FUNDAMENTALS OF TOOL DESIGN

Third Edition

David T. Reid
Publications Administrator

Published by
Society of Manufacturing Engineers
Publications Development Department
One SME Drive
P.O. Box 930
Dearborn, MI 48121-0930

Based on Fundamentals of Tool Design, Second Edition
Edward G. Hoffman, Ph.D., Editor

FUNDAMENTALS OF TOOL DESIGN

Copyright © 1991
Society of Manufacturing Engineers

Third Edition

Second Printing

Library of Congress Catalog Card Number: 91-66441

International Standard Book Number: 0-87263-412-4

Manufactured in the United States of America

SME would like to extend special thanks to the following contributors and reviewers of *Fundamentals of Tool Design*, Third Edition, for their invaluable assistance and devotion to accuracy in the application of design principles.

CONTRIBUTORS

William E. Boyes (retired)
Mack Trucks Inc.

Duane D. Dunlap, Ph.D.
Purdue University

Edward G. Hoffman, Ph.D.
Hoffman & Associates

Paul Jacobs, Ph.D.
3D Systems, Inc.

Elsayed A. Orady, Ph.D.
University of Michigan-Dearborn

David A. Smith
Smith & Associates

REVIEWERS

James Albrecht
Dayton Progress

Charles DeRoller, Ph.D.
Rochester Institute of Technology

David M. Johnson
Ford Motor Co.

Donald H. Koch
Flexible Fixturing Systems, Inc.

Alvin G. Neumann
Neumann & Associates

Thanks also to the contributors to *Fundamentals of Tool Design*, Second Edition: Paul Dalla Guardia, *Hoffman & Associates*; Dr. David Goestsch, *Okalossa-Walton Junior College*; Earl Harp, *Hewlett-Packard Corporation*; Martin Kuklinski, *Falk Corporation*; Paul A. Longo, *U.S. Army*; John Mitchell, *Carboloy Systems*; Paul Suksi, *Western Tool and Die*; Thomas Ury, *AMPEX Corporation*.

Revised and updated from *Fundamentals of Tool Design*, First Edition

originally published by Prentice-Hall, Inc.
Englewood Cliffs, New Jersey
1962

PREFACE

This book provides to engineers and students a fundamental text on one of the most significant areas of manufacturing—tool design. Its coverage ranges from simple jigs and fixtures and cutting tools to the latest advances in computer aided design and rapid prototyping.

Since its initial publication in 1962, *Fundamentals of Tool Design* has proven to be an outstanding source for information in the area of tool design. From classrooms to the design shops of major corporations around the world, this published resource has demonstrated its value for almost 30 years.

Written by tool designers for tool designers, this real-world book presents useful information on presses and dies, new design ideas, practical tips for geometric dimensioning and tolerancing, and information on modular tooling and automated tool handling.

I would like to acknowledge the contribution of the authors who worked not only on this important third edition update of *Fundamentals of Tool Design*, but also the authors who produced both the first and second editions of this highly successful book. Their efforts in the area of sound tool design are evident in the pages that follow.

Thomas J. Drozda, CMfgE, PE
Director
SME Publications

TABLE OF CONTENTS

1

TOOL DESIGN

Tool design is a specialized area of manufacturing engineering which comprises the analysis, planning, design, construction, and application of tools, methods, and procedures necessary to increase manufacturing productivity. To carry out these responsibilities, today's tool designer must have a working knowledge of machine shop practices, toolmaking procedures, machine tool design, manufacturing procedures and methods, as well as the more conventional engineering disciplines of planning, designing, engineering graphics and drawing, and cost analysis.

Responsibilities of a Tool Designer

Typically, tool designers are responsible for a wide variety of special tools. Whether these tools are an end product or merely an aid to manufacturing, the tool designer must be familiar with the following:

1. Cutting tools, toolholders, and cutting fluids.
2. Machine tools, particularly modified or special types.
3. Jigs and fixtures.
4. Gages and measuring instruments.
5. Dies for sheet metal cutting and forming.
6. Dies for forging, upsetting, cold finishing, and extrusion.
7. Fixtures and accessories for welding, riveting, and other mechanical fastening.

In addition, the tool designer must be familiar with other engineering disciplines such as metallurgy, electronics and computers, and machine design, insofar as they affect the design of special tools.

In most cases the size of the company which employs the tool designer, or the type of product will determine the exact duties of each designer. Larger companies with several product lines may employ many tool designers. Here, each designer may have an area of specialization, such as die design, jig and fixture design, gage design, or any similar specialized tool design area. In smaller companies, however, one tool designer may have to do all tool designs as well as other tasks in manufacturing.

Objectives of Tool Design

The main objective of tool design is to increase production while maintaining quality and lowering costs. To this end, the tool designer must realize the following objectives, or goals:

1. Reduce the overall cost of manufacturing a product by producing acceptable parts at the lowest cost.
2. Increase the production rate by designing tools that will produce parts as quickly as possible.
3. Maintain quality by designing tools which will consistently produce parts with the required precision.
4. Reduce the cost of special tooling by making every design as cost effective and efficient as possible.
5. Design tools that will be safe and easy to operate.

Every design must be created with these objectives in mind. No matter how well a tool functions or how well it produces parts, if it costs more to make the tool than it saves in production, its usefulness is questionable. Likewise, if a tool cannot maintain the desired degree of repeatability from one part to the next, it's of no value in production.

The Design Process

While the specifics of designing each type of tool will be discussed in the subsequent chapters of this text, there are a few basic principles and procedures that should be introduced at this point. The design process consists of five basic steps:

1. Statement and analysis of the problem.
2. Analysis of the requirements.
3. Development of initial ideas.
4. Development of design alternatives.
5. Finalization of design ideas.

While these five steps are separated for this discussion, in practice, each of these steps actually overlaps the others. For example, when stating the problem, the requirements must also be kept in mind to properly define and determine the problem or task to be performed. Likewise, when determining the initial design ideas, the alternative designs are also developed. So, like many other aspects of manufacturing, tool design is actually an ongoing process of creative problem solving.

Statement of the Problem

The first step in the design of any tool is to define the problem as it exists without tooling. This may simply be an assessment of what the proposed tool is expected to do, such as drill four holes. Or, it may be an actual problem which has been encountered in production where tooling may be beneficial, such as low-volume production caused by a bottle neck in assembly. Once the exact extent of the problem has been determined, the problem can be analyzed and resolved throughout the remaining steps of the design process.

Analysis of the Requirements

After the problem has been isolated, the specific requirements such as function, quality, cost, due date, and other related specifics can be used to determine the specific parameters within which the designer must work. Every tool that is designed (1) must perform certain functions; (2) must meet certain minimum precision requirements; (3) must keep the cost to a minimum; (4) must be available when the production schedule requires it; (5) must be operated safely; (6) must meet various other requirements such as adaptability to the machine on which it is to be used, and have an acceptable working life. *Figure 1-1* shows a method of applying these criteria to the process of choosing a tool design. Rarely, if ever, will one tool design be best in each of these five areas. The tool designer's task here is to weigh all these factors and select the tool that best meets these criteria and the task to be performed.

Development of Initial Ideas

Initial design ideas are normally conceived after an examination of the preliminary data. This data consists of the part print, process sheet, engineering notes, production schedules, and any other related information. While evaluating this information, the designer should take notes to insure nothing is forgotten during the initial evaluation. Should the designer need more information than that furnished with the design package, the planner responsible for the tool request should be consulted to determine the required additional information. In many cases, the designer and planner must jointly develop parameters during initial design.

Development of Design Alternatives

During the initial concept phase of design, many ideas will occur to the designer. As these ideas are thought out, they should also be written down so they are not lost or forgotten. There are always several ways to do any job. As each method is developed and analyzed, the information should be added to the list shown in *Figure 1-1*.

Create	Analyze in terms of these criteria				
Alternatives	*Function*	*Quality*	*Cost*	*Date*	*Auxiliary*
A	x	x	x	x	x
B	x	x	x	x	x
C	x	x	x	x	x
.	.	.	.	.	.
.	.	.	.	.	.
.	.	.	.	.	.
n	x	x	x	x	x

Figure 1-1. Basic pattern for tool analysis.

Complete Design Ideas

Once the initial design ideas and alternatives are determined, the tool designer must analyze each element of the design to determine the best possible way to proceed toward the final tool design. As was stated earlier, rarely is one tool alternative a clear favorite over any other. Rather, the tool designer must evaluate the strong points of each alternative and weigh this value against the weak points of the design. For instance, one tool design may have a high production rate, but the cost of the tool may be very high. On the other hand, a second tool may have a medium production rate and will cost much less to build. In this case, the value of production over cost must be evaluated to determine the best design for the job.

If the job is a long-term production run, the first tool may pay for itself in increased production volume. If, however, the production run is short or is a one-time run, the second tool may work best by sacrificing production speed for a reduced tool cost. Seldom will one tool be able to meet all the expectations of the tool designer. In most cases, the best design is a compromise of the basic criteria of function, quality, cost, due date, safety and other requirements.

Economics of Design

The tool designer must know enough economics to determine, for example, whether temporary tooling would suffice though funds are provided for more expensive permanent tooling. He or she should be able to check the design plan well enough to initiate or defend a planning decision to write off the tooling on a single run as opposed to write-off distributed against probable future reruns. The tool designer should have an opinion backed by economic proof of certain changes that would make optimum use of the tools.

Many aspects of manufacturing economics are beyond the scope of this book, but excellent literature is available.

Analysis of Small Tool Costs

The following analyses are applicable where production rates are small and fixed:

Let:

N = number of pieces manufactured per year,
C = first cost of fixture,
I = annual allowance for interest on investment, percent,
M = annual allowance for repairs, percent,
T = annual allowance for taxes, percent,
D = annual allowance for depreciation, percent,
S = yearly cost of setup,
a = saving in labor cost per unit,
t = percentage of overhead applied on labor saved,
V = yearly operating profit over fixed charges,
H = number of years required for amortization of investment out of earnings.

For simplification, the following values are used in the following examples:

$$a = 0.03, t = 50\%, \text{cost of each setup} = \$10$$
$$I = 6\%, T = 4\%, M = 10\%, D = 50\%$$
$$I + T + D + M = 70\%$$

Number of pieces required to pay for fixture:

$$N = \frac{C(I + T + D + M) + S}{a(1 + t)} \tag{1-1}$$

Example: If a fixture costs $400 and one run is made per year,

$$N = \frac{(400 \times 0.70) + 10}{0.03 \times 1.5} = 6445 \text{ pieces}$$

Therefore, on two yearly runs of 6445 pieces, the fixture will pay for itself.

If six runs per year are made,

$$N = \frac{(400 \times 0.70) + (6 \times 10)}{0.03 \times 1.5} = 7556 \text{ pieces}$$

If a fixture must pay for itself in a single run, $H = 1$ and $D = 100\%$ and $(I + D + T + M) = 120\%$,

$$N = \frac{(400 \times 1.20) + 10}{0.045} = 10,889 \text{ pieces}$$

Economic investment in fixtures for given production:

$$C = \frac{Na(1 + t) - S}{I + T + D + M} \tag{1-2}$$

Example: For a single run of 10,900 pieces with an estimated saving of three cents per piece,

$$C = \frac{(10,900 \times 0.045) - 10}{1.20} = \$400$$

If 7550 pieces are made in six runs per year and the fixture is to pay for itself in two years and $(I + D + T + M) = 70\%$,

$$C = \frac{(7550 \times 0.045) - 60}{0.70} = \$400$$

Number of years required for a fixture to pay for itself:

$$H = \frac{C}{Na(1 + t) - C(I + T + M) - S} \tag{1-3}$$

Example: In the preceding example, $C = \$400$; then,

$$H = \frac{400}{(7550 \times 0.045) - (400 \times 0.20) - 60} = 2 \text{ years}$$

Profit from improved fixture designs:

$$V = Na(1 + t) - C(I + T + D + M) - S \tag{1-4}$$

The previous equations have assumed a break-even cost.

Example: Assume that C = $250 instead of $400. Then with 7550 pieces per year and six runs per year.

$$V = (7550 \times 0.045) - (250 \times 0.70) - 60 = \$105 \text{ per year}$$

Equation 1-2 can be used for comparing costs of alternative fixtures having different setup and labor costs.

Example: In a cost of a 2000-piece run, and D = 100%, assume (1) a fixture with S = $10 and a = $0.03 and (2) a fixture with S = $15 and a = $0.05.

$$C_1 = \frac{(2000 \times 0.03 \times 1.5) - 10}{1.20} = \$66.67$$

$$C_2 = \frac{(2000 \times 0.05 \times 1.5) - 15}{1.20} = \$112.50$$

Tooling Economics in Combined Operations. Analysis may sometimes show that operations can advantageously be combined. The total cost of tooling, the production costs, or both can thus be reduced. *Figure 1-2* illustrates a case where the cost of combined tools was less than the total cost of the separate tools otherwise required. The combined operation was done at the speed of the blanking operation above.

Costs	Blanking operation alone	Forming operation alone	Total blank and form	Combined operations
Tools	$40.00	$30.00	$ 70.00	$50.00
Setup	2.00	2.00	4.00	3.00
Maintenance	2.00		2.00	2.00
Processing	4.00	30.00	34.00	4.00
Total cost	$48.00	$62.00	$110.00	$59.00

Figure 1-2. Cost of combined versus separate operations.

Process-cost Comparisons. During process planning, many possible methods of manufacturing may be reduced to a few based upon alternate process steps, use of available equipment, or combined operations. Under these conditions, a comparison of costs for different tools and process steps may reveal the combination that will result in the lowest total cost per part.

Let:

N_t = total number of parts to be produced in a single run
N_b = number of parts for which the unit costs will be equal for each of two compared methods Y and Z ("break-even point")
T_y = total tool cost for method Y
T_z = total tool cost for method Z
P_y = unit tool process cost for method Y
P_z = unit tool process cost for method Z
C_y, C_z = total unit cost for methods Y and Z, respectively.

Then:

$$N_b = \frac{T_y - T_z}{P_z - P_y} \tag{1-5}$$

$$C_y = \frac{P_y N_t + T_y}{N_t} \tag{1-6a}$$

$$C_z = \frac{P_z N_t + T_z}{N_t} \tag{1-6b}$$

Example: The aircraft-flap nose rib shown in *Figure 1-3* of 0.02″ (0.5 mm) 2024-T Alclad was separately calculated to be formed by Hydropress, drop hammer, Marform, steel draw die, and hand forming. For such reasons as die life, equipment available, and handwork required, the choice narrowed down to Hydropress versus steel draw die. With Hydropress, the flanges had to be fluted, and for piece quantities over 100, a more expensive steel die costing $202 had to be used.

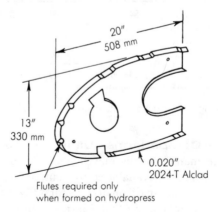

Figure 1-3. General specifications for aircraft-flap nose rib.

Actual die and processing costs for both methods are listed in *Figure 1-4*. P_y and P_z are processing costs, and T_y and T_z are die costs, for the steel draw die and Hydropress methods, respectively.

Figures in the last column of *Figure 1-4* were not stated in the original report but can properly be extrapolated on the basis of apparent stability of P_y and P_z at $N_t = 500$, and assuming their stability at higher production.

N_t	5	25	50	100	500	780†
P_y	$ 3.00	$ 1.18	$ 1.11	$ 1.05	$ 1.05	$ 1.05
P_z	4.40	2.05	1.96	1.85	1.85	1.85
T_y	810.00	810.00	810.00	810.00	810.00	810.00
T_z	103.00	103.00	103.00	103.00	202.00	202.00
C_y	165.00	33.60	17.30	9.15	2.67	2.12
C_z	25.00	6.17	4.02	2.88	2.25	2.12

† Extrapolated.

Figure 1-4. Cost comparison of methods for producing aircraft-flap nose rib.

On the basis of listed figures at $N_t = 500$, and from *Equation 1-5*, the production at which total unit costs C_y and C_z will be the same for both methods is,

$$N_b = \frac{810 - 202}{1.85 - 1.05} = 760 \text{ pieces}$$

As an alternate method for calculating the break-even point between two machines, e.g., a turret lathe and an automatic, the Warner & Swasey Company uses a formula based on known or estimated elements that make up production costs. The break-even-point formula is,

$$Q = \frac{pP(SL + SD - sl - sd)}{P(1 + d) - p(L + D)} \tag{1-7}$$

Where:

Q = quantity of pieces at break-even point,
p = number of pieces produced per hour by the first machine,
P = number of pieces produced per hour by the second machine,
S = setup hours required on the second machine,
s = setup hours required on the first machine,
L = labor rate for the second machine, in dollars,
l = labor rate for the first machine, in dollars,

D = hourly depreciation rate for the second machine (based on machine-hours for the base years period),

d = hourly depreciation rate for the first machine (based on machine-hours for the base years period).

Example: Assume that it is desired to find value Q when the various factors are as follows:

p = 10 pieces per hour,
P = 30 pieces per hour,
S = 6 hour,
s = 2 hour,
L = $1.50 per hour,
l = $1.50 per hour,
D = $1.175 per machine-hour,
d = $0.40 per machine-hour (10-year period).

Then, substituting these values in the formula,

$$Q = \frac{10 \times 30(6 \times 1.5 + 6 \times 1.175 - 2 \times 1.5 - 2 \times 0.4)}{30(1.5 + 0.4) - 10(1.5 + 1.175)}$$

Answer: Q = 121 pieces, or the quantity on which the cost is the same for either machine.

Effect of Tool Material on Minimum-cost Tool Life. The tool designer can often influence the choice of cutting-tool materials. Certainly the tool designer must know the tool-life economics involved.

Typical tool-life curves are shown in *Figure 1-5* for high-speed steel, sintered carbide, and oxide tools. The wear characteristics, as denoted by the n values, show that economic life T_c for carbide tools is shorter than that for HSS tools (T_c equals 15 minutes for carbide vs. 35 minutes for HSS). It is even

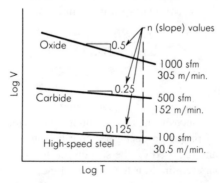

Figure 1-5. Tool-life comparison for various tool materials.

more important to use higher speeds and shorter tool life for oxide tools, since T_c equals five minutes.

From *Figure 1-6*, it is evident that most economic metal removal demands both high horsepower and speeds when cutting with carbide and oxide tools.

Tool material	$1/n - 1$* values	Tool Life, min T_c	T_p	Minimum-cost cutting speed, V_c, sfpm
High-speed steel	7	35	14	107
Tungsten carbide	3	15	6	701
Oxide	1	5	2	3500

* The term $1/n - 1$ is derived from differentiation of the tool-life equation $VT^n = C_t$, where $V =$ cutting speed, $T =$ actual cutting time between sharpenings, and C_t is a constant numerically equal to the cutting speed that gives a 1-minute tool life under the actual cutting conditions.

Figure 1-6. Performance data for various tool materials.

Economic Lot Sizes

Economic lot sizes are calculated to obtain the minimum unit cost of a given part or material. This minimum is reached when the costs of planning, ordering, setting up, handling, and tooling equal the costs of storage of finished parts. These costs may be equated and the lot size determined by mathematical calculation. Depending on the number of variables to consider, the formula can range from one of relative simplicity to one that is relatively complex.

By assuming the number of pieces required per month as constant, the inventory increasing until the lot ordered is completely sent to stock, and decreasing uniformly with use, a relatively simple formula for calculation of economic lot size can be devised.

$$L = \sqrt{\frac{24mS}{kc(1 + mv)}} \tag{1-8}$$

Where:

c = value of each piece, dollars,
k = annual carrying charge per dollar of inventory,
L = lost sizes, pieces,
m = monthly consumption, pieces,
S = setup cost per lot, dollars,
v = ratio of machining time to lot sizes, months per piece.

Labor, material, and other costs not related to lot size are also omitted. In actual practice, reasonable values for S, m, and c are difficult to obtain. These values are generally determined by the standards, sales, or methods departments.

Formulas for calculating economic lot sizes can never prove out exactly because of the assumptions upon which many of the factors may or must be based. Such a formula should be regarded as just a useful guide, to be applied with mature judgment.

Break-Even Charts

Break-even charts are perhaps most widely used to determine profits based on anticipated sales. They have other uses, however, such as for selecting equipment or for measuring the advisability of increased automation.

To determine which of two machines is most economical, the fixed cost and variable cost of each machine are plotted as shown in *Figure 1-7*. (The fixed cost is the cost of the machine, and the variable costs are obtained by multiplying the number of pieces produced by the unit-piece cost.) The total cost is composed of the sum of the fixed and variable costs. For example: Assume the initial cost for machine A is $1500, and the unit production cost

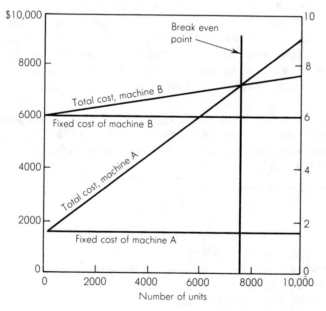

Figure 1-7. Break-even chart for machine selection; choice is based upon volume production.

on the machine is $0.75 each. For the other machine *B*, the initial cost is $6000 and the unit production cost $0.15 each.

The chart shows that it is more economical to purchase machine *A* if production never exceeds 7500 pieces. For higher production quantities, machine *B* is more economical.

Tool Drawings

A tool designer must have a strong background in drafting, dimensioning, and mechanical drawing to properly present design concepts to the people who will make the proposed tool. Often, computer aided design (CAD) technology is used for this purpose (see Chapter 12). For the most part, toolmakers and diemakers do not require the same type of drawings as do the less experienced machine operators in a production department. For this reason, tool drawings can be drawn much simpler and faster to keep design cost to a minimum. The following are basic points to remember when creating tool drawings.

1. Draw and dimension with due consideration for someone using the drawing to make the item in the toolroom.
   a. Do not crowd views or dimensions.
   b. Analyze each cut to be sure it can be done with standard tools.
2. Use only as many views as necessary to show all required detail.
3. Surface roughness must be specified.
4. Tolerances and fits peculiar to tools need special consideration. It is not economical as a rule to tolerance both details of a pair of mating parts as is required on production part detailing. In cases where a hole and a plug are on different details to be made and mated, the fit tolerance should be put on the male piece and the hole should carry a nominal size. This allows a toolmaker to ream the hole with a nominal-size tool and grind the plug to fit it, although nominal may vary several thousandths from exact.
5. The stock list of any tool drawings should indicate all sizes required to obtain the right amount for each detail. It is necessary to allow material for finishing in almost all cases, although some finished stock is available today. As far as possible, stock sizes known to be on hand should be used, but in all cases, available sizes should be specified. A proper, finished detail is dependent upon starting with the right material.
6. Use notes to convey ideas that cannot be communicated by conventional drawing. Heat treatments and finishes are usually identified as specification references rather than being spelled out on each drawing.
7. Secondary operations such as surface grinding, machining of edges, polishing, heat treating, or similar specifications should be kept to a

minimum. Only employ these operations when they are important to the overall function of the tool; otherwise these operations will only add cost, not quality to the tool.

8. Apply tolerances realistically. Overly tight tolerances can add a great deal of additional cost with little or no added value to the tool. The function of the detail should determine the specific tolerance, not a standard title block tolerance value.

Note: If tooling is to be duplicated in whole or in part for other locations or if multiple tools are required, drawings should be well detailed.

Tooling Layout

The actual work of creating on paper the assembly design of equipment or tools for manufacturing processes should be done within the general framework of the following rules. (This will insure the details of the tool will work and not interfere with any other part of the tool.)

1. Lay out the part in an identifying color (red is suggested).
2. Lay out any cutting tools. Possible interference or other confining items should be indicated in another identifying color (blue suggested). Use of the cutting tool should not damage the machine or the fixture.
3. Indicate all locating requirements for the part. There are three locating planes: use three points in one, two points in the second, and only one point in the third plane. This is called the 3-2-1 locate system. Do not locate on the parting line of castings or forgings. All locators must be accessible for simple cleaning of chips and dirt. For a more detailed explanation, see Chapters 5 and 6.
4. Indicate all clamping requirements for the part. Be careful to avoid marking or deforming finished or delicate surfaces. Consider the clamping movements of the operator so injury to the hands or unsafe situations are eliminated. Be sure it is possible to load and unload the part.
5. Lay out the details with due considerations to stock sizes, so as to minimize machining requirements.
6. Use full scale in the layout if possible.
7. Indicate the use of standard fixture parts (shelf items) whenever possible.
8. Identify each different item or detail of any design by the use of balloons with leaders and arrows pointing to the detail in the view that best shows the outline of the item. These should not go to a line that is common to other details.

Safety as Related to Tool Design

Safety laws vary greatly from one state to another. All states have some laws requiring protective guards and devices to safeguard workers.

Safety should be designed into the tooling. One of the first and least expensive requirements should be that of breaking all sharp edges and corners. More minor injuries result from this cause than from most others. Cutting should never be performed against a clamp, because of vibration and tool chatter. Instead, parts should be nested against pins in order to take the cutter load. Rigidity and foolproofing should always be built into the tooling. Make drill jigs large enough to hold without the danger of spinning. Small drill jigs should always be clamped in a vise or against a bar or backstop. Install plexiglass guards around all milling and flycutting operations where chips endanger workers or work areas. High-speed open cutters on production milling, drilling, turning, and jig borers are difficult to safeguard because of the varied size and location of the workpieces to be machined.

In guarding punch presses, no one type of guard is practical for all operations. Ring guards work well on small punching setups where the ring meets an obstruction and the downward motion of the ram is stopped. Larger die sets and presses can be better protected by gate guards which must be positioned after the work is loaded and then interlocked with the clutch mechanism. Barrier guards protect by preventing hands and arms from being placed inside the work area. These generally are telescoping perforated sheet metal coverings with an opening to feed the stock. Sweep guards actually sweep clear the danger area as the press ram descends and even provides protection during an accidental descent of the ram. Wire-cage guards are still another way to protect the operator's hands; the operator feeds the work through a small opening. This type of guard is useful for secondary operations. Another method of protection is to provide space between the punch and die too small for the operator's hand to enter; this distance should not exceed 3/8″ (10 mm).

Limit switches can be used extensively to protect both worker and product. In punches and dies, limit switches can detect a misfeed or buckling of stock, and check the position of parts in assembling. Photoelectric equipment to protect the operator's hands or body operates when the set beam is interrupted. It is good practice for all punch presses and air or hydraulically operated tooling to be installed with a double-button interlocking protection system, requiring both bottons to be activated before the tool can be used. Other interlocking systems can include two valves in conjunction with a hand valve designed to prevent the tooling or press from operating in case one of the buttons is locked in a closed position. The device should be located in such a position or guarded in such a manner that the operator cannot operate the tooling or press while in the danger area.

Feed mechanisms should be provided for all high-speed punching, machining, or assembling operations. The safety function of a feeding device is to provide a means of moving the part into the nest by gravity or mechanical action so that there is no necessity for the operator to place his hands in the danger zone. High-speed presses equipped with automatic feeds operate at such speeds that it would be impractical as well as hazardous for an operator to attempt to feed the stock.

Tools and additions to machinery involving electrical equipment must be grounded. Portable electrical equipment should be checked periodically because of rough handling, and all equipment should be grounded to prevent injuries to personnel. Locks, which hold electrical switches open while tools or punches and dies are being repaired or set, help to prevent accidental damage. Local regulations can be found in each state's electrical code book covering all applications.

Tooling for various industries requires different treatment to insure safe operations. For example, special electrical controls and motors are required for industries handling explosive material. Tooling material for the chemical industry should be designed to withstand corrosive actions, and electrical equipment should be properly sealed. Careful analysis must be given to tooling for each type of application to provide maximum protection and long life.

Machining of various plastics sometimes generates abrasive dust particles or poisonous fumes, and exhaust systems are required in such cases. Safe procedures should be practiced when handling certain metals. Areas around magnesium machining should be kept clean and tools kept sharp to reduce fire hazard. Metal powders present a hazard of combustion during grinding operations. Data can be found in chemical, electrical, and industrial handbooks to help in tooling these special problems.

Tooling and additions to machines must be designed so that the operator does not have to lean across a moving cutter or table. All adjustments and clamping should be easily accessible from the front or from the operator's position. Consider body geometry when designing tooling—not only in terms of safety but also production.

Tooling involving welding must be guarded to prevent severe burns or eye injury from high-intensity arc welding rays. Safety glasses should be worn during all machining, grinding, and buffing operations.

All belts, chain drives, gears, sprockets, couplings, keys, and pulleys should be well guarded by sheet metal and panel guards. These guards should be strong enough to support and protect in the event that someone or something falls against the guard, or in case the belt or chain should break. Always provide an adequate factor of safety in the design of all tools and tooling applications.

Safety standards are available covering all types of industrial applications. Personnel responsible for plant safety should be familiar with them. Some of the agencies handling such information are as follows: American Standards Association; National Bureau of Standards; U.S. Department of Commerce; National Safety Council; National Board of Fire Underwriters; Association of Casualty and Surety Companies; and the Occupational Safety and Health Administration (OSHA).

Tooling is not restricted solely to machining operations. Automatic assembling and inserting equipment may be classified as either a tool or a machine. One of today's complex tools might contain any combination of electric motors, air cylinders, hydraulic equipment, conveyors, and precision indexing tables. Safety is of the utmost concern when a production tool involves such operating mechanisms.

Material Handling at the Workplace

It is beyond the scope of this text to go into detail regarding the bulk handling of materials and parts through the factory, or the many principles that underly the design of a workplace. Full texts are devoted to this subject-matter area. It is the objective of this brief discussion to emphasize the important part tool design plays in the total cost of an operation from the standpoint of materials handling.

To pick up a workpiece, place it into a tool, clamp the part, unclamp, and remove and set aside after machining may consume more time than the actual machining. Such operations, however essential, contribute nothing of value to the product, but their performance is paid for at the same rate as the productive effort. Consequently, tooling should be designed to reduce the nonproductive time and costs.

Once a tool is designed and built, the methods of handling materials into and out of it are fixed. It is essential then, that the original methods planning be carried out with care. The following list of principles* should be applied to help ensure maximum motion economies. It will be noted that the principles are arranged under three general headings: (1) use of the human body; (2) arrangement of the workplace; and (3) design of tools and equipment.

　1. Use of the human body
　　　a. Both hands should begin and end their basic divisions of accomplishment simultaneously and should not be idle at the same instant, except during rest periods.
　　　b. The motions made by the hands should be made symmetrically

* Niebel, B. W., *Motion and Time Study*, Richard D. Irwin, Inc., Homewood, IL, 1958.

and simultaneously away from and toward the center of the body.

c. Momentum should be employed to assist the worker wherever possible, and should be reduced to a minimum if it must be overcome by muscular effort.

d. Continuous curved motions are preferable to straightline motions involving sudden and sharp changes in direction.

e. The least number of basic divisions should be used, and these should be confined to the lowest possible classifications. These classifications, summarized in ascending order of time and fatigue expended in their performance, are:

   (1) Finger motions
   (2) Finger and wrist motions
   (3) Finger, wrist, and lower arm motions
   (4) Finger, wrist, lower arm, and upper arm motions
   (5) Finger, wrist, lower arm, upper arm, and body motions.

f. Work that can be done by the feet should be arranged so that it is done simultaneously with work being done by the hands.

2. Arrangement and conditions of the workplace

a. Fixed locations should be provided for all tools and materials to permit the best sequence and to reduce or eliminate the need to search, select, and find.

b. Gravity bins and drop delivery should be used to reduce times for reaching and moving. Ejectors also should be provided wherever possible to remove finished parts automatically.

c. All materials and tools should be located within the normal area in both the vertical and horizontal planes.

d. A comfortable chair should be provided for the operator, and the height should be arranged so the work can be efficiently performed by the operator alternately standing and sitting.

e. Proper illumination, ventilation, and temperature should be provided.

f. Visual requirements of the workplace should be considered so that eye fixation demands are minimized.

g. Rhythm is essential to the smooth and automatic performance of an operation. Work should be arranged to permit an easy and natural rhythm wherever possible.

3. Design of tools and equipment

a. Make multiple cuts whenever possible by combining two or more tools in one, or by arranging simultaneous cuts from both feeding devices if available (cross slide and hex turret).

b. All levers, handles, wheels, and other control devices should be readily accessible to the operator and should be designed to give the best possible mechanical advantage.

   c. Use fixtures to hold parts in position.

   d. Always investigate the possibility of powered or semiautomatic tools, such as power nut and screw drivers, and speed wrenches.

For the manipulation of materials in and out of presses, there are many types of mechanical feeding and ejecting devices. These may be used alone or in combination with manual feeding and ejecting methods. Good design practice calls for full consideration of available handling systems.

TOOL DESIGN

Review Questions

1. Name five design aspects with which the tool designer must be concerned.
2. List the seven areas of tool design that a designer must be familiar with.
3. List the four principle objectives which must be realized in any tool design.
4. How many steps are there in the basic design process?
5. List these steps and briefly describe each.
6. How many years would it take and how many pieces would have to be made to reach the break-even point if the fixture cost $500? Given: $a = 0.02$; $M = 20\%$; $t = 50\%$; $D = 40\%$ $I + T + D + M = 75\%$
   $S=\$15$ = yearly cost of setup.
7. Using the values given in question six, how many pieces must be run to break even in one run?
8. When constructing tool drawings, should they be drawn in the same manner as production drawings? Why?
9. What is the main purpose of a tooling layout?
10. What are the three general categories of principles that should be applied to ensure motion economics when handling materials?

TOOL DESIGN

Answers to Review Questions

1. a. Function
   b. Quality
   c. Cost
   d. Due date
   e. Auxiliary requirements
2. a. Cutting tools
   b. Machine tools
   c. Jigs and fixtures
   d. Gages and measuring instruments
   e. Dies for sheet metal
   f. Dies for forging
   g. Fixtures for welding and assembly
3. a. Reduce the overall cost of manufacturing a product by producing acceptable parts at the lowest cost.
   b. Increase the production rate by designing tools that will produce parts as quickly as possible.
   c. Maintain quality by designing tools which will consistently produce parts with the required precision.
   d. Reduce the cost of special tooling by making every design as cost-effective and efficient as possible.
4. Five steps.
5. a. Statement of the problem—identify exactly what is to be done.
   b. Analysis of the requirements—identify the parameters within which the task must be accomplished. These parameters are the five basic aspects of tool design.
   c. Development of initial ideas—begin to formulate the initial design ideas.
   d. Development of design alternatives—determine several methods of performing the task to be done.
   e. Finalize design ideas—select the one most acceptable and complete the tool design.
6. $H = \dfrac{1}{D} = \dfrac{1}{40} = 2\,1/2$ years

   $I + T + D + M = 75\%$

From *Equation 1-1*

$$N = \frac{500(0.75) + 15}{0.02(1 + 0.50)} = 13,000 \text{ parts}$$

The fixture will pay for itself in 2 1/2 years after an average yearly run of 13,000 parts.

7. If a fixture is to pay for itself in one run, $H = 1$ and $D = 100\%$

$$I + D + T + M = 0.10 + 1 + 0.05 + 0.20 = 1.35$$

Substituting in *Equation 1-1*

$$N = \frac{500(1.35) + 15}{0.20(1.00 + 0.50)} = 23,000 \text{ parts}$$

8. No, because toolmakers and diemakers do not require as much detail as do production machine operators. Simplifying the tool drawing also helps reduce the design cost of a tool. Unless the tool is to be built outside, or several tools are to be built.

9. To insure the arrangement of details will work and not interfere with any other part of the tool.

10. a. Use of the human body.
    b. Arrangement and conditions of the workplace.
    c. Design of tools and equipment.

2

TOOL MATERIALS

The specific material selected for a particular tool is normally determined by the mechanical properties necessary for the proper operation of the tool. These materials should be selected only after a careful study and evaluation of the function and requirements of the proposed tool. In most applications, more than one type of material will be satisfactory, and a final choice will normally be governed by material availability and economic considerations.

The principal materials used for tools can be divided into three major categories: ferrous materials, nonferrous materials, and nonmetallic materials. Ferrous tool materials have iron as a base metal and include tool steel, alloy steel, carbon steel, and cast iron. Nonferrous materials have a base metal other than iron and include aluminum, magnesium, zinc, lead, bismuth, copper, and a variety of alloys. Nonmetallic materials are those materials such as woods, plastics, rubbers, epoxy resins, ceramics, and diamonds that do not have a metallic base. To properly select a tool material, there are several physical and mechanical properties you should understand to determine how the materials you select will affect the function and operation of the tool.

Physical and Mechanical Properties

Physical and mechanical properties are those characteristics of a material which control how the material will react under certain conditions. Physical properties are those properties which are natural in the material and cannot be permanently altered without changing the material itself. These properties include: weight, color, thermal and electrical conductivity, rate of thermal expansion, and melting point.

Weight. The weight of a material is a measure of the amount of substance contained in a specific volume, such as pound per foot or kilogram per meter.

Color. The color of a material is the natural tint contained throughout the material. For example, steels are normally a silver-gray color and copper is usually a reddish-brown.

Thermal and Electrical Conductivity. The thermal and electrical conductivity of a material is the measure of how fast or slow a specific material conducts heat or electricity. Aluminum and copper, for example, have a high rate of thermal and electrical conductivity, while nickel and chromium have a low rate of conductivity.

Rate of Thermal Expansion. The rate of thermal expansion is a measure of how a material expands when exposed to heat. Materials such as zinc and lead have a high rate of expansion, while carbon and silicon expand very little when heated.

Melting Point. The melting point of a material is the point where a material changes from a solid to a liquid state. Materials such as tantalum and tungsten have a very high melting point while lead and bismuth have a very low melting point.

The mechanical properties of a material are those properties which can be permanently altered by thermal or mechanical treatment. These properties include strength, hardness, wear resistance, toughness, brittleness, plasticity, ductility, malleability, and modulus of elasticity.

Strength. Strength is the ability of a material to resist deformation. The most common units used to designate strength are pounds per square inch (psi) and kiloPascals (kPa). The principal categories of strength you will be most concerned with when designing tools are tensile strength, compressive strength, shear strength, and yield strength.

Tensile Strength. This property of materials is the value obtained by dividing the maximum load observed during tensile testing by the specimen's cross-sectional area before testing (*Figure 2-1*).

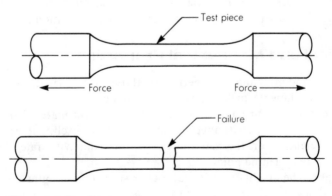

Figure 2-1. Tensile strength.

Tensile strength is an important property to consider when designing large fixtures or other tooling. It is of lesser importance in tools and dies except where soft or medium hard ferrous or nonferrous materials are used.

If a steel elongates slightly before it breaks, a reasonably accurate tensile figure can be obtained. However, if the tool material is so hard that it breaks before it begins to elongate, the specimen will rupture in test long before the true strength is obtained.

The tensile tests successfully made on tool steel involve the use of drawing temperatures much higher than actually used on tools. Tool steels used for hot work, fatigue, or impact applications are usually used at lower hardness levels. Tensile properties can be obtained for tool steels.

Compressive Strength. Compressive force plays an important part in tool design. It is the maximum stress that a metal, subject to compression, can withstand without fracture (*Figure 2-2*).

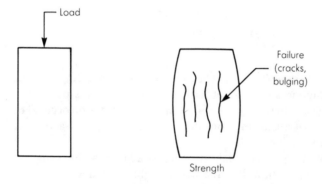

Figure 2-2. Compressive strength.

This test is used on hardened tool steels, especially at high hardness levels. For all ductile materials, the specimens flatten out under load, and there is no well-marked fracture. For these materials, compressive strength is usually equal to tensile strength.

Shear Strength. The shear strength of materials is significant, especially in designing machines and members subject to torsion. It may be defined as the value of stress necessary to cause rupture in torsion (*Figure 2-3*).

For most steels, except tool and other highly alloyed steels, the shear strength lies between 50 and 60% of the yield strength; hence the yield strength in tension serves fairly well as an index of shear strength.

Hardened tool steels are considered to be brittle because they deform very little before fracture in tension or bending, yet when subjected to torsion they

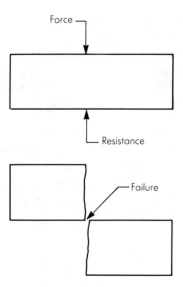

Figure 2-3. Shear strength.

exhibit considerable ductility. The torsion method of testing offers advantages for the "toughness testing" of hard tool steels. This is particularly true in those instances where very small amounts of deformation occur. Toughness is the property of absorbing considerable energy before fracture. It involves both ductility and strength.

Yield Strength. This is the property of a material that generally limits its strength in application. It is the stress level at which a material will show a permanent elongation after the load is released.

To compare the elastic properties of steel, both soft and hard, a definite amount of permanent elongation is used as the criterion of yield strength. This is generally 0.2% of the two-inch gauge length used. Heat treatment is used to improve the yield strength.

Hardness. Hardness is the ability to resist penetration, or the ability to withstand abrasion. It is an important property in selecting tool materials. Hardness alone does not determine the wear resistance or abrasion resistance of a material. In alloy steels, especially tool steels, the resistance to wear or abrasion varies with the alloy content.

Rockwell Hardness. This is the most widely used method for measuring the hardness of steel. The test is made using a dead weight acting through a series of levers to force a penetrator into the surface of the metal being tested. A micrometer dial gauge tells the depth to which the penetrator sinks. The softer the metal being tested, the deeper it will penetrate with a given load. The dial gauge does not read directly in depth of penetration, but shows arbitrary scales of "Rockwell numbers". A variety of loadings can be used, each designated by a different letter, and the relative hardness or softness is measured.

Two types of penetrators are used: a diamond cone known as a brale, for hard materials such as a hardened tool steel, and a hardened steel ball for testing soft materials.

Brinell Hardness. This method of hardness measurement is much older than the Rockwell. It operates similarly to the Rockwell ball-test principle. In the Brinell machine a much larger steel ball is forced into the material being tested under a load of 3000 kilograms (approximately 6500 pounds). Instead of measuring the penetration, diameter of the impression in the test piece is measured with a small hand microscope with a lens calibrated in millimeters. The measured diameter is converted by means of a table into a Brinell hardness number.

The Brinell hardness measurement is most useful on soft and medium hard materials. On steels of high hardness, the impression is so small it is difficult to read; therefore, the Rockwell test is used more commonly for such materials. A comparison of the designations for each system, as well as other hardness tests, is shown in *Figure 2-4*.

Wear Resistance. Wear resistance is the property of a material which allows it to resist wearing by abrasive action. This property is normally linked to hardness; so if a material is hard, it is usually wear resistant as well.

Toughness. Toughness is a property which permits a material to resist fracture or deformation from sudden loading or shocks. In most cases, a hard material is also tough. However, if the hardness goes beyond a certain point (normally Rockwell C57), toughness is replaced by brittleness.

Brittleness. Brittleness is the property of a material which causes it to break or fracture easily when sudden loads are applied. Most materials which are extremely hard are also very brittle. While brittleness is often thought of as a negative property in most materials, in some applications, such as shear pins, brittleness is a desirable property.

Plasticity. Plasticity is the property of a material which allows it to be extensively deformed without fracture. Two general categories of plasticity are ductility and malleability.

Ductility. Ductility is the property of a material which allows it to be stretched or drawn with a tensional force without fracture or rupture.

Malleability. Malleability is the property of a material which permits it to be hammered or rolled without fracture or rupture.

Modulus of Elasticity (Bending). This is a measure of the stiffness of a material. It is indicated by the slope of the line generated below the elastic limit during tensile testing. The modulus cannot be materially altered by heat

Figure 2-4. Approximate relations among various hardness scales for steel*

Diam, mm (3,000-kg 10-mm ball)	Brinell Ball — Steel	Brinell Ball — Tungsten-carbide	Vickers or Firth	Rockwell Standard — C 150 kg (diamond Brale)	Rockwell Standard — B 100 kg (1/16-in ball)	Rockwell Superficial — 30-N 30 kg (diamond Brale)	Rockwell Superficial — 30-T 30 kg (1/16-in ball)	Knoop	Sclero-scope	Monotron	Herbert pendulum time (steel ball)	Tensile strength, 1,000 psi
2.25	745	840	1050	68		86		822	100	122	85	368
2.30	712	812	960	66		84		787	95	111	80	352
2.35	682	794	885	64		83		753	91	103	76	337
2.40	653	760	820	62		81		720	87	96	72	324
2.45	627	724	765	60		79		688	84	91	67	311
2.50	601	682	717	58		78		657	81	86	63	298
2.55	578	646	675	57		76		642	78	82	60	287
2.60	555	614	633	55	*120*	74		610	75	78	56	*276*
2.65	534	578	598	53	*119*	72		578	72	74	53	*266*
2.70	*514*	555	567	52	*119*	70		563	70	71	51	*256*
2.75	495	525	540	50	*117*	69		533	67	68	48	*247*
2.80	477	514	515	49	*117*	68		520	65	66	47	*238*
2.85	461	477	494	47	*116*	67		493	63	63	44	*229*
2.90	444	460	472	46	*115*	66		480	61	61	41	*220*
2.95	429	432	454	45	*115*	65		467	59	59	39	*212*
3.00	415	418	437	44	*114*	64		455	57	57	37	*204*
3.05	401	401	420	42	*113*	63		432	55	55	35	*196*
3.10	388	388	404	41	*112*	62		421	54	53	34	*189*
3.15	375	375	389	40	*112*	61		410	52	51	33	*182*
3.20	363	364	375	38	*110*	60		388	51	50	32	*176*
3.25	352	352	363	37	*110*	59		378	49	48	30	*170*
3.30	341	341	350	36	*109*	58		368	48	47	29	*165*
3.35	331	330	339	35	*109*	57		359	46	45	28	*160*
3.40	321	321	327	34	*108*	56		350	45	44	27	*155*
3.45	311	311	316	33	*108*	55		341	44	42	26	*150*
3.50	302	302	305	32	*107*	54		332	43	41	24	*150*
3.55	293		296	31	*106*	53		323	42	40	24	*146*
3.60	285		287	30	*105*	52		315	40	39	23	*142*
3.65	277		279	29	*104*	51		307	39	38	23	*138*
3.70	269		270	28	*104*	50		299	38	37	22	*134*
3.75	262		263	26	*103*	49		285	37	36	22	*131*
3.80	255		256	25	*102*	48		278	37	35	21	*128*
3.85	248		248	24	*102*	47		272	36	34	21	*125*
												122

Figure 2-4. (Continued)

3.90	241	241	23	100	46	85	266	35	33	20	*119*
3.95	235	235	22	99	45	84	258	34	32	20	*116*
4.00	229	229	21	98	44	83	250	33	31	19	*113*
4.05	223	223	20	97	43	82	243	32	30	19	*110*
4.10	217	217	*18*	96	42	82	237	31	29	19	*107*
4.15	212	212	*17*	95	40	81	237	31	29	18	*104*
4.20	207	207	*16*	94	39	80	230	30	28	18	*101*
4.25	202	202	*15*	93	38	79	224	30	28	18	99
4.30	197	197	*13*	92	37	78	218	29	27	17	97
4.35	192	192	*12*	91	36	78	213	28	26	17	95
4.40	187	187	*10*	90	35	77	209	28	26	17	93
4.45	183	183	*9*	89	34	77	205	27	25	17	93
4.50	179	179	*8*	88	33	76	201	26	25	16	91
4.55	174	174	*7*	87	32	76	197	26	24	16	89
4.60	170	170	*6*	86	31	75	193	25	23	16	87
4.65	166	166	*4*	85	30	74	189	25	23	16	85
4.70	163	163	*3*	84	29	73	186	24	22	16	83
4.75	159	159	*2*	83	28	73	182	24	22	15	82
4.80	156	156	*1*	82	27	72	179	23	21	15	80
4.85	153	153		81		71	176	23	21	15	78
4.90	149	149		80		71	173	22	20	15	76
4.95	146	146		79		70	170	22	20	15	75
5.00	143	143		78		69	167	21	19	14	74
5.05	140	140		77		68	164	21	19	14	72
5.10	137	137		76		68	161	21	19	14	71
5.15	134	134		74		67	158	20	18	14	71
5.20	131	131		73		67	154		18	14	70
5.25	128	128		72		66	152		18		68
5.30	126	126		71		66	150		17		66
5.35	124	124		70		65	148		17		65
5.40	121	121		69		64	146		17		64
5.45	118	118		68		63	144		16		63
5.50	116	116		67		63	142		16		62
5.55	114	114		66		62	140		16		61
5.60	112	112		65		61	139		15		60
5.65	109	109		64		60	137		15		59
5.70	107	107		62		59	135		14		58
5.75	105	105		61			132				56
5.80	103	103					130				55

*Figures in *italics* should be used only as a guide.

treatment. This is an important property to take into consideration when designing long tools and machine parts. Section size must then be taken into consideration.

Modulus of Elasticity (Torsion). This modulus corresponds to the modulus of elasticity in the tensile test except that it is measured in a torsion test. The modulus is the ratio of the unit shear stress to the displacement caused by the stress per unit length in the elastic range. It is a usable value when designing shafts, taps, twist drills, or other tools working in torsion. Values should be used as a guide when available.

Ferrous Tool Materials

Many ferrous materials may be used in tool construction. Typically, materials such as carbon steel, alloy steel, tool steel, and cast iron are widely used for jigs, fixtures, dies, and similar special tools. These materials are supplied in several different forms. The most common types used for tools are hot rolled, cold rolled, and ground.

When steel is hot rolled at a mill, a layer of decarburized slag, or bark, covers the entire surface of the metal. This bark should be removed when the part being made is to be hardened. If, however, the metal is to be used in an unhardened condition, the bark may be left on the part. When ordering hot rolled materials, the designer must make allowance for the removal of the bark.

Cold rolled steels are generally used for applications when little or no machining or welding are required. Cold rolled bars are reasonably accurate and relatively close to size. When rolled, these steels develop internal stresses which could warp or distort the part if it were extensively machined or welded. Cold rolled bars are distinguished from hot rolled bars by the bright, scale-free surface.

Steels are also available in a ground condition. These materials are held to very close tolerances and are available commercially in many sizes and shapes. These materials are normally used where a finished surface is required without additional machining. The two standard types of ground materials are to-size and over-size. To-size materials are ground to a specific size, such as 0.25" (6.35 mm), 0.50" (12.7 mm) or any similarly standard size. Over-size materials are normally ground 0.015" (0.4 mm) over the standard size.

The following is a brief description of the most common applications of these ferrous tool materials.

Carbon Steels. Carbon steels are used extensively in tool construction. Carbon steels are those steels which only contain iron and carbon, and small amounts of other alloying elements. Carbon steels are the most common and least expensive type of steel used for tools. The three principal types of carbon

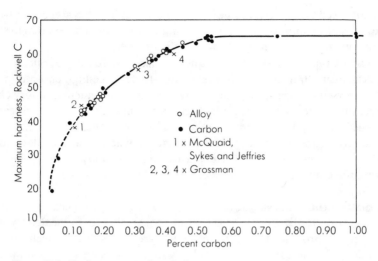

Figure 2-5. Carbon content in relation to hardness possible.

steels used for tooling are low carbon, medium carbon, and high carbon steels. Low carbon steel contains between 0.05 and 0.30% carbon. Medium carbon steel contains between 0.30 and 0.70% carbon. And high carbon steel contains between 0.70 and 1.50% carbon. As the carbon content is increased in carbon steel, the strength, toughness, and hardness also increase when the metal is heat treated (*Figure 2-5*).

Low carbon steels are soft, tough steels that are easily machined and welded. Due to their low carbon content, these steels cannot be hardened except by case hardening. Low carbon steels are well suited for the following applications: tool bodies, handles, die shoes, and similar situations where strength and wear resistance are not required.

Medium carbon steels are used where greater strength and toughness is required. Since medium carbon steels have a higher carbon content they can be heat treated to make parts such as studs, pins, axles, and nuts. Steels in this group are more expensive as well as more difficult to machine and weld than low carbon steels.

High carbon steels are the most hardenable type of carbon steel and are used frequently for parts where wear resistance is an important factor. Other applications where high carbon steels are well suited include drill bushings, locators, and wear pads. Since the carbon content of these steels is so high, parts made from high carbon steel are normally difficult to machine and weld.

Alloy Steels. Alloy steels are basically carbon steels with additional elements added to alter the characteristics and bring about a predictable change in the mechanical properties of the alloyed metal. Alloy steels are not normally used for most tools due to their increased cost, but some have found favor for special applications. The alloying elements used most often in steels are manganese, nickel, molybdenum, and chromium.

Another type of alloy steel frequently used for tooling applications is stainless steel. Stainless steel is a term used to describe high chromium and nickel-chromium steels. These steels are used for tools which must resist high temperatures and corrosive atmospheres. Some high chromium steels can be hardened by heat treatment and are used where resistance to wear, abrasion, and corrosion are required. Typical applications where a hardenable stainless steel is sometimes preferred are plastic injection molds. Here the high chromium content allows the steel to be highly polished and prevents deterioration of the cavity from heat and corrosion.

Tool Steels. Tool steels are alloy steels which are produced primarily for use in cutting tools. Proper selection of tool steels is complicated by their many special properties. The five principal properties of tool steels are heat resistance, abrasion resistance, shock resistance, resistance to movement or distortion in hardening, and cutting ability.

Because no one steel can possess all of these properties to the optimum degree, hundreds of different tool steels have been developed to meet the total range of service demands.

The steels listed in *Figure 2-6* will adequately serve 95% of all metal stamping operations. The list contains 31 steels, nine of which are widely applied and readily available. The other steels represent slight variations for improved performance in certain instances, and their use is sometimes justified because of special considerations.

Tool steels are identified by letter and number symbols. All the steels listed, except those in the *S* and *H* groups, can be heat treated to a hardness greater than Rockwell C62 and, accordingly, are hard, strong, wear resistant materials. Frequently hardness is proportional to wear-resistance, but this is not always the case, because the wear resistance usually increases as the alloy content, and particularly the carbon content, increases.

The toughness of steels, on the other hand, is inversely proportional to the hardness and increases markedly as the alloy content or the carbon content is lowered.

Figure 2-7 lists the basic characteristics, *Figure 2-8* the hardening and tempering treatments, and *Figure 2-9* shows typical applications of the various steels listed.

The general nature and application of the various standard tool steel classes are as follows:

W: Water-Hardening Tool-Steels. This group includes plain carbon (W1) and carbon vanadium (W2). Carbon steels were the original tool steels. Because of their low cost, abrasion-resisting and shock-resisting qualities,

Steel types*	Average Composition, percent							
	C	Mn	Si	Cr	W	Mo	V	Other
W1	1.00							
W2	1.00						0.25	
O1	0.90	1.00		0.50	0.50			
O2	0.90	1.60						
O7	1.20			0.75	1.75	0.25		
A2	1.00			5.00		1.00		
A4	1.00	2.00		1.00		1.00		
A5	1.00	3.00		1.00		1.00		
A6	0.70	2.00		1.00		1.00		
D2	1.50			12.00		1.00		
D3	2.25			12.00				
D4	2.25			12.00		1.00		
D6	2.25		1.00	12.00	1.00			
S1	0.50			1.50	2.50			
S2	0.50		1.00			0.50		
S4	0.50	0.80	2.00					
S5	0.50	0.80	2.00			0.40		
H11	0.35			5.00		1.50		
H12	0.35			5.00	1.50	1.50	0.40	
H13	0.35			5.00		1.50	1.00	
H21	0.35			3.50	9.00			
H26	0.50			4.00	18.00		1.00	
T1	0.70			4.00	18.00		1.00	
T15	1.50			4.00	12.00		5.00	5.00 Co
M2	0.85			4.00	6.25	5.00	2.00	
M3	1.00			4.00	6.00	5.00	2.40	
M4	1.30			4.00	5.50	4.50	4.00	
L2	0.50			1.00			0.20	
L3	1.00			1.50			0.20	
L6	0.70			0.75				1.50 Ni
F2	1.25				3.50			

* W, water-hardening; O, oil-hardening, cold-work; A, air-hardening, medium-alloy; D, high-carbon high-chromium, cold-work; S, shock-resisting; H, hot-work; T, tungsten-base high-speed; M, molybdenum-base high-speed; L, special-purpose, low-alloy; F, carbon-tungsten, special-purpose.

Figure 2-6. AISI identification and classification of tool steels.

AISI Steel No.	Non-deforming properties	Safety in hardening	Toughness	Resistance to softening effect of heat	Wear resistance	Machin-ability
W1	Poor	Fair	Good	Poor	Fair	Best
W2	Poor	Fair	Good	Poor	Fair	Best
O1	Good	Good	Fair	Poor	Fair	Good
O2	Good	Good	Fair	Poor	Fair	Good
O7	Good	Good	Fair	Poor	Fair	Good
A2	Best	Best	Fair	Fair	Good	Fair
A4	Best	Best	Fair	Poor	Fair	Fair
A5	Best	Best	Fair	Poor	Fair	Fair
A6	Best	Best	Fair	Poor	Fair	Fair
D2	Best	Best	Fair	Fair	Good	Poor
D3	Good	Good	Poor	Fair	Best	Poor
D4	Best	Best	Poor	Fair	Best	Poor
D6	Good	Good	Poor	Fair	Best	Poor
S1	Fair	Good	Good	Fair	Fair	Fair
S2	Poor	Fair	Best	Fair	Fair	Fair
S4	Poor	Fair	Best	Fair	Fair	Fair
S5	Fair	Good	Best	Fair	Fair	Fair
H11	Best	Best	Best	Good	Fair	Fair
H12	Best	Best	Best	Good	Fair	Fair
H13	Best	Best	Best	Good	Fair	Fair
H21	Good	Good	Good	Good	Fair	Fair
H26	Good	Good	Good	Best	Good	Fair
T1	Good	Good	Fair	Best	Good	Fair
T15	Good	Fair	Poor	Best	Best	Poor
M2	Good	Fair	Fair	Best	Good	Fair
M3	Good	Fair	Fair	Best	Good	Fair
M4	Good	Fair	Fair	Best	Best	Poor
L2	Fair	Fair	Good	Poor	Fair	Fair
L3	Fair	Poor	Fair	Poor	Fair	Good
L6	Good	Good	Good	Poor	Fair	Fair
F2	Poor	Poor	Poor	Fair	Best	Fair

Figure 2-7. Comparison of basic characteristics of steels used for press tools.

AISI Tool-steel	Preheat Temp., °F (°C)	Rate of heating for hardening	Hardening temp., °F (°C)	Time at temp., min.	Quenching medium	Tempering temp., °F (°C)	Depth of hardening	Resistance to decarburizing
W1		Slow	1425-1500 (744-816)	10-30	Brine or water	325-550 (163-288)	Shallow	Best
W2		Slow	1425-1550 (744-843)	10-30	Brine or water	325-550 (163-288)	Shallow	Best
O1	1200 (649)	Very slow	1450-1500 (778-816)	10-30	Oil	325-500 (163-260)	Medium	Good
O2	1200 (649)	Very slow	1400-1475 (760-802)	Do not soak	Oil	325-600 (163-316)	Medium	Good
O7	1200 (649)	Slow	1575-1625 (857-885)	10-30	Oil	350-550 (177-288)	Medium	Good
A2	1450 (788)	Very slow	1700-1800 (927-982)	30	Air	350-700 (177-371)	Deep	Fair
A4	1250 (677)	Slow	1450-1550 (788-843)	15-30	Air	300-500 (149-260)	Deep	Very good
A5	1250 (677)	Slow	1450-1550 (788-843)	15-30	Air	300-500 (149-260)	Deep	Very good
A6	1250 (677)	Slow	1500-1600 (816-871)	15-30	Air	300-500 (149-260)	Deep	Very good
S1		Slow to 1400	1650-1750 (899-954)	10-30	Oil	500-600 (260-316)	Medium	Fair
S2		Slow	1525-1575 (829-857)	10-30	Brine or water	350-700 (177-371)		Fair
S4		Slow	1550-1650 (843-899)	10-30	Brine or water	350-700 (177-371)		Poor
S5			1600-1700 (871-927)		Oil	350-700 (177-371)	Medium	Poor
H11	1400 (760)	Slow	1800-1850 (982-1010)	15-60	Air	900-1200 (482-649)	Deep	Good
H12	1400 (760)	Slow	1800-1850 (982-1010)	15-60	Air	900-1200 (482-649)	Deep	Good
H13	1400 (760)	Slow	1800-1850 (982-1010)	15-60	Air	900-1200 (482-649)	Deep	Good
H21	1550 (843)	Medium	2000-2200 (1093-1204)	5-15	Air, oil	1000-1200 (538-649)	Deep	Good
H26	1550 (843)	Medium	2000-2200 (1093-1204)	5-15	Air, oil	1000-1200 (538-649)	Deep	Good
T1	1500-1600 (816-871)	Rapid from preheat	2150-2300 (1177-1260)	Do not soak	Air, oil or salt	1025-1200 (552-649)	Deep	Good
T15	1500-1600 (816-871)	Rapid from preheat	2125-2270 (1163-1243)	Do not soak	Air, oil or salt	1025-1200 (552-649)	Deep	Fair
M2	1500 (816)	Rapid from preheat	2125-2225 (1163-1218)	Do not soak	Air, oil or salt	1025-1200 (552-649)	Deep	Poor
M3	1450-1550 (788-843)	Slow	2125-2225 (1163-1218)	Do not soak	Air, oil or salt	1025-1200 (552-649)	Deep	Poor
M4	1450-1550 (788-843)	Slow	2125-2225 (1163-1218)	Do not soak	Air, oil or salt	1025-1200 (552-649)	Deep	Poor
L2		Slow	1550-1700 (843-927)	15-30	Oil	350-600 (177-316)	Medium	Good
L3		Slow	1425-1500 (774-816)	10-30	Brine or water	300-800 (149-427)	Medium	Good
L3		Slow	1500-1600 (816-871)	10-30	Oil	300-800 (149-427)	Medium	Good
L6		Slow	1450-1550 (778-843)	10-30	Oil	300-1000 (149-538)	Medium	Fair
F2	1200 (649)	Slow	1525-1625 (829-885)	15-30	Brine or water	300-500 (149-260)	Shallow	Good

Figure 2-8. Hardening and tempering treatments for press tools.

Application	Suggested AISI tool-steels	Rockwell C hardness range
Arbors	L6, L2	47-54
Axle burnishing tools	M2, M3	63-67
Boring bars	L6, L2	47-54
Broaches	M2, M3	63-67
Bushings (drill jig)	M2, D2	62-64
Cams	A4, O1	59-62
Centers, lathe	D2, M2	60-63
Chasers	M2	62-65
Cutting tools	M2	62-65
Dies, blanking	O1, A2, D2	58-62
Dies, bending	S1, A2, D2	52-62
Dies, coining	S1, A2, D2	52-62
Dies, cold heading:		
Solid	W1, W2	56-62
Insert	D2, M2	57-62
Dies, hot heading	H12, H13	42-48
Dies, lamination	D2, D3	60-63
Dies, shaving	D2, M2	62-64
Dies, thread rolling	D2, A2	58-62
Die casting:		
Aluminum	H13	42-48
Form tools	M2, M3	63-67
Lathe tools	M2, T1	63-65
Reamers	M2	63-65
Shear blades:		
Light stock	D2, A2	58-61
Heavy stock	S1, S4	52-56
Rolls	A2, D2	58-62
Taps	M2	62-65
Vise jaws	L2, S4	48-54
Wrenches	L2, S1	40-50

Figure 2-9. Applications of tool steels.

ease of machinability, and ability to take a keen cutting edge, the carbon grades are widely applied. Both W1 and W2 steels are shallow-hardening and are readily available.

O: Oil-Hardening Tool Steels. Types O1 and O2 are manganese oil-hardening tool steels. They are readily available and inexpensive. These steels have less internal molecular movement than the water-hardening steels, and are of equal toughness with water-hardening steels when the latter are hardened throughout. Wear resistance is slightly better than that of water-hardening steels of equal carbon content. Steel O7 has greater wear resistance because of its increased carbon and tungsten content.

A: Air-Hardening Die Steels. Type A2 is the principal air-hardening tool steel. It has minimum movement in hardening and has higher toughness than the oil-hardening die steels, with equal or greater wear resistance. Steels A4, A5, and A6 can be hardened from lower temperatures, but have lower wear resistance and better distortional properties.

D: High-Carbon High-Chromium Die Steels. Type D2 is the principal steel in this class. It finds wide application for long-run dies. It is deep-hardening, fairly tough, and has good resistance to wear. Steels D3, D4, and D6, containing additional carbon, have very high wear resistance and lower toughness. Steels D2 and D4 are air-hardened.

S: Shock-Resisting Tool Steels. These steels contain less carbon and have higher toughness. They are applied where heavy cutting or forming operations are required, and where breakage is a serious problem. Steels S1, S4, and S5 are readily available. Steels S4 and S5 are more economical than S1.

H: Hot-Work Die Steels. These steels combine red hardness with good wear resistance and shock resistance. They are air-hardening and on occasion are used for cold-work applications. They have relatively low carbon content and intermediate to high alloy content.

T and M: Tungsten and Molybdenum High-Speed Steels. Steels T1 and M2 are equivalent in performance and have good red hardness and abrasion resistance. They have higher toughness than many of the other die steels. They may be hardened by conventional methods or carburized for cold-work applications. Steels M3, M4, and T15 have greater cutting ability and resistance to wear. They are more difficult to machine and grind because of their increased carbon and alloy contents.

L: Low-Alloy Tool Steels. Steels L3 and L6 are used for special die applications. Other L steels find application where fatigue and toughness are important considerations, such as in coining or impression dies.

F: Finishing Steels. Steel F2 is of limited use but occasionally applied where extremely high wear resistance in a shallow-hardening steel is desired.

Cast Iron. Cast iron is essentially an alloy of iron and carbon, containing from 2-4% carbon, 0.5 to about 3.00% silicon, 0.4 to approximately 1% manganese, plus phosphorus and sulphur. Other alloys may be added depending on the properties desired.

The high compressive strength and ease of casting of the gray irons are utilized in large forming and drawing dies to produce such items as automobile panels, refrigerator cabinets, bath tubs, and other large articles. Conventional methods of hardening result in little distortion.

Alloying elements are added to contract graphitization, to improve mechanical properties, or to develop special characteristics.

Nonferrous Tool Materials

Nonferrous tool materials are used to some degree as die materials in special applications, and generally for limited-production requirements. On the other hand, in jig and fixture design some nonferrous materials are used extensively where magnetism or tool weight are important factors. Another area where nonferrous materials are finding increased use is for cutting tools. Here alloys and compositions of nonferrous materials are used extensively to machine the newer, exotic, high-strength metals. The following is a brief description of the typical nonferrous materials used for special purpose tooling.

Aluminum. Aluminum is a nonferrous metal which has been used for special tooling for quite some time. The principal advantages in using aluminum are high strength-to-weight ratio, nonmagnetic properties, and relative ease in machining and forming. Pure aluminum is corrosion resistant, but not well suited for use as a tooling material except in very limited low strength applications. Aluminum alloys, while not as corrosion resistant as pure aluminum, are much stronger and are well suited for many special tooling applications. The alloys most frequently used for tooling applications are aluminum/copper (2000 series) and aluminum/zinc (7000 series). Depending on composition, some aluminum alloys are weldable and some can be heat treated. One form of aluminum alloy finding increased use today is aluminum tooling plate. This material is available in sheets and bars made to very close tolerances. Aluminum tooling plate is very useful for a wide variety of tooling applications. From supports and locators to base plates and tool bodies, aluminum tooling plate provides a lightweight alternative to steel. Other variations of aluminum frequently used for tooling are aluminum extrusions and cast bracket materials. In most cases these materials can be used as is with little or no machining required.

Magnesium. Magnesium, like aluminum, is a lightweight yet strong tooling material. Magnesium is lighter than aluminum and has a very good strength-to-weight ratio. Magnesium is commercially available in sheets, bars, and extruded forms. The only disadvantage to using magnesium is its potential fire hazard. When specifying magnesium as a tooling material make sure those who are to make the various parts are well acquainted with the precautions which must be observed when machining this material.

Bismuth Alloys. Bismuth alloys have several different uses in special tools. One application is as a matrix material for securing punch and die parts in a die assembly, and as cast punches and dies for short-run forming and drawing operations. Another frequent application of these alloys is for cast workholders. Here the material is melted and poured around the part and once cool, the part is removed and the cast nest is used to hold subsequent parts for machining.

One of the principal advantages of bismuth alloys is their very low melting temperature. Many alloy compositions will melt in boiling water. In addition to acting as a reusable nesting material, these low melt alloys are also useful when machining parts with very thin cross sections, such as turbine blades. In these applications, the material is cast around the thin sections and acts as a support during machining. Once the machining is complete, the material is melted off the part and can be reused.

Carbides. Carbides are a family of tool materials made from the carbides of tungsten, titanium, tantalum or a combination of these elements. They are powder metals consisting of the carbide with a binder—usually cobalt—hot pressed or sintered to desired shapes. The most common carbide material used for special tools is tungsten carbide. All carbides are characterized by high hardness values and resistance to wear making them an excellent choice for cutting tools. The specific grades and characteristics commonly used to classify carbides are shown in *Figure 2-10* and *Figure 2-11*. *Figure 2-10* shows the nonofficial C-classification system and *Figure 2-11* shows the International Standards Organization (ISO) system.

Nonmetallic Tool Materials

Nonmetallic tool materials are chiefly used where the production of parts is limited and where the cost of using tool steels or similar materials would not

Code	Application	Carbide characteristics
C-1	Roughing	Medium-high shock resistance Medium-low wear resistance
C-2	General-purpose	Medium shock resistance Medium wear resistance
C-3	Finishng	Medium-low shock resistance Medium-high wear resistance
C-4	Preicsion finishing	Low shock resistance High wear resistance
C-5	Roughing	Excellent resistance to cutting temperature Shock and cutting load Medium wear resistance
C-50	Roughing and heavy feeds	Same as above
C-6	General-purpose	Medium-high shock resistance Medium wear resistance
C-7	Finishing	Medium shock resistance Medium wear resistance
C-70	Semifinishing and finishing	High cutting-temperature resistance Medium wear resistance
C-8	Precision finishing	Very high wear resistance Low shock resistance

Toughness increases from bottom to top, hardness from top to bottom.

Figure 2-10. JIC Carbide-classification code.

Main machining group	Color marking	Application group	Operations and working conditions
P: steel, cast steel, long-chipping malleable	Blue	P01	High-precision turning and boring, high cutting speeds, small chip cross section, dimensional accuracy, good surface finish, and vibration-free machining
		P10	Turning, copy turning, thread cutting and milling, high cutting speeds, and small to medium chip cross section
		P20	Turning, copy turning, milling, medium cutting speeds, and medium chip cross section; planing with small chip cross section
		P30	Turning, milling, planing, medium to low cutting speeds, medium to large chip cross section, also under unfavorable conditions
		P40	Turning, planing, milling, shaping, low cutting speeds, large chip cross section, high rake angles, unfavorable conditions; also automatic turning
		P50	Where highest demands are made on toughness of carbide: turning, planing and shaping, low cutting speeds, large chip cross section, and high rakes under unfavorable conditions. Automatic turning
M: steel, cast steel, austenitic manganese steel, cast-iron alloys, austenitic steels, malleable and spheroidal cast iron, free-cutting mild steel	Yellow	M10	Turning, medium-high cutting speeds, small to medium chip cross sections
		M20	Turning, milling, medium cutting speeds, and medium chip cross section
		M30	Turning, milling, planing, medium cutting speeds, medium to large chip cross sections
		M40	Turning, form turning, parting off and recessing, particularly for automatics
K: cast iron, chilled cast iron, short-chipping malleable cast iron, hardened steel, non-ferrous metals, non-metallic materials	Red	K01	Turning, precision turning and precision boring, finish milling, and scraping
		K10	Turning, milling, boring, countersinking, reaming, scraping, and broaching
		K20	Turning, milling, planing, countersinking, scraping, reaming, and broaching under tougher conditions than K10
		K30	Turning, milling, planing, shaping under unfavorable conditions, high rakes
		K40	Turning, milling, planing, shaping under unfavorable conditions, high rakes

Cutting speed and wear resistance increase from bottom to top, feed and carbide toughness from top to bottom.

Figure 2-11. ISO Carbide-classification system.

be economically practical. In many cases where nonmetallic tool materials are used for special tools they are used in conjunction with steel parts such as bushings or blades rather than by themselves. However, in other applications, these nonmetallic materials may be used alone. The principal nonmetallic materials used for special tooling are wood, composition materials, plastics, epoxy resins, rubber, urethane, ceramics, and diamonds.

Wood. Wood is a material frequently used for low cost, limited production tooling. Typical applications include jig plates with inserted steel bushings and backing and support parts for steel rule dies. When working with wood the designer must anticipate the problems inherent in this material. For example, wood has a tendency to swell and warp. However, by selecting a relatively stable type of wood and properly positioning the parts, *Figure 2-12,* these problems can be minimized or eliminated.

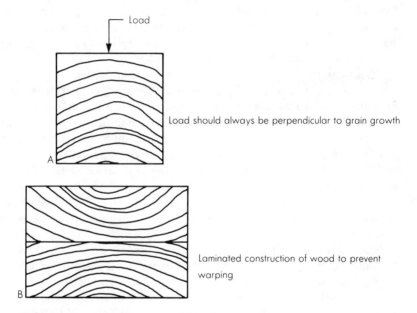

Figure 2-12. Using wood properly.

Some variations of wood products often used in tooling applications include hardboard, densified wood, plywood, and particle board. Hardboard is basically a material made of compressed wood fiber. Typical uses for this material include forming punches and dies, form blocks for rubber forming, and for stretch dies. Densified woods are woods which have been impregnated with a phenolic resin and laminated. After lamination, the assembled parts are compressed to about 50% of the original thickness of the wood layers. Densified woods are used for basically the same applications as hardboard. Plywood may be used for special tooling in either its natural condition or as a clad plywood. Clad plywood has a thin sheet of metal applied to help it resist wear and damage in use. Plywood may be used for tumble jigs, steel rule dies, milling fixtures, and stretch forming dies. When using plywood for any tooling application, it is best to use an exterior plywood grade to prevent the wood plies from separating. Particle board is a composition material made from wood chips and epoxy resins. These materials are available in several grades and serve basically the same functions as plywood. When using any type of wood or wood byproduct, it is always a good precaution to coat the tool with a lacquer or shellac to preserve the wood and to help reduce swelling and warpage.

Composition Materials. Composition are materials normally made from a filler, or base material, and some type of resin acting as a binder. Two typical composition materials used for special tooling are phenolic and Bakelite. Both materials are used for the same applications as wood, but are more stable and less susceptible to moisture.

Plastics. Plastics are some of the newer tooling materials, and are used where the operations are not severe and the production run is short to medium. Plastics are resistant to chemicals, moisture, and temperature. They are inexpensive and facilitate tool repair and modification. In most cases, plastics can be machined with the same tools and equipment as metals and can be easily adapted to toolroom uses. Plastics have been recently developed to withstand high heat, abrasion, and some have a nonstick surface that makes an excellent sliding surface. Some newer "space-age" plastics have tensile and shear strength equal to steels. With constant research and development in the plastics industry, plastics will play an increasingly important role as a tool material.

Epoxy Resins. Epoxy resins are mainly used for casting and laminating. Castable resins are used for jig plates, workholders, silk screen fixtures, duplicating patterns, and for large forming dies. In addition to the resin, a filler material is added to the mixture to increase its strength and to provide better dimensional stability. Typical filler materials include glass beads, metal beads, and metal filings. Cast resins are strong and relatively lightweight. When properly cast, these resins require little or no machining. Laminated resins are used for large, stretch forming dies and checking fixtures. These materials are generally laminated over a wooden frame.

In either case, epoxy resins combine low cost, ease of modification, and shortened lead times into a single tooling material. They work well with intricate or complex part shapes, and depending on the filler material, will normally last a long time.

Rubber. Today, rubber is used less than it was in earlier times. This is due in large part to the newer and better materials which have been developed. But rubber is still used for specialized drawing, blanking, and bulging die operations as well as protective elements for other special tools.

Urethane. Urethane is a material that is becoming widely used for special tooling. It is available in solid bars that can be machined to suit a specific application or cast into almost any desired form. Urethane is noncompressive and acts as a liquid when force is applied. That is, when force is applied, urethane displaces the force equally in all directions. By containing and redirecting these displaced forces, urethane can be used to form complex shapes without marring the workpiece material. When used as a clamp pad, urethane transfers all the clamping forces without damaging the workpiece surfaces. Urethane is also used as a stripper in some larger, low production, blanking dies.

Urethane does not shrink an appreciable amount and can be used to duplicate parts exactly. Lack of shrinkage makes urethane ideal for nests for ultrasonic fixtures or molds for model parts. Urethane is also used for embossing or shallow forming dies. When used in forming, only the die is made from urethane, the punch is normally made from steel or a similar material.

Ceramics. Ceramics, or oxide cutting tools, are basically aluminum oxide materials. Ceramics have a high compressive strength, high red hardness, high abrasive resistance, low heat conductivity, and good resistance to galling and welding. They are harder than carbides, but will chip or break easily when bending or twisting loads are applied. Ceramics should not be used for interrupted cuts since they have a low resistance to shock loads. For this reason, the machine selected to use ceramic cutting tools must be extremely rigid. Ceramic cutting tools are used to machine cast iron, carbon steels, low alloy steels, and for finishing hard steels (Rockwell C60 and 65) at high speeds. They may also be used for machining carbon, graphite, fiberglass, and other highly abrasive materials.

Diamonds. Diamonds are the hardest substance known, but have only a limited use as a tool material. Industrial diamonds are either synthetic or natural and are used for turning tools, grinding wheels, and for grinding wheel dressers. Diamonds are frequently used for turning plastics, precious metals, nonferrous metals, and general finishing operations with light cuts, fine feeds, and high cutting speeds. Diamond powder is used for lapping and polishing. These powders, or flours, create a very smooth finish and high luster especially on ferrous materials.

Other Materials. Cubic Boron Nitride (CBN) is a product similar to diamond. It is used for grinding wheels and cutting tools. CBN can be used on ferrous materials. It is a granular material compacted in a binder.

Heat Treating

The purpose of heat treatment is to control the properties of a metal or alloy through the alteration of the structure of the metal or alloy by heating it to definite temperatures and cooling at various rates. This combination of heating and controlled cooling determines not only the nature and distribution of the microconstituents, which in turn determine the properties, but also the grain size.

Heat treating should improve the alloy or metal for the service intended. Some of the various purposes of heat treating are as follows:

1. To remove strains after cold working.
2. To remove internal stresses such as those produced by drawing, bending, or welding.
3. To increase the hardness of the material.
4. To improve machinability.
5. To improve the cutting capabilities of tools.
6. To increase wear-resisting properties.
7. To soften the material, as in annealing.
8. To improve or change properties of a material such as corrosion resistance, heat resistance, magnetic properties, or others as required.

Treatment of Ferrous Materials

Iron is the major constituent in the steels used in tooling, to which carbon is added in order that the steel may harden. Alloys are put into steel to enable it to develop properties not possessed by plain carbon steel, such as the ability to harden in oil or air, increased wear resistance, higher toughness, and greater safety in hardening.

Heat treatment of ferrous materials involves several important operations which are customarily referred to under various headings, such as normalizing, spheriodizing, stress relieving, annealing, hardening, tempering, and case hardening.

Normalizing. Normalizing involves heating the material to a temperature of about 100° to 200° F (55 to 110° C) above the critical range and cooling in still air. This is about 100° F (55° C) over the regular hardening temperature.

The purpose of normalizing is usually to refine grain structures that have been coarsened in forging. With most of the medium-carbon forging steels, alloyed and unalloyed, normalizing is highly recommended after forging and before machining to produce more homogeneous structures, and in most cases, improved machinability.

High-alloy air-hardened steels are never normalized, since to do so would cause them to harden and defeat the primary purpose.

Spheroidizing. Spheroidizing is a form of annealing which, in the process of heating and cooling steel, produces a rounded or globular form of carbide—the hard constituent in steel.

Tool steels are normally spheroidized to improve machinability. This is accomplished by heating to a temperature to 1380-1400° F (749-760° C) for carbon steels and higher for many alloy tool steels, holding at heat one to four hours, and cooling slowly in the furnace.

Stress Relieving. This is a method of relieving the internal stresses set up in steel during forming, cold working, and cooling after welding or machining. It is the simplest heat treatment and is accomplished merely by heating to 1200° to 1350° F (649° to 732° C) followed by air or furnace cooling.

Large dies are usually roughed out, then stress-relieved and finish-machined. This will minimize change of shape not only during machining but during subsequent heat treating as well. Welded sections will also have locked-in stresses owing to a combination of differential heating and cooling cycles as well as to changes in cross section. Such stresses will cause considerable movement in machining operations.

Annealing. The process of annealing consists of heating the steel to an elevated temperature for a definite period of time and, usually, cooling it slowly. Annealing is done to produce homogenization and to establish normal equilibrium conditions, with corresponding characteristic properties.

Tool steel is generally purchased in the annealed condition. Sometimes it is necessary to rework a tool that has been hardened, and the tool must then be annealed. For this type of anneal, the steel is heated slightly above its critical range and then cooled very slowly.

Finished parts may be annealed without surface deterioration by placing them in a closed pot and covering with compounds that will combine with the air present to form a reducing atmosphere. Partially spent carburizing compound is widely used, as well as cast iron chips, charcoal, and commercial neutral compounds.

Hardening. This is the process of heating to a temperature above the critical range, and cooling rapidly enough through the critical range to appreciably harden the steel. (See *Figure 2-8* for specific treatment.)

A simplified theory of hardening steel is that iron has two distinct and different atomic arrangements, one existing at room temperature (or again near the melting point), and one above the critical temperature. Without this phenomenon it would be impossible to harden iron-base alloys by heat treatment.

What happens in the heat treatment of die steels is represented graphically in *Figure 2-13*. Starting in the annealed condition at *A*, the steel is soft, consisting of an aggregate of ferrite and carbide. Upon heating above the critical temperature to *B,* the crystal structure of ferrite changes, becomes austenite, and dissolves a large portion of the carbide. The new structure, austenite, is always a prerequisite for hardening. By quenching it (cooling rapidly to room temperature), the carbon is retained in solution, and the structure known as martensite (*C* in diagram) results. This is the hard matrix structure in steels. It is initially high stressed, for the change from austenite

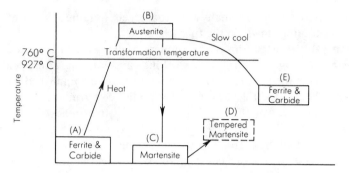

Figure 2-13. Transformations in the hardening of steel.

involves some volumetric expansion against natural stiffness of the steel, so it must be reheated to an intermediate temperature (*D*) to soften it slightly and relieve those internal stresses which unduly embrittle the steel.

If quenching is not rapid enough, the austenite reverts to ferrite and carbide (*E*) and high hardness is not obtained. The rate at which quenching is required to produce martensite depends primarily on the alloy content. Low-alloy die steels are water or oil hardened, while highly alloyed steels usually can be hardened in air, i.e., quenched at a much slower rate. The high alloys make the reaction more sluggish.

Tempering. This is the process of heating quenched and hardened steels and alloys to some temperature below the lower critical temperature to reduce internal stresses setup in hardening. Thus, the hard martensite resulting from the quenching operation is changed in tempering in the direction of the equilibrium properties, the degree being dependent on the tempering temperature and rate of cooling. (See *Figure 2-8* for specific treatment.)

Case Hardening. The addition of carbon to the surface of steel parts and the subsequent hardening operations are important phases in heat treating. The process may involve the use of molten sodium cyanide mixtures, pack carburizing with activated solid material such as charcoal or coke, gas or oil carburizing, and dry cyaniding.

Whether a solid carbonaceous packing material is used, or a liquid gas, the objective is to produce a hard, wear-resistant surface with a core of such hardness or toughness as is best suited for the purpose. The carbon content of the surface is raised to 0.80-1.20% and the case depth can be closely controlled by the time, the temperature, and the carburizing medium used. Pack carburizing is generally done at 1700° F (927° C) for eight hours to produce a case depth of 1/16″ (1.6 mm). Light cases up to 0.005″ (0.13 mm), can be obtained in liquid cyanide baths, and case depths to 1/32″ (0.8 mm) are economically practical in liquid carburizing baths.

Usually, low-carbon and low-carbon alloy steels are carburized. The normal carbon range is 0.10 to 0.30% carbon, though higher carbon-content steels may be carburized as well.

Treatment of Nonferrous Materials

The heat treatment of nonferrous metals and alloys closely approximates that of steel except that the temperature ranges used are lower, and hardening is accomplished by the precipitation of hard metallic compounds or particles.

Nonferrous metals and alloys that are not heat-treatable harden by cold work only.

For the heat-treatable alloys of aluminum, hardening is accomplished by precipitation. When an alloy is water-quenched from the "hardening heat" it is very soft; this is known as the solution treatment. Hardness is accomplished by aging, which follows the quenching operation. The aging temperature for some aluminum alloys is room temperature; others may require an elevated temperature of 290° to 360° F (143° to 182° C), depending on the alloy. As a rule, the lower the aging temperature, the longer the time required for the alloy to reach full hardness.

Beryllium copper is a precipitation-hardening alloy and is usually furnished by the manufacturer in the very soft solution-treated condition. It has excellent forming properties in this condition. Formed parts are hardened by aging at 560° to 620° F (293° to 327° C) for two hours at heat. A hardness of 38 to 42 Rockwell C can be expected.

All other brass and bronze alloys are hardenable only by cold working and may be softened to various degrees by stress relieving or annealing.

TOOL MATERIALS

Review Questions

1. What is meant by the term "physical properties of a material"?
2. What is a mechanical property of a material?
3. How are the strength values of a material normally expressed?
4. What are the two most common systems used to specify the hardness of a part?
5. What is the ability to resist shocks and sudden loads called?
6. What makes a material either ferrous or nonferrous?

7. Which carbon steel is used mainly for tool bodies?
8. What are the five principal properties of tool steels?
9. What is cast iron?
10. What hazard must be considered when using magnesium?
11. Which aluminum alloys are the most commonly used for tooling?
12. What two systems are used to grade carbide?
13. List three variations of wood products sometimes used for special tooling.
14. What are the two principal applications of epoxy resins used for tooling?
15. What should be added to epoxy resins to increase their strength and dimensional stability?
16. Which nonmetallic material is frequently used for forming dies because of its noncompressive properties?
17. What type of cut should not be taken with a ceramic cutting tool?
18. What heat treating process makes the metallic carbides in a metal form into small rounded globules?
19. What process is used to remove the internal stresses created during a hardening operation?
20. In hardenable aluminum, how is hardening accomplished after the part is heated and quenched?

TOOL MATERIALS

Answers to Review Questions

1. Those properties which are natural in the material.
2. Those properties which can be altered, or modified.
3. In units of pounds per square inch (psi) or in kiloPascals (kPa).
4. Rockwell and Brinell.
5. Toughness.
6. The base metal must be iron in ferrous and a metal other than iron in nonferrous.
7. Low carbon steel.
8. Heat resistance, abrasion resistance, shock resistance, resistance to movement or distortion in hardening, and cutting ability.
9. An alloy of iron and carbon.
10. Fire.
11. Aluminum/copper and aluminum/zinc.
12. Joint Industrial Council (JIC) and International Standards Organization ISO).

13. Hardboard, densified wood, plywood, or particle board. (ANY THREE)
14. Casting and laminating.
15. Filler material.
16. Urethane.
17. Interrupted.
18. Spheriodizing.
19. Tempering.
20. Aging.

3

CUTTING TOOL DESIGN

Physics of metal cutting provide the theoretical framework by which we must examine all other elements of cutting tool design. We have workpiece materials from a very soft, buttery consistency to very hard and shear resistant. Each of the workpiece materials must be handled by itself; the amount of broad information that is applicable to each workpiece material is reduced as the distinctions between workpiece characteristics increase. Not only is there a vast diversity of workpiece materials, but there is also a variety of shapes of tools and tool compositions.

The tool designer must match the many variables to provide the best possible cutting geometry. There was a day when trial and error was normal for this decision, but today, with the ever-increasing variety of tools, trial and error is far too expensive.

The designer must develop expertise in applying data and making comparisons on the basis of the experience of others. For example: tool manufacturers and material salesmen will have figures their companies have developed. The figures are meant to be guidelines; however, a careful examination of the literature available will provide an excellent place from which to start, and be much cheaper than trial and error.

The primary method of imparting form and dimension to a workpiece is the removal of material by the use of edged cutting tools. An oversize mass is literally carved to its intended shape. The removal of material from a workpiece is termed generation of form by machining, or simply machining.

Form and dimension may also be achieved by a number of alternative processes such as hot or cold extrusion, sand casting, die casting, and precision casting. Sheet metal can be formed or drawn by the application of pressure. In addition to machining, metal removal can be accomplished by chemical or electrical methods. A great variety of workpieces may be produced without resorting to a machining operation. Economic considerations, however, usually dictate form generation by machining, either as the complete process or in conjunction with another process.

Elements of Machining. Material removal by machining involves interaction of five elements: the cutting tool, the toolholding and guiding device, the workholder, the workpiece, and the machine. The cutting tool may have a single cutting edge or may have many cutting edges. It may be designed for linear or rotary motion. The geometry of the cutting tool depends upon its intended function. The toolholding device may or may not be used for guiding or locating. Toolholder selection is governed by tool design and intended function.

The physical composition of the workpiece greatly influences the selection of the machining method, the tool composition and geometry, and the rate of material removal. The intended shape of the workpiece influences the selection of the machining method and the choice of linear or rotary tool travel. The composition and geometry of the workpiece to a great extent determines the workholder requirements. Workholder selection also depends upon forces produced by the tool on the workpiece. The workholder must hold, locate, and support the workpiece. Tool guidance may be incorporated into the workholding function.

Successful design of tools for the material removal processes requires, above all, a complete understanding of cutting tool function and geometry. This knowledge will enable the designer to specify the correct tool for a given task. The tool, in turn, will govern the selection of toolholding and guidance methods. Tool forces govern selection of the workholding device. Although the process involves interaction of the five elements, everything begins with and is based on what happens at the point of contact between the workpiece and the cutting tool.

Single Point Tools

The Basic Tool Angles

Cutting tools are designed with sharp edges to minimize rubbing contact between the tool and workpiece. Variations in the shape of the cutting tool influence tool life, surface finish of the workpiece, and the amount of force required to shear a chip from the parent metal. The various angles on a tool compose what is often termed the tool geometry. The tool signature or nomenclature is a sequence of alpha and numeric characters representing the various angles, significant dimensions, special features, and the size of the nose radius. This method of identification has been standardized by the American National Standards Institute for carbide and for high speed steel, and is

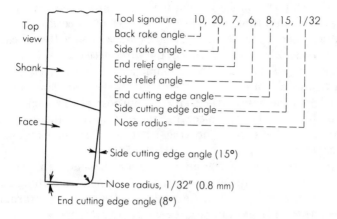

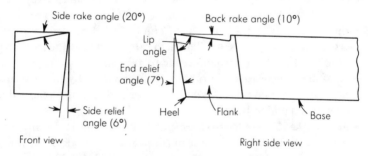

Figure 3-1. A straight-shank, right-cut, single-point tool, illustrating the elements of the tool signature as designated by the ANSI. Positive rake angles are shown.

illustrated in *Figure 3-1*, together with the elements that make up the tool signature.

Back Rake Angle. This is the angle between the face of the tool and a line that is parallel to the base of the toolholder. It is measured in a plane that is parallel to the side cutting edge and perpendicular to the base. Variations in the back rake angle affect the direction of chip flow and cutting force. As this angle is increased while other conditions remain constant, tool life will increase slightly and the cutting force required will decrease. Cutting edge strength decreases dramatically as positive back rake angles are increased above 5°. Similarly, cutting edge strength increases as back rake becomes negative and is optimized around -5°.

Side Rake Angle. This angle is defined as the angle between the tool face and a plane parallel to the tool base. It is measured in a plane perpendicular to both the base of the holder and the side cutting edge. Variations in this angle have the largest effect on cutting force, and to some extent, affect direction of chip flow. As the angle is increased, forces are reduced about 1% for each degree of positive side rake as less tearing of the workpiece occurs. Negative side rake increases edge strength and is recommended for most steels.

End Relief Angle. This is the angle between the end flank and a line perpendicular to the base of the tool. The purpose of this angle is to prevent rubbing between the workpiece and the end flank of the tool. An excessive clearance or relief angle reduces the strength of the tool, so the angle should not be larger than necessary, typically in the 5-7° range.

Side Relief Angle. This is the angle between the side flank of the tool and a line drawn perpendicular to the base. Comments regarding end relief angles are applicable also to side clearance or relief angles. For turning operations, the side relief angle must be large enough to prevent the tool from advancing into the workpiece before the material is machined away. Angles of 5-7° are sufficient for feed ratio under 0.030″ (0.8 mm) per revolution. Threading of low pitch threads requires up to 25° clearance.

End Cutting Edge Angle. This is the angle between the edge on the end of the tool and a plane perpendicular to the side of the tool shank. The purpose of the angle is to avoid rubbing between the edge of the tool and the workpiece. As with end relief angles, excessive end cutting angles reduce tool strength with no added benefits.

Lead Angle (Side Cutting Edge Angle). This is the angle between the straight cutting edge on the side of the tool and the side of the tool shank. This side edge provides the major cutting action and should be kept as sharp as possible. Increasing this angle tends to widen and thin the chip, and influences the direction of chip flow. An excessive side cutting edge angle redirects feed forces in the radial direction which may cause chatter. As the angle is increased from 0° to 45°, workpiece entry is moved away from the vulnerable tip (radius) of the tool to a stronger more fully supported part of the tool, usually resulting in increased tool life. However, these benefits will usually be lost if chatter occurs, so an optimum maximum angle should be sought.

Nose Radius. The nose radius connects the side and end cutting edges and dramatically affects tool life, radial force, and surface finish. Sharp-pointed tools have a nose radius of zero. Increasing the nose radius from zero avoids high heat concentration at a sharp point. Improvements in tool life and surface finish usually result as nose radius is increased up to 1/16″ (1.6 mm). There is, however, a limit to radius size that must be considered. Chatter will result if the nose radius is too large; an optimum maximum value should be sought.

Tool Signature. A comparison of *Figure 3-1* and *3-2* illustrates the difference between a right and left hand tool. Right hand tools are the most popular.

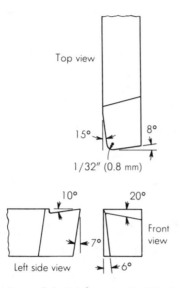

Figure 3-2. A left-cut tool. All other
aspects are identical with *Figure 3-1*.

Figures 3-3, 3-4, and *3-5* give the recommended angles for single-point
tools of high-speed steel, carbide, and cast alloys respectively.

Chip Groove. Pressed-in chip grooves provide for chip control and force
reductions with unique grooves designed for specific applications. Rapid
advances in chip groove design are resulting from research into the effects of
changes in groove geometry. *Figure 3-7* lists several standard chip groove
geometries.

Chip grooves can be ground into the cutting surface of high-speed steel and
brazed carbide tools. *Figure 3-6* shows typical dimensions used. Press
technology has obsoleted the need to hand grind carbide tools because
indexable inserts have the grooves pressed in.

Chip Formation

The majority of metal cutting operations involve the separation of small
segments or chips from the workpiece to achieve the required shape and size
of manufactured parts. Chip formation involves three basic requirements: (1)
there must be a cutting tool that is harder and more wear-resistant than the
workpiece material; (2) there must be interference between the tool and the
workpiece as designated by the feed and depth of cut, and (3) there must be a
relative motion or cutting velocity between the tool and the workpiece with
sufficient force to overcome the resistance of the workpiece material. As long

Material	Side-relief angle, deg.	Front-relief angle, deg.	Back-rake angle, deg.	Side-rake angle, deg.
High-speed, alloy, and high-carbon tool steels and stainless steel	7 to 9	6 to 8	5 to 7	8 to 10
SAE steels:				
1020, 1035, 1040	8 to 10	8 to 10	10 to 12	10 to 12
1045, 1095	7 to 9	8 to 10	10 to 12	10 to 12
1112, 1120	7 to 9	7 to 9	12 to 14	12 to 14
1314, 1315	7 to 9	7 to 9	12 to 14	14 to 16
1385	7 to 9	7 to 9	12 to 14	14 to 16
2315, 2320	7 to 9	7 to 9	8 to 10	10 to 12
2330, 2335, 2340	7 to 9	7 to 9	8 to 10	10 to 12
2345, 2350	7 to 9	7 to 9	6 to 8	8 to 10
3115, 3120, 3130	7 to 9	7 to 9	8 to 10	10 to 12
3135, 3140	7 to 9	7 to 9	8 to 10	8 to 10
3250, 4140, 4340	7 to 9	7 to 9	6 to 8	8 to 10
6140, 6145	7 to 9	7 to 9	6 to 8	8 to 10
Aluminum	12 to 14	8 to 10	30 to 35	14 to 16
Bakelite	10 to 12	8 to 10	0	0
Brass, free-cutting	10 to 12	8 to 10	0	1 to 3
Red, yellow, bronze—cast, bronze— commercial	8 to 10	8 to 10	0	-2 to -4
Bronze, free-cutting	8 to 10	8 to 10	0	2 to 4
Hard phosphor bronze	8 to 10	6 to 8	0	0
Cast iron, gray	8 to 10	6 to 8	3 to 5	10 to 12
Copper	12 to 14	12 to 14	14 to 16	18 to 20
Copper alloys:				
Hard	8 to 10	6 to 8	0	0
Soft	10 to 12	8 to 10	0 to 2	0
Fiber	14 to 16	12 to 14	0 to 2	0
Formica	14 to 16	10 to 12	14 to 16	10 to 12
Nickel iron	14 to 16	10 to 12	6 to 8	12 to 14
Micarta	14 to 16	10 to 12	14 to 16	10 to 12
Monel and nickel	14 to 16	12 to 14	8 to 10	12 to 14
Nickel silvers	10 to 12	10 to 12	8 to 10	0 to -2
Rubber, hard	18 to 20	14 to 16	0 to -2	0 to -2

Figure 3-3. Recommended angle for high-speed steel single-point tools.

as these three conditions exist, the portion of the material being machined that interferes with free passage of the tool will be displaced to create a chip.

Many possibilities and combinations exist that may fulfill such requirements. Variations in tool material and tool geometry, feed and depth of cut, cutting velocity, and workpiece material have an effect not only upon the

Material	Normal end-relief, deg.	Normal side-relief, deg.	Normal back-rake, deg.	Normal side-rake, deg.
Aluminum and magnesium alloys	6 to 10	6 to 10	0 to 10	10 to 20
Copper	6 to 8	6 to 8	0 to 4	15 to 20
Brass and bronze	6 to 8	6 to 8	0 to -5	+8 to -5
Cast iron	5 to 8	5 to 8	0 to -7	+6 to -7
Low-carbon steels up to SAE 1020	5 to 10	5 to 10	0 to -7	+6 to -7
Carbon steels SAE 1025 and above	5 to 8	5 to 8	0 to -7	+6 to -7
Alloy steels	5 to 8	5 to 8	0 to -7	+6 to -7
Free-machining steels SAE 1100 and 1800 series	5 to 10	5 to 10	0 to -7	+6 to -7
Stainless steels, austenitic	5 to 10	5 to 10	0 to -7	+6 to -7
Stainless steels, hardenable	5 to 8	5 to 8	0 to -7	+6 to -7
High-nickel alloys (Monel, Inconel, etc.)	5 to 10	5 to 10	0 to -3	+6 to +10
Titanium alloys	5 to 8	5 to 8	0 to -5	+6 to -5

Figure 3-4. Recommended angles for carbide single-point tools.

Material	Back-rake angle, deg.	Side-rake angle, deg.	Side-relief angle, deg.	Front-relief angle, deg.	Side-cutting-edge angle, deg.	End-cutting-edge angle, deg.
Steel	8-20†	8-20†	7	7	10	15
Cast steel	8	8	5	5	10	10
Cast iron	0	4	5	5	10	10
Bronze	4	4	5	5	10	10
Stainless steel	8-20†	8-20†	7	7	10	15

* Stellite 98M2-turning tools.
† Angle depends on grade and type of steel. Boring tools use the same rake but greater relief to clear the work.

Figure 3-5. Cutting angles for cast alloy tools.*

formation of the chip, but also upon cutting force, cutting horsepower, cutting temperatures, tool wear and tool life, dimensional stability, and the quality of the newly created surface. The interrelationship and the inter-dependence among these "manipulating factors" constitute the basis for the

	Feed	0.006-0.012 (0.15-0.30)	0.013-0.017 (0.33-0.43)	0.018-0.025 (0.46-0.64)	0.028-0.040 (0.71-1.02)	Over 0.040 (1.02)
Depth of cut	R	0.010-0.025 (0.25-0.64)	0.035-0.065 (0.89-1.65)	0.035-0.065 (0.89-1.65)	0.035-0.065 (0.89-1.65)	0.035-0.065 (0.89-1.65)
	T	0.010 (0.25)	0.015 (0.38)	0.020 (0.51)	0.030 (0.76)	0.030 (0.76)
1/64-3/64 (0.4-1.2)	W	1/16 (1.6)	5/64 (2.0)	7/64 (2.8)	1/8 (0.1)	
1/16-1/4 (1.6-6.4)	W	3/32 (2.4)	1/8 (0.1)	5/32 (4.0)	3/16 (4.8)	3/16 (4.8)
5/16-1/2 (7.9-12.7)	W	1/8 (3.2)	5/32 (4.0)	3/16 (4.8)	3/16 (4.8)	3/16 (4.8)
9/16-3/4 (14.3-19.0)	W	5/32 (4.0)	3/16 (4.8)	3/16 (4.8)	3/16 (4.8)	3/16 (4.8)
Over 3/4 (19.0)	W	3/16 (4.8)	3/16 (4.8)	3/16 (4.8)	3/16 (4.8)	1/4 (6.4)

Figure 3-6. Dimensions for parallel- and angular-type chip breakers, in. (mm).

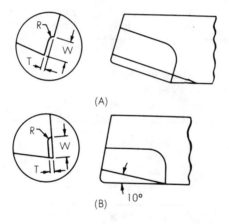

(A)

(B)   10°

Figure 3-7. Chip grooves ground into tools.

study of machinability—a study which has been popularly defined as the response of a material to machining.[1]

Types of Chips. *Figure 3-8* illustrates the necessary relationship between cutting tool and the workpiece for chip formation in several common machining processes. Although it is apparent that different general shapes and sizes of chips may be produced by each of the basic processes, all chips regardless of process are usually classified according to their general behavior during formation.

The three most common types of chips are illustrated by the photomicrographs of *Figure 3-9*: (a) discontinuous or segmental, (b) continuous, without built-up edge, and (c) continuous with built-up edge. The type of chip is

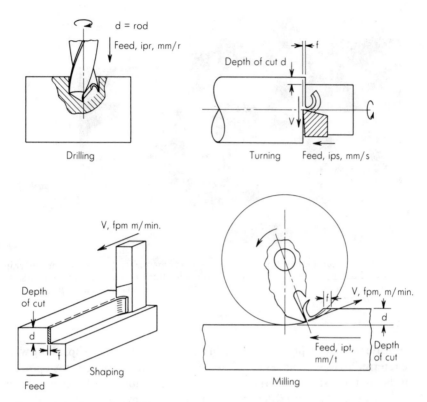

Figure 3-8. Examples of feed depth and velocity relationships for several chip-formation processes.

generally a function of the work material and the cutting conditions. A discontinuous chip is typical of the more brittle materials such as cast iron, while continuous chips are typical of ductile materials such as steel.

Mechanism of Chip Formation.[2,3,4] Observations during metal cutting reveal several important characteristics of chip formation: (1) the cutting process generates heat; (2) the thickness of the chip is greater than the thickness of the layer from which it came; (3) the hardness of the chip is usually much greater than the hardness of the parent material; and (4) the above relative values are affected by changes in cutting conditions and in properties of the material to be machined, to produce chips that range from small lumps to long continuous ribbons.

These observations indicate that the process of chip formation is one of deformation or plastic flow of the material, with the degree of deformation dictating the type of chip that will be produced.

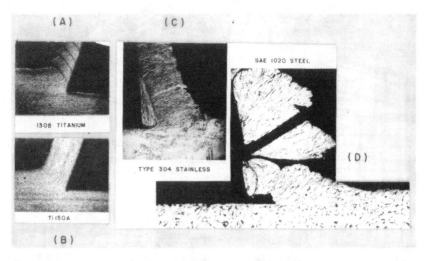

Figure 3-9. Examples of types of chips: (*A*) segmental chip. (*B*) continuous chip without built-up edge. (*C*) continuous chip with built-up edge. Part (*D*) shows the "piling up" or heavy distortion common to soft, ductile materials which have a large capacity for plastic flow.

Plastic flow takes place by means of a phenomenon called slip, along what are referred to as slip planes. The capacity for plastic flow depends upon the number of these slip planes that are available. The number of planes, in turn, depends upon the crystal lattice structure of the material and upon prior treatment. When the resisting stresses in a material exceed its elastic limit, a permanent relative motion occurs between those adjacent slip planes which are most favorably oriented in the direction of the applied force. Once this motion or slip takes place, these particular planes are strengthened and resist further deformation in preference to other, now weaker, planes that are available. This strengthening is called work or strain hardening, and is characteristic of all steels but most dramatically exhibited in stainless steels.

As the tool advances into the workpiece, frictional resistance to flow along the tool face and resultant work hardening cause deformation by shear to take place ahead of the tool along a shear plane. This plane extends from the vicinity of the cutting edge toward the free surface of the workpiece at some shear angle θ, as illustrated in *Figure 3-10*. If the workpiece material is brittle and has little capacity for deformation before fracture, when the fracture shear stress is reached, separation will take place along the shear plane to form a segmental chip. Ductile materials, however, contain sufficient plastic flow capacity to deform along the shear plane without rupture. Strain hardening permits a transfer of slip to successive shear planes, and the chip tends to flow in a continuous ribbon along the face of the tool and away from

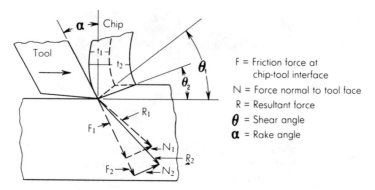

Figure 3-10. Effect of friction force upon shear angle and upon amount of chip distortion. Force polygons show effect of friction force upon magnitude and direction of resultant force.

the work surface. The chip is highly worked, and much harder than the material from which it is taken.

Figure 3-11 shows the relative distortion in grid specimens of brittle and ductile materials. The ductile specimen (B) shows evidence of great chip distortion; the grid lines are completely obliterated. In contrast, the grid lines on the brittle segment (A) are plainly visible and show little distortion. The rate of deformation in continuous-chip formation is extremely high since the shear plane is the boundary between relatively undisturbed and highly disturbed material.

The effect of a change in shear is shown in *Figure 3-10.* For a given depth of cut, a smaller shear angle θ_2 causes greater chip cross section and, therefore, greater distortion than in the case of the larger shear angle θ_1. For a given material, the shear angle is a function of the tool rake angle and the coefficient of friction along the tool face. Materials that have fewer slip planes and low work-hardening capacity generally show higher shear angles than the more ductile materials, and have a lower ratio of chip thickness to undeformed chip thickness. Under ideal conditions, this ratio will approach 1.5 and can be checked to a rough extent by measuring the thickness of the chip and comparing it to the feed rate.

Built-up Edge. Consideration of chip flow along the face of the tool in the formation of continuous chips is of prime importance. If the friction force that resists the passage of the chip along the tool face is less than the force necessary to shear the chip material, the entire chip will pass off cleanly, as shown in *Figure 3-9B*. This ideal case of chip formation may be approached, but is seldom realized. It is generally associated with materials of high strength and of low work-hardening capacity and with low coefficients of

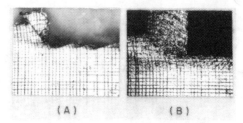

Figure 3-11. Grid specimens showing amount of distortion in chip and adjacent area in (*A*) brittle material, and (*B*) ductile material. Grid lines 0.003 in. apart. Line of demarcation between chip and parent material is the shear plane. Wavy surface in (*A*) due to chatter.

friction—factors which lead to large shear angles. High cutting speeds are also favorable.

In most cases, however, it is virtually impossible to prevent some amount of seizure between the chip and the tool face. Unless surfaces are perfectly flat, contact is made along the high spots over only a fraction of the total area. As the chip passes over the tool face, cutting forces give rise to extremely high unit pressures, sufficient to form pressure welds. If these welds are stronger than the ultimate shear strength of the material, that portion of the chip which is welded to the tool shears off as the chip is displaced and becomes what is called a built-up edge. Continuous chips with built-up edges are illustrated in *Figures 3-9C* and *3-9D*.

The built-up edge is common to most metal cutting operations and is particularly evident in machining aluminum and some stainless steels. The edge builds up to a point where it eventually breaks down, part of it going off with the chip, and part of it being deposited on the work surface. This characteristic occurs at rapid intervals and is exhibited in *Figure 3-12*. Any change in cutting conditions that reduces or eliminates the built-up edge will usually improve surface quality (*Figure 3-13*). Built-up edge affords some protection to the cutting edge to reduce wear, and a small amount may be desirable. The problem then becomes one of size control through the effects of the various manipulating factors.

Effect of Manipulating Factors

Certain manipulating factors provide some control of the metal cutting characteristics. Some effects of these factors are illustrated in *Figure 3-12* through *3-16*. The results shown are derived from turning cuts on AISI 1020 hot-rolled steel with sharp tools. The results cannot be listed as all-inclusive,

Figure 3-12. Underside of a chip that had seized to the face of the cutting tool in the light area along the cutting edge. Part of the built-up edge is shown in the process of passing off with the chip. Part of it was also being forced over the cutting edge of the tool and would eventually be deposited on the work surface.

since effects of tool wear have not been considered. However, they do represent the general trends of most metalcutting operations even though built-up edge, surface roughness, and chip shapes may not be the same for each. *Figure 3-17* summarizes surface roughness for *Figures 3-14* through *3-16*.

Hss tool		Carbide tool
60 fpm	1020 steel	350 fpm
f = 0.012 ipr	d = 0.125 in.	f = 0.006 ipr

Figure 3-13. Surfaces produced on hot rolled AISI 1020 steel under conditions that resulted in continuous chip formation with built-up edge (left), and without built-up edge (right).

In studying the various examples, one should keep in mind the relationship between tool, chip, and surface appearance. The size and the degree of brittleness of the chip may be a good indication of the severity of the cutting operation. The back of the chip, which is in contact with the tool face, gives a fairly good indication of the built-up edge condition. In the absence of built-up edge, the back of the chip should be clean, smooth, and highly burnished. The workpiece surface should be correspondingly good. As the size of the built-up edge increases, more and more markings are evident on the chip. Generally the workpiece is affected in the same manner.

Velocity. Velocity affects temperature, which in turn effects the cutting process. At low velocities, the temperature at the tool point is below the recrystallization temperature of the material. As a result, work hardening in the chip is retained and the workpiece material is not softened due to failure to reach the yield strength temperature of the material. If the velocity increases to the point where the cutting temperature is above the yield strength temperature of the material, the chip material at the interface tends to soften and machine much more efficiently. Higher shear angles occur at higher velocities and an ideal chip thickness of 1.5 times the feed can be approached. Excessive velocity will cause the tool to fail rapidly since speed has the greatest effect on tool life.

Chip form or shape at high velocities can be very troublesome on ductile materials. The reduced resistance to chip flow and the resultant increase in shear angle gives a thinner, less distorted chip, but one which becomes longer and straighter as velocity increases. Chip grooves specifically designed for thin chips can be employed to deal with this problem.

Size of Cut. Changes in the size of cut effectively change the cross-sectional area of chip contact (*Figures 3-18A* and *3-18B*). How this area is changed determines the effect upon the cutting process. An increase in depth of cut for a constant feed merely lengthens the contact, but does not change the thickness; the force per unit length remains the same. However, an increase in feed for a given depth widens the area of contact and changes the force per unit length. This results in greater chip distortion and reduced tool life although increased feed reduces the machining cycle.

Several factors affect surface quality to a greater degree than may be predicted. Lack of rigidity will permit greater deflections as a result of higher forces. Increases in feed and depth of cut may then cause chatter, poor surface quality, and loss of dimensional stability. Deep turning cuts on relatively small diameters have a greater percentage change in velocity along the length of the cutting edge. This might result in erratic tool life behavior with poor surface quality.

The effect upon the chip is also much more pronounced with increases in feed than with increases in depth. Because of the greater distortion at high

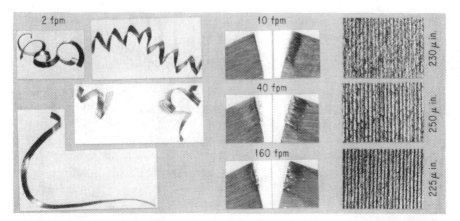

Figure 3-14. Examples of chips, tools, and surfaces to show effect of velocity on chip formation, built-up edge, and surface quality. Approximate relative magnifications: chips, 1X; tools, 4X; surface, 10X. Note that velocity has practically no effect upon direction of chip flow as seen by markings on tool face. The relatively small effect of velocity on surface finish is due to the fact that practical permissible velocities with HSS tools are not high enough to completely eliminate built-up edge. Tool material—HSS. Tool shape—8, 21, 6, 6, 6, 15, 0. Work material—SAE 1020. Cutting fluid—dry.

feed rates, the chips tend to break up more readily. Chip control is essential for operator safety since long continuous chips can wrap around the rotating part and be very dangerous.

Effect of Tool Geometry. For given cutting conditions, changes in tool geometry have two direct effects on chip formation: (1) effect upon shear angle, and (2) effect upon chip thickness. The two are related in that a change in one usually affects the other.

The effects of side cutting edge angle and of nose radius can be explained in terms of the effect upon chip thickness. *Figure 3-18C* shows that an increase in the side cutting edge angle reduces the chip thickness for a given feed by a factor of the cosine of the angle. This, in effect, reduces the chip contact width to thin out the built-up edge. An increase in nose radius has the same general effect as seen in *Figure 3-18D*. The shape of the contact area changes, but at the point of contact between the machined surface and the tool, the chip is very thin. In comparison, the feed marks and resultant surface finish are much smoother than those left by a sharp-nosed tool.

The effects of changes in rake angles are shown in *Figure 3-19*. The lower rake angles decrease the shear angle, cause greater chip distortion, and increase the resistance to chip flow. Lower rake angles (negative) produce rougher and more work-hardened surfaces. At low or negative rake angles, the chip is so highly distorted that it facilitates chip control by breaking the

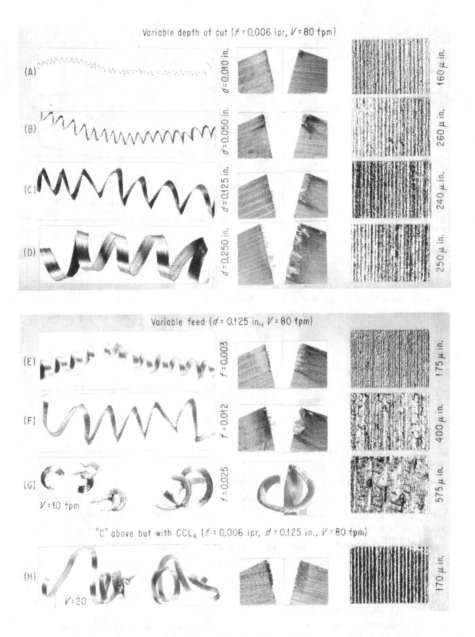

Figure 3-15. Examples showing effect of feed and depth of cut on chip form, built-up edge, and surface quality. Note how direction of chip flow changes with size of cut. Tool material—HSS. Tool shape—8, 21, 6, 6, 6, 15, 0. Work material—SAE 1020. Cutting fluid—dry.

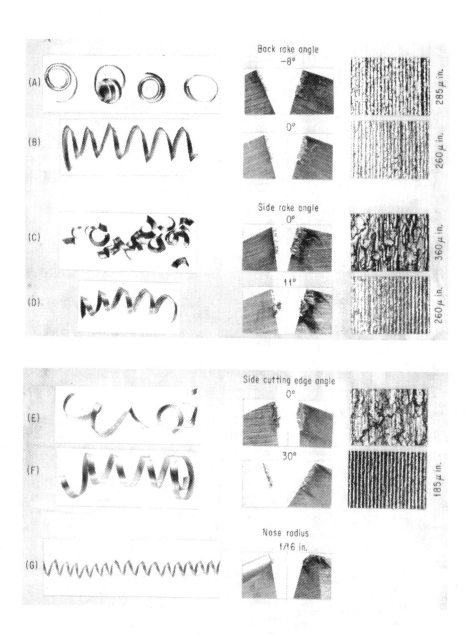

Figure 3-16. Effect of changes in tool geometry upon chip form, built-up edge, and surface quality. Note also the effect upon direction of chip flow. Tool material—HSS. Work material—SAE 1020. Cutting fluid—dry. Basic tool shape—8, 21, 6, 6, 6, 15, 0.

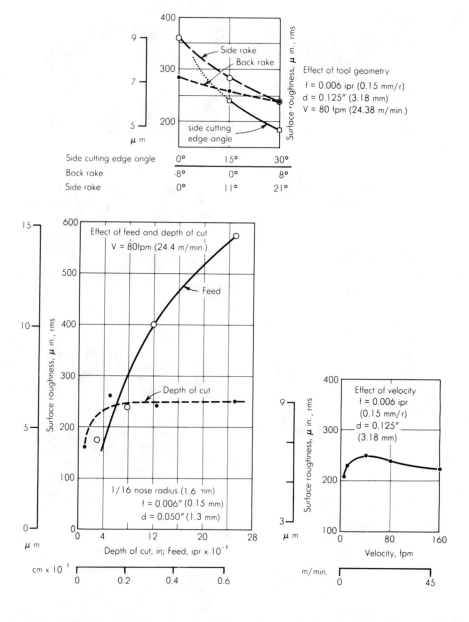

Figure 3-17. Summary of surface roughness results given in *Figures 3-14* through *3-16.* Standard tool geometry—8,21,6,6,6,15,0. HSS tools. Work material—AISI 1020 hot rolled steel.

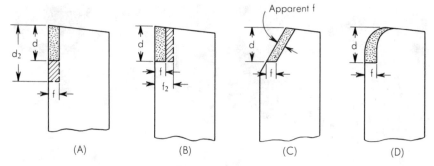

Figure 3-18. Effect of size of cut and changes in tool geometry upon chip thickness: (A) change in depth, (B) change in feed, (C) effect of side cutting edge angle, and (D) effect of nose radius. Crosshatched portions represent increase in contact area.

chip into six short lengths. The side rake angle has much more effect than does the back rake angle. The effects on three-dimensional cutting must be described on the basis of the effective rake angle in the direction of chip flow.

Tool Material. One effect of tool material lies in its ability to sustain high cutting velocities, as for example between high-speed steel and carbides. The effect of high velocity has already been described. Another factor is the coefficient of friction between chip and tool material. Usually, this is of little consequence with high-speed steel tools because the coefficient of friction does not change appreciably among the various grades. Sintered carbide tools and ceramic tools are made with different compositions and these tools may behave quite differently for similar cutting conditions.

Cutting Fluids. Ideally, if a cutting fluid provides lubrication between the chip and the tool, the coefficient of friction will be reduced and the shear angle will be increased. However, effective lubrication may be difficult to achieve except possibly at very low cutting speeds. Lubricant effects vary with cutting conditions and with work materials. At high speed, fluids act principally as coolants, but may effectively lubricate the tool-chip interfacial zone providing more efficient machining which often results in increased tool life and improved surface finish. Constant, even flow is essential when cutting fluids are applied with carbide tools to prevent thermal shocking and resultant fracture.

Workpiece Materials. Brittle materials form discontinuous chips, and can be machined with flat top inserts. Having no chip groove pressed in, these stronger, flat top inserts are more economical and last longer, particularly when applied to machining cast iron. Cutting forces are usually lower than they would be for a ductile material of corresponding strength because of

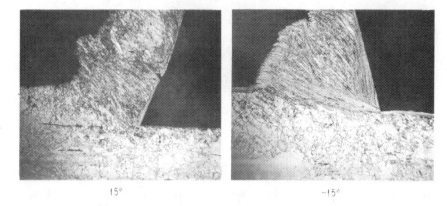

15°   −15°

Figure 3-19. Photomicrographs showing effect of rake angle upon shear angle, chip distortion, built-up edge, and work hardening of machined surface. Shaper tools set for some depth of cut. Apparent difference in depth due to higher separating force and greater tool deflection with negative rake tool. Work material—304 stainless steel.

generally large shear angles and lower resistance along the tool face.

Ductile materials produce continuous chips. Continuous chips can be dangerous to operators and machines, so the problem has to be dealt with. With low friction and high cutting velocities, particularly with materials of low work-hardening capacity, a thinner, less distorted chip is produced. Pressed-in chip grooves are required for low feed rates in order to break up the chip. High frictional resistance to flow, low shear angles, and materials of high work-hardening capacity are associated with large distortions during cutting, and are not as big a problem to break up the chip. Additions of lead, sulfur, and phosphorus to low-carbon steels help to break up chips, reduce built-up edge, and improve surface quality.

Tool Wear

For the sake of recognition and understanding of the fundamentals of metal cutting, the effects of changes in the manipulating factors have been described without regard to their influence upon such criteria as tool wear and tool life. Yet, there is no known tool material that can completely resist contact and rubbing at high temperatures and at high pressures without some changes from its original contours over a period of time. It becomes necessary, therefore, to think of the effect of the manipulating factors not only upon the cutting process itself, but upon the performance of the cutting tool, which may, in turn, itself affect the cutting process.

Tool Failure. Failure of the cutting tool has occurred when it is no longer capable of producing parts within required specifications. The point of failure, together with the amount of wear that determines this failure, is a function of the machining objective. Surface quality, dimensional stability, cutting forces, cutting horsepower, and production rates may alone, or in combination, be used as criteria for tool failure. It may, for instance, take very little wear to affect surface quality, although the tool itself could continue to remove metal with little, if any, loss of efficiency. In contrast, only a few thousandths of an inch of wear on a wide form tool might cause such a large increase in thrust or feeding forces that it would result in a loss of dimensional stability, or require excessive power in addition to a loss of surface quality.

Type of Tool Wear. Tool failure is associated with some form of breakdown of the cutting edge. Under proper operating conditions, this breakdown takes place gradually over a period of time. In the absence of rigidity, or because of improper tool geometry that gives inadequate support to the cutting edge, the tool may fail by mechanical fracture or chipping under the load of the cutting forces. This is not truly a wear phenomenon for it can be eliminated or at least minimized by proper design and application.

As a result of direct contact with the work material, there are three major regions on the tool where wear can take place: (1) face, (2) flank, and (3) nose (*Figure 3-20*).

Face Wear. The face of the tool is the surface over which the chip passes during its formation. Wear takes the form of a cavity or crater which has its origin not along the cutting edge but at some distance away from it and within the chip contact area. As wear progresses with time, the crater gets wider, longer, and deeper and approaches the edges of the tool.

This form of wear is usually associated with ductile materials which give rise to continuous chips. If crater wear is allowed to proceed too far, the cutting edge becomes weak as it thins out, and breaks down suddenly. Usually, there is some preliminary breakthrough of the crater at the nose and at the periphery prior to total failure of the cutting edge. These preliminary breaks serve as focal points for the development of notches along the flank. In general, crater wear develops faster than flank wear on ductile materials and is the limiting factor in determination of tool failure.

Flank Wear. Although crater wear is most prominent in the machining of ductile materials, flank wear is always present regardless of work and tool material, or even of cutting conditions. The flank is the clearance face of the cutting tool, along which the major cutting edge is located. It is the portion of the tool that is in contact with the work at the chip separation point and that resists the feeding forces. Because of the clearance, initial contact is made along the cutting edge. Flank wear begins at the cutting edge and develops into a wider and wider flat of increasing contact area called a wear land.

Materials that do not form continuous chips promote little if any crater

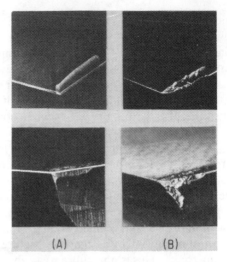

Figure 3-20. Representative wear patterns on face, flank, and nose of cutting tool, typical of chip-removal process on ductile materials. Crater on face of tool in (*A*) started well back of cutting edge. In (*B*) crater wear had progressed to point where weak cutting edge broke down under cutting forces.

wear, and flank wear becomes the dominant factor in tool failure. In the case of most form tools and certain milling cutters, the wear land is in direct contact with the finished surface, and usually becomes the basis for failure even on ductile materials, particularly if surface finish specifications are the controlling factors in the process. Quite often, flank wear is accompanied by a rounding of the cutting edge, particularly in the machining of abrasive materials. This results in large increases of cutting and feeding forces which, if carried too far, could lead to tool fracture.

Nose Wear. Nose wear is similar to and is often considered a part of flank wear. There are times when it should be considered separately. Nose wear sometimes proceeds at a faster rate than flank wear, particularly when one is working on rather abrasive materials and using small nose radii. In finish turning operations, for example, excessive wear will affect finished part dimensions as well as surface roughness. Where sharp corners are specified on the part drawing, the rounding or flattening of the nose can cause out of tolerance conditions long before flank wear itself becomes a factor.

Mechanism of Tool Wear. Evidence indicates that wear is a complex phenomenon and is influenced by many factors. The causes of wear do not always behave in the same manner, nor do they always affect wear to the same degree under similar cutting conditions. The causes of wear are not fully understood. In recent years, great strides have been made by various researchers. Even though there is some disagreement regarding the true mechanisms by which wear actually takes place, most investigators feel that there are at least five basic causes of wear:

1. Abrasive action of hard particles contained in the work material.
2. Plastic deformation of the cutting edge.
3. Chemical decomposition of the cutting tool contact surfaces.
4. Diffusion between work and tool material.
5. Welding of asperities between work and tool.

The relative effects of these causes are a function of cutting velocity or cutting temperatures and are shown in *Figure 3-21*. Investigations have also been made on other possible causes such as oxidation and electrochemical reactions in the tool work contact zone.

The most important factor influencing tool wear is cutting temperature. Of the five basic causes of wear, temperature has considerable effect in all but one. Cutting temperatures are important for two basic reasons: (1) most tool materials show rapid loss of strength, hardness, and resistance to abrasion above some critical temperature, and (2) the rate of diffusion between work and tool materials rises very rapidly as temperature increases past the critical.

Analytical and experimental methods have been used[6,7] to show that the average peak temperatures at the tool-chip interface occur near the point where the chip leaves the tool surface (*Figure 3-22*). Crater wear appears

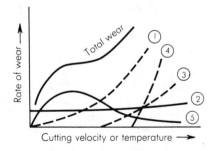

Figure 3-21. Relative effects of various causes of tool wear: (1) abrasive wear, (2) plastic deformation of cutting edge, (3) chemical decomposition, (4) diffusion, (5) welding of asperities.

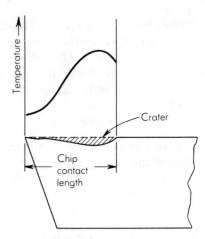

Figure 3-22. Temperature distribution
along tool-chip contact length.[6]

greatest at this point. The relationship between crater wear rate and average
tool-chip interface temperature is shown in *Figure 3-23*. The rate of wear
increases very rapidly beyond a critical temperature. Flank temperatures
were found to be maximum near the tool point as shown in *Figure 3-24*.

The significance of temperature effects upon wear are associated with tool
material properties. High-speed steel tools begin to lose their properties very
rapidly at approximately 1100°F (593°C). Carbides show less drastic

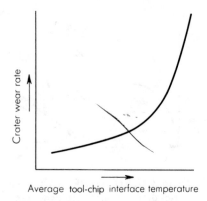

Figure 3-23. Relationship between rate
of crater wear and average tool-chip
interface temperature.[6]

sensitivity to temperature up to about 1600° F (871° C). Chemical decomposition and diffusion will not occur at any appreciable extent until the critical temperatures are reached.

Wear By Abrasive Action. This mechanism may be partly explained by the fact that hard particles (sand inclusions, carbides, etc.) in the workpiece material literally gouge or dislodge particles from the tool, causing continuous wear under any cutting condition. The rate of wear is thus dependent upon the number, size, distribution, and the hardness of the particles in the work material as well as the hardness of the cutting tool and the workpiece. At higher cutting speeds, even some of the softer constituents may contribute to the gouging action as a result of higher impact values and reduced tool resistance to abrasion.

Plastic Deformation of the Cutting Edge. This wear mechanism is believed to take place at all ranges of cutting temperatures; it arises from the high unit pressures imposed on the tool. This results in a slight depression and bulging of the edge, similar to that shown in *Figure 3-25*. The net effect is greater tool pressure and increased cutting temperature resulting in further

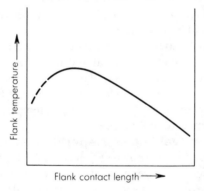

Figure 3-24. Distribution of flank temperature along flank-work contact length.[6]

deformation and concluding in edge wipe out. This mode of failure is common when machining hardened materials at high speeds.

Chemical Distortion of the Cutting Tool. This wear mechanism occurs through localized chemical reactions at the tool-workpiece interface. These reactions are temperature-dependent and result in weakening the bond between minute tool segments and the segments surrounding them. This may occur either through formation of weaker compounds, or in the case of carbide tools, by a dissolving action of the bond between the binder and

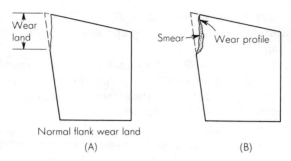

Figure 3-25. Section of tool to show effect of smear and other causes on type of flank wear.

individual carbide particles. As a result of this weakening effect, the particles are pulled out from the main body of the tool by the chip or work as it moves past the contact surfaces. Once the critical temperature for this chemical action is reached, the rate of wear is relatively rapid.

Diffusion. This complex wear phenomenon between the work and the tool results in a rapid breakdown of tool material once critical temperature is reached. Much has yet to be learned about this cause of wear, but basically it involves a change of composition at the tool-chip interface. There is an alloying effect which weakens the bond of the tool particles and permits them to be pulled out by the chip as it sloughs off. Carbon transfer from the tool material to the workpiece is enhanced at higher temperatures and this greatly contributes to premature tool failure.

Wear Through Welding of Asperities. This mechanism parallels that of the built-up edge. As shown in *Figure 3-21*, the greatest rate of wear by this mechanism occurs at lower cutting velocities or temperatures. A built-up edge forms because of a high resistance to chip flow along the tool face, which causes a portion of the chip to shear off as it moves past the tool. This action is most prominent when cutting temperatures are below the recrystallization temperature for the material. Work hardening is retained and the built-up edge is harder and stronger than the rest of the chip. This same situation exists for wear through welding of asperities.

The asperities on a tool are brittle and relatively weak in bending or tension. If welding takes place between the chip and the asperities because of the extremely high unit pressures, the work-hardened chip material is strong enough to pull these asperities off. However, if the temperature is near or beyond the recrystallization temperature, then the bond between the chip or built-up edge and the tool is no weaker than the material adjacent to it, because work hardening has not been retained. Therefore, the rate at which these asperities are pulled out diminishes.

If cutting conditions are such that the resulting temperatures approach or go beyond the critical temperature for a given tool material, the reduced resistance of the tool, and the increasing tendency for alloying between work and tool material, cause a high rate of wear and very rapid failures. When cutting temperatures are low, the processes of wear by abrasion and by welding at the asperities become most prominent.

Effects of Manipulating Factors Upon Tool Wear. The effects of the manipulating factors upon tool wear are concerned either with modifications that influence the cutting process directly for given tool and workpiece materials, or with inherent properties of materials that resist or promote wear. For a given tool and workpiece material combination, cutting temperatures are influenced most by cutting speed, and to a lesser extent, feed and depth of cut (*Figure 3-26*). Adjustments in speed or feed, or both, will affect tool wear. It may be possible to substitute another tool material or a coated tool material that has inherently better temperature-resistant properties to maintain original or even higher production rates with less sensitivity to temperature failure. The cost of the second material may be higher than that of the first, but it may be more than justified by higher production rates at increased operating temperatures.

Changes in tool geometry that result in higher shear angles, less chip distortion, lower frictional resistance, and thinner chips will lower cutting forces and decrease cutting temperatures, and thus contribute to a reduction in the rate of tool wear for given cutting conditions. Within practical and design limitations, rake, relief, nose radii, and side cutting edge angles should be matched to the application providing the most free-cutting strong geometry that directs the cutting forces in the most rigid section of the workpiece. Heat transfer characteristics may also be adversely affected if the point of the tool is too thin as a result of high relief and rake angles. The heat at the point does not dissipate as rapidly, and higher temperatures prevail.

Workpiece materials that have relatively high hardness, high shear strength, high coefficient of friction, and high work-hardening capacities, and contain hard constituents, promote more rapid wear for certain cutting conditions. Materials such as titanium or stainless steels, which have poor

Figure 3-26. Relative effects of velocity, feed, and depth of cut on cutting temperatures for a given work material and tool geometry.

thermal conductivity, do not dissipate heat from the cutting zone as rapidly as others, and temperature failures are more common.

Effect of Wear on Machinability Criteria

The study of machinability involves certain criteria which play a prominent part in the evaluation of the cutting process. Machinability ratings for a given material are entirely relative in that one material is used as a base. The ratings can vary not only among the machining processes, but also with the criterion used in the evaluation for a given process. Though many data are available about numerous materials, erroneous conclusions may result if the data are not interpreted or applied properly.

Many of the available data, particularly with respect to cutting forces, specific power requirements, and surface quality, are based upon the results of sharp-tool investigations. These investigations serve a valuable purpose in analysis of the cutting process and in determination of initial levels of performance, but they give no indication as to how long the initial level of performance will be maintained when tools begin to wear. Some materials that are given very high machinability ratings with sharp tools are extremely sensitive to tool wear. Performance may drop off rapidly with time. On the other hand, very similar materials may have lower initial levels of performance but are less sensitive to tool wear, maintain the original levels for longer periods of time, and actually receive higher production performance ratings.

Another sort of data that warrants some caution in direct application is the tool-life curve. These curves (see *Figure 3-27*) are usually based on accelerated-wear tests. They are actually plots of cutting velocity versus cutting time, or of cubic inches of metal removal before failure for a given size of cut and otherwise constant cutting conditions. Failure can be identified in several ways. A preselected amount of flank wear (0.030″ [0.8 mm] for example) may indicate failure in tests with carbide and ceramic tools or quality of a machined surface may be used to denote failure of any tool material. Whenever wear is used in some form as the criterion of failure, the results can be plotted as a generally acceptable straight line on log-log coordinates.

For various reasons, mostly economic, the points used to establish the tool-life curve come from comparatively short tool lives, usually less than one hour of cutting time. If the velocity for some longer tool life is desired for a practical application, these curves are usually extended and the velocity for the desired tool life is extrapolated as indicated by the dotted line in *Figure 3-27*. This practice can sometimes lead to very unsatisfactory results.

High cutting velocities result in high cutting temperatures. When short

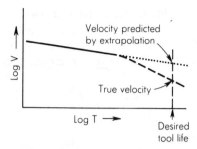

Figure 3-27. Tool-life curve showing error that can be introduced in predicting velocity for longer tool lives by extrapolation of typical accelerated-wear results (solid line). Lower velocities affect the rate and type of tool wear to change slope of curve. Slopes approaching 0 indicate greater sensitivity to temperature while slopes approaching 1 indicate effect of abrasive wear.

tool lives are encountered, the cutting temperature is in or near the critical temperature ranges, which promote rapid tool wear by diffusion and chemical decomposition. In this region, small changes in velocity or temperature may have relatively large effects upon wear rate and, therefore, on tool life. The solid line in *Figure 3-27* represents a typical plot.

The further the cutting temperature is removed from the critical, the less effective becomes the wear by diffusion and chemical decomposition, and the more prominent becomes the wear by abrasion and welding at the asperities. Since the total wear rate decreases at this point, there should be a lesser effect upon tool life for the same increment change in cutting velocity. The absolute slope of the curve should increase as represented by the dashed line in *Figure 3-27.* The actual velocity for a specified tool life may be considerably lower than the predicted velocity by extrapolation from a curve based on short-time tests.

Whether all workpiece materials and tool materials exhibit this kind of behavior to any predictable degree is not fully known at this time. There is considerable evidence that accelerated-wear tests can give misleading information and that care should be exercised to make proper use of this information. Some materials are much more sensitive to tool wear than others. The general wear trends are similar, although not to the same degree.

Numerous changes may take place during the life of a cutting tool. Whether these changes are important or not depends upon the machining objective. When cutting ductile materials, increases in crater and flank wear literally cause a change in the true tool geometry. This change in tool geometry usually affects chip formation (*Figure 3-28*). Chip form changes can be very sudden, particularly when the crater starts to break through the cutting edge.

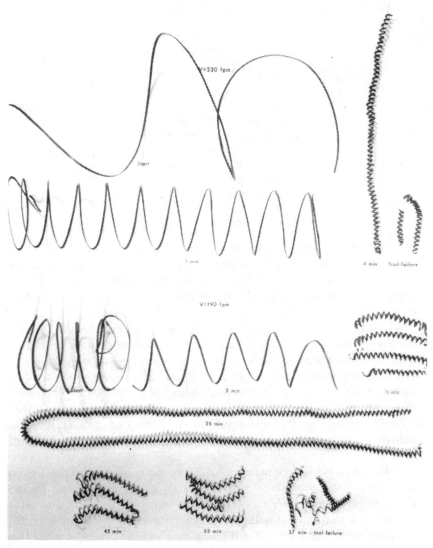

Figure 3-28. Effect of tool wear upon chip form and shape. Times represent points at which changes occurred. Tool geometry—8, 14, 6, 6, 6, 0, 0.010. Tool material—HSS. Cutting fluid—oil with sulfur additive. Work material—C 1045.

One of the most notable effects of tool wear is the change in surface quality. During the test on AISI 1045 steel at 190 fpm, from which the chip samples in *Figure 3-28* were taken, there were five distinct changes in surface appearance before total tool failure occurred at 57 minutes. *Figure 3-29* lists these changes. The changes in surface quality reflect the unstable character of the cutting edge as wear progresses. With materials that are notorious for formation of built-up edge, surface roughness usually reaches unsatisfactory proportions long before actual cutting tool failure. In the previous example, if surface finish was the criterion of failure, tool life would have been about 11 minutes rather than the 57 minutes of actual cutting time. *Figure 3-30* shows an actual application of failure on the basis of surface quality in the finish turning of large axles of AISI 1045 steel.

Tool wear has a definite effect upon cutting forces; the ratio of increase can be beyond expectation. In production operations, feeding or thrust forces in straight forming can rise from as low as two to as high as 40 times the sharp-tool values, depending upon the type of flank encountered. In one test,* the feeding force rose from a sharp-tool value of 11 lb. to 505 lb. (49 to 2246N) for a flank wear land of 0.0084" (0.213 mm). In another test on a material of the same commercial grade but from another source, the feeding force rose from an initial value of 16 lb. (71N) to only 180 lb. (800N) for the same width of wear land. The cutting conditions were exactly the same in each case, but there was sufficient difference in material behavior (even though both materials were within commercial specifications for the grade) to cause flank wear patterns as illustrated in (A) of *Figure 3-25*.

On the basis of the condition of the flanks, one might suspect that tangential cutting forces or cutting power would, or should, show the same general characteristics as the feeding forces. Actually, in the tests cited, there was not only comparatively little difference in power requirements for the operation in spite of the large difference in feeding force, but power requirements increased by a factor of only one-half in each case.

Although a large increase in feeding force may not appreciably affect power requirements, the effect upon tool or workpiece deflections can be quite pronounced. In the example cited above, the formed diameters increased by 0.050 and 0.028" (1.27 and 0.71 mm) respectively, with the higher and lower forces. Because of clearances in machine tool assemblies, the typical force-deflection characteristics of a tool mounted on a slide controlled through a series of links can be represented by the curve shown in *Figure 3-31*.

* University of Michigan Research Institute Project No. 2575, Reynolds Metals Company.

Cutting time, min.	Surface appearance
0-2	Clean, only minor traces of built-up edge
2-11	Dull and streaky
11-25	Numerous large deposits of built-up edge; unsatisfactory
24-45	Numerous small deposits of built-up edge; very streaky
45-57 (Failure)	Partially burnished at 45 minutes to very highly burnished at failure

Figure 3-29. Effects of tool wear on surface quality.

Initially, a small change in force can result in a rather large deflection. As the play between parts is taken up and elastic resistance increases, a comparable change in force results in a smaller deflection of the tool. Thus, it is apparent that a range of feeding force from a very low value to a high value would absorb the greatest deflection range. If the initial feeding force is high, most of the tool deflection occurs prior to any effects created by tool wear, and the change in dimensions is lower in magnitude.

The previous examples illustrate certain effects of tool wear, but more than that, they illustrate the ever-present variations and difficult-to-explain results that complicate metal cutting practice.

Tool Life

The types and mechanisms of tool failure have been previously described. It was shown that excessive cutting speeds cause a rapid failure of the cutting edge; thus, the tool can be declared to have had a short life. Other criteria are sometimes used to evaluate tool life:

1. Change of quality of the machined surface.
2. Change in the magnitude of the cutting force resulting in changes in machine and workpiece deflections causing workpiece dimensions to change.
3. Change in the cutting temperature.
4. Costs, including labor costs, tool costs, tool changing time (cost), etc.

The selection of the correct cutting speed has an important bearing on the economics of all metalcutting operations. Fortunately, the correct cutting speed can be estimated with reasonable accuracy from tool-life graphs or from the Taylor Tool Life Relationship, provided that necessary data are obtainable.

The tool-life graph is shown in *Figure 3-32.* The logarithm of tool life in minutes is plotted against the logarithm of cutting speed in feet per minute. The resulting curve is very nearly a straight line in most instances. For

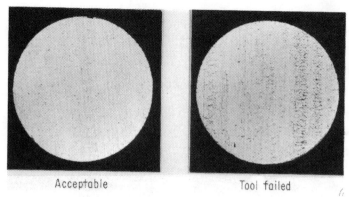

Acceptable   Tool failed

Figure 3-30. Example of tool failure based upon surface finish. Fax film reproductions show tearing and built-up-edge deposits which rendered surface unacceptable for functional requirements. Wear land at failure was only a few thousandths of an inch wide. Tool material—HSS. Work material—AISI 1045 steels.

practical purposes, it can be considered a straight line. This curve is expressed by the following equation (unless otherwise mentioned, equations and figures are for U.S. customary units only):

$$VT^n = C \qquad (3\text{-}1)$$

Where:

V = cutting speed, feet per minute
T = tool life, minutes
C = a constant equal to the intercept of the curve and the ordinate or the cutting speed—actually it is the cutting speed for a one-minute tool life
n = slope of the curve $\left(n = \tan \phi = \dfrac{\log V_1 - \log V_2}{\log T_2 - \log T_1} \right)$

    Example 1. A 2″ diam bar of steel was turned at 284 rpm and tool failure occurred in 10 min. The speed was changed to 232 rpm and the tool failed in 60 min. of cutting time.

    Assume a straight-line relationship exists, what cutting speed should be used to obtain a 30-min. tool life (V_{30})?

    Solution. Calculating the cutting speed,

$$V_1 = \frac{\pi DN}{12} = \frac{\pi\,(2)\,(284)}{12} = 149 \text{ fpm}$$
$$V_2 = \frac{\pi\,(2)\,(232)}{12} = 122 \text{ fpm}$$

The slope n of the tool-life curve can be determined from the equation

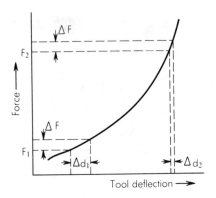

Figure 3-31. Effect of forces upon ma-
chine-tool component deflections.

shown above. If a log-log calculator is available, a faster method of
solution is shown as follows:

$$V_1 T_1^n = V_2 T_2^n = C$$

$$\frac{V_1}{V_2} = \left(\frac{T_2}{T_1}\right)^n$$

$$\frac{149}{122} = \left(\frac{60}{10}\right)^n$$

$$1.22 = 6^n$$

$$n = 0.11$$

From Eq. (3-1),

$$C = 149(10)^{0.11} = 149(1.2888)$$

$$= 192$$

Thus,

$$VT^{0.11} = 192$$

$$V_{30} = \frac{192}{(30)^{0.11}} = \frac{192}{1.455}$$

$$V_{30} = 132 \text{ fpm} \quad \text{Answer}$$

These values are for the particular feed, depth of cut, and tool geometry
shown. Significant changes in the tool geometry, depth of cut, and feed will
change the value of the constant C and may cause a slight change in the

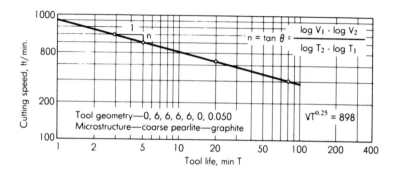

Figure 3-32. Tool life vs. cutting speed. Tool material—Kennametal carbide. Tool geometry—0, 6, 6, 6, 6, 0, 0.050. Work material—gray cast iron, 195 Bhn.

exponent n. In general, n is more a function of the cutting tool material. The value of n for the common cutting tool materials is as follows:

$$\text{H.S.S.: } n \cong 0.1 \text{ to } 0.15$$
$$\text{Carbides: } n \cong 0.2 \text{ to } 0.25$$
$$\text{Ceramics: } n \cong 0.6 \text{ to } 1.0$$

Equation (3-2) incorporates the effect of the size of cut:

$$K = VT^n f^{n_1} d^{n_2} \tag{3-2}$$

Where:

K = constant of proportionality
f = feed, inches per revolution
d = depth of cut, inches
n_1 = exponent of feed (average value = 0.5 to 0.8)
n_2 = exponent of depth of cut (average value = 0.2 to 0.4)

The optimum cutting speed for a constant tool life is more sensitive to changes in the feed than to changes in the depth of cut. The tool life is most sensitive to changes in the cutting speed, less sensitive to changes in the feed, and least sensitive to changes in the depth of cut. This relationship is shown by *Figures 3-33* and *3-34*.[1]

Example 2. The following equation has been obtained when machining AISI 2340 steel with high-speed steel cutting tools having a 8, 22, 6, 6, 6, 15, 3/64 tool signature:

$$2.035 = VT^{0.13} f^{0.77} d^{0.37}$$

A 100-min. tool life was obtained using the following cutting condition:

$$V = 75 \text{ fpm}, \qquad f = 0.0125 \text{ ipr}, \qquad d = 0.100''$$

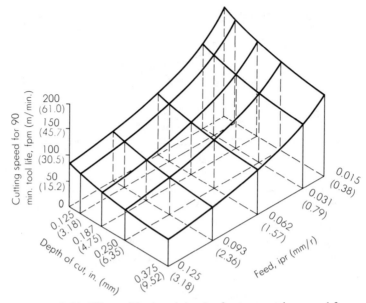

Figure 3-33. Effect of feed and depth of cut on cutting speed for 90-min tool life.[1] Workpiece material—gray cast iron. Tool material—HSS.

Calculate the effect upon the tool life for a 20% increase in the cutting speed, feed, and depth of cut, taking each separately. Calculate the effect of a 20% increase in each of the above parameters taken together:

a. When:

$$V = 1.2 \times 75 = 90 \text{ fpm}$$

$$T^{0.13} = \frac{K}{Vf^{n_1}d^{n_2}} = \frac{2.035}{90(0.0125)^{0.77}(0.100)^{0.37}}$$

$$= \frac{2.035}{90(0.034)(0.426)} = 1.56$$

$$T = 1.56^{1/0.13} = 1.56^{7.7}$$

$$= 31 \text{ min} \quad \text{Answer}$$

b. When:

$$f = 0.0125 \times 1.2 = 0.015 \text{ ipr}$$

$$T^{0.13} = \frac{2.035}{75(0.015)^{0.77}(0.100)^{0.37}} = \frac{2.035}{75(0.039)(0.426)}$$

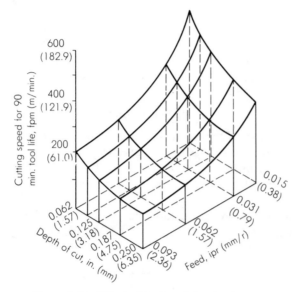

Figure 3-34. Effect of feed and depth of cut on cutting speed for 90-min tool life.[1] Workpiece material—gray cast iron. Tool material—HSS.

$$T = (1.63)^{7.7}$$

$$= 43 \text{ min} \quad \text{Answer}$$

c. When:

$$d = 1.2 \text{ x } 0.100 = 0.120''$$

$$T^{0.13} = \frac{2.035}{75 \,(0.0125)^{0.77}\,(0.120)^{0.37}} = \frac{2.035}{75(0.034)(0.456)}$$

$$T = (1.745)^{7.7}$$

$$= 74 \text{ min} \quad \text{Answer}$$

d. When:

$$V = 90 \text{ fpm}$$

$$f = 0.015 \text{ ipr}$$

$$d = 0.120''$$

$$T^{0.13} = \frac{2.035}{90(0.039)(0.456)}$$

$$T = (1.3)^{7.7}$$

$$= 7.5 \text{ min} \quad \text{Answer}$$

Tool life is very sensitive to changes in cutting tool geometry. However, tests varying the geometry do not generally yield curves that are consistent enough to interpret in general mathematical terms such as those already given. Tool life is also very sensitive to the microstructure and the hardness of the workpiece. An approximate equation relating tool life to Brinell hardness number (Bhn) is as follows:

$$K = VT^n f^{n_1} d^{n_2} \text{Bhn}^{1.25} \qquad (3\text{-}3)$$

The microstructure of the metal has a more pronounced effect on tool life than hardness alone. It is possible to have two pieces of steel at the same hardness but with different microstructures. The two pieces will yield a different tool life and surface finish when machined under the same cutting conditions.

Tool life is also sensitive to the tool material and the use of cutting fluids. The following general equation and *Figures 3-35* and *3-36* take these factors into consideration:

$$V = \frac{K_1}{d^{0.37} f^{0.77}} \quad \text{x} \quad \sqrt[6]{\frac{60}{T}} \quad \text{x } C.F. \qquad (3\text{-}4)$$

Metal to be cut	K_1 For 18-4-1 high-speed-steel tool and tool life of		
	60 min without cutting fluid, or 480 min with cutting fluid	60 min with cutting fluid	480 min without cutting fluid
Light alloys	25.0		
Brass (80-120 Bhn)	6.7		
Cast brass	4.2		
Cast steel	1.5	2.1	1.1
Carbon steel:			
SAE 1015	3.0	4.2	2.1
SAE 1025	2.4	3.3	1.7
SAE 1035	1.9	2.7	1.3
SAE 1045	1.5	2.1	1.1
SAE 1060	1.0	1.4	0.7
Chrome-nickel steel	1.6	2.3	1.1
Cast iron:			
100 Bhn	2.2	3.0	1.5
150 Bhn	1.4	1.9	1.0
200 Bhn	0.8	1.1	0.5

Figure 3-35. Numerical values for K_1.

Type	Approximate composition, %						C.F.
	W	Cr	V	C	Co	Mo	
14-4-1	14	4	1	0.7 -0.8	---	---	0.88
18-4-1	18	4	1	0.7 -0.75	---	---	1.00
18-4-2	18	4	2	0.8 -0.85	---	0.75	1.06
18-4-3	18	4	3	0.85-1.1	---	---	1.15
18-4-1+ 5% Co	18	4	1	0.7 -0.75	5	0.5	1.18
18-4-2+ 10% Co	18	4	2	0.8 -0.85	10	0.75	1.36
20-4-2+ 18% Co	20	4	2	0.8 -0.85	18	1.0	1.41
Sintered carbide	---	---	---	---	---	---	Up to 5

Figure 3-36. Correction factors for compositions of tool material.

Where:

V = cutting speed, fpm

k_1 = proportionality constant

$\sqrt[6]{\dfrac{60}{T}}$ = a factor which will correct the cutting speed from that obtained for a basic 60-minute tool life to the cutting speed for the desired tool life.

T = tool life, minutes

C.F. = corrections factor for the tool material. (18-4-1 HSS = 100)

Figure 3-35 lists values for K_1. *Figure 3-36* lists correction factors (*C.F.*) for different tool materials. *Figure 3-37* lists values for $d^{0.37}$ and $f^{0.77}$.

Example 3. A 5″ (127 mm) diameter bar of SAE 1020 steel is to be turned to a 4″ (102 mm) diameter in one cut, using cutting fluid. A feed of 0.020 ipr

d	$d^{0.37}$	d	$d^{0.37}$	f	$f^{0.77}$	f	$f^{0.77}$
0.01	0.182	0.25	0.598	0.001	0.004	0.025	0.059
0.02	0.235	0.30	0.640	0.002	0.008	0.030	0.067
0.04	0.305	0.35	0.678	0.004	0.017	0.035	0.075
0.06	0.353	0.40	0.712	0.006	0.019	0.040	0.084
0.08	0.393	0.45	0.744	0.008	0.024	0.045	0.092
0.10	0.427	0.50	0.774	0.010	0.029	0.050	0.099
0.14	0.482	0.75	0.899	0.014	0.037	0.075	0.135
0.18	0.530	1.00	1.000	0.018	0.045	0.100	0.170
0.22	0.571	---	---	0.022	0.053		

Figure 3-37. Numerical values for $d^{0.37}$ and $f^{0.77}$.

(0.51 mm/rev) will provide a satisfactory surface finish and a tool life of 180 minutes is desired.

Determine the cutting speed which will accomplish the cut if an 18-4-3 HSS tool is to be used.

K_1 = 3.75 (interpolate from *Figure 35*, 60 min with cutting fluids, basic)

d = 0.500; $d^{0.37}$ = 0.774 (from *Figure 3-37*)

f = 0.020; $f^{0.77}$ = 0.049 (interpolate from *Figure 3-37*)

T = 180 min

C.F. = 1.15 (from *Figure 3-38*)

$$V = \frac{3.75}{0.774 \times 0.049} \times \sqrt[6]{\frac{60}{180}} \times 1.15$$

$$= 99 \times 0.33^{.166} \times 1.15$$

$$= 99 \times 0.833 \times 1.15$$

$$= 95 \text{ fpm}$$

The recommended lathe rpm = $N = \dfrac{12V}{\pi D} = \dfrac{12 \times 95}{\pi \times 5}$

$$= 72.5 \text{ rpm}$$

Cutting Forces

Orthogonal cutting (*Figure 3-38*) is defined as two-dimensional cutting in which the cutting edge is perpendicular to the direction of motion relative to the workpiece and the cutting edge is wider than the chip. The forces arising from this type of cutting action are shown in *Figure 3-39*. F_c and F_N are the cutting and normal force, respectively. These components of the resultant force R can be measured by means of a dynamometer. F is the force required to overcome the friction between the chip and the face of the tool. If N is the coefficient of friction and α is the rake angle,

$$N = \tan\beta$$
$$= \frac{F}{N} = \frac{F_N - F_c \tan\alpha}{F_c - F_N \tan\alpha}$$

$$F_T = F_c \cos\phi - F_N \sin\phi$$

F_T cannot be calculated from the static mechanical properties of the material cut, since the strain rate encountered in metal cutting is much greater than that encountered in any conventional test. The shear angle ϕ must be determined before the shearing force and the shear flow stress can be determined from F_c and F_N. One method of approximating the shear angle is by measuring the thickness of the chip. Optimum shear angles occur when N is in the range of 1.5.

Let:

r = cutting ratio
t_1 = depth of cut or feed
t_2 = thickness of the chip

$$r = \frac{t_1}{t_2}$$

Then:

$$\tan \phi = \frac{f \cos \alpha}{1 - r \sin \alpha}$$

The shearing force can be determined from F_c and F_N by the geometry of the force system in *Figure 3-39*.

Figure 3-40 shows the resultant force acting on the cutting tool in a three-dimensional cut and the components of this force which can be measured with a three-component dynamometer. A typical relationship of the magnitude of these forces is shown in *Figure 3-41*.

Although some variation of tangential cutting force with respect to changes in speed may occur at low cutting speeds, the cutting force can be considered to be independent of cutting speed within the practical ranges of cutting speeds normally used. The effect of feed and the depth of cut on the cutting force is illustrated in *Figure 3-42*.[1] The following equation is obtained from *Figure 3-42*:

$$F_T = cf^{n_3} d^{n_4} \tag{3-5}$$

Where:

F_T = tangential cutting force, pounds
f = feed, inches per revolution
d = depth of cut, inches
c = constant of proportionality
n_3 = slope of F_T vs. f graph (typical values = 0.05 - 0.98)
n_4 = slope of F_T vs. d graph (typical values = 0.90-1.4)

The cutting-tool geometry has a considerable effect upon the cutting forces. The most pronounced effect is due to variations in the rake angle. This effect is shown in *Figure 3-43* for orthogonal cutting conditions. In three-dimensional cutting, the true rake angle is the particular rake angle on the face of the tool along which the chip is sliding.

The true rake angle can be closely approximated by the following equations (*Figure 3-44*):

$$\tan\theta = \frac{d}{NR + (d - NR) \tan SCEA} \tag{3-6}$$

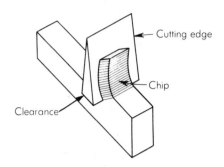

Figure 3-38. Orthogonal cutting.

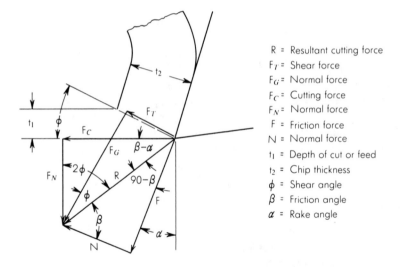

R = Resultant cutting force
F_T = Shear force
F_G = Normal force
F_C = Cutting force
F_N = Normal force
F = Friction force
N = Normal force
t_1 = Depth of cut or feed
t_2 = Chip thickness
ϕ = Shear angle
β = Friction angle
α = Rake angle

Figure 3-39. Forces acting on a continuous chip in orthogonal cutting.

Where:

θ = chip flow angle
NR = nose radius
$SCEA$ = side cutting edge angle

This equation is correct for tools having a zero-degree rake angle. For most normal rake angles, it will be correct within a few degrees of error. For very large rake angles, significant error can be introduced. Equation (3-6) can be used for most practical applications.

Equation (3-7) expresses the true rake angle corresponding to the chip flow angle.

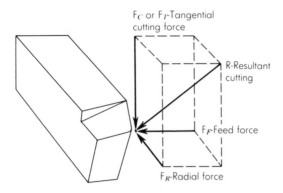

Figure 3-40. Three components of cutting force.

$$\tan\rho = \tan\gamma \ \sin\phi + \tan\theta \ \cos\phi \qquad (3\text{-}7)$$

Where:

ρ = true rake angle
γ = side rake angle
θ = chip flow angle
ϕ = shear angle

The location of the maximum rake angle is found from the following expression:

$$\tan\theta_{max} = \frac{\tan\gamma}{\tan\theta} \qquad (3\text{-}7a)$$

Figure 3-45 is a graphical representation of the effect of feed on the several forces for various depths of cut. The effects of cutting speed on cutting force (*Figure 3-46*) and tool forces (*Figure 3-47*) are also illustrated.

Power Requirements

The horsepower at the cutting tool is defined by,

$$hp_c = \frac{F_T V}{33,000} \qquad (3\text{-}8)$$

Where:

hp_c = horsepower at the cutting tool
V = cutting speed, feet per minute
F_T = tangential cutting force component

The radial cutting force does not contribute to horsepower. Although the

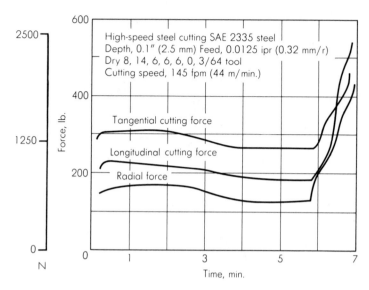

Figure 3-41. Three components of cutting force through tool life.[1] Tool material—HSS. Work material—SAE 2335 steel. Tool geometry—8, 14, 6, 6, 6, 0, 3/64.

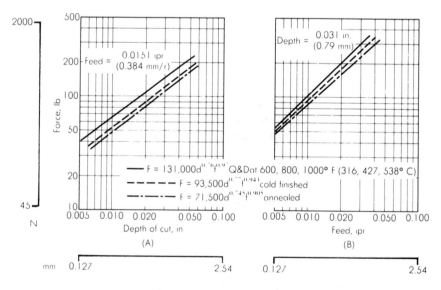

Figure 3-42. (*A*) Cutting force vs. depth of cut.[1] Tool material—HSS. Tool geometry—6, 11, 6, 6, 6, 15, 0.01. Work material—SAE 1045 steel. (*B*) Cutting force vs. feed.[1] Depth—0.031 inches.

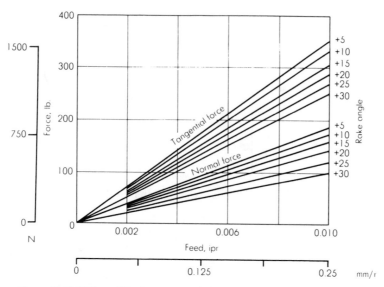

Figure 3-43. Effect of feed on tangential and normal cutting forces for a range of rake angles when turning nickel steel.

feed force can be considerable, the feeding velocity is generally so low that the horsepower required to feed the tool can be neglected.

Substituting Eq. (3-5) in (3-8),

$$hp_c = \frac{Cf^{n_3}d^{n_4}V}{33,000} \tag{3-9}$$

There are losses in the machine which must be considered when estimating the size of the electric motor required:

$$hp_g = \frac{hp_c}{eff} + hp_t \tag{3-10}$$

Where:

hp_g = gross or motor horsepower

hp_t = tare horsepower, the power required to run the machine at no-load conditions

eff = the mechanical efficiency of the machine

Specific Power Consumption or Unit Horsepower. The specific power consumption, W_p, or unit horsepower, is defined as the horsepower required to cut a material at a rate of one cubic inch per minute.

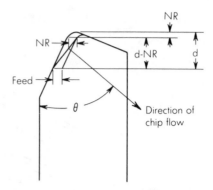

Figure 3-44. Approximate chip flow direction.

$$W_p = \frac{hp_c}{\text{cubic inches per minute removed by cut}} \qquad (3\text{-}11)$$

For turning,

$$W_p = \frac{hp_c}{12Vfd} \qquad (3\text{-}11a)$$

$$= \frac{F_T}{396,000fd} \qquad (3\text{-}11b)$$

Note that Wp is independent of the cutting speed.

$$W_p = \frac{Cf^{n_3}d^{n_4}}{396,000fd} = \frac{C}{396,000f^{1-n_3}d^{1-n_4}} \qquad (3\text{-}11c)$$

from which,

$$\frac{C}{396,000} = C' = W_{p_1}\,(f_1)^{1-n_3}(d_1)^{1-n_4} = W_{p_2}(f_2)^{1-n_3}(d_2)^{1-n_4}$$

or,

$$W_{p_2} = W_{p_1}\left(\frac{f_1}{f_2}\right)^{1-n_3}\left(\frac{d_1}{d_2}\right)^{1-n_4} \qquad (3\text{-}12)$$

Equation (3-12) defines the effect of feed and depth of cut on the specific power consumption. This equation may be reduced to the following form since the value of n_4 is often nearly equal to one.

$$W_{p2} = W_{p1}\left(\frac{f_1}{f_2}\right)^{1-n_3}$$

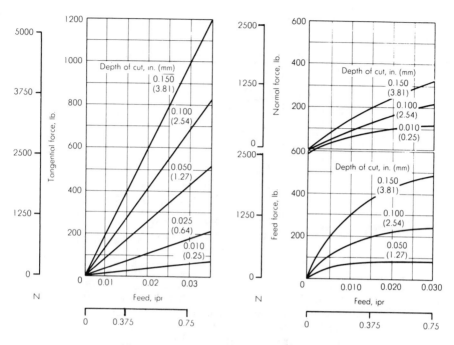

Figure 3-45. Effect of feed on tangential feed and normal forces for various depths of cut. Tool material—HSS. Tool shape—8, 14, 6, 6, 6, 15, 3/64. Work material—SAE 3135 steel, annealed. Cutting speed—50 fpm (dry).

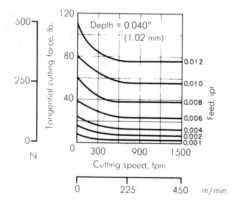

Figure 3-46. Effect of cutting speed on cutting force for various feeds when cutting bronze with a tungsten carbide tool. Depth of cut—0.040 inches.

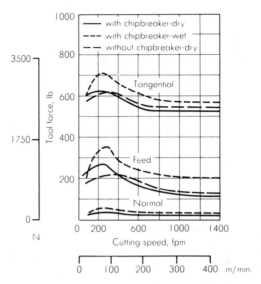

Figure 3-47. Effect of cutting speed on tool forces. Tool material—tungsten carbide. Tool shape—10, 10, 8, 8, 7, 0, 0.015. Work material—AISI C 1118 steel. Depth of cut—0.150 inches.

Typical values of specific power consumption are given in *Figure 3-48*.

Example 1. An alloy steel having a hardness of 250 Bhn is to be machined in a milling machine. The depth of cut is to be 0.250″, the feed is 0.005″ per tooth, and the cutting speed is 300 fpm. The milling cutter has 12 teeth and is 10″ in diameter. The width of the cut is 5″. The specific power consumption is 0.70 hp/cu in./min.

$$hp_c = W_p \text{ x cu. in. per min}$$

$$= (W_p) \left(\frac{12V}{\pi D}\right)(n)(f)(w)(d)$$

$$= (0.70) \left[\frac{12(300)}{\pi(10)}\right](12)(0.005)(5)(0.250)$$

$$= 6.03$$

Assume:

$$hp_t = 0.5 \text{ and } eff = 0.8$$

$$hp_g = \frac{hp_c}{eff} + hp_t = \frac{6.03}{0.8} + 0.5$$
$$= 8.05$$

Use 10-hp motor.

Dry Cutting; Depth, 1/8 in. (3.2 mm); Feed, 1/64 ipr (0.4 mm/r)			
Material cut	Tool shape	Brinell hardness no.	HP_c/in.^{3} per min. (kW/cm^3 per sec.)
Plain carbon steel		126	0.59-0.66 (1.6-1.8)
		179	0.70-0.79 (1.9-2.2)
		262	0.85-0.95 (2.3-2.6)
Free-cutting steel	8, 14, 6, 6, 6, 0, 1/16	118	0.36-0.39 (1.0-1.1)
		179	0.44-0.48 (1.2-1.3)
		229	0.50-0.54 (1.4-1.5)
Alloy steel		131	0.46-0.57 (1.3-1.6)
		179	0.55-0.68 (1.5-1.9)
		269	0.67-0.83 (1.8-2.3)
		429	1.10-1.90 (3.0-5.2)
Cast iron		140	0.22-0.32 (0.6-0.9)
		179	0.45-0.68 (1.2-1.9)
		256	0.85-1.30 (2.3-3.5)
Leaded brass		33	0.18-0.27 (0.5-0.7)
		76	0.22-0.31 (0.6-0.8)
		131	0.25-0.35 (0.7-1.0)
Unleaded brass	8, 14, 6, 6, 6, 15, 0	50.9	0.54 (1.5)
Pure copper		40.4	0.88 (2.4)
Magnesium alloys		32	0.084-0.10 (0.2-0.3)
		49	0.094-0.11 (0.26-0.3)
		68	0.10-0.12 (0.3-0.33)
	8, 14, 6, 6, 6, 15, 0	55	0.28 (0.8)
		159	0.26 (0.7)
Aluminum alloys	20, 40, 10, 10, 10, 15, 0	32	0.12 (0.33)
		94	0.15 (0.4)
		115	0.17-0.21 (0.5-0.6)
		153	0.20 (0.55)
Monel metal	8, 14, 6, 6, 6, 15, 3/64	147	0.58-0.75 (1.6-2.0)
		160	1.35 (3.7)

Figure 3-48. HP_c/CIM values for various materials.

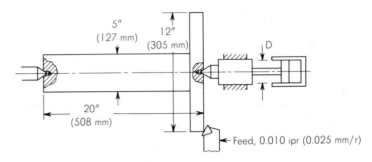

Figure 3-49. Forces in a turning operation.

Example 2. A steel part, shown in *Fig. 3-49*, is to be machined between centers in a lathe equipped with an air-operated tailstock spindle. The maximum depth of cut is to be 0.100", the feed is 0.010 ipr, and the cutting speed is 300 fpm. The cutting-tool geometry is 10, 10, 6, 6, 10, 15, 0.030. The feeding force is assumed to be two-thirds of the tangential cutting force. The following equation applies to this material.

$$F_T = Cf^{0.8}d \tag{3-13}$$

Calculate the diameter D of the air cylinder required to hold the work between centers against the cutting forces if the minimum pressure in the air cylinder may reach 60 psi. Both centers rotate; thus the friction between the work and the centers need not be considered.

$W_p = 0.70$ (from *Figure 3-48*)

$$W_{P_2} = W_{P_1} \left(\frac{f_1}{f_2} \right)^{1-0.8} = 0.70 \left(\frac{0.0156}{0.010} \right)^{0.2}$$

$\qquad = 0.765 \; hp_c / \text{cu in.} / \min$

$F_T = 396{,}000 \; W_p fd = 396{,}000(0.765)(0.010)(0.100)$

$\qquad = 303 \; \text{lb}$

$F_F = 2/3 \times 303 = 202 \; \text{lb}$

$$\tan\theta = \frac{d}{NR + (d - NR)\tan SCEA}$$

$$\qquad = \frac{0.100}{0.030 + (0.100 - 0.030)\tan 15}$$

$\theta = 64°01'$; say $64°$

$F_R = F_F \tan(90 - \theta) = 202 \tan 26°$

$\qquad = 99 \; \text{lb}$

Taking moments about the headstock center (*Figure 3-50A*),

$$\Sigma \quad M = 0 = 6F_F - 20F_R + 20F_X$$
$$F_X = 38.4 \text{ lb}$$

Resolving the forces at the cutting tool in a vertical plane normal to the work axis (*Figure 3-50B*).

$$R = (F_T{}^2 + F_X{}^2)^{1/2} = 307 \text{ lb}$$

Resolving R about the tailstock center (*Figure 3-50C*).

$$F_Y = \frac{R}{\tan 60} = 177 \text{ lb}$$

$$D_{CYL} = \frac{4F_Y}{\pi 60} = 3.76 \quad \text{Answer}$$

Basic Principles of Multiple-Point Tools

Multiple-point cutting tools are basically a series of single-point tools mounted in or integral with a holder or body and operated in such a manner that all the teeth (tools) follow essentially the same path across the workpiece. The cutting edges may be straight or may be in the form of various contours which are to be reproduced upon the workpiece. Multiple-point tools may be either linear-travel or rotary. With linear-travel tools, the relative motion between the tool and workpiece is along a straight-line path. The teeth of rotary cutting tools revolve about the tool axis. The relative motion between the workpiece and a rotary cutting tool may be either axial or in a plane normal to the tool axis. In some cases, a combination of the two motions is used. Certain form-generating tools involve a combination of linear travel and rotary motions.

Figures 3-51 and *3-52* illustrate two types of milling cutters and indicate the difference in nomenclature between the angles of these cutting tools as compared to single-point tools. *Figure 3-52* shows the peripheral cutting edge angle as zero degrees with positive and negative directions indicated.

Whether a cutting tool is single-point or is one component of a milling cutter, the various angles must provide the most efficient cutting action. Theoretical considerations may dictate larger angles, but actual cutting experience may dictate smaller angles for greater tool strength without chatter. Advantages from increasing any angle must always be considered together with the effect on tool strength.

Cutting Processes. The cutting processes for multiple-point tools are essentially similar to those for single-point tools. Linear-travel tools produce a series of chips that are similar to those produced by single-point tools on

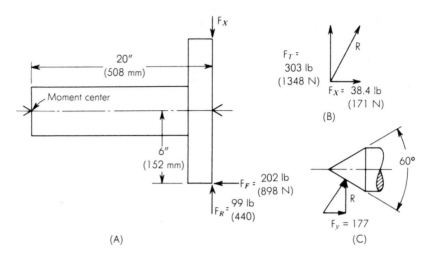

Figure 3-50. Forces in turning operations.

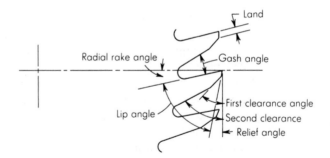

Figure 3-51. A solid plain milling cutter showing the basic tooth angles and the tooth land.

planing cuts. Milling cutters produce chips that vary in thickness because of the nature of the tooth path as illustrated in *Figure 3-53*. The chips produced by axial-feed tools tend to be conical because the varying diameters across the cutting edge cause a difference in the distance travelled by different portions of the cutting edge. Aside from these differences, studies have shown that there is no fundamental difference between the metal-formation process involved in forming chips with these tools and that involved in using single-point tools on turning or planing cuts.

The twist drill is unique in that it involves two different metal-deformation processes. The main cutting lips produce conventional chips as illustrated in the chip section shown in *Figure 3-55*. Aside from some curvature of the

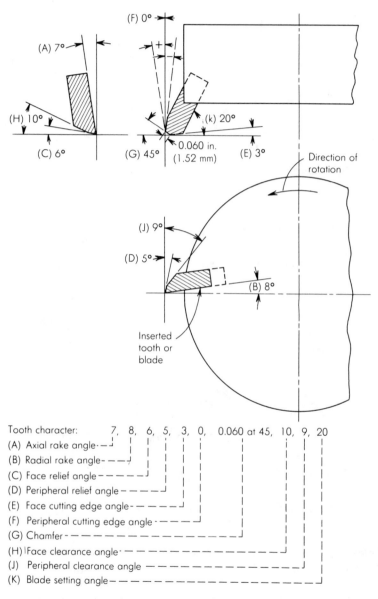

Tooth character: 7, 8, 6, 5, 3, 0,   0.060 at 45, 10, 9, 20
(A) Axial rake angle
(B) Radial rake angle
(C) Face relief angle
(D) Peripheral relief angle
(E) Face cutting edge angle
(F) Peripheral cutting edge angle
(G) Chamfer
(H) lFace clearance angle
(J) Peripheral clearance angle
(K) Blade setting angle

Figure 3-52. A face milling cutter with inserted teeth. All pertinent tooth angles are included.

finished surface, the chip is quite similar to the chip produced by a single-point tool. The metal deformation under the chisel edge is much more

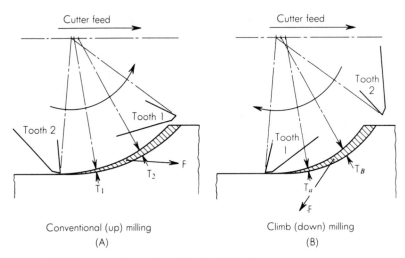

Figure 3-53. A comparison of undeformed chip shapes.

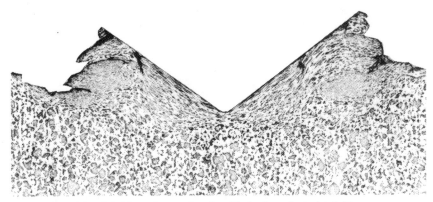

Figure 3-54. Metal deformation under the chisel edge of a twist drill.

complex. At the exact center of the hole, the only motion of the drill is axial, so the deformation resembles that produced by an indenting punch. As the radius increases, the rotation of the drill becomes important and the chisel-edge wedge appears to both cut and extrude the metal. An analysis of this region of the drill based upon such an assumption fits experimental data. *Figure 3-54* shows a diametral section normal to the chisel edge. The complex metal deformation shown is a major factor in the high thrust forces required by twist drills with conventional drill points.

Design Considerations. The most important single factor affecting

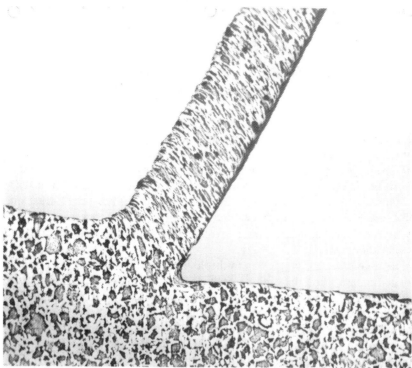

Figure 3-55. Section showing conventional chip formation by a single-point tool.

the performance of any cutting tool is the attainment of a high degree of rigidity in the entire machining system. This includes the cutting tool, the machine tool, the fixture, and the workpiece. A lack of rigidity in any of the system elements can largely nullify the benefits of high rigidity in the other elements. This interrelationship is all too often overlooked by fixture designers so that workpieces are not adequately supported at the point of cutting.

With adequate workpiece support and machine tool rigidity, an increase in the rigidity of the cutting tool can effect large improvements in tool life. Such improvements are particularly evident in the case of the super-strength alloys used in aircraft and missile production. When drilling such alloys, improvements in tool life by a factor of 50 or more have been obtained by using short, heavy-web drills in place of conventional drills.

In the design of multiple-point tools, there must always be some compromise with maximum rigidity. Most multiple-point tools are required to carry the chips generated for some distance before they can be ejected. Adequate chip space must be provided to avoid jamming, which can cause tool breakage. The amount and shape of the chip space provided depend upon the material and the nature of the cut. If the chips are discontinuous, less chip room is required and closer tooth spacing can be used. More chip room and wider tooth spacing are required for high tensile continuous chips. With some tools, it is possible to incorporate some form of chip breaker on the cutting face to produce smaller chips and thus improve the chip-conveying ability of the tool.

With tools such as milling cutters and broaches, tooth spacing can also affect smoothness of operation and the accuracy of the work. If too many teeth are in contact with the work at the same time, the tool part, or the machine may be overloaded. Overloading may cause breakage of the tool or driving key. The deflection due to the increase in loading may cause a decrease in accuracy. With milling cutters it is generally desirable that at least one tooth be in contact with the work at all times. This keeps the tool and workpiece under load at all times avoiding the vibration that occurs from varying shock loads.

The rake and relief angles on multiple-point cutting tools affect both the tool performance and the strength of the teeth. High rake angles usually make cutting freer and more efficient. High relief angles reduce the rubbing that occurs on the flank of the tool. These angles, however, cannot be increased without limit. As either the rake angle or the relief angle is increased, cutting-edge strength is reduced. In addition, while high relief angles reduce the temperature resulting from flank friction, they do cause a greater change in size for the development of a wear land of given size. The selection of rake and relief angles therefore requires a compromise.

Operating Considerations. In setting up the operating conditions for any multiple-point cutting tool, there are three important variables that can be adjusted. These are the cutting speed, the feed per cutting edge, and the cutting fluid. Of these, the feed is the most important and should be established first. With some tools, such as taps and broaches, the feed per tooth is determined by the design of the tool and can be changed only by tool modification. With most other tools, the feed is determined by selecting an appropriate machine setting. It has long been demonstrated that it is more efficient to remove metal in the form of thick chips than in thin ones, so the maximum possible feed per tooth should be used. The maximum feed per tooth is limited by the following factors: (1) cutting-edge strength, (2) rigidity and allowable deflection, (3) the surface finish required, and (4) tool chip space.

The feed cannot be increased without limit because some point will be reached at which either the tool or the machine will be overloaded. Overload can cause breakage either immediately or with accumulative increase in cutting forces due to dulling. As the feed is increased, the cutting force increase causes greater deflection between the tool and the workpiece. Deflection can become so large that it is impossible to hold the accuracy required. The surface finish produced on the workpiece usually deteriorates as the feed is increased. This may necessitate the use of light feeds. With some metals, the greater volume of chips produced with heavy feeds may overload the chip space in the tool. In other cases, heavier feeds will produce broken chips which can be handled better than continuous chips resulting from light feeds.

In peripheral milling, the actual chip thickness is affected by the depth of cut. Maximum undeformed chip thickness equal to the feed per tooth is obtained when the width of cut equals one-half the cutter diameter. With shallow finishing cuts the maximum undeformed chip thickness is much less than the cutter advance per tooth. In the extreme case of peripheral milling, with a shallow width of cut, the undeformed chip thickness may be so low that there is no chip load, which causes the tool to burnish or workharden the material rather than cut it. To prevent this on these shallow cuts, feed rates of three to four times the normal should be used.

After the maximum allowable feed has been established, the cutting speed should be considered. At a constant feed per tooth, there is relatively little change in cutting forces as the cutting speed is increased. In the normal operating range, speed has relatively little effect on surface finish, but sometimes a large increase of speed (usually made possible by a change of tool material) will yield an improvement in surface finish by increasing the temperature such that the workpiece is softened, resulting in less tearing. The principal effect of increasing cutting speed is to produce parts faster at constant feed per tooth. This increase in production normally justifies the tool life reduction resulting from operating at higher temperatures. The optimum cutting speed is that which will permit production of parts at the lowest cost per piece. This requires analysis of all of the costs including machining cost, tool changing cost, and cutting tool acquisition and maintenance costs.

The use of the correct cutting fluid can result in substantial improvements in a machining operation. The proper cutting fluid can permit higher feeds and increased speeds as well as attainment of better surface finishes. Cutting fluids should be directed to the exact point where the cutting is being done and should be applied in a constant, even flow. No machining operation should be set up without some consideration of cutting fluids.

Forces and Power Requirements

Knowledge of cutting forces is essential to machine and fixture design, and for the determination of the power requirements. Force and power predictions for multiple-point tools are more complex than those for a single-point tool.

Varying numbers of teeth may be in contact with the work, the chip size can vary in different parts of the cut, and the orientation of the cutting teeth may not be constant with respect to the workpiece. The best method to determine forces on such tools is by actual measurement on the machine to be used or on a simulated setup in a metalcutting laboratory. Since such measurements are sometimes impossible or inconvenient to make, methods for making reasonable force and power estimates may be of value.

One approach is to consider the multiple-point tool equivalent to a series of single-point tools, then estimate the contribution of each tool and sum these to arrive at the resultant forces. The following methods of calculation are preferable.

Power Requirements for Drilling and Reaming. For rotary axial feed tools, such as twist drills, core drills, and reamers, reasonably accurate estimates of forces and power can be made through the use of formulae developed experimentally and analytically by Shaw and Oxford.[8]

The torque and thrust for a twist drill operating in an alloy steel with a hardness of 200 Bhn can be represented by the following formulae:

$$M = 23,300 f^{0.8} d^{1.8} \left[\frac{1 - \left(\frac{c}{d}\right)^2}{\left(1 + \frac{c}{d}\right)^{0.2}} + 3.2 \left(\frac{c}{d}\right)^{1.8} \right] \quad (3\text{-}14)$$

$$T = 42,600 f^{0.8} d^{1.8} \left[\frac{1 - \left(\frac{c}{d}\right)}{1 + \left(\frac{c}{d}\right)^{0.2}} + 2.2 \left(\frac{c}{d}\right)^{0.8} \right] + 19,300 c^2 \quad (3\text{-}15)$$

Where:

  M = torque, inch-pounds
  T = thrust, pounds
  d = drill diameter, inches
  c = chisel edge length, inches (approximately 1.15 times the web thickness for normal sharpening)
  f = feed per revolution, inches

For drills of regular proportions the ratio c/d can be set equal to 0.18 and the equation simplified to,

$$M = 25,200 f^{0.8} d^{1.8} \quad (3\text{-}16)$$

$$T = 57,500 f^{0.8} d^{0.8} + 625 d^2 \quad (3\text{-}17)$$

Note: There is a difference in the two power requirements for drilling, because there is a difference in the configuration of the point. A reduced web area will take far less power to drive because of the rubbing action of the flat created by the chisel point.

For reamers or core drills, which are used for enlarging existing holes, the effects of the chisel-edge region can be eliminated and the equations reduced to the following:

$$M = 23,300 k f^{0.8} d^{1.8} \left[\frac{1 - \left(\dfrac{d_1}{d} \right)^2}{\left(1 + \dfrac{d_1}{d} \right)^{0.2}} \right] \tag{3-18}$$

$$T = 42,600 k f^{0.8} d^{0.8} \left[\frac{1 - \dfrac{d_1}{d}}{\left(1 + \dfrac{d_1}{d} \right)^{0.2}} \right] \tag{3-19}$$

Where:

d = drill diameter, inches
d_1 = diameter of hole to be enlarged, inches
f = feed, ipr
k = a constant depending upon the number of flutes

The constant k is necessary since, for a given feed per revolution, the number of flutes affects the feed per tooth. It has been pointed out that it is more efficient to remove metal in the form of thick chips than thin ones, so tools with a large number of flutes will require proportionately more energy because of the thinner chips. Values of k for different numbers of flutes are tabulated below (for U.S. customary units).

Number of flutes	Constant, k	Number of flutes	Constant, k
1	0.87	8	1.32
2	1.00	10	1.38
3	1.08	12	1.43
4	1.15	16	1.51
6	1.25	20	1.59

Whereas the thrust forces can be substantial and thus can have a large influence upon the required strength and rigidity, the power required in feeding the tool axially is very small (less than 2% of the total power requirements), and can usually be disregarded. The cutting power is a function of the torque and rotational speed and can be computed from the following equation:

$$hp = \frac{MN}{63,025} \qquad (3\text{-}20)$$

Where:

  hp = horsepower
  N = speed, rpm
  M = tool torque, inch-pounds

Note: To convert to SI Units:

$$hp \times 746 = \text{watt (W)}$$

In using these formulae for estimating purposes, an allowance of at least 25% should be made for increases due to tool dulling, and further allowances should be made for the efficiency of the machine drive train. For other materials, the cutting forces vary in about the same ratio as noted for single-point tools.

When it is not possible to estimate forces and power directly, a fairly good estimate can be made by considering the cutting energy. It has been found that the removal of one cubic inch (16.4 cubic centimeters) of an alloy steel with a hardness of about 200 Bhn at normal feeds requires the expenditure of about 500,000 inch-pounds (56429 Joule) of energy. In terms of engineering units, this can be stated as 1.25 horsepower per cubic inch per minute. Thus, if the maximum rate of metal removal is computed in terms of cubic inches per minute, this value can be multiplied by 1.25 to give a reasonable estimate of the horsepower required at the cutting tool.

For SI metric conversion:

$$\text{Unit horsepower (hp/in.}^3/\text{min)} \times 2.73$$
$$= \text{unit power (kW/cm}^3/\text{s)}$$

If a rotary tool is involved, the tool torque M, in inch-pounds (newton, metres—N · m), can be estimated from the following equation:

$$M = \frac{63,025\ hp}{N} \qquad (3\text{-}21)$$

Where:

  hp = tool horsepower
  N = tool rotational speed, rpm

Note: To convert to SI Units:

$$M \times 0.11298 = \text{newton} \cdot \text{metre (N} \cdot \text{m)}$$

In the case of linear-travel tools, the force in the cutting direction can be estimated from the following formulae:

$$F = \frac{33,000 \ hp}{V} \tag{3-22}$$

Where:

F = cutting force, pound
hp = tool horsepower
V = cutting speed, feet per minute

Note: To convert to SI Units:

$$F \times 4.448 \ = \ \text{Newtons}$$

Equations 3-21 and 3-22 must be applied with caution since they will be useful in estimating torque or cutting force only if an accurate estimate of the maximum rate of metal removal is made. If they are applied to the average rate of metal removal, they will indicate only average torque or average force, and this could be considerably below the peak forces in the case of intermittent cutting as might be encountered in milling or broaching. Further, all of the equations presented here ignore the even higher peak forces resulting from impact or vibration during cutting. Because of these limitations, the equations should not be used as a basis for the design of critical elements in the machine or fixtures.

Power Requirements for Milling. Milling machines, like other machine tools, are not 100% efficient. That is, it takes higher motor horsepower than the horsepower required at the cutting tool to run the milling machine in a machining operation. The reason is due to frictional losses, gear-train inefficiencies, spindle speeds that are too high for the particular machine, mechanical condition, etc. Consequently, the power of milling must include the machine power losses and the power actually used at the cutter. Efficient use of power at the cutter is influenced by cutter speed, design, cutting edge geometry, and by workpiece material.

The total horsepower required at the cutter (hp_c) is given by the equation:

$$hp_c = \frac{cim}{K} \tag{3-23}$$

Where:

hp_c = horsepower at the cutter
cim = metal removal rate, cubic inches per minute
K = a factor reflecting the efficiency of the metalcutting operation

Note: To convert to SI Units:

$$hp_c \times 746 = \text{watt (W)}$$

The K factor varies with type and hardness of material; also for the same material, it varies with the feed per tooth, increasing as the chip thickness increases. Time-consuming trials are required to determine the quantities involved, because in each case the K factor represents a particular rate of metal removal and not a general or average rate.

To make available a quick approximation of the total power requirements, a milling-machine selector table has been devised (*Figure 3-56*), which estimates the metal removed in cubic inches per minute for various rated horsepowers of various machines under constant load conditions.

The metal removal rates in *Figure 3-56* may be considered as products of the K factor (*Figure 3-57*) and over-all efficiencies listed in *Figure 3-56*. All K constants given here are for dull cutters; hence no allowance for increase in horsepower due to dulling need be made. All values apply to average milling speeds and rake angles recommended for the various materials, and 0.010" (0.25 mm) feed per tooth.[8]

Linear-Travel Tools

Broaches. The most common multiple-point linear-travel tool is the broach. Broaches are used for producing either external or internal surfaces. The surfaces produced may be flat, circular, or of quite intricate profile, as

Rated hp of machine	3	5	7.5	10	15	20	25	30	40	50
Overall machine efficiency, per cent	40	48	52	52	52	60	65	70	75	80
Material	Max metal removal, in.^{3}/min. (cm^3/min.)									
Aluminum	2.7 (44)	5.5 (90)	8.7 (143)	12 (197)	18 (295)	27 (442)	37 (606)	48 (787)	69 (1131)	91 (1491)
Brass, soft	2.4 (39)	4.7 (77)	7.5 (123)	10 (164)	16 (262)	24 (393)	32 (524)	41 (672)	60 (983)	79 (1295)
Bronze, hard	1.7 (28)	3.3 (54)	5.3 (87)	7.3 (120)	11 (180)	17 (279)	23 (213)	30 (492)	43 (705)	56 (918)
Bronze, very hard	0.78 (13)	1.6 (26)	2.5 (41)	3.4 (56)	5.3 (87)	7.8 (128)	11 (180)	15 (246)	20 (328)	26 (426)
Cast iron, soft	1.6 (26)	3.2 (52)	5.2 (85)	7.1 (116)	11 (180)	16 (262)	22 (361)	28 (459)	41 (672)	54 (885)
Cast iron, hard	1 (16)	2 (33)	3.3 (54)	4.6 (75)	7 (115)	11 (164)	14 (229)	18 (295)	26 (426)	35 (574)
Cast iron, chilled	0.78 (13)	1.6 (26)	2.5 (41)	3.4 (56)	5.3 (87)	7.8 (128)	10 (164)	13 (213)	19 (311)	26 (426)
Malleable iron	1 (16)	2.1 (34)	3.4 (56)	4.7 (77)	7.3 (120)	11 (180)	14 (229)	18 (295)	26 (426)	36 (590)
Steel, soft	1 (16)	2 (33)	3.3 (54)	4.6 (75)	7 (115)	11 (164)	14 (229)	18 (295)	26 (426)	35 (574)
Steel, medium	0.78 (13)	1.6 (26)	2.5 (41)	3.4 (56)	5.3 (87)	7.8 (128)	10 (164)	13 (213)	19 (311)	26 (426)
Steel, hard	0.56 (9)	1.1 (18)	1.8 (29)	2.5 (41)	3.9 (64)	5.7 (93)	7.7 (126)	10 (164)	14 (229)	19 (311)

Data courtesy of Kearney & Trecker Corp.

Figure 3-56. Milling-machine selector table.

Aluminum and magnesium 2.5-4.0
Bronze and brass, soft 1.7-2.5
Bronze and brass, medium 1.0-1.4
Bronze and brass, hard 0.6-1.0
Cast iron, soft 1.5
Cast iron, medium 0.8-1.0
Cast iron, hard 0.6-0.8
Malleable iron and cold-drawn steel, SAE 6140 .... 0.9
Cold-drawn steel, SAE 1112, 1120, and 1315 .... 1.0
Forged and alloy steel, SAE 3120, 1020, 2320, and 2345, 150-300 Bhn 0.63-0.87
Alloy steel, 300-400 Bhn 0.5
Stainless steel, AISI 416, free-machining 1.1
Stainless steel, austenitic, AISI 303, free-machining 0.83
Stainless, steel, austenitic, AISI 304 .... 0.72
Monel metal 0.55
Copper, annealed 0.84
Tool-steel 0.505
Nickel 0.525
Titanium 0.75

Figure 3-57. Value of K factor for various materials for (U.S. customary units).

viewed in a section normal to the tool travel. A broach is essentially a series of single-point tools following each other in the axial direction along a tool body or holder. Successive teeth vary in size or shape in such a manner that each following tooth will cut a chip of the proper thickness. The basic elements of broach construction are illustrated in *Figure 3-58*.

The spacing and shape of broach teeth are determined by the length of the workpiece and the chip thickness per tooth as well as by the type of chips formed. The chip space between the broach teeth must be sufficient to take care of the volume of chips generated. Broach teeth are provided with rake and relief angles in the same manner as other cutting tools. Standard broaching nomenclature designates the rake angle as the face angle and the relief clearance as the back-off angle. The rake angles fall in the same range as used for other tools, but the back-off angles are normally quite low, in the range between 0.5° and 3.5°. Low back-off angles are used on broaches to minimize the loss of size in resharpening. Final finishing teeth are often provided with an unrelieved land behind the cutting edge to assure proper sizing of the workpiece. Sometimes noncutting burnishing teeth follow the final cutting teeth.

Internal broaches (*Figure 3-59*) are either pulled or pushed through the work. Strength considerations limit the design of such broaches. Surface broaches (*Figure 3-60*) are ordinarily carried on a large guided ram; here strength is not so critical since the cutting tool can be transferred to the ram at many points along the broach length.

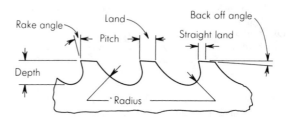

Figure 3-58. Basic elements of broach construction.

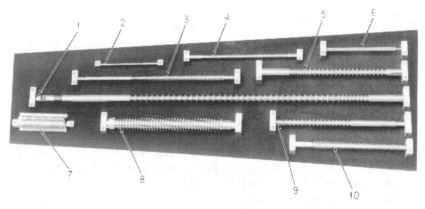

Figure 3-59. Internal broaches: (1) round pull-type, (2) round push-type, (3) involute spline pull, (4) round push-type, (5) push-type with overlapping teeth, (6) two-spline type with round body, (7) shaving shell for pump rotor, (8) helical spline push-type, (9) round pull-type with overlapping teeth, (10) combination round and serration type. (*Courtesy, American Broach & Machine, Sundstrand Machine Tool Co.*)

Broaches are commonly made of high-speed steel as solid units, but carbide-tipped and inserted-blade broaches are sometimes economical. This is particularly true in the case of surface broaches which are better adapted to mounting carbide indexable inserts. Broaches can be used to cut helical internal forms if the broaching machine is equipped to rotate the broach at the proper lead-rate as it passes through the work.

Gear Shaper Cutters. A gear shaper cutter is a tool that looks, to some extent, like a gear, the teeth of which are relieved to provide cutting edges. Typical gear shaper cutters are illustrated in *Figure 3-61*. Gear shaper cutters are reciprocated in the axial direction, but at the same time are rotated in timed relationship to the gear being generated. The rotational speed is low

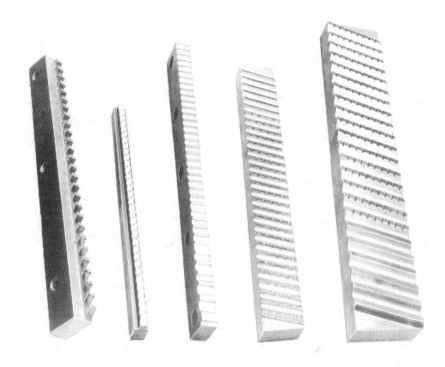

Figure 3-60. Surface broach sections showing tooth detail and chip breakers on taper teeth. (*Courtesy, American Broach & Machine, Sundstrand Machine Tool Co.*)

compared to the speed of reciprocation, so the shaper cutter is principally a linear-travel tool. The feed per tooth is determined by the rate at which the cutter is fed into depth in the gear blank and to some extent by the speed of rotation. Gear shaper cutters can, within limits, be designed for forms other than gears, such as splines, sprockets, etc.

Axial-Feed Rotary Tools

Twist Drills. In its most basic form, a twist drill (*Figure 3-62*) is made from a round bar of tool material. It has a pair of helical flutes which form the cutting surfaces and act as chip conveyors. Relief is provided behind the two cutting edges or lips. The intersection of the two relief surfaces across the web between the two flutes is known as the chisel edge. The lands between the flutes are cut away to a narrow margin to reduce the area of contact between the lands and the wall of the hole and/or guide bushing. The metal cut away to form the margin is known as body diameter clearance.

Twist drills can be provided with a variety of shanks, but the straight shank

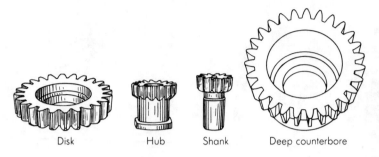

Figure 3-61. Standard gear shaper cutters.

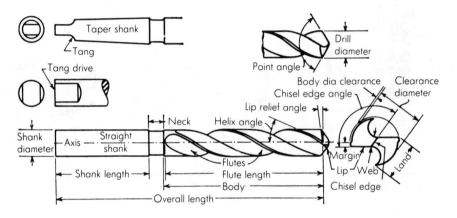

Figure 3-62. Nomenclature for twist drills.

and the Morse taper shank are the most common. Twist drills are ordinarily made of high-speed steel, but certain sizes and types are available, made from solid carbide or with carbide cutting tips. In 1978, indexable carbide inserted drills were introduced which greatly improved the metal removal rates for drilling holes from 1″ to 3″ (25.4 mm to 76.2 mm) diameter.

Standard drills are available in a wide variety of designs; a few are shown in *Figure 3-63*, which may be used for various materials and services. The most common variations are those of the helix angle and the web thickness. Though materials usually require the use of rigid drills with heavy webs and reduced helix angles. Free-machining materials can be cut with drills having higher helix angles and lighter webs which provide for more efficient cutting and better chip ejection.

Reamers and Core Drills. Twist drills produce holes; core drills and reamers are hole enlarging and finishing tools. The cutting ends of these tools

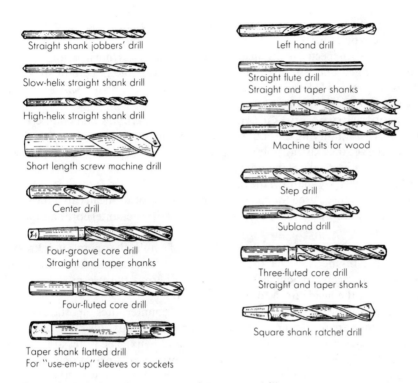

Straight shank jobbers' drill

Slow-helix straight shank drill

High-helix straight shank drill

Short length screw machine drill

Center drill

Four-groove core drill
Straight and taper shanks

Four-fluted core drill

Taper shank flatted drill
For "use-em-up" sleeves or sockets

Left hand drill

Straight flute drill
Straight and taper shanks

Machine bits for wood

Step drill

Subland drill

Three-fluted core drill
Straight and taper shanks

Square shank ratchet drill

Figure 3-63. Conventional and special-purpose drills.

have relieved chamfers which extend to a small enough diameter to permit their entry into the hole being enlarged or finished.

Core drills are designed principally to enlarge existing holes, and provide a greater degree of accuracy and better finish than a two-flute twist drill. These tools usually have three or four moderately deep helical flutes. The lands between the flutes are given body diameter clearance to produce a margin similar to that on a twist drill. Typical core drills are shown in *Figure 3-63*. Core drills are usually made of high-speed steel, but in some sizes they can be furnished with carbide tips. Sometimes they are made in a two-piece construction so that only the cutting end need be replaced.

Reamers are designed principally for hole sizing where only a moderate amount of stock is to be removed from the hole walls. The stock is removed by a larger number of flutes than are common with drills. Reamed holes usually have superior accuracy and finish. Because of the greater number of flutes and cutting edges, the margins on reamers are much narrower than those on twist drills and core drills, which minimizes galling of the margins. Reamers have little or no starting taper, which improves their ability to accept guidance by bushings. The flutes on reamers may be either straight or helical depending upon the type of work to be finished.

Solid, high-speed steel construction of reamers is most common, but carbide-tipped and solid carbide reamers are available in certain sizes and styles. *Figure 3-64* shows various commercial types of reamers. Also shown are expansion and taper reamers.

Counterbores and Countersinks. Counterbores and countersinks are tools for modifying the ends of holes. They are often provided with pilots which engage the existing hole to improve alignment, and are commonly used to provide a seat in a plane normal to the hole axis. Where the flat seating surface is relatively shallow, the tools used are often called spotfacers. Deeper seats, such as those used for recessing the heads of socket-head cap screws, are called counterbores. *Figure 3-65* illustrates various counterbores and spotfacers.

When a conical seat is required at the end of a hole, the operation is known as countersinking and the tools are called countersinks. Such seats are required to receive machine centers and to permit the use of flat-headed screws with the heads flush to the drilled surface.

Except for the angle of the cutting edges, counterbores, spotfacers, and countersinks are of similar construction. They are commonly made with two or more flutes which usually have a right-hand helix. The flutes are usually shorter than those on drills because the seats produced are relatively shallow. When pilots are provided, they may be either integral with the tool or removable. The tools themselves are often made quite short and are used as readily replaceable tips in the holder which is mounted in the machine spindle. In some cases, one holder will drive a fairly wide range of sizes of cutters.

The combined drill and countersink is a specialized tool used for producing the center holes required for lathe and other machine centers. This tool combines a short two-flute drill with a two-fluted countersink so that the entire drilling and countersinking operation can be performed in one pass on a single setup. Such tools are usually made in double-end construction as shown in *Figure 3-66*. When one end becomes dull, the tool can be reversed in the chuck to provide another cutting end.

Multiple-Diameter Tools. Multiple-diameter holes can be produced or finished in a single operation. Common situations involve drilling and counterboring or drilling and chamfering a hole on the same setup. In high-production operations, this can result in lower machining costs.

The simplest way to make a multiple-diameter tool is to start with a tool of the size that will produce the largest diameter and then grind down the cutting end of the tool to the size required for the small diameter. The cutting portions of the tool are provided with appropriate relief. This is known as step construction.

Another style of multiple-diameter tool employs what is known as subland construction. With this type of tool, separate tool lands are provided for each cutting diameter. The small-diameter lands run between the large-diameter lands. Subland tools are more expensive because of the additional manufacturing operations required in producing the additional cutting surfaces.

Step and subland drills are included in *Figure 3-63*. Multiple-diameter reamers (*Figure 3-64*) and counterbores (*Figure 3-65*) are also commercially available. Where the length of the small-diameter portion of the tool is relatively long in terms of the usable tool length, the simpler step construction is usually preferred. Besides lower acquisition costs, such tools provide more chip space in the flutes.

Where the small-diameter step is relatively short, the use of subland tools may be advantageous. In step construction, the usable sharpening life is limited to something less than the length of the small-diameter step. After this is used up, it is necessary to cut off the tool and completely remanufacture the step end. In subland construction, the two diameters can be sharpened for the entire length of the tool and the length of the step is not critical. The restricted flute space due to the presence of the additional lands sometimes limits the depth of holes to which they may be applied. With any multiple-diameter, it is important that the difference in diameters be relatively small. With a large difference in diameter it is not possible to maintain optimum cutting speeds on both diameters, and there is some difficulty in maintaining adequate strength in the tool.

Milling Cutters. Milling cutters are cylindrical cutting tools with cutting teeth spaced around the periphery (*Figure 3-67*). *Figure 3-51* shows the basic tooth angles of a solid plain milling cutter. A workpiece is traversed under the cutter in such a manner that the feed of the workpiece is measured in a plane perpendicular to the cutter axis. The workpiece is plunged radially into the cutter, and sometimes, in rare cases, there is also an axial feed of the cutter which results in a generated surface on the workpiece. Milling-cutter teeth intermittently engage the workpiece with the chip thickness being determined by the motion of the workpiece, the number of teeth in the cutter, the rotational speed of the cutter, the cutter lead angle and the overhang of the cutter on the workpiece.

There are two modes of operation for milling cutters. In conventional (up) milling the workpiece motion opposes the rotation of the cutter (*Figure 3-68A*), while in climb (down) milling the rotational and feed motions are in the same direction (*Figure 3-68B*). Climb milling is preferred wherever it can be used since it provides a more favorable metal-cutting action and generally yields a better surface finish. Climb milling requires more rigid equipment and there must be no looseness in the workpiece feeding mechanism since the cutter will tend to pull the workpiece.

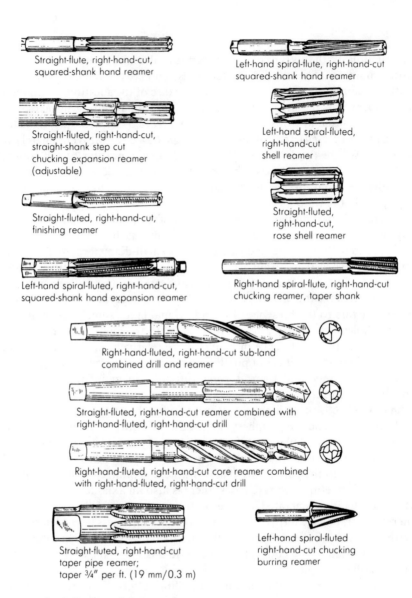

Straight-flute, right-hand-cut,
squared-shank hand reamer

Left-hand spiral-flute, right-hand-cut
squared-shank hand reamer

Straight-fluted, right-hand-cut,
straight-shank step cut
chucking expansion reamer
(adjustable)

Left-hand spiral-fluted,
right-hand-cut
shell reamer

Straight-fluted, right-hand-cut,
finishing reamer

Straight-fluted,
right-hand-cut,
rose shell reamer

Left-hand spiral-fluted, right-hand-cut,
squared-shank hand expansion reamer

Right-hand spiral-flute, right-hand-cut
chucking reamer, taper shank

Right-hand-fluted, right-hand-cut sub-land
combined drill and reamer

Straight-fluted, right-hand-cut reamer combined with
right-hand-fluted, right-hand-cut drill

Right-hand-fluted, right-hand-cut core reamer combined
with right-hand-fluted, right-hand-cut drill

Straight-fluted, right-hand-cut
taper pipe reamer;
taper ¾" per ft. (19 mm/0.3 m)

Left-hand spiral-fluted
right-hand-cut chucking
burring reamer

Figure 3-64. Commercial types of reamers.

Indexable milling cutters have precision ground carbide inserts positioned around the cutter body and they are held in by pins or wedges which can be released for indexing. Some milling cutters may have either profile-sharpened or form-relieved teeth. Profile-sharpened cutters are those which are

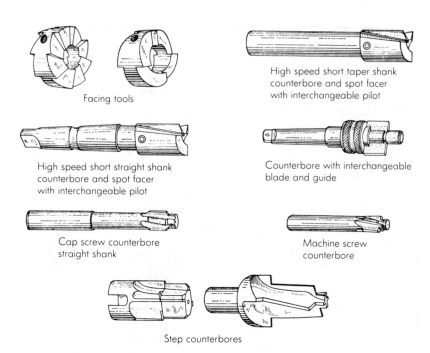

Facing tools

High speed short taper shank
counterbore and spot facer
with interchangeable pilot

High speed short straight shank
counterbore and spot facer
with interchangeable pilot

Counterbore with interchangeable
blade and guide

Cap screw counterbore
straight shank

Machine screw
counterbore

Step counterbores

Figure 3-65. Types of counterbores and spotfacers.

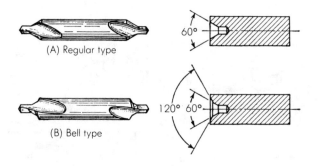

(A) Regular type

60°

(B) Bell type

120° 60°

Figure 3-66. Combined drills and countersinks.

sharpened on the relief surface using a conventional cutter grinding machine. Form-relieved cutters are made with uniform radial relief behind the cutting edge. They are sharpened by grinding the face of the teeth. The profile style provides greater flexibility in adjusting relief angles for the job to be done, but it is necessary that any form on the cutter be reproduced during each resharpening. In the form-relieved style, the relief angle cannot be

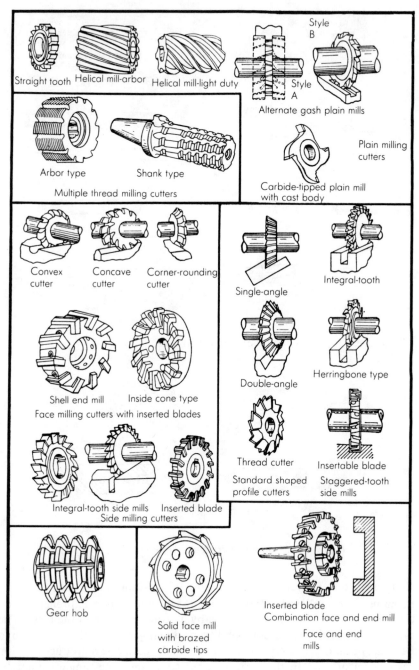

Figure 3-67. Common types of milling cutters.

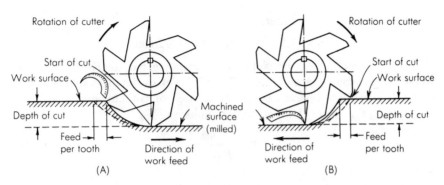

Figure 3-68. (*A*) "Out-cut," "conventional," or "up" milling; also called "feeding against the cutter." (*B*) "in-cut," "climb," or "down" milling; also called "feeding with the cutter."

changed since it is fixed in the manufacture of the cutter. However, the form-relieved construction is well adapted to cutters with intricate profiles since the profile is not changed by resharpening.

Most large milling cutters are provided with an axial hole for mounting on an adapter or arbor, and usually have a drive key slot. Certain small-diameter cutters and some cutters for specialized applications are made using an integral shank construction where the cutting section is at the end of a straight or tapered shank which fits into the machine tool spindle or adapter. Also, some large facing cutters are designed to mount directly on the machine tool spindle nose.

Number of Teeth in a Milling Cutter. The number of teeth (n) may be expressed as[10]:

$$n = \frac{F}{F_t N} \tag{3-24}$$

Where:

F = feed rate, in. per min.
F_t = feed per tooth, in. (chip thickness)
N = cutter speed rpm

Equation (3-25) has been found satisfactory for high-speed steel cutters such as plain, side, and end mills, for both production and general-purpose use.

$$n = 19.5\sqrt{R} - 5.8 \tag{3-25}$$

Where:

R = cutter radius, in.

Example: For a 4″ diam cutter,

$$n = 19.5\sqrt{2} - 5.8 = 21.8 \text{ or } 22 \text{ teeth}$$

Equation (3-26)[11] was developed for calculating the number of teeth (n) when using carbide-tipped cutters. Since it involves power available at the cutter, it is helpful in permitting full utilization of available power without overloading the motor (U.S. customary units only).

$$n = \frac{K\,hp_c}{F_tNdw} \tag{3-26}$$

Where:

    K = a constant related to the workpiece material. It may be given the values of 0.650 for average steel, 1.5 for cast iron, and 2.5 for aluminum (conservative, taking into account the dulling of the cutter in service)

    hp_c = horsepower available at cutter

    w = width of cut, in.

    d = depth of cut, in.

End Mills. End mills are shank-type milling cutters which are usually designed with some form of relieved end teeth (*Figure 3-69*). This construction allows them to do some end cutting as the majority of the cutting takes place on the periphery. Indexable end mills generally use radius corner inserts, but can be furnished with parallel land wiper inserts for generating finishes down to 50 μ in. (1.27 μ m). End mills are usually considered in a separate category from other milling cutters.

High-speed steel end mills have helical flutes with a flute helix angle between 20° and 40°. Flutes are usually several diameters long, but both longer and shorter designs are available. Tools with two and four flutes are most common, but the larger sizes are available with more flutes. The relieved end teeth permit their use in keywaying or pocketing operations with a minimum of contact with the surface at the bottom of the recess. Most two-flute end mills have end teeth which extend from the periphery to the center of the tool. This permits their use as drills to sink into depth before starting a cut. Some four-flute end mills are also made with two diametrically opposed teeth cutting to the tool center. The corners at the intersection of the peripheral and end teeth can be provided with radii to minimize the formation of stress risers in the work. When this radius becomes equal to the cutter radius, the end mill is said to have a ball end. Such tools are widely used in forming recesses and cavities in dies. An operation of this sort is often called die sinking.

Taps. The tap is essentially a screw that has been fluted to form cutting edges. The cutting end of the tap has a relieved chamfer which forms the

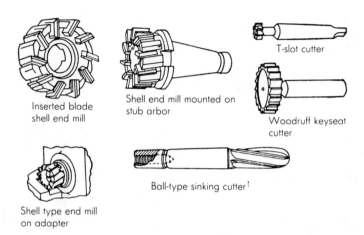

Inserted blade
shell end mill

Shell end mill mounted on
stub arbor

T-slot cutter

Woodruff keyseat
cutter

Shell type end mill
on adapter

Ball-type sinking cutter[1]

Figure 3-69. End mills.

cutting edges and permits it to enter an untapped hole. If succeeding chamfered teeth are followed along the thread helix, it will be seen that the effect of the chamfer is to make each tooth have a slightly larger major diameter than the preceding one. Thus, the major feed involved in tapping is a radial feed, although the feed is built into the tool rather than controlled by the machine as is the case with most other rotary tools. The actual feed per tooth depends on the chamfer angle and the number of flutes. A reduction in the chamfer angle or an increase in the number of flutes will reduce the feed per tooth.

The most common type of tap is the straight fluted hand tap (*Figure 3-70*). This type of tap has a straight shank with a driving square at the shank end. In spite of the hand tap designation, these taps are generally used in tapping machines. Hand taps are usually available with three different chamfer lengths of nominally 1 1/4, 4, and 8-thread pitches, which are designated respectively as bottoming, plug, and taper taps. Taps are also available with helical flutes and other modifications to suit them to specific applications. Most taps used in production operations are made of high-speed steel. Some solid carbide and carbide-tipped taps are used in work materials of high abrasiveness or hardness. However, these are more susceptible to breakage or chipping.

Hobs. A hob is a generating, rotary cutting tool which has its teeth arranged along a helical thread (*Figure 3-71*). The hob and the workpiece are rotated in timed relationship to each other, while the hob is fed axially or tangentially across or radially into the workpiece. The cutting edges of the hob teeth lie in a helicoid which is usually conjugate to the form produced on the workpiece. The teeth on a hob resemble those on a form-relieved milling

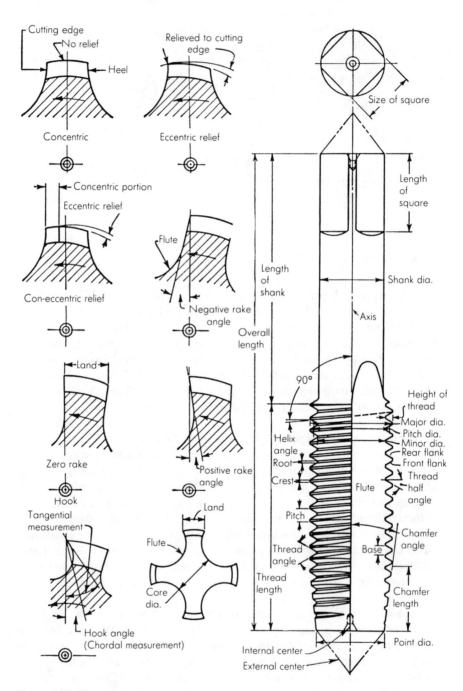

Figure 3-70. Tap nomenclature.

cutter in that they are designed to be sharpened on the rake face. Most hobs are designed for generating gear teeth or splines. Certain other evenly spaced forms on the periphery of a cylindrical workpiece can also be hobbed. There are some limitations on the shape of forms that can be generated by hobs.

Most hobs are manufactured from high-speed steel. Carbide-tipped hobs are sometimes used for hobbing low-strength, abrasive, nonmetallic materials.

Control of the Causes of Tool Wear and Failure

The design of cutting tools is not a pure science, involving only computations to be carried out in functional isolation. Of itself, a cutting tool is only a piece of metal of special shape and construction, although frequently a very expensive piece of metal.

Good tool design is accomplished with consideration of many inseparably related factors: the composition, hardness, condition, and shape of the workpiece material; the rate and volume of the specified production; the type, motions, power, and speed of the machine tool to be used; the toolholders and workholders available or to be designed; the specified accuracy and surface of the finished workpiece, and many other factors, common or specific to the particular operation.

The following discussion is, therefore, very much the business of the good tool designer. He cannot escape responsibility by saying that some of the factors leading to undue tool wear, or to tool failure, are the responsibility of the process or methods functions. He should know, or anticipate, the possible application difficulties that lie ahead for the tool he is about to design, and then accommodate the difficulties by tool design, or consult with the other manufacturing functions as to changing certain conditions, or both.

Rigidity. Setup rigidity is vital to the maintaining dimensional accuracy of the cut surface, since the tool shifts into or out of the cut with the accumulation of static deflections and take-up of loose fits. Rigidity also maintains surface-finish quality, avoiding the marks made by elastic vibration and free play of loose fits and backlash. In the control of vibration, rigidity of the part and cutting tool can make the difference between success or failure of the machining operation.

Increasing mass reduces vibration amplitude and resonant frequency, while dampening reduces amplitude by dissipating vibratory energy as frictional heat. Since each part of the cutting system (i.e. the machine, the fixture, the tool and the workpiece) can affect the mode and amount of vibration, most should be made oversize and broadly supported. This provides design latitude for those members having more severe cost and space limitations. Designers should be generous with rigidity, anticipating fast, efficient cuts.

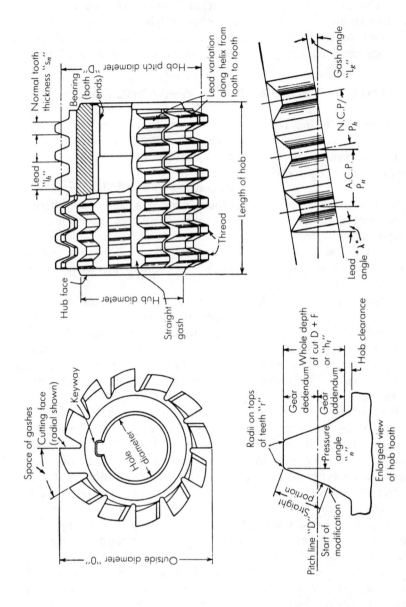

Figure 3-71. Nomenclature of hobs.

Strength. The strength of each member can be considered separately and related to the magnitude and application of the forces it will transmit. It should clearly be sufficient to prevent breakage or deformation beyond the elastic limit when the operation is performed correctly. The designer must also consider overloads and damage that may be encountered, providing abundant strength wherever economically possible. In particular, generous size and material specification should give good working life to areas subject to abrasive wear and work hardening under impact loads. But chatter, packed chips, or binding due to setup misalignment can multiply normal operating forces many times. Tool failure, mechanical malfunctions, and operating errors threaten destructive casualties even with costly overdesign.

Weak Links. A common practice is to protect the structural chain with weak links in anticipation of casualties, and to confine or limit the possible damage. Suitable low strength with high rigidity is illustrated by the common soft shear pin. But these weak links must be strong enough to withstand normal operation and overload if possible. The permanent members are made unquestionably stronger by comparison and are protected by location. Indentical design criteria make weak links an ideal combination with wearing details. They should be comparatively cheap, with duplicates widely stocked or readily produced, and of a form of easy, accurate replacement mounting. Mass-produced, delicate workpieces are of themselves natural weak links, though cutting-tool inserts are the typical wear and breakaway members. Indexable insert cutting tools, using replaceable backing seats, are ideal examples of wear and damage protection.

Force Limitations. Operating forces may be obviously limited by a weak-link member, as in the case of a delicate workpiece. A machine tool such as a hydraulic-stroke planer or broaching machine may be related in terms of force, with the ratings understated but subject to measurement and control. Some saws, mills, and grinders regulate the feeding force instead of the cutting force in the direction of cutting velocity. This causes the machine to stop feeding if an accident occurs or if forces begin to exceed a certain safe level.

Speed, Feed, and Size. A machine tool's speed and feed ranges, its cutting tool adaptor capacity, and its working clearance establish restrictions on tool design and production rate. The effect of forces is indirect but inevitable. A milling cutter with few teeth contacting a delicate workpiece exerts only a few times the cutting forces of one tooth, while high speed permits rapid completion of the cut. Limited cutter diameter and a low speed range would require more teeth and more force or a longer cutting time. Variable infeed rates can be used to speed rough stock removal and then minimize distortion while finishing, as in the case of the spark-out of a grinding wheel. The important thing is to design around limitations and take full advantage of flexibility.

Related Force Components. Total cutting force is usually resolved into three mutually perpendicular components. Force in the direction of feeding motion of a turning, boring, facing, plunge forming, or parting tool corresponds to radial force on a peripheral-cutting milling cutter tooth or abrasive wheel or belt contact area. It is commonly taken as from one-fourth up to three-fourths of the tangential force F_T for sharp tools, the larger fraction being appropriate for extremely heavy feeds. The straight-line cutting tools, namely teeth of saw blades or broaches and planer or shaper tools, develop this force component in the direction of feed into the work. The radial force component for turning or boring tools is normal to the finished work surface in facing, planing, and shaping, and in an axial direction for peripheral milling. It may be negative, pulling into the work as a result of large positive back rake, but it is usually a pushing force of relatively low value.

The third and most significant force component is the tangential force F_T which acts on the top of the tool tangent to the direction of rotation of the part or tool. Carbide turning tools typically have 1000 pounds (4450 N) of tangential force in general purpose machining applications.

Chip Disposal. One sure way to overload a cutting tooth is to block the path of the chip flowing across its face so that the chip is recut. Single-point tools cutting ductile work frequently employ a pressed-in chip breaker to curl an otherwise stringy chip so that it will break in the form of a figure nine and fall away. If groove design is too weak for the size of cut being taken, it can cause edge chipping, or breakage. Small diameter coarse pitch milling cutters commonly have ample chip spaces, provided that the chips are thrown or washed out between successive passes through the cut. On the other hand, large diameter cutters taking full width cuts must carry the chip a half rotation before the chip can exit. These cuts require large chip slots. It is difficult to remove work materials like soft steel or copper alloys and titanium, whose chips tend to weld onto the tool face. Chip disposal in milling slots may demand high positive rake angles and climb milling instead of conventional cutter rotation to eject the chips. Complex selection and application methods have been developed for tapping and deep hole drilling where chip clogging, misalignment, and runout can readily break tools. Tool design should provide space for chip flow and means of disposal, which may well be the solution to many problems of tool chipping.

Uneven Motions. Another sure way to overload a cutting tooth is to increase the feed rate drastically beyond its structural or chip-disposal capacity. Machine structural deflection accomplishes this in the example of a drill breaking as it breaks through the work. As the heavy thrust of the chisel edge is relieved, structural members spring back toward their unstressed shape, and the drill lips plunge into the work for an oversize bite. Feed mechanisms may employ air or hydraulic fluid whose compression is elastic;

or gearing and a leadscrew nut fit may introduce backlash. Machine way motion becomes jumpy at slow speeds ("slip-stick" motion), even with heavy lubrication. A milling cutter at slow feed may actually rub until pressure builds up. It then may dig into the work and surge ahead. Adding to the difficulty, the sudden change in cutting torque adds to the pounding caused by teeth entering the cut.

Torsional vibration and backlash tend to develop in a rotary drive train. Should cutter rotation become so erratic that it momentarily stops, carbide teeth will generally break at once by being bumped into the work. With some teeth gone, the entire cutter may fail progressively as each successive tooth is unable to carry the extra load left by the preceding damaged teeth.

Chatter. The rapid, elastic vibration that sometimes appears between tool and work is easily detected by marks on the work surface and by the sound that gives it the name "chatter." Chatter is the momentary separation of the tool and workpiece and the immediate banging back into contact at an audible frequency. It is a danger signal of impending possible chipping or fracture. The remedy is to eliminate uneven motion and loose fits. Chatter is less likely with few teeth moving at high velocity taking thick chip loads, and having high rake and ample relief angles. A negative rake angle may prevent pulling into the cut. In grinding, harder action or broader contact helps withstand bumping. As an extreme simplification, chatter can be combatted with lower cutting forces while looseness and backlash cannot. Like all other problems in machining, chatter can be greatly reduced by proper tool design.

References

1. Boston, O. W., *Metal Processing*, 2nd ed., John Wiley & Sons, Inc., New York, 1951.
2. Shaw, M. C., *Metalcutting Principles*, 3rd ed., Massachusetts Institute of Technology, 1954.
3. Ernst, H., "Physics of Metalcutting," *Machining of Metals*, American Society of Metals, 1938.
4. Ernst, H., and Merchant, M. E., *Chip Formation, Friction and Finish*, Cincinnati Milling Machine Company.
5. Opitz, I. H., "Present-Day Status of Chip-Formation Research," *Microtechnic*, No. 4 (1960).
6. Chao, B. T., and Trigger, K. J., "Temperature Distribution at the Tool-Chip Interface in Metalcutting," *Trans. ASME*, 77 (1955).

7. Chao, B. T., and Trigger, K. J., "Temperature Distribution at the Tool-Chip and Tool-Work Interface in Metalcutting," *Trans. ASME*, 80 (1958).
8. Shaw, M. C., and Oxford, C. J. Jr., (1) "On the Drilling of Metals,"(2) "The Torque and Thrust in Milling," *Trans. ASME*, 79:1 (Jan. 1957).
9. "Milling Cutters, Nomenclature, Principal Dimensions, etc.,"American Standard ASA B5.3—1959, American Standard Association, New York.
10. *Tool Engineers Handbook*, 2nd ed., American Society of Tool and Manufacturing Engineers, McGraw-Hill Book Co., Inc., New York, 1959.
11. *A Treatise on Milling and Milling Machines*, 3rd ed., The Cincinnati Milling Machine Co., 1951.

CUTTING TOOL DESIGN

Review Questions

1. What five elements interact in the machining process?
2. What factor will greatly influence the cutting tool material and geometry?
3. Can tool guidance be incorporated in the tool function?
4. What factors are influenced by the shape of a cutting tool?
5. What are the single-point tool angles?
6. What are the advantages of increasing the nose radius?
7. What normal tool angles should a single-point tool have for turning AISI 3120 steel?
8. What tool angles should be specified for a single-point tool for turning aluminum?
9. What is the primary function of a relief angle?
10. What is the rake angle requirement for ductile work materials; for brittle materials?
11. What are the machining factors used to evaluate machinability?
12. Name the three most common types of chips.

13. What type of chip is generally produced by the machining of brittle materials?
14. Upon what does the capacity of a material for plastic flow depend?
15. What causes a built-up edge?
16. What effect does a negative side rake angle have on cutting forces?
17. The part shown in *Figure A* is to be turned and bored in a turret lathe. Design the appropriate boring and turning tools to turn the 2.500 diameter and bore the 1.750 and 1.500 diameters.
18. Design a boring bar employing a standard adjustable boring tool cartridge for boring the 2.750″ diameter hole shown in *Figure B*.

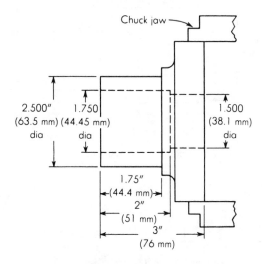

Chuck jaw

2.500″ (63.5 mm) dia   1.750 (44.45 mm) dia   1.500 (38.1 mm) dia

1.75″ (44.4 mm)
2″ (51 mm)
3″ (76 mm)

Figure A. Part for which a multiple tool is to be designed. The tool is to turn and bore the part on a turret lathe.

19. You must remove an area 3 1/2″ wide x 6 3/8″ long x 1/16″ to 3/4″ deep the long way. How long will it take using a 1″ carbide (of your choice, but note your choice) and a 50 horsepower machine?
20. Design a pull broach to put a 3/16″ x 3/16″ keyway in a soft steel pulley.
21. Design a simple form cutter to cut the profile in *Figure C*.

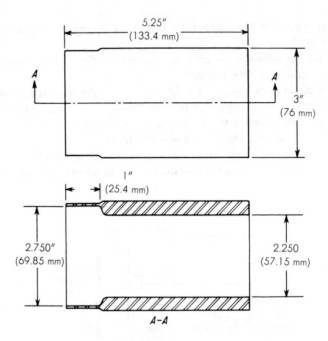

Figure B. Workpiece for Problem 18.

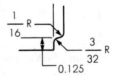

Figure C.

CUTTING TOOL DESIGN

Answers to Review Questions

1. The cutting tool, the tool holding or guiding device, the workholder, the workpiece, and the machine.
2. Intended function and design.
3. Yes, tool guidance may be incorporated in the workholding function.
4. Tool life and surface finish of the workpiece.
5. Side rake angle, back rake angle, end relief angle, side relief angle, end cutting edge angle and side cutting angle. (Nose radius may also be included.)
6. Avoids high heat concentration at a sharp point, improves tool life and surface finish and effects a slight reduction in cutting forces.
7. Side relief angle 7°–9°, front relief angle 7°–9°, back rake angle 8°–10°, and side rake angle 10°–12°.
8. 18°–20°.
9. Relief angles are clearance angles to keep the tool from rubbing.
10. Ductile work materials 0°–20°, brittle work materials 0°–7°.
11. Power requirements. Rigidity.
12. Discontinuous or segmental, continuous without built-up edge, and continuous with built-up edge.

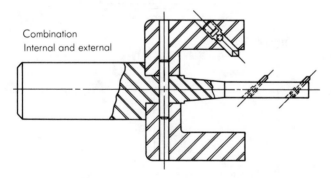

Combination
Internal and external

Answer Figure A.

13. Discontinuous or segmental.
14. The number of slip-planes that are available.
15. Seizure between the chip and tool face.
16. Increases the cutting forces.
17. (See *Answer Figure A.*)
18. (See *Answer Figure B.*)
19. Four passes at full depth 7 1/2" long at 1 1/4" per minute = 24 minutes and three moves will be approximately 26 minutes.
20. (See *Answer Figure C.*)
21. (See *Answer Figure D.*)

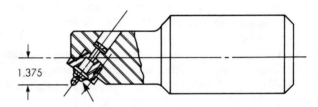

Answer Figure B.

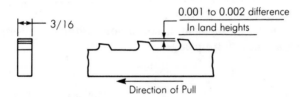

Answer Figure C.

Answer Figure D.

4

WORKHOLDING PRINCIPLES

The term workholder includes all devices that hold, grip, or chuck a workpiece to perform a manufacturing operation. The holding force may be applied mechanically, electrically, hydraulically, or pneumatically. This section considers workholders used in material-removing operations. Workholding is one of the most important elements of machining processes.

Figure 4-1 illustrates almost all the basic elements that are present in a

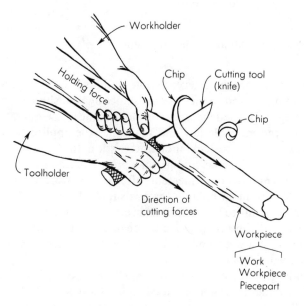

Figure 4-1. Principles of workholders.

material-removing operation intended to shape a workpiece. The right hand is the toolholder, the left hand is the workholder, the knife is the cutting tool, and the piece of wood is the workpiece. Both hands combine their motion to shape the piece of wood by removing material in the form of chips. The body of the person whose hands are shown may be considered a machine that imparts power, motion, position, and control to the elements shown. Except for the element of force multiplication, these basic elements may be found in all of the forms of manufacturing setups where toolholders and workholders are used.

Figure 4-2 shows a pair of pliers or tongs used to hold a rod on which a point has to be ground or filed. This simple workholder illustrates the element

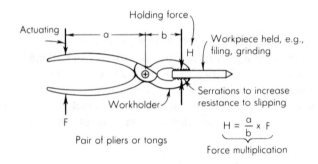

Figure 4-2. Multiplication of holding force.

of force multiplication by a lever action, and also shows serrations on the parts contacting the rod to increase resistance against slippage.

Figure 4-3 shows a widely used workholder, the screw-operated vise. The screw pushes the movable jaw and multiplies the applied force. The vise remains locked by the self-locking characteristic of the screw, provides means of attachment to a machine, and permits precise placement of the work.

A vise with a number of refinements often used in workholders is depicted in *Figure 4-4*. The main holding force is supplied by hydraulic power, the screw being used only to bring the jaws in contact with a workpiece. The jaws may be replaceable inserts profiled to locate and fit a specific workpiece as shown. Other, more complicated jaw forms are used to match more complicated workpieces.

Another large group of workholders are the chucks. They are attached to a variety of machine tools and are used to hold a workpiece during turning, boring, drilling, grinding, and other rotary operations. Many types of chucks

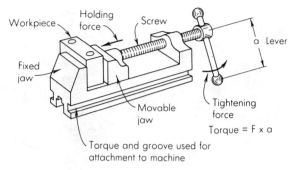

Figure 4-3. Elementary workholder (vise).

are available. Some are tightened manually with a wrench, others are power-operated by air or hydraulic means or by electric motors. On some chucks, each jaw is individually advanced and tightened, while others have all jaws advance in unison. *Figure 4-5* shows a workpiece clamped in a four-jaw independent chuck. The drill, which is removing material from the workpiece, is clamped in a universal chuck.

Purpose and Function of Workholders

A workholder must position or locate a workpiece in a definite relation to the cutting tool and must withstand holding and cutting forces while

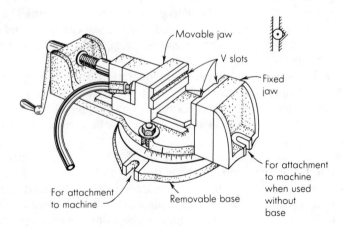

Figure 4-4. Vise with hydraulic clamping.

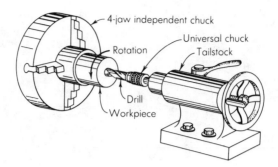

Figure 4-5. Holding (chucking) a round workpiece.

maintaining that precise location. A workholder is made up of several elements, each performing a certain function. The locating elements position the workpiece; the structure, or tool body, withstands the forces; brackets attach the workholder to the machine; and clamps, screws, and jaws apply holding forces. Elements may have manual or power activation. All functions must be performed with the required firmness of holding, accuracy of positioning, and with a high degree of safety for the operator and the equipment.

General Considerations

The design or selection of a workholder is governed by many factors, the first being the physical characteristics of the workpiece. The workholder must be strong enough to support the workpiece without deflection. The workholder material must be carefully selected with the workpiece in mind so that neither will be damaged by abrupt contact, e.g., damage to a soft copper workpiece by hard steel jaws.

Cutting forces imposed by machining operations vary in magnitude and direction. A drilling operation induces torque, while a shaping operation causes straight-line thrust. The workholder must support the workpiece in opposition to the cutting forces and will generally be designed for a specific machining operation.

The workholder establishes the location of the workpiece relative to the cutting tool. If the operation is to be performed at a precise location on the workpiece, locating between the workpiece and workholder must be equally precise. If the cutting tool must engage the workpiece at a specified distance from a workpiece feature such as a line or plane of the workpiece, then the workholder or workholding fixture must establish the line or plane at the specified distance. The degree of precision required in the workholder will usually exceed that of the workpiece because of cumulative error.

The strength and stiffness of the workpiece will determine to what extent it must be supported for the machining operation. If the workpiece design is such that it could be distorted or deflected by machining forces, the workholder must support the affected area. If the workpiece is sufficiently rigid to withstand the machining forces, workholder support at the edge of the workpiece may be adequate. The strength of the workholder is determined by the magnitude of the machining forces and the weight of the workpiece.

Production requirements will greatly influence workholder design. If a great number of workpieces are to be processed, the cost of an elaborate workholder might well be offset by savings due to increased hourly production made possible by the elaborate workholder, since the cost of the workholder will be prorated against the great number of workpieces. High production rates and volume can therefore justify expensive fixturing. Conversely, if only one or two workpieces are to be machined, the operation will usually be performed with standard toolroom equipment and little to no fixturing costs can be justified. Production schedules may limit the time available for workholder acquisition and may compel the use of standard equipment.

Safety requirements must always dictate workholder design or selection. A workholder must not only withstand normal cutting forces and workpiece weight but may also have to withstand large momentary loads. In machining a cast workpiece, the cutting tool might strike an oxide inclusion causing instantaneous multiplication of force. The tool might cut through the inclusion; the tool might break; or the machine might stall. If the workholder broke, however, the tool might impart motion to the workpiece. A workpiece in uncontrolled motion is a missile. The workholder must also be designed to protect workers from their own negligence. Where possible, a shield should be interposed between the worker and the tool.

A workholder should be designed to receive the workpiece in only one position. If a symmetric workpiece can be clamped in more than one position, it is probable that a percentage of workpieces will be incorrectly clamped and machined. Workholders should be designed to prevent incorrect placement and clamping.

It is advisable to use standard workholders and commercially available components whenever possible. Not only can these items be purchased for less than the cost of making them, but they are generally stronger and accurate enough.

Many workholders are used in industry that are not used on material-removing operations. Workholders may be used for the inspection of workpieces, assembly, welding, and so on. There may be very little difference in their basic design and their appearance. Quite often a standard commercial design may be used in one application for a turning operation and for the same or another workpiece in an inspection operation.

Locating Principles

To insure successful operation of a workholding device, the workpiece must be accurately located to establish a definite relationship between the cutting tool and some points or surfaces of the workpiece. This relationship is established by locators in the workholding device which position and restrict the workpiece to prevent its moving from its predetermined location. The workholding device will then present the workpiece to the cutting tool in the required relationship. The locating device should be so designed that each successive workpiece, when loaded and clamped, will occupy the same position in the workholding device. Various methods have been devised to effectively restrict the movement of workpieces. The locating design selected for a given workholding device will depend on the nature of the workpiece, the requirements of the metal-removing operation to be performed, and other restrictions on the workholding device.

Workpiece Surfaces

One major consideration involved in selecting locators for a workpiece is the shape of the locating surface. All workpiece surfaces can be divided into three basic categories, insofar as location is concerned: flat surfaces, cylindrical surfaces, and irregular surfaces. While most workpieces are a combination of different surfaces, the designer must identify the specific surfaces which are to be used to locate any part.

Flat surfaces, as their name implies, are those surfaces which, regardless of their position, have a flat bearing area for the locators. Typical examples of flat surfaces include edges, flanges, steps, faces, shoulders, and slots.

Cylindrical surfaces are those surfaces which are located on a circumference or diameter. Typical examples of cylindrical surfaces include internal (concave) surfaces of holes or external (convex) surfaces of turned cylinders.

Irregular surfaces provide neither a flat or cylindrical locating surface. Typical examples of irregular surfaces include many cast or forged workpieces.

Since irregular locating surfaces are so inclusive—that is, these surfaces include all locating surfaces which are neither flat nor cylindrical—every locating surface can be grouped into one, and only one, of these three surface categories.

Types of Location

Basic workpiece location can be divided into three fundamental categories: plane, concentric, and radial. In many cases, more than one category of location may be used to locate a particular workpiece. However, for the purpose of identification and explanation, each will be discussed individually.

Plane Location. Plane location is normally considered the act, or process, of locating a flat surface. But, many times, irregular surfaces may also be located in this manner. Plane location is simply locating a workpiece with reference to a particular surface, or plane (*Figure 4-6*).

Figure 4-6. Plane location.

Concentric Location. Concentric location is the process of locating a workpiece from an internal or external diameter (*Figure 4-7*).

Radial Location. Radial location is normally a supplement to concentric location. With radial location (*Figure 4-8*), the workpiece is first located concentrically and then a specific point on the workpiece is located to provide a specific fixed relationship to the concentric locator.

Combined Location. Most workholders use a combination of locational methods to completely locate a workpiece. The part shown in *Figure 4-9* is an example of all three basic types of location being used to reference a workpiece.

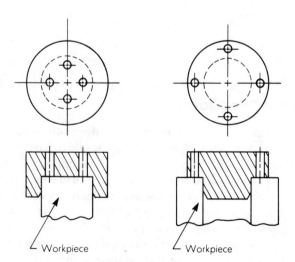

Figure 4-7. Concentric location.

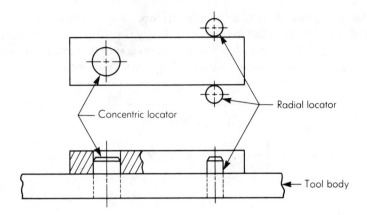

Figure 4-8. Radial location.

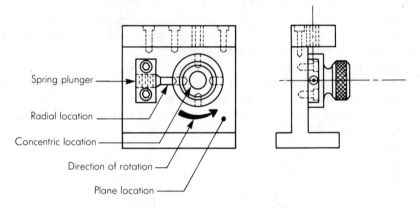

Figure 4-9. Plane, concentric, and radial location.

Degrees of Freedom

A workpiece in space, free to move in any direction, is designed around three mutually perpendicular planes and may be said to have twelve modes or degrees of freedom. It may move in either of two opposed directions along three mutually perpendicular axes, and may rotate in either of two opposed directions around each axis, clockwise and counterclockwise. Each direction of movement is considered one degree of freedom. The twelve degrees of freedom as they apply to a rectangular prism are shown in *Figure 4-10*. *Figure 4-10B* shows three views of the prism in orthographic projection, with all twelve degrees of freedom indicated in their respective positions. To accurately

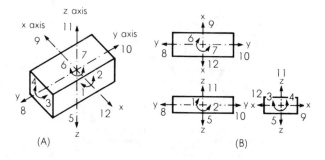

Figure 4-10. Twelve degrees of freedom.

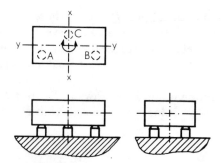

Figure 4-11. Three pins arrest five degrees
of freedom.

locate a workpiece, it must be confined to restrict it against movement in all of
the twelve degrees of freedom except those called for by the operation. When
this condition is satisfied, the workpiece is accurately and positively confined
in the workholding device.

3-2-1 Method of Locating. A workpiece may be positively located by
means of six pins positioned so that collectively they restrict the workpiece in
nine of its degrees of freedom. This is known as the 3-2-1 method of location.
Figure 4-11 shows the prism resting on three pins, *A, B,* and *C.* The faces of the
three pins supporting the prism form a plane parallel to the plane that
contains the *X* and *Y* axes. The prism cannot rotate about the *X* and *Y* axes
and it cannot move downward in the direction of freedom five. Therefore,
freedoms 1, 2, 3, 4, and 5 have been restricted.

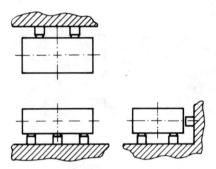

Figure 4-12. Five pins arrest eight
degrees of freedom.

In *Figure 4-12*, two additional pins *D* and *E* whose faces are in a plane
parallel to the plane containing the *X* and *Z* axes prevent rotation of the prism
about the *Z* axis. It is not free to move to the left in the direction of freedom 9.
Therefore, freedoms 6, 7, and 9 have been restricted, and the prism cannot
rotate.

Finally, with the addition of pin *F* as shown in *Figure 4-13*, freedom 8 is
restricted. Thus by means of six locating points, three in a base plane, two in a

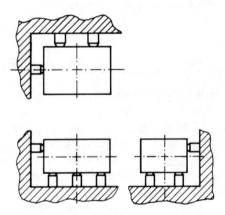

Figure 4-13. Six pins arrest nine degrees
of freedom.

vertical plane, and one in a plane perpendicular to the first two, nine degrees of freedom have been restricted.

Three degrees of freedom, 10, 11, and 12, still remain unrestricted. The addition of three more pins, one for each remaining freedom, would completely restrict movement of the prism. The pins would then entirely enclose the workpiece. This is not practical since it would prevent loading of the workpiece into the workholding device. The remaining three freedoms may be restricted with clamps, which also serve to resist the forces generated by the metal-removing operation being performed on the workpiece. Any combination of three clamping devices and locating pins may be used if this is more suitable to the design of a particular workholding device.

Concentric and Radial Methods of Locating. A workpiece which is located concentrically and radially is restricted from moving in eleven degrees of freedom. *Figure 4-14* shows a typical workpiece which is located concentrically. The base and center pin restrict nine degrees of freedom as shown. The base restricts any downward movement as well as rotation around the X and Y axes. The center locating pin prevents any movement in either a traverse or longitudinal direction along the X and Y axes. Located in this manner, the part is only free to move vertically or radially around the Z axis.

To prevent movement around the Z axis, a radial locator is positioned as shown in *Figure 4-15*. In this position, both degrees of freedom around the Z axis are restricted. The only possible movement this part can make is vertically, up the Z axis. This degree of freedom is restricted by the clamping device.

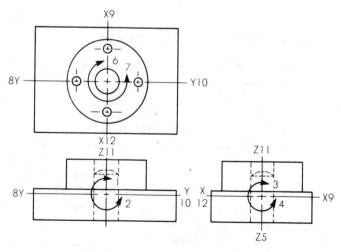

Figure 4-14. Base and center pin restrict nine degrees of freedom.

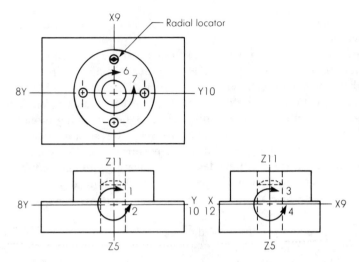

Figure 4-15. Base, center pin, and radial locator restrict 11 degrees of freedom.

Basic Locating Rules

To function properly, locators must be positioned correctly, properly designed, and accurately sized. To do all this and still permit easy loading and unloading of the tool requires forethought when planning the locational elements of a workholder. The following are a few basic principles every designer should keep in mind when planning part location.

Position and Number of Locators

Locators and part supports should always contact a workpiece on a solid, stable point. When possible, the surface should be machined to insure accurate location.

Locating points should be chosen as far apart as possible on any one workpiece surface. Thus, for a given displacement of any locating point from another, the resulting deviation decreases as the distance between the points increases.

The most satisfactory locating points are those in mutually perpendicular planes. Other arrangements are possible but not desirable. Two disadvantages result from locating from other than perpendicular surfaces: (1) the consequent wedging action tends to lift the workpiece, (2) the displacement of a locating point or a particle (chip or dirt) adhering to it introduces a correspondingly larger error. In *Figure 4-16* the introduced error *T* is

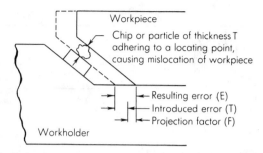

Figure 4-16. Magnification and projection of error.

projected to become the resulting error *E*. The projection factor *F* is zero when the locating surfaces are perpendicular, and increases as the angle between them becomes more acute.

The number of locators used to reference a part normally depends on the part itself. But no more points than necessary should be used to secure location in any one plane. The 3-2-1 principle determines the minimum number required. More can be used, but the additional points should be used only if they serve a useful purpose, and care must be taken that they do not impair the location function.

Redundant Locators

Always avoid redundant, or duplicate locators on any part. Redundant location occurs when more than one locator is used to locate a particular surface or plane of a workpiece. *Figure 4-17* shows examples of redundant location. The principal objection to using more than one locator, or series of locators, to reference one location is the variance in part sizes. Any variation in the part size, even within the tolerance, can cause the part to be improperly located or bind between the duplicate locators. Besides these obvious problems, it is not cost effective to use more locators than necessary.

Locational Tolerances

Locational tolerance is one point which must always be considered when specifying locators for any workholder. As a general rule, locational tolerance should be approximately 20% to 50% of the part tolerance. Making tolerance excessively tight only increases costs. Likewise, overly large tolerances can shorten the life of a workholder. The designer must balance the cost against the expected life of the tool and the required accuracy of the parts to

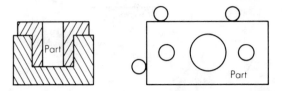

Figure 4-17. Redundant locators.

determine a locational tolerance that will provide the required number of parts without excessive tooling costs.

Foolproofing

Foolproofing is the process of positioning locators so a part will only fit in the workholder in its proper position. This is accomplished by a number of different means. The simplest and most cost effective method is positioning a foolproofing pin to prevent incorrect loading (*Figure 4-18*). In any case where a part has the possibility of being loaded improperly, use a foolproofing device.

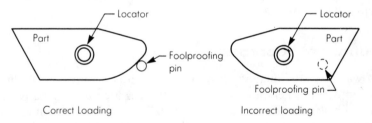

Figure 4-18. Foolproofing.

Basic Types of Locators

Locators are made in a wide variety of shapes and sizes to accommodate the large range of workpiece configurations. In addition, commercial locators are available in many styles to suit their ever-increasing use. To properly design and specify an appropriate locator, the designer first must be familiar with different types of locators commonly used in jig and fixture applications.

External Locators. External locators are those devices which are used to locate a part by its external surfaces. These locators are normally classified as either locators or supports. Locators are those elements which prevent movement in a horizontal plane. Supports are locating devices which are positioned beneath the workpiece and prevent downward movement of the part as well as rotation around the horizontal axes.

The two basic forms of external locators or supports are fixed or adjustable. Fixed locators are solid locators that establish a fixed position for the workpiece. Typical examples of fixed locators include integral locators, assembled locators, locating pins, vee locators, and locating nests.

Adjustable locators are movable locators which are frequently used for rough cast parts or similar parts with surface irregularities. Examples of adjustable locators are threaded locators, spring pressure locators, and equalizing locators. Adjustable locators are used in conjunction with fixed locators to permit variations in part sizes while maintaining a fixed relative position of the part against the fixed locators.

Integral Locators. Integral locators are locators which are machined into the body of the workholder (*Figure 4-19*). In most instances this type of

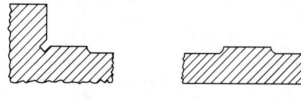

Figure 4-19. Integral locators.

locating, or supporting device is the least preferred. The principal objections to using integral locators are the added time required to machine the locator and the problem of replacing the locator if it wears or becomes damaged. Another drawback to using integral locators is the additional material required to allow for the machining of the locator.

Assembled Locators. Assembled locators are similar to integral locators in that they must both be machined. However, these locators have the advantage of being replaceable. Assembled locators may be used as locators for supports (*Figure 4-20*), and since they are not part of the major tool body, using assembled locators does not require additional material for the tool body. Assembled locators are frequently made of tool steel and hardened to prevent wear.

Locating Pins. Locating pins are the simplest and most basic form of locating element. These locators may be made in-house from steel drill rod or purchased commercially. Commercial locating pins are available in several styles and types (*Figure 4-21*). Standard hardened dowel pins are another form of commercial component frequently used for locating devices. Due to

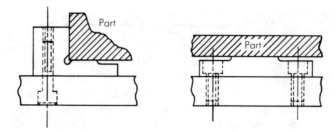

Figure 4-20. Assembled locators.

their simplicity, easy application, and replaceability; round pins are the most commonly used form of locating device.

The location and number of locating pins is generally determined by the size, shape, and configuration of the part. But in most cases the 3-2-1 principle is applied. In this principle there are three pins placed under the part to control five degrees of freedom and three more pins placed perpendicular to the base to control four more degrees. The remaining two degrees of freedom are controlled by the clamping element.

Consideration other than location of the workpiece will often affect the number of locating pins used. The workholding device must be designed to clamp the workpiece securely and support it to resist the forces generated by machining. If the operation performed applies considerable force, the workpiece may spring out of shape. Thus, the locating elements must be designed also to provide adequate support for the workpiece against the forces acting upon it. Many workpieces are essentially flat, or have a flat surface that can be used for locating purposes. These are commonly located

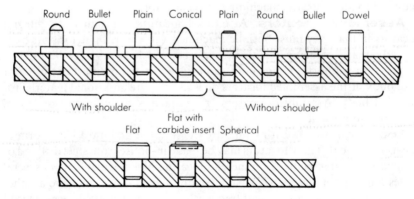

Figure 4-21. Locating pins.

by placing the workpiece on a plane surface to restrict it in five freedoms as shown in *Figure 4-22*. The addition of six locating pins, *A, B, C, D, E,* and *F* will restrict it in six more degrees of freedom. The workpiece can move only in an upward direction. This final degree of freedom is restrained by a suitable clamp having a plane surface parallel to the one on which the workpiece is placed. When this method of location is used, a great deal of planning must be done before specifing the position of the locators. Since the workpiece is confined on all sides, the designer must specify locator positions that will not interfere with each other. The locators must be positioned to eliminate the possibility of duplicate location.

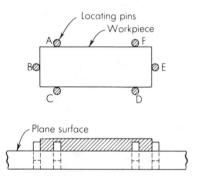

Figure 4-22. Simple workholder made of plane surface and pins.

If the workpiece must be held in a vertical position, the same principle of clamping and supporting between two plane surfaces may be used. This again will restrict motion in six degrees of freedom. Of the six remaining freedoms, only five must be restricted by locating pins as shown in *Figure 4-23*. The four pins *A, B, C,* and *D* restrict freedom of motion downward, to the left and right, and both clockwise and counterclockwise around axis *X*. Gravity may be used to locate the workpiece and restrict freedom of movement during the machining or other operation to be performed. Here again, the locators must be positioned to minimize the chance of duplicate location.

V Locators. A cylinder, like the prism, also has twelve degrees of freedom. The cylinder in *Figure 4-24* is free to move in two opposed directions along each axis, and to rotate both clockwise and counterclockwise around each axis. To accurately locate a cylindrical workpiece, it must be confined to restrict its motion in each of its twelve freedoms.

Figure 4-25 shows a cylinder placed in the intersection of two perpendicular planes. The base plane is parallel to the *X* and *Z* axes, and the vertical plane is

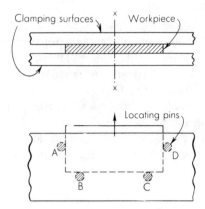

Figure 4-23. Vertical locating with pins.

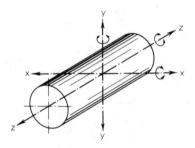

Figure 4-24. Twelve degrees of freedom of a cylindrical workpiece.

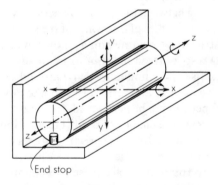

Figure 4-25. Seven degrees of freedom arrested by V locator with stop pin.

parallel to the Y and Z axes. The horizontal plane restricts movement in the two rotational freedoms around the X axis and the downward freedom along the Y axis. The vertical plane restricts the two rotational freedoms around the Y axis and the leftward movement along the X axis. The pin that forms the end stop restricts one freedom, i.e., forward movement along the Z axis. This corresponds to the basic 3-2-1 method of location used for the prism, but it restricts movement only in seven freedoms. The cylinder can move backward along the Z axis; in addition, it is free to rotate clockwise and counterclockwise around the Z axis.

Rotation around the Z axis can be restrained by clamping friction applied against the V formed by the two planes. This does not locate in a definite angular position about the Z axis and therefore cannot be considered true locating. No provision has been made to accurately locate a particular point on the cylindrical surface.

Locating a cylinder in a V places its longitudinal axis in true location. This is often sufficient for the operation to be performed. In addition, the basic principle of V location can be applied to workpieces that are not true cylinders but do contain cylindrical segments.

A single V locator provides two points for locating the points where the cylindrical end of the workpiece is tangent to both sides of the V. The equivalent of three points in a base plane and a radial locator are required for complete location of the workpiece. In *Figure 4-26*, a workpiece with two cylindrical ends is confined by means of two V locators. The movable V locator serves only to locate one point, the center of the cylindrical portion.

The included angle between the two surfaces of a V locator governs the

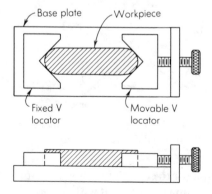

Figure 4-26. Workholder with multiple V locators.

positions of circular sections of varying diameters (*Figure 4-27*). A V with an included angle of $2X$ locates a circle of radius R_1 with center A and a circle of radius R_2 with center B. The distance between centers $= C$. By similar triangles:

$$\frac{R_1}{OA} = \frac{R_2}{OB} = \frac{R_2}{OA + C}$$

but $OA = R_1$ cosecant X

$C =$ cosecant $X(R_2 - R_1)$

and $D_2 = 2R_2$; $D_1 = 2R_1$

therefore, $$C = \frac{\text{cosecant } X(D_2 - D_1)}{2}$$

Consequently, the distance between positions of any two diameters in a V varies as one-half the included angle of the V. The smallest variation occurs when $X = 90°$, where cosecant $X = \dfrac{0.5}{2}$. However, as X approaches $90°$,

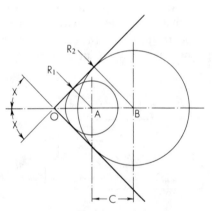

Figure 4-27. Positions of circular sections
of varying diameter in a V locator.

there is less inclination for the circular section to seat positively in the V and more difficulty in retaining it. Note that with $X = 90°$, $2X = 180°$ or a straight line. The best compromise is achieved with $X = 45°$ and the included angle of the V is $90°$.

Irregularities in the circular section of a workpiece, or a chip lodged between the workpiece and the V locating surface can introduce errors of location. The included angle of the V has a definite influence on the effect of such displacement. In *Figure 4-28*, *E* is the displacement caused by a rough surface or chip. The circular section of the workpiece may be considered to rest on another side of the V indicated by a dotted line. The displaced side of the V, shown by the solid line, forms a new V with the opposite side to define the displaced position of the circular section. The change in the position of the center of the workpiece is identical to the shift in the apex from the original to the new V. The original V is *BOC* and the new V is *AO'C*. The axis of the workpiece is displaced by the distance *00'*. For a constant displacement *E*, *00'* is a minimum when $2X = 90°$.

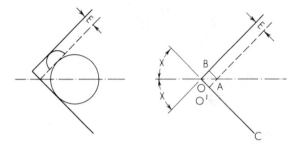

Figure 4-28. Influence of the included angle or errors of location.

Consideration must be given to the axis of the rotating tool and its relationship with the position of the V locator. The V locates the longitudinal axis of the cylindrical workpiece. When work is done perpendicular to this axis, the position of the V locator should be arranged to keep displacement of the workpiece to a minimum. In *Figure 4-29A*, a cylindrical workpiece is placed in a V locator so that a hole can be drilled perpendicular to the longitudinal axis. Any variation in diameter of the workpiece will cause a displacement in the location of the vertical axis. The drill bushing, however, remains in its original position and the drilled hole will deviate from its required position by the amount of the displacement. In *Figure 4-29B*, the V locator is so positioned that its axis is parallel to that of the drill bushing. Variation in diameter of the workpiece will cause no displacement of the vertical axis and the drilled hole will not deviate from its required position.

Locating Nests. The nesting method of locating features a cavity in the workholding device into which the workpiece is placed and located. If the cavity is the same size and shape as the workpiece, this is an effective means of locating. *Figure 4-30* illustrates a nest which encloses the workpiece on its

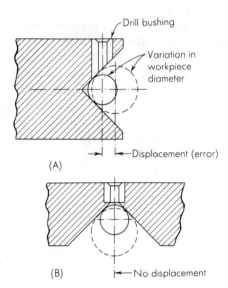

Figure 4-29. Minimizing error by proper placement of a V locator.

bottom surface and around the entire periphery. The only degree of freedom remaining is in an upward direction. A similar nest can be used to locate cylindrical workpieces. Cavity nests are used to locate a wide variety of workpieces regardless of the complexity of their shape. All that is necessary is to provide a cavity of the required size and shape. No supplementary locating devices such as pins are normally required.

The cavity nest possesses some disadvantages. Since the workpiece is completely surrounded, it is often difficult to lift it out of the nest. This is particularly true when no portion of the workpiece projects out of the nest to afford a good grip for unloading. The workholding device can, of course, be turned over, and the workpiece shaken out. When the workpiece tends to stick, an ejecting device can be incorporated in the workholder. This, however, adds time to the processing. Another disadvantage is that the operation performed may produce burrs on the workpiece which tend to lock it into the nest. In this case, the workpiece must be pried out or an ejector must be provided. Chips from the cutting operation may lodge in the nest and must be removed before loading the next workpiece. Any chips remaining may interfere with the proper positioning of the next workpiece.

To avoid the disadvantage of the cavity type of nest, partial nests are often used for locating. Flat members, shaped to fit portions of the workpiece, are fastened to the workholding device to confine the workpiece between them. *Figure 4-31* shows two partial nests, each confining one end of the bow-

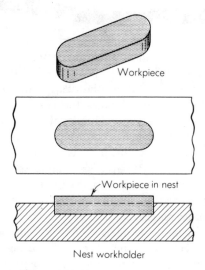

Figure 4-30. Nest-type workholder.

shaped workpiece. Each nest is fastened to the flat supporting surface of the workholder by means of two screws. Accurate positioning of the nests is ensured by dowels which prevent each nest from shifting from its required position.

Partial nests eliminate many of the disadvantages of cavity nesting. Since they do not have the entire contour of the workpiece, they require less time to

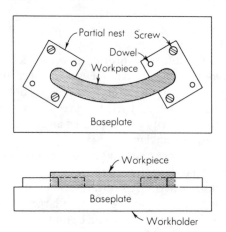

Figure 4-31. Workholder with partial nests.

make. Since they do not completely confine the workpiece, it is easily lifted out of the workholder. Normally, an ejecting device is not required. The cavity in each nest is open at one end to permit easy removal of chips.

With the development of plastic casting materials, making cavities for locating workpieces has been simplified. The casting material can be poured around a workpiece prototype. After solidification and cure, the workpiece is removed. The resulting cavity becomes the nest for locating workpieces during the production operation. This method is much simpler than machining the cavity from a solid piece of material and requires considerably less time, particularly when the workpiece is complex.

Many varieties of plastic materials are available, some with fillers of solid material, including steel or aluminum, to increase strength. Those with metallic fillers may be sawed, milled, drilled, and tapped after curing to a solid state. This machinability makes it easier to fasten the poured plastic cavity to the workholding device and eases alteration to accommodate future changes in the workpiece. Selection of the proper plastic depends on the forces generated by the operation to be performed, the quantity of parts to be made, and the effects of shrinkage of the material as it solidifies and cures.

Since plastic materials do not possess the strength of steels, plastic cavities are often reinforced with steel members. They increase resistance to tool forces, but do not materially increase the wear resistance of the plastic nesting surface. Consequently, plastic cavities generally are not used when large quantities of parts must be made. Also, if the plastic material used has a high degree of shrinkage, the resulting cavity may be too small to adequately locate the work. This, however, is not a critical factor since continuing improvements in the formulation of plastic compounds have consistently reduced inherent shrinkage.

Since the workpiece itself is required for the preparation of a nesting cavity, this type of workholder cannot be completed in advance of the production process. Of course, a prototype workpiece may be used, but making the prototype costs time and money so that the primary advantage of the cast locating cavity is lost. Therefore, the cavity is not usually made until production parts are available. This, however, does not usually cause a serious delay, since a cast cavity can be poured and cured in a few hours and be ready for use the following day.

A minimum-size workpiece must not be used for making the cavity since it will not accept parts that vary toward the maximum dimension. When the contours of the workpiece permit, its surfaces can be built up to increase vital locating dimensions to insure casting a cavity large enough to adequately locate a maximum-size part. Often this can be done by applying strips of thin masking tape to the proper surfaces of the workpiece.

Experience has developed the casting process of making locating cavities to the extent that a complete workholding device can be poured in one piece. *Figure 4-32* shows a plate type drill jig made this way. The workpiece containing the required holes is laid on a flat surface. Round pins are pressed into the holes and project upward to locate the drill bushings whose outer periphery is serrated to insure a firm grip in the cast plastic. A dam may be made from four rectangular bars placed around the workpiece. The pins are then removed, and the drill jig is ready for use.

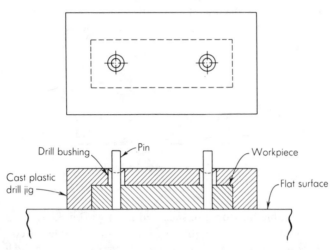

Figure 4-32. Cast plastic plate-type drill jig.

Other materials frequently used for casting locating nests are low-melt alloys. These bismuth, lead, tin, and antimony alloys are ideal for many difficult workholding and locating problems. Quite often, low-melt alloys are used for applications where plastics cannot be used. One principal advantage to using low-melt alloys rather than plastic is their ability to be reused.

In use, the part is first positioned in a container. The container can be made of almost any material. Sheet metal, low-carbon steel, and aluminum are typical examples of material which could be used. Once positioned, the molten alloy is poured into the container and allowed to solidify.

The part is then removed, and the nest is cleaned and all burrs removed. If the part has a configuration that prevents its removal from the cast alloy, the part is machined in the encasing alloy, and when finished, the alloy is melted off the part.

Frequently, additional materials such as metal filings, ball bearings, or similar materials are added to the alloy to add wear resistance and to permit the nest to hold up under longer production runs.

In any case, both plastic and low-melt alloys provide cost-effective alternatives to expensive machined nests. The low initial costs and adaptability of these materials make them very suitable for a wide range of workholding applications.

Adjustable Locators. Adjustable locators are widely used for applications where the workpiece surface is irregular, or where large variations between parts make solid locators impractical. The principal type of adjustable locator is the threaded style (*Figure 4-33*). In some cases, this type of locator can also be used as a clamping device rather than a locator, as shown. However, for the most part, this type of locator is used simply as a locator.

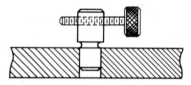

Figure 4-33. Threaded adjustable locator.

When adjustable locators are specified for a workholder, the position of the locator is not as critical as with solid locators, so the relative cost is greatly reduced. Frequently, adjustable locators are actually used as solid locators by simply adding a locknut, or screw, to secure the adjusting screw (*Figure 4-34*).

Adjustable Supports. Adjustable supports are simply adjustable locators that are positioned beneath the workpiece. The primary variations of adjustable supports include threaded, spring, and equalizing. Threaded supports are used along with solid supports to permit easy leveling of irregular parts in the workholder (*Figure 4-35*). Spring supports are also used with solid supports to level the workpiece, but rather than using threads to

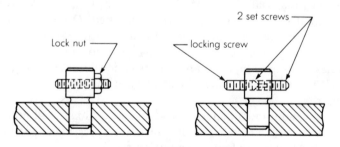

Figure 4-34. Adjustable locators with locknut or screw.

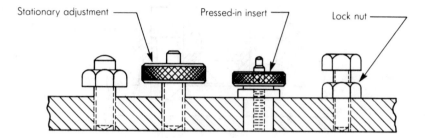

Figure 4-35. Threaded type adjustable supports.

elevate the locator, a secondary threaded element, such as a thumbscrew, is used to lock the position of the spring support (*Figure 4-36*). Equalizing supports are used to insure constant contact of the supports and workpiece. These supports are normally self-adjusting. That is, as one is depressed, the other rises (*Figure 4-37*).

Sight Locators. Sight locators are an effective means of locating sand

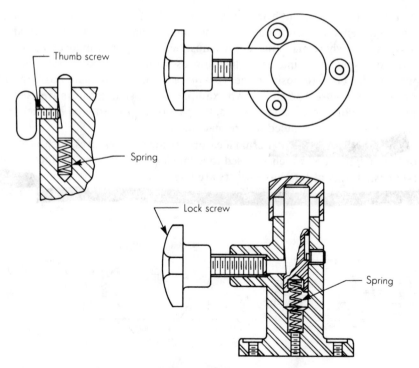

Figure 4-36. Spring type adjustable supports.

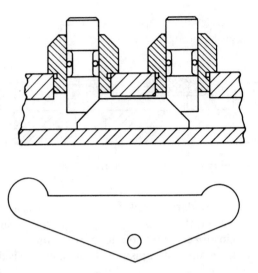

Figure 4-37. Equalizing type adjustable supports.

castings and similar rough and irregular parts for first-operation machining. These elements, while not locators in the conventional sense, are well suited to applications where machined details must be in an approximate area, rather than at a specific point. Sight location uses lines, slots, and holes in the workholder body to position the workpiece in an approximate position for machining. *Figure 4-38* shows two examples where sight locators are used to position a workpiece. In most cases, the part is simply centered between the sight locators and clamped before machining.

Internal Locators. Internal locators are locating features, such as holes or bored diameters, which are used to locate a part by internal surfaces. The two basic forms of internal locators are fixed size and compensating. Fixed size locators are made to a specific size to suit a certain hole diameter. Typical

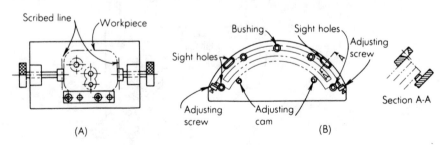

Figure 4-38. Sight locators: (*a*) by scribed lines; and (*b*) by sighting holes.

examples of this type of internal locator include machined locators, commercial pin locators and relieved locators. Compensating locators are generally used to centralize the location of a part or to allow for larger variations in hole sizes. The two typical forms of compensating locators are conical and self-adjusting.

Machined Internal Locators. Machined internal locators are made to suit special size hole diameters. In most cases, machined locators are made for larger hole diameters. The exact form and shape of these locators are normally determined by the part to be located. They are generally machined to size and then attached to the tool by screws and dowels (*Figure 4-39*). In cases where small-diameter locating pins are required, materials such as drill rod or commercial drill blanks are frequently purchased in the desired diameter and then cut to the required length. Both drill rod and drill blanks are normally available in most standard sizes.

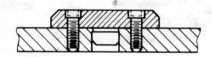

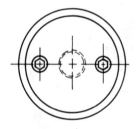

Figure 4-39. Machined internal locator.

When round plugs are used in holes for locating there is a tendency to stick when a close-fitting workpiece is applied. A plug of diameter d extending from a faceplate, (*Figure 4-40A*) will not stick in the hole of diameter D of a workpiece if the plug length $1 = \sqrt{2WC}$ where W is the outside diameter of the workpiece concentric with the hole and $C = (D - d)$.

A projection of diameter d on a workpiece will not stick in a locating hole, (*Figure 4-40B*), of diameter D, if the hole is relieved or countersunk so that the length of engagement between hole and workpiece is $E = 1 - m \leqq \sqrt{WC}$.

A method of reducing the tendency for workpieces to stick on a locating plug extending from a faceplate is to relieve the plug by cutting away three equal segments, (*Figure 4-40C*). A workable value for the angle σ is 15°, which results in $m = 0.35d$. A plug cut away in that manner will not stick if its length $1 - 2.4\,(2a + 0.85d)(D - d)$. The disadvantage is that a workpiece can be displaced in three directions on a relieved plug, and the extra error in inches of location as compared with a full round plug is $0.207\,(D - d)$, or about a 20% loss in locating accuracy.

An aligning groove may be put on the end of a plug of diameter d of unlimited length l to keep it from sticking when inserted in a hole of diameter D, (*Figure 4-40D*). Workable dimensions for such a design are $l_1 = \quad 2AC_2$, $l = \mu d$, and $B = 0.95d$, where C_2 is the clearance between the plug pilot and hole $= (D - A)$, and μ is the coefficient of friction between plug and hole surfaces, usually about 0.15 to 0.25 for steel. The pilot diameter may be made as convenient between size d and $A = (2d_2 / D) - d$.

Commercial Pin Locators. Commercial pin locators are made in two general styles, plain and shouldered (*Figure 4-41*). The ends of these locators are made in either round, flat, or bullet shapes and facilitate easy loading and unloading of parts. These locating pins are normally made between 0.0005

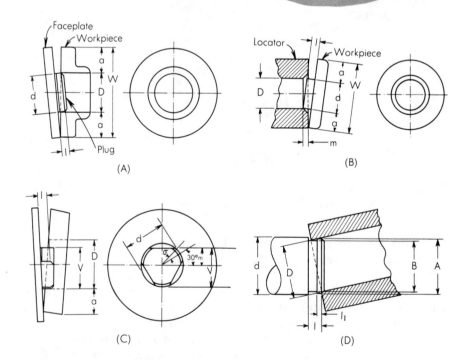

Figure 4-40. Nonsticking locator design.

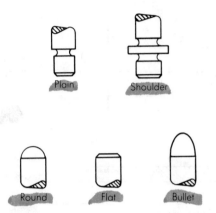

Figure 4-41. Commercial pin locators.

and 0.002" (0.013 and 0.05 mm) undersize to prevent jamming and binding in the located hole. The installed end of these pins is generally 0.0625" (1.588 mm) smaller than the location end to prevent improper installation.

Commercial locating pins are also made in press fit and slip fit styles. Press fit pins are those which are installed directly into the tool body. Slip fit pins are used along with liner bushing which are installed into the tool body. A lockscrew is also used to hold the pin in place.

Relieved Locators. Relieved locators, as their name implies, are designed to minimize the area of contact between the workpiece and the locating pin. This reduces the chances of the locator sticking or jamming in the part. *Figure 4-42* shows several examples of relieved locators. The most commonly used form of relieved locator is the diamond pin.

Diamond pins are used for radial location in conjunction with round locating pins. It is possible to accurately locate a workpiece with two round pins, but allowances must be made for the variations encountered in hole sizes and locations. For instance, the distance between holes *A* and *B* (*Figure 4-43*) will vary to the extent of tolerance *X*. Similarly, the distance between pins *A*

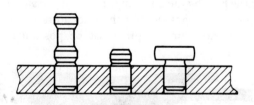

Figure 4-42. Relieved locators.

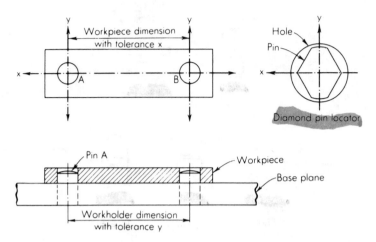

Figure 4-43. Radial location with internal pins or plugs.

and *B* in the workholder has a tolerance *Y*. For accurate location, there should be an allowance between pin *A* and hole *A* of only a few ten-thousandths of an inch. But if pin *B* is a complete cylinder (same as pin *A*) the allowance between pin *B* and hole *B* must be at least as great as the sum of tolerances *X* and *Y*. This is necessary for the pins to engage holes within the permissible tolerance *X*. The diameter of pin *B* can be calculated using the formula *B* = minimum hole diameter minus the sum of the center-to-center pin tolerance. Extreme cases occur when both hole and pin center-to-center dimensions are at maximum or minimum conditions. As a result, there will be a large allowance between the hole and pin at *B* in the *Y* direction. This will permit an undesirable amount of radial rotation around the axis *A* and defeats the purpose for which pin *B* is intended.

To achieve more accurate radial location, *B* may be a diamond pin as shown in the inset in *Figure 4-43*. It is relieved on two sides to allow for variations in the *C* direction and has two cylindrical portions to locate the hole in the *Y* direction. The minimum radial movement of the workpiece occurs when the diameter of the cylindrical portion of the pin is smaller than the diameter of the hole by the allowance necessary to slip the minimum size hole over the pin. When positioning these locators in the tool body, the bearing surface of the diamond pin must be positioned to restrict the movement of the part.

In some cases, the part may be completely located using diamond pins. As shown in *Figure 4-44*, the part is completely located by using two diamond pins placed to restrict the rotational movement of the part. When used in this fashion, the pins must be positioned to restrict the movement permitted by the other pin.

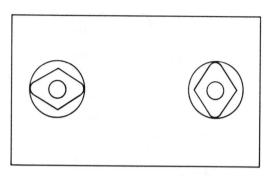

Figure 4-44. Locating completely with diamond pins.

One other style of locating pin that will correct slight differences between locating holes is the floating locating pin (*Figure 4-45*). This pin provides precise location in one axis while allowing up to 0.125" (3.18 mm) movement in the perpendicular axis. The body of the locator is pressed into the tool body, and is referenced to the tool body and to both the fixed and movable axes with a roll pin.

The floating pin locator generally works like a diamond pin. Due to the increased movement, however, this pin should be used for parts with somewhat looser locational tolerances of the mounting holes. The floating locating pin permits greater variation than that typically allowed by a diamond pin.

As shown in *Figure 4-46*, the floating locating pin is used along with a round locating pin. The part is first positioned on the round locator. It is then located on the floating pin locator. This pin, when positioned as shown, prevents any radial movement about the round locator. It also compensates for differences of up to 0.125" (3.18 mm) in mounting hole positions.

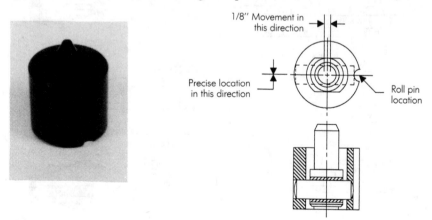

Figure 4-45. Floating locating pin. (*Courtesy, Carr Lane Manufacturing Co.*)

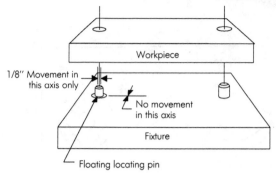

Figure 4-46. Floating locating pin used in combination with round locating pins. (*Courtesy, Carr Lane Manufacturing Co.*)

Conical Locators. Conical locators are centralizing locators that compensate for variations in part sizes as well as centering a part in the workholder. The most efficient types of conical locators are spring-loaded or threaded (*Figures 4-47A* and *4-47B*). Conical locators, while normally used for internal location, may also be used for external location with a conical cup (*Figure 4-47C*).

Self-Adjusting Locators. Self-adjusting locators are used in applications such as sand casting, where there is great variation in the size of the holes to be located. These locators can be made in a wide variety of styles. *Figure 4-48* shows one example of a self-adjusting locator that could be used for internal location.

Spring Locating Pins. Unless the parts are properly positioned against the locators, errors will result, no matter how well a locating system is designed

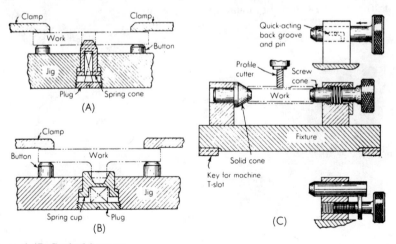

Figure 4-47. Conical locators.

Figure 4-48. Self-adjusting locator.

or made. One type of locating device that helps reduce these locational errors is the spring locating pin (*Figure 4-49*).

These pins are designed to push the part against the fixed locators. This will ensure proper contact during the clamping operation. Although these spring locating pins are not actually locating devices, they help reduce locational errors by correctly positioning the part against the locators. In addition, these pins can eliminate the need for a third hand when positioning and clamping some parts. The small size and compact design of these pins makes them very useful for smaller parts or confined space. A protective rubber seal around the contact pin helps seal out debris and coolant.

Spring locator pins may be installed directly in a hole or mounted in an eccentric liner (*Figure 4-50*). The liner permits adjustment of the pins to suit parts with looser tolerances (*Figure 4-51*).

When positioning a flat plate, shown in *Figure 4-52*, the first step is to position the part over the workholder. The part is then placed against the solid locator and pushed down against the spring pin. When seated, the spring locating pins push the workpiece against the solid locator to ensure proper contact. These locating pins are well suited for a variety of applications and part shapes.

Spring Stop Buttons. *Figure 4-53* illustrates spring stop buttons, another spring-loaded workholding device. They work much like the spring pins, but are designed for larger parts or where more force is needed. The spring stop buttons are made with three different contact faces. The first is a spherical button contact. The other two have flat contacts. The flat face contacts are made with or without a tang (*Figure 4-54*).

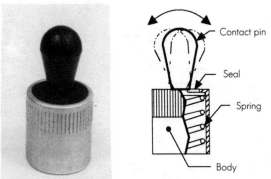

Figure 4-49. Spring locating pin. (*Courtesy, Carr Lane Manufacturing Co.*)

Figure 4-50. Spring locating pin mounted in an eccentric liner. (*Courtesy, Carr Lane Manufacturing Co.*)

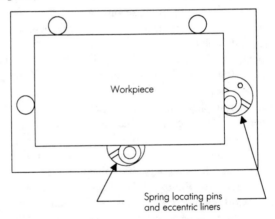

Figure 4-51. Eccentric liner permits adjustment of the spring locating pins for loosely toleranced parts. (*Courtesy, Carr Lane Manufacturing Co.*)

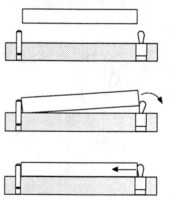

Figure 4-52. Positioning a flat plate. (*Courtesy, Carr Lane Manufacturing Co.*)

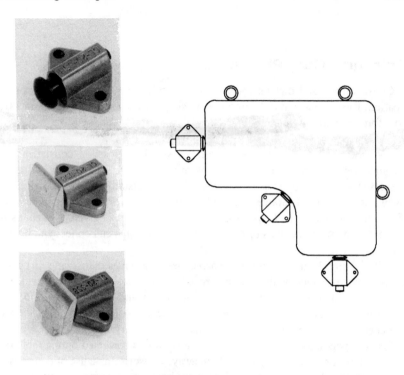

Figure 4-53. Spring stop buttons. (*Courtesy, Carr Lane Manufacturing Co.*)

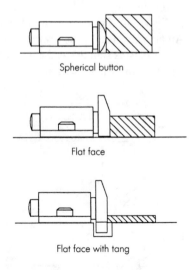

Spherical button

Flat face

Flat face with tang

Figure 4-54. Spring stop buttons showing spherical and flat contacts. (*Courtesy, Carr Lane Manufacturing Co.*)

Chip and Burr Problems

Chips, burrs, and dirt on locating surfaces cause wear and disturb proper location. Every means must be provided to keep locating surfaces and points free from foreign matter. There are three general methods of chip and dirt control: (1) make locators easy to clean, (2) make them self-cleaning, and (3) protect them.

For ease of cleaning, make locators as small as possible consistent with adequate wearing qualities; rest buttons are popular for this reason. Jigs and fixtures should be as open as possible so that supports are readily accessible and visible. Raise supports above surrounding surfaces so chips fall or can be swept off readily, as shown in *Figure 4-55*. Provide easy exit or passage avenues for chip ejection. Avoid pockets or obstructions where chips can collect and be hard to clear.

Self-cleaning locators can have sharp edges or grooves that scrape dirt off of the workpiece surface as they slide across it. Relief around locating surfaces is essential as a means of escape for unwanted chips and dirt (*Figure 4-56*). For corner relief, suitable recessor grooves are provided so dirt and chips do not pack into corners and burrs do not bear against locating surfaces.

Fixed wipers may push chips along as a fixture is traversed on the table of a machine tool, or coolant may flush them away. Indiscriminate use of compressed air for blowing chips about has its drawbacks because chips can be quite harmful when blown into ways and other bearing surfaces of machine tools. Shields and guards control and gather blown chips.

The drill jig shown in *Figure 4-57* automatically cleans the jig and locating points prior to loading. An air valve is actuated by contact of the jig top plate in the full-open position. The air blast is shut off as the top plate is lowered for each cycle. The chips are stopped by the shield.

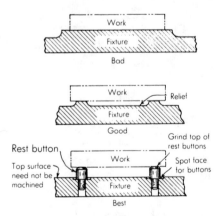

Figure 4-55. Raised workpiece supports.

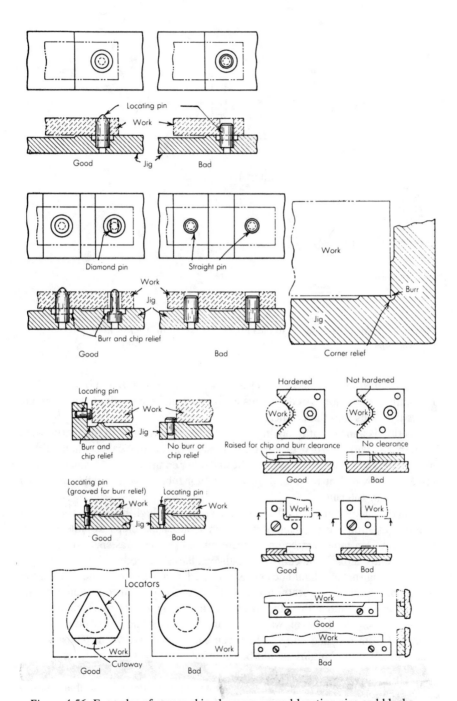

Figure 4-56. Examples of proper chip clearance around locating pins and blocks.

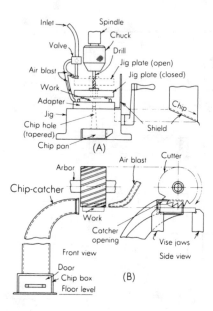

Figure 4-57. Air blast used (*a*) to clean drill jig and (*b*) in front of milling cutter to remove chips.

An air blast in front of the cutter in view *B* of *Figure 4-57* removes the chips in a milling operation. A continuous air flow of sufficient volume and pressure blows the chips into the catcher.

Suction is a means of removing light, discontinuous chips from the tool work area, particularly in dry grinding. It is sometimes applicable when milling nonmetallic materials, such as phenolic and other compositions. A suitable mesh screen is provided at the mouth of the suction tube to keep clothes and other objects from entering.

Gravity slides can be used to control chips in a high-production setup for milling cast iron parts. Machine vibration induces the chips to slide into the trough between table and column where hinged pushers, fastened to the table, push the chips along. The chips accumulate in barrels at both ends of the trough.

A locating surface should be entirely covered by a workpiece when the fixture is loaded. If part of the locator is exposed, dirt and chips can collect on it during cutting and may move over the locator as the workpiece is removed. Bearing surfaces, such as those of ways, and indexing mechanisms should also be protected from dirt and chips, which can cause excessive wear. Thus, slides, indexing pins, and buttons should be enclosed.

A burr raised on the work at the start of a cut is termed a ''minor'' burr, and that at the end of a cut is a ''major'' burr. Jigs should be designed so that removal of the workpiece is not hindered by burrs. In *Figure 4-58*, work removal does not

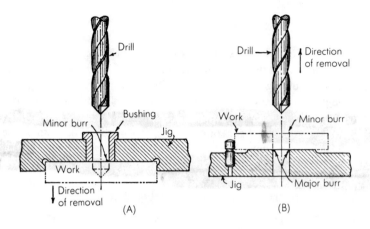

Figure 4-58. Work removal does not tend to shear the burrs; the direction of workpiece removal eliminates sticking of minor and major burrs.

tend to shear the burrs since it parts the work and jig directly. In contrast, if the work must be slid across the burr, sticking complicates removal. To avoid this, suitable clearance grooves or slots should be provided (*Figure 4-59*).

Clamping Principles

For a specific operation, the selection of general clamping—simple hand-operated clamps, quick-acting hand-operated clamps, power-operated clamps, etc.—should primarily be a function of operation analysis. This selection is based on an effort to balance the cost of the clamp against the cost of the operation to obtain the lowest possible total cost of both fixture and operation.

Sound judgment by the tool designer in the application of these specific clamping principles to the job at hand is essential. In general, clamping arrangements should be as simple as possible. Complicated arrangements tend to lose their effectiveness as the parts become worn, necessitating excessive maintenance, which might readily offset the savings of a faster operation.

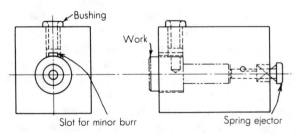

Figure 4-59. Use of burr relief for the minor burr.

The purpose of a clamp is to exert a force to press a workpiece against the locating surfaces and hold it there in opposition to the actions of cutting or other processing forces. Clamping forces should be directed within the locating area, preferably through heavy sections of the workpiece directly upon locating spots or supports. Cutting forces should be taken by the fixed locators in a jig or fixture as much as possible, but generally some components of, or moments set up by, the cutting forces must be counteracted by clamping forces. To be effective, a clamp should be designed to exert a minimum force equal to the largest force imposed upon it in the operation.

The following design and operational factors should be considered:

1. Simple clamps are preferred because complicated ones lose effectiveness as they wear.
2. Some clamps are more suitable for large and heavy work, others for small pieces.
3. Rough workpieces call for a longer travel of the clamp in the clamping range, but clamps may be made to dig into rough surfaces to hold them firmly.
4. The type of clamp required is determined by the kind of operation to which it is applied. A clamp suitable for holding a drill jig leaf may not be strong enough for a milling fixture.
5. Clamps should not make loading and unloading of the work difficult, nor should they interfere with the use of hoists and lifting devices for heavy work.
6. Clamps that are apt to move on tightening, such as plain straps, should be avoided for production work.
7. The anticipated frequency of setups may influence the clamping means. For example, the use of hydraulic clamps, even if simple and of low cost, might be inadvisable if frequent installation and removal of piping and valves is necessary.

Tool Forces

A clear understanding of the direction and magnitude of cutting forces may eliminate the need to restrain all 12 degrees of freedom of a workpiece. *Figure 4-60* shows how two pins and a table absorb the torque and thrust of a drilling operation. Although the workpiece is free to turn in a direction opposite to the torque, this freedom is insignificant unless a force is applied in that direction. No such force will be encountered in the planned drilling operation, and the remaining freedom need not be restrained. If, however, as part of the drilling operation, the spindle rotation is reversed for tool removal, such force will be encountered, and the freedom must be restrained.

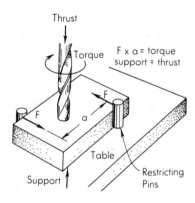

Figure 4-60. Pin-type drill fixture resisting torque and thrust.

Theoretically, there is no need to hold the workpiece down, as this is accomplished by the thrust of the drill. When the drill breaks through the thickness of the workpiece, an upward force may be created by interaction between the drill flutes and material remaining around the periphery of the hole. If there is no restraint in this upward direction, the workpiece may be lifted above the pins, creating a very dangerous condition. An upward force may also be produced when a drill or reamer gets lodged in a workpiece and the tool is to be withdrawn.

Figure 4-61 shows how a workpiece must be restrained for tapping. Torque in both directions must be absorbed and the lifting pull of the tap must be resisted by some form of a holddown, such as a clamp (not shown). For leadscrew tapping, torque resistance is still required, but no holddown is needed because the leadscrew, having the same lead as the tap, eliminates all thrust. When two holes are tapped simultaneously on a two-spindle setup, each tap prevents workpiece rotation by the torque exerted by the other tap. Without a leadscrew, a holddown would still be needed to prevent accidental lifting of the workpiece by the spindles.

Figure 4-62 shows another instance where the cutting force holds the workpiece, in this case against the support plate of a broaching machine. The broach is guided in the support plate and, to some extent, in the workpiece. The broach in turn also holds the workpiece. The cutting tool and cutting force both contribute to the workholding operation.

Once the designer of a workholder has identified the possible direction and magnitude of the forces, he has two ways to restrain the workpiece to counteract these forces. One utilizes the strength and rigidity of some part of the workholder against which the workpiece rests or is forced by a clamp, screw, or wedge. The other way utilizes friction between the workpiece surface contacting under pressure a surface of the workholder.

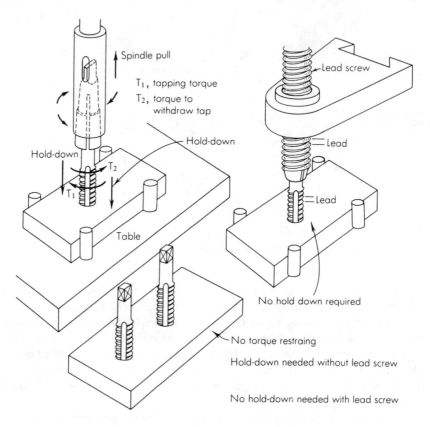

Figure 4-61. Designing a workholder to resist torque and thrust in a tapping operation.

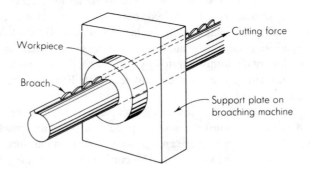

Figure 4-62. Workholder for broaching operation.

Figure 4-63 shows a workpiece held in a vise. The horizontal component of the cutting force is absorbed by the solid jaw of the vise. The vertical component is resisted by friction between the workpiece and the jaws. *Figure 4-64* shows the cutting force absorbed only by the friction between the jaws and the workpiece.

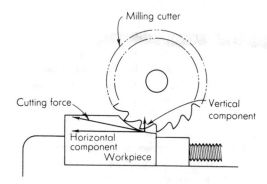

Figure 4-63. Cutting force resisted by solid jaw of vise.

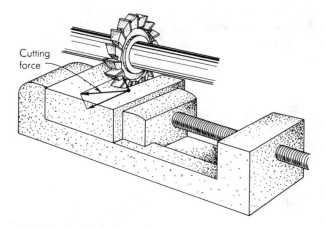

Figure 4-64. Cutting force resisted only by friction.

Wherever possible, cutting forces should be opposed by the structure of the workholder and preferably by the strongest and most rigid parts of it. If necessary, a movable element may be used to absorb cutting forces, but only if properly designed for strength and rigidity.

Clamping Forces

Complete analysis of the tool forces in a proposed operation will disclose which of the 12 degrees of freedom must be restrained and to what extent.

Quite often, tool forces are of such magnitude and direction that a workpiece may be dislodged or moved from its required location. If the locating elements of a fixture cannot assure adequate restraint, it may be necessary to clamp the workpiece against them.

Clamps hold a workpiece against a locator. Perhaps the most common application is the bench vise, where a movable jaw exerts pressure on a workpiece, thereby holding it in a precise location determined by a fixed jaw. The bench vise uses a screw to convert actuating force into holding force. *Figure 4-65* shows a number of commonly used mechanical methods for transmitting a multiplying force.

The clamping forces applied against the workpiece must counteract the tool forces. Having accomplished this, further force is unnecessary and may be detrimental. The physical characteristics of the workpiece greatly influence clamping pressure. Hard vise jaws can crush a soft, fragile workpiece. The clamping pressure must hold, but not damage, deform, or impose too great a load on the workpiece.

The direction and magnitude of clamping pressure must be consistent with the purpose of the operation. An example is the boring of a precise round hole, with the workpiece clamped in a heavy vise. Excessive clamping pressure can

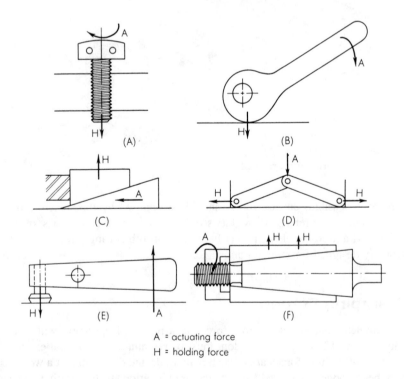

A = actuating force
H = holding force

Figure 4-65. Mechanical methods of transmitting and multiplying force: (*A*) screw, (*B*) cam, (*C*) wedge, (*D*) toggle linkage, (*E*) lever, (*F*) combined screw and wedge.

compress the workpiece. The bored hole may be perfect in size and roundness while the workpiece is compressed. The release of clamping pressure might permit the workpiece to return to its normal rather than compressed condition, and the hole might then be offsize and elongated. Another example of excessive or misdirected clamping pressure may be found in a cutoff operation. If a workpiece clamped in a vise is to be cut between the jaws, in a direction parallel to the jaws, the removal of metal by the saw blade permits the remaining metal to flow into a cut area. This metal movement can effectively wedge and stop the saw blade.

Clamping pressure should not be directed toward a cutting operation, but should, wherever possible, be parallel to it. Clamping pressure should never be great enough to change any dimension of the workpiece.

Positioning Clamps

Clamps must be positioned to contact a workpiece at its most rigid point. When possible, clamps should be located over a locator or support. In cases where a part cannot be clamped over a locator, a secondary support must be installed to counteract the clamping forces and prevent damage to the part. As shown in *Figure 4-66*, the flanged part is located by its center. If the part was clamped as shown in (a), the part would distort. Therefore, a secondary support must be added (b). This additional support provides the required backing to prevent distorting the part. Remember, when adding a secondary support, allow enough space between the support and the part to prevent redundant locating.

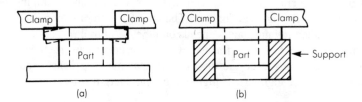

Figure 4-66. Positioning clamps.

When clamping large rings, or similar parts, the location of the clamping devices is very important to prevent warping or distorting the part. *Figure 4-67* shows two methods of clamping this type of workpiece. If the part were clamped as shown at (a), the clamping forces could easily distort the part. However, if the clamps are positioned as shown at (b), the chance of part distortion is greatly minimized.

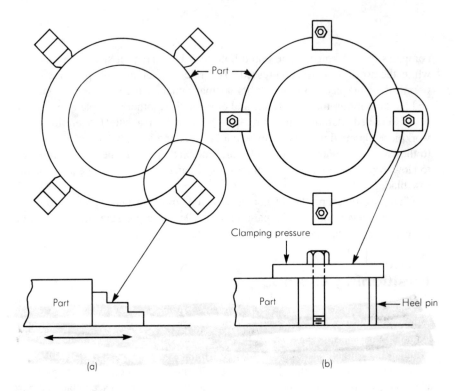

Figure 4-67. Clamping large rings.

Basic Types of Clamps

There are several basic types and styles of clamps and clamping devices commonly used for jigs and fixtures. The specific type of clamp selected for a particular application normally depends on the type of tool, part shape and size, and the operation to be performed. Other considerations such as speed of operation and permanence are also factors that must be considered for long production runs or high-speed, high-volume production runs.

Rigid versus Elastic Workholdings

Workholders may be either rigid or elastic. Since there are no absolutely rigid bodies and materials, rigid means the holding elements are preset to a fixed position. *Figure 4-68* shows a screw holding a workpiece as an example of rigid workholding. There is some elasticity in the screw and nut, but it is not intentionally provided. *Figure 4-69* illustrates a pressure-supported piston, which bears down on and holds the workpiece as an example of elastic workholding.

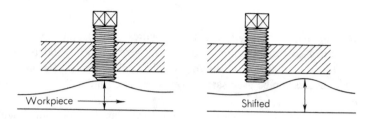

Figure 4-68. Rigid workholding.

In *Figure 4-68*, a sideways shift will bring the screw out of contact with a workpiece that is not of uniform thickness, causing complete loosening. The workpiece, however, resists any upward force against the screw because the screw is a self-locking, mechanically irreversible element. In *Figure 4-69*, the piston clamp will continue to exert holding force in case of a sideways shift. An upward force is resisted only by hydraulic pressure exerted on the piston.

Figure 4-70 shows examples of elastic workholding using self-contained, hydraulically operated clamping cylinders. The workholder may be clamped by hydraulic pressure and released by spring pressure (*View A*), or may be held by spring pressure and released by hydraulic pressure (*View B*).

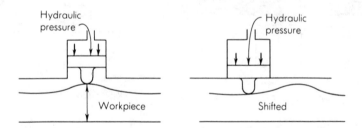

Figure 4-69. Elastic workholding.

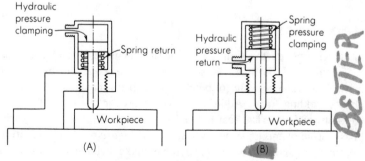

Figure 4-70. Elastic workholding with self-contained hydraulically operated clamp cylinders.

Air pressure may be used, but usually requires considerably larger piston areas to obtain a sufficient holding force. The size of the cylinders may be reduced by the use of force-multiplying mechanical elements. Before elastic workholding devices can be used safely, the work forces and their direction must be determined.

Figure 4-71 compares a hydraulic mandrel for elastic workholding with a mechanical mandrel for rigid workholding. Hydraulic pressure is produced by a screw-piston arrangement. Line *A-A* traces a path through a solid workpiece, a solid expandable shell, and an elastic layer of a hydraulic compound. A similar path *A-A* on the split-collet expanding mandrel passes through only consecutive layers of rigid metal. This does not mean that the elastic hydraulic mandrel is not as positive as the rigid mandrel. In fact, the opposite may be true. The hydraulic mandrel's torsional stiffness may be greater, and it may possess many other desirable features such as higher inherent precision and less vulnerability to dirt.

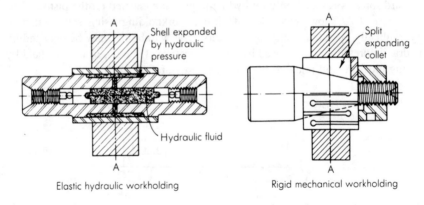

Elastic hydraulic workholding   Rigid mechanical workholding

Figure 4-71. Elastic and rigid mandrels.

Strap Clamps

Strap clamps are the simplest and least expensive type of clamping device used for jigs and fixtures. As shown in *Figure 4-72*, the basic strap clamp consists of a bar, a heel pin (or block), and either a threaded rod or cam lever to apply the holding force. Additional accessories for this type of clamp include hand knobs and spherical seat nut and washer sets (*Figure 4-73*). Variations of the basic strap clamping system are shown in *Figure 4-74*. The positions of the force, fulcrum, and workpiece are altered to make differences in the level classes of the strap clamps. Other variations of this basic type of clamp include the latch strap, sliding strap, and the hinged strap clamp (*Figure 4-75*).

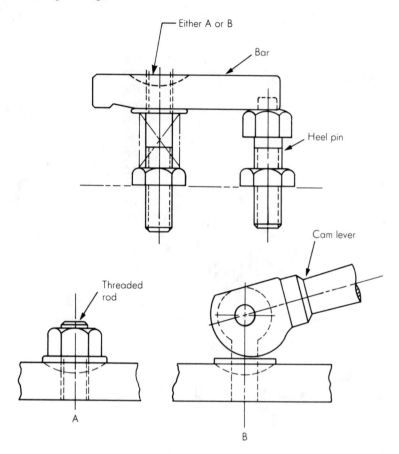

Either A or B

Bar

Heel pin

Cam lever

Threaded rod

A

B

Figure 4-72. Strap clamps.

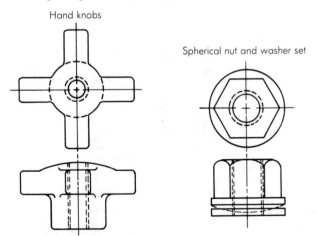

Hand knobs

Spherical nut and washer set

Figure 4-73. Accessories for strap clamps.

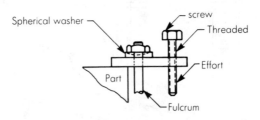

First class lever action

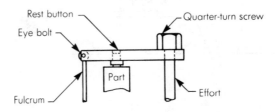

Second class lever action

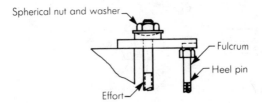

Third class lever action

Figure 4-74. Different classes of strap clamps.

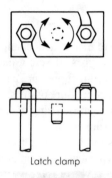

Latch clamp

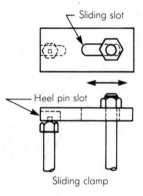

Sliding clamp

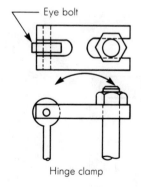

Hinge clamp

Figure 4-75. Variations of the strap clamp.

Screw Clamps

Screw clamps are a type of mechanical clamp that uses a screw thread to apply the holding force. The two types of screw clamps generally used for jig and fixture work are classified as direct pressure and indirect pressure clamps. Direct pressure clamps use the direct pressure of the screw thread to hold the workpiece. Typical examples are hook clamps, swing clamps, and quick-acting hand knobs (*Figure 4-76*). Indirect pressure clamps use the screw thread in combination with a secondary device to transmit the holding force of the thread. The arrangement shown in *Figure 4-77* is a typical application of an indirect pressure screw clamp. One additional benefit of using indirect pressure screw clamps is that the holding force can be magnified by simply increasing the leverage of the holding member. This will permit the holding force to be two, three, or more times greater than the actuating force.

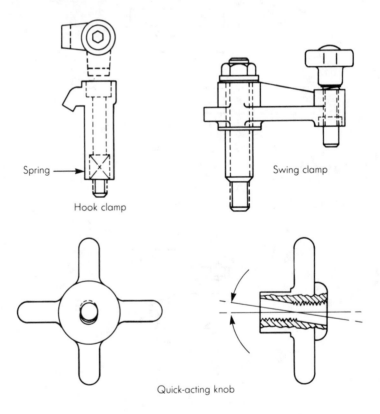

Spring

Hook clamp

Swing clamp

Quick-acting knob

Figure 4-76. Direct pressure screw clamps.

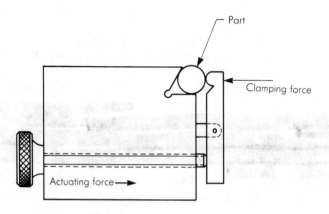

Figure 4-77. Indirect pressure clamp.

Cam Clamps

Cam action clamps are frequently used for fast-operating clamping devices. The three principle types of cam action clamps used for workholders are the flat eccentric cam, flat spiral cam, and the cylindrical cam (*Figure 4-78*). Flat eccentric cams operate on a high center principle and must be positioned exactly at the high center to hold properly. Flat spiral cams, on the other hand, have a locking range, which permits them to hold at any point within the range along the

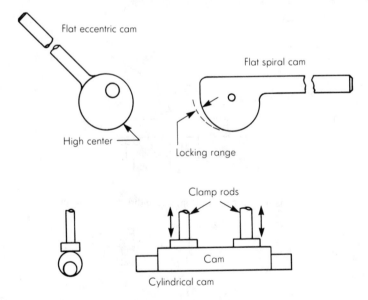

Figure 4-78. Cam clamps.

cam surface. Of these two styles, the flat spiral cam is the easiest and safest to use. All commercial cam action clamps are made with this flat spiral cam design.

Both flat eccentric and flat spiral cams can be used for direct pressure or indirect pressure applications (*Figure 4-79*). Indirect pressure is the most efficient and safest design for jig and fixture work. Direct pressure cam clamps have a tendency to loosen during machining. Since indirect pressure cams do not contact the workpiece, there is less chance for the vibration of the machining operation to loosen the cam.

Cylindrical cams are also used for workholding applications. With these clamps, the cam surface is generated on a cylindrical surface. *Figure 4-80* shows two applications where a cylindrical cam provides a good advantage.

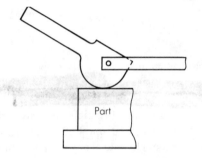

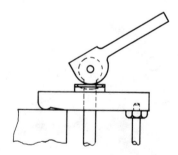

Figure 4-79. Spiral cams used for direct (top) and indirect (bottom) pressure.

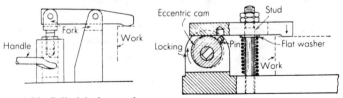

Figure 4-80. Cylindrical cam clamps.

Toggle Clamps

Toggle action clamps are commercially available clamps made with four general clamping motions: hold-down, push-pull, squeeze, and straight pull. The main advantages of using toggle clamps are their fast clamping and release actions, their ability to move completely clear of the workpiece, and their high ratio of holding force to actuation force. Several variations of toggle clamps are available to suit almost every workholding application. *Figure 4-81* shows several different styles of toggle clamps commonly used for jig and fixture work.

For all their advantages, standard toggle clamps have always caused problems because of their limited range of movement and inability to compensate for different thicknesses. Once set to a clamping height, the standard toggle clamp can only suit very slight changes in workpiece thicknesses. Larger variations

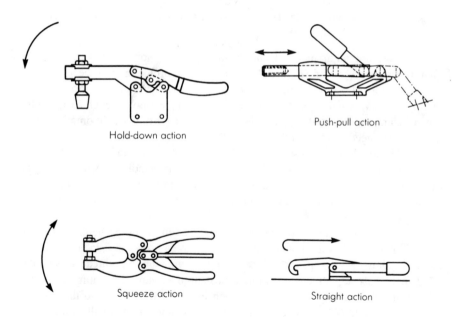

Hold-down action

Push-pull action

Squeeze action

Straight action

Figure 4-81. Toggle clamps.

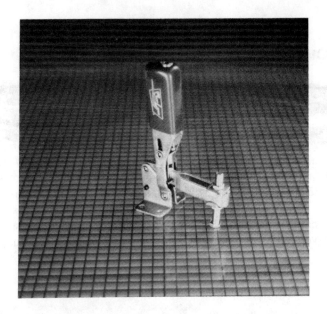

Figure 4-82. Automatic toggle clamp. (*Courtesy, Carr Lane Manufacturing Co.*)

usually require adjustment of the clamp spindle. Now, however, with the automatic toggle clamp (*Figure 4-82*), both these problems have been solved.

The standard toggle clamp design uses a "four-bar linkage" arrangement with fixed pivots and levers to produce the clamping action. Though adequate for many clamping operations, this design does not permit the automatic height adjustment needed for some parts. Rather than using only fixed points, the automatic toggle clamp uses a self-adjusting feature to readjust the clamp to different workpiece heights. With this clamp, the handle, normally a fixed component, has a variable length. The self-adjusting, self-locking wedge arrangement within the handle alters the pivot length to suit a variety of workpiece heights.

The clamp arm can accommodate differences in clamping heights of up to 15°. This results in a total automatic-adjusting range of over 1.25″ (31.8 mm). Added adjustment is permitted by manually moving the threaded spindle. Together, these adjustments result in a substantial amount of clamping capacity. To set up these clamps, the spindle extension must first be set to the average workpiece height (*Figure 4-83A*). Once set, the clamp automatically adjusts to a considerable range of workpiece variations (*Figure 4-83B* and *4-83C*).

Adjustments to the clamping force applied by the clamp are made by turning the screw located in the end of the handle. This adjustment is made with a standard screwdriver. The holding capacity of this clamp is 750 lbs. (3,336 N). The total movement of the vertical handle is 60° and the arm moves a total of 105° to permit easy workpiece loading and unloading.

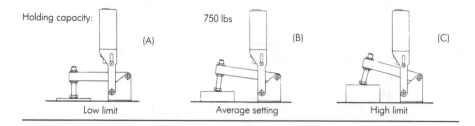

Figure 4-83. Automatic adjustment to workpiece variations. (*Courtesy, Carr Lane Manufacturing Co.*)

Wedge Action Clamp

Wedge action clamps use the basic principle of the inclined plane to securely hold and clamp a workpiece. The two principal types of wedge clamps are flat wedges and conical wedges. The flat wedge clamp (*Figure 4-84*) works as a flat cam to provide the holding force. As shown, the clamp is tightened and released by swinging the lever around the fulcrum pivot and contacting the inclined wedge against the spherical head pin. Conical wedges, or mandrels, are normally used in two styles, the solid mandrel and the expansion mandrel, (*Figure 4-85*). The wedging action of the conical surface directly or indirectly holds the workpiece.

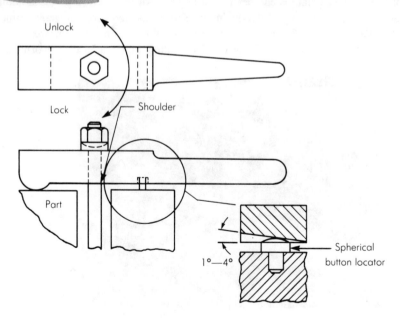

Figure 4-84. Flat wedge clamp.

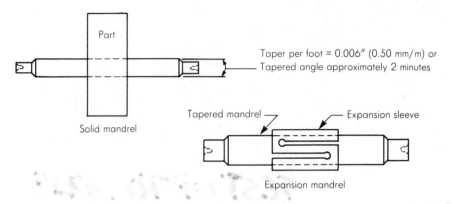

Part

Taper per foot = 0.006" (0.50 mm/m) or
Tapered angle approximately 2 minutes

Solid mandrel

Tapered mandrel

Expansion sleeve

Expansion mandrel

Figure 4-85. Conical wedges.

Specialty Clamps

In addition to the standard clamp variations, there is also a group of clamping devices that are not quite as easily classified. Rather than using one of the more common clamping methods, these clamps use a variety of nonstandard clamping actions. Incorporating the concept of a universal clamping system for today's multipurpose machine tools offers an almost endless range of possible applications. Unlike the limited purpose clamping devices usually found on machine tools today, these clamps can perform a majority of the necessary workholding operations while substantially reducing the setup time and part changeover time.

The Mono-Bloc™ clamp, as shown in *Figure 4-86*, is one of the newer clamp designs. It consists of a self-contained clamping unit that uses a worm and

Figure 4-86. Mono-Bloc™ clamp. (*Courtesy, Royal Products*)

worm-wheel arrangement to perform the clamping action (*Figure 4-87*). This allows the clamps to have a wide clamping range and complete adjustability. These clamps are available in three different types or grades. The most common is the standard-duty model, as shown. Light-duty and heavy-duty models are also available.

The standard-duty clamp is available with two different clamp arm lengths, and has an extension arm that can be installed on the basic clamp. The clamping range is increased by a series of available 3″ (76.2 mm) riser blocks, which are installed under the base of the clamp and can be used to build the clamp to a maximum height of 12″(304.8 mm). Both the basic clamp and the riser blocks are assembled with a single screw. This allows the clamp assembly to be swiveled to any angle necessary to clamp the workpiece.

The Terrific 30™ clamp (*Figure 4-88*) is another form of universal clamp. It is a retractable clamp that operates from a single point. By simply turning the

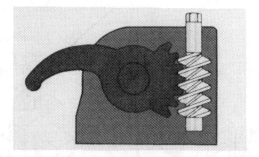

Figure 4-87. Worm and worm-wheel clamp arrangement. (*Courtesy, Royal Products*)

Figure 4-88. Terrific 30™ clamp. (*Courtesy, Royal Products*)

handle 1½ turns, both the extension and clamping action are achieved (*Figure 4-89*). Turning the handle in the opposite direction releases and retracts the clamp. This clamp may be mounted either directly to a machine table or on a series of riser elements. It also uses a single mounting screw, which allows the clamp to be positioned wherever necessary.

Edge clamping small or multiple parts may be difficult. Often, standard edge gripping clamps are too large to suit the workpiece or setup. This usually forces a design compromise. The workholder must either use the larger standard clamps, thus reducing the number of parts clamped, or expensive special purpose clamps must be custom made to hold the workpieces. An alternative is the MITEE-BITE™ clamping system (*Figure 4-90*).

This clamping system has the design advantages of space-saving edge clamps with the cost savings of standard clamps. These clamps combine the security of a screw clamp and the speed of a cam action clamp into a single clamping device. The main element is the eccentric socket head cap screw and hexagonal

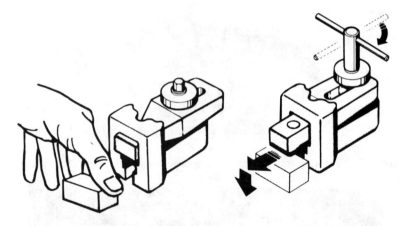

Figure 4-89. Extension of clamp arm to hold workpiece. (*Courtesy, Royal Products*)

Figure 4-90. MITEE-BITE™ clamping system. (*Courtesy, MITEE-BITE Products Co.*)

brass clamping element, shown in *Figure 4-91*. The cam-action clamping movement is achieved by turning the screw inside the clamping element. The hexagonal clamping element is made of brass so there is little chance of damaging the clamped workpiece. For special purpose applications, stainless steel clamping elements are also available. The cam-action of this system results in a significant mechanical advantage that can apply up to 4000 lbs. (17,792 N) of holding force directly to the workpiece.

These clamps may be mounted directly to the tool body, or in a tee nut for use on a machine table. For applications where the workpiece must be elevated off the mounting surface, a riser clamp variation (*Figure 4-92*) is also available. The riser clamp incorporates the benefits of the original basic clamp unit with the added flexibility of a small toe clamp. Rather than applying only lateral clamping force to hold the workpiece, the workpiece is held with both lateral and downward forces.

The basic clamps in the riser clamp set are mounted on a base element that permits the clamps to be mounted directly in the tee slots of the machine table and provides the 10° clamping angle of the clamping element. This base element also establishes the mounting surface that holds the workpiece 0.5″ (12.7 mm) off the machine table. The clamping element in the riser clamp is made of steel and has two clamping surfaces that can hold the workpiece. The first is a smooth surface that holds the workpiece without leaving clamp marks. The second is serrated to provide a better grip on the clamped surface. A clamp bar that works with the clamp to locate the opposite side of the workpiece during the clamping is included with the basic clamp. It also has an end stop unit mounted at the end of the bar (*Figure 4-93*). This stop can be set to suit the desired location of the workpiece.

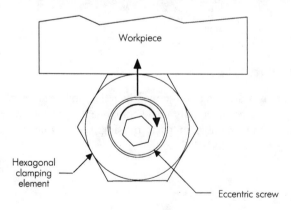

Figure 4-91. Eccentric socket head cap screw inside hexogonal clamping element. (*Courtesy, MITEE-BITE Products Co.*)

Figure 4-92. Riser clamp variation of MITEE-BITE clamp. (*Courtesy, MITEE-BITE Products Co.*)

Figure 4-93. Clamp bar with end stop unit.

Chucks and Vises

Chucks and vises are some of the most popular workholders used for jig and fixture work. These commercially made components allow a single workholder to service an infinite number of parts by simply changing or modifying the holding and locating surfaces of the tool.

Chucking Operations

Chucking operations can be distinguished by the location at which they are performed on the workpiece. Common methods shown in *Figure 4-94* include (A) external chucking, (B) internal chucking, (C, D, K, L) endwise chucking,

(*E, F, G, H*) holding or driving on centers, and (*J*) combinations of these methods. Although different workholders may be used, the differences are primarily in name and appearance rather than function or mechanical principles employed.

Motion is imparted to the workpiece either by friction or by positive means. In chucking, the jaw elements of the workholder bear on but do not positively engage the workpiece. The friction between the jaws and the workpiece rotates or drives the workpiece. Positive means for driving the workpiece can be provided by making use of its structure. Cutouts, keyways, gear teeth, or splines on a workpiece can be mated with matching elements on the workholder to give a positive drive. One distinguishing element can be found should failure occur: in a positive drive something breaks, while in a friction drive something slips.

Chucking Nomenclature. A chuck is a workholder generally used for gripping the outside or end of a workpiece, and it is usually attached to a machine tool spindle. An arbor is a workholder generally used for internal chucking, holding, or gripping, and it is usually attached to a machine tool spindle. A mandrel is a workholder that, like an arbor, is used for internal chucking, holding, or gripping. A mandrel is not generally as precise as an arbor, and is usually held between centers instead of being attached to the spindle.

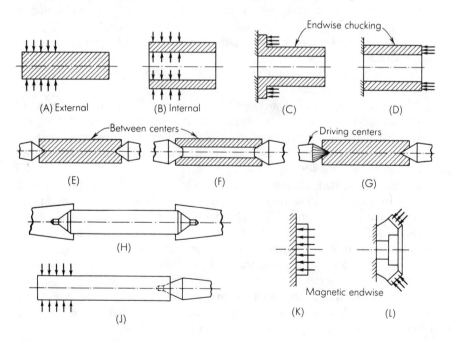

Figure 4-94. Location and clamping of round workpieces.

Range is the amount of variation in workpiece diameter that can be accommodated by workholder expansion and contraction. Interference is the amount by which the chucking diameter in its holding condition (without the workpiece) differs from the diameter of the workpiece to be held. Chucking area is the area of a workholder that can be used for holding. The holding elements of workholders are called by various names depending on the type of workholder. They are the jaws on some chucks; the fingers or prongs on some collets; the expanding wedges or inserts on some arbors; the expanding or contracting cylinders or sleeves on some special arbors and mandrels, and so on. These elements of the workholders are made to deflect by the actuating mechanism that transmits the force needed to operate the workholder. This force in turn may be applied manually, or by power (pneumatic, hydraulic, or electric). The holding elements are backed up by the actuating elements; these again are backed up by the operating mechanisms and forces.

Lathe Chucks. Lathe chucks consist of a body with inserted workholding jaws that slide radially in slots and are actuated by various mechanisms such as screws, scrolls, levers, and cams, alone and in a variety of combinations. The number of jaws varies. Chucks in which all the jaws move together are self-centering and are used primarily for round work. Two-jaw chucks operate somewhat like a vise, and may be used for round and for irregular-shaped workpieces by the use of suitably shaped jaws or jaw inserts.

The accuracy of a chuck deteriorates with usage owing to wear, dirt, and deformation caused by excessive tightening. Independent jaw chucks permit each jaw to move independently for chucking irregular-shaped workpieces or to center a round workpiece.

The jaws of most lathe chucks can be reversed to switch from external to internal chucking. Jaws may be adapted to fit workpiece shapes that are not round. The means of attaching a lathe chuck to different machine tools have been well standardized, so that chucks made by different manufacturers can be easily interchanged (*Figure 4-95*).

In addition to their standard jaws, lathe chucks may also be fitted with a variety of special purpose jaws to accommodate different types of workpiece surfaces and configurations. The principal types of chuck jaws used for these purposes are called *soft jaws* and are generally made of cast aluminum. The two standard forms of soft jaws are regular soft jaws and pie-type soft jaws (*Figure 4-96*). Regular soft jaws resemble standard jaws since they are made to the same width as the jaw carrier in the chuck. Pie-type soft jaws are much larger and completely cover the face of the chuck. In use, these jaws are first attached to the chuck, and the chuck is tightened. The desired form or shape is then machined into the soft jaws forming a type of partial or full nest. Once machined to a size slightly less than the part size, the normal operation of the chuck securely holds and locates the parts to be machined.

Figure 4-95. Lathe chucks.

Figure 4-96. Soft jaws. (*Courtesy, Carr Lane Manufacturing Co.*)

Another type of soft jaw sometimes used for large chucks uses an insertable element in either the top or end of the jaw (*Figure 4-97*). This arrangement reduces the cost of replacing the entire set of jaws and reduces the space required to stock and store machine jaw inserts. Since only the small insert is machined, the major body of the jaw can be used repeatedly for any number of different inserts. *Figure 4-98* shows a few examples of how these inserts may be machined to suit a variety of workpieces.

Solid Arbors and Mandrels. The solid, slender taper mandrel (*Figure 4-99*) is about the simplest possible workholder for round workpieces. Its main characteristic is a slightly tapered chucking surface with a taper of 0.004" to 0.006" per foot (0.33-0.50 mm/m). The workpiece diameter must be smaller than the largest diameter of the mandrel. The workpiece is forcibly pushed endwise onto the mandrel. This produces a gripping force around the hole in the workpiece, decreasing axially in relation to the interference produced between the outer diameter of the mandrel and the inner diameter of the workpiece. The driving torque that can be transmitted depends on the radial gripping forces and the tangential friction forces produced. The resistance against axial slipping depends on the axial friction forces produced.

It is not always easy to obtain the same driving power or to position the workpiece to a definite stop when trying to control the resulting interference between workpiece and mandrel. Pressing the workpiece on the mandrel requires an arbor press. The procedure is slow and may damage the finish of the workpiece bore and score the mandrel. Workpieces with accurate round and straight bores are held with great accuracy on this simple mandrel. If the bore is not round and straight, the workpiece and mandrel will mutually distort under the forces used to press on the workpiece.

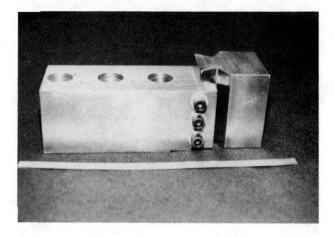

Figure 4-97. Insertable element for soft jaws. (*Courtesy, Starwood Enterprises, Inc.*)

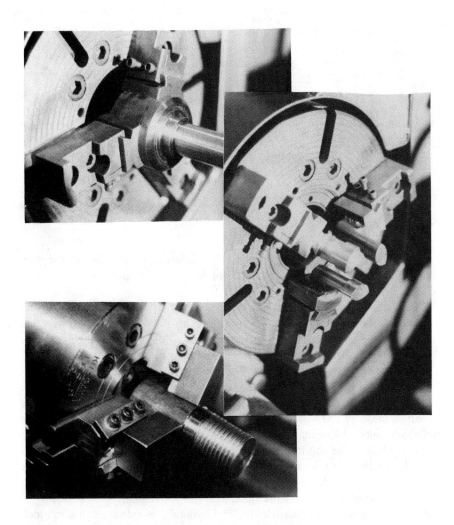

Figure 4-98. Inserts may be machined. (*Courtesy, Starwood Enterprises, Inc.*)

Straight Mandrel. This mandrel resembles the previous mandrel but has a straight (untapered) chucking area. To produce the required pressfit, the outer diameter of this mandrel is made larger than the bore of the workpiece by an amount called the *interference*. The amount of permissible interference is determined by the wall thickness, the diameter, and the material of the workpiece. It must be controlled to avoid exceeding the elastic limits of the

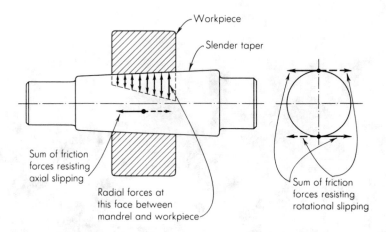

Figure 4-99. Solid mandrel.

workpiece and the mandrel. Exceeding such limits could produce a change in bore diameter, especially in materials of low strength and low modulus of elasticity.

The obtained driving torque together with the axial resistance to slipping depends on the interference and is not easy to control. Possible mandrel wear and damage to the workpiece offset the advantages of the simple low-cost design. The range of interference-fit mandrels is limited.

Combination of Slender Taper and Straight Mandrel. This mandrel has a short tapered chucking length followed by a straight length. The straight area fits snugly into the workpiece without interference, and helps prealign the bore. The tapered surface provides the driving area. To obtain reliable results for torque, accuracy, and axial location, hole tolerances must be carefully held. Like the taper and straight mandrels, this type also requires a pressfit condition between the workpiece and the chucking surface. The larger diameter of the mandrel is forced into the smaller bore of the workpiece. This axial pressing often damages the workpiece and workholder.

Solid Expanding Mandrels. The hydraulic mandrel and the roll-lock type of arbor (*Figure 4-100*) produce a shrinkfit by expanding tubular shells that are not split lengthwise. The internal actuating mechanism expands the shell into the workpiece bore, and the amount of pressfit produced can be controlled by stops limiting the amount of expansion. The hydraulic type expands by hydraulic pressure; the roll-lock type expands by the gradual rolling and wedging action of straight rollers between the tapered inner diameter of the shell and a tapered plug, which is turned by a wrench.

Mandrels with a chucking area that is not weakened by axial slots generally have only small ranges. The solid mandrel will permit a resulting interference of

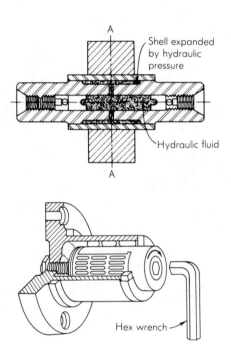

Figure 4-100. Elastic hydraulic workholding (top). Roll-lock expanding solid mandrel (below).

0.001-0.002″ per inch (0.001-0.002 mm per mm) of diameter. The hydraulic and roll-lock workholders expand from 0.002-0.003″ per inch (0.002-0.003 mm per mm) of diameter. The obtainable accuracy is approximately 10% to 20% of the range of the workholder. A mandrel expanding 0.002″ (0.05 mm) may then hold a round piece within a 0.0002-0.0004″ (0.005-0.01 mm) total indicator reading or 0.0001-0.0002″ (0.003-0.005 mm) eccentricity.

Split-Collet and Bushing Workholders. Solid expanding mandrels have a very small range. To increase range and to hold workpieces with larger diameter variations, split collets and bushings are used. These are basically slotted shells of various shapes. The slots permit greater flexibility to increase for internal chucking and to decrease for external chucking. The more flexibility that is provided, the greater the range will be. The shell is split, and acts as a spring (*Figure 4-101*). *Figure 4-102* shows a very popular collet in which the range is obtained by cantilever deflection. The cantilevers are produced by splitting the collet from one end only and leaving a solid ring on the other end. *Figure 4-103* shows a high-range collet where great flexibility is obtained by imbedding loose individual collet jaws in a suitable rubber compound. Many varieties of these chucking elements (*Figures 4-101, 4-102,* and *4-103*) form the

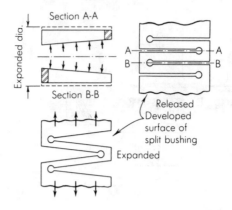

Figure 4-101. Expansion of a split bushing.

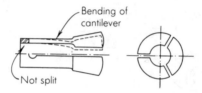

Figure 4-102. Split collet.

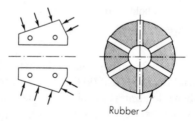

Figure 4-103. High-range type collet.

basis of various workholders. They are used for internal and external chucking, and are actuated in most cases by cones or tapers acting on corresponding surfaces of the chucking elements. Some workholders use small tapers which are self-locking since a force must be applied to disengage the mating tapers. A taper of less than 10° is usually self-locking (*Figure 4-104*).

Above 12° to 16°, the taper becomes self-releasing and requires force only to engage the mating tapers (*Figure 4-105*). Split collets and bushings, made with care, give satisfactory results for most applications. They are hard to maintain where extreme accuracies are required, and are vulnerable to the entry of dirt and to wear on the sliding surfaces.

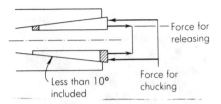

Figure 4-104. Collet with self-locking taper.

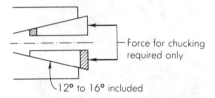

Figure 4-105. Collet with self-releasing taper.

Standard 5*C* collets, step collets, and step chucks may serve as workholders for a wide variety of parts. As shown in *Figure 4-106*, 5*C* collets are available in a wide range of sizes and standard shapes. In addition, unmachined, soft, emergency collets, step collets, and step chucks may also be machined to suit any particular part shape or configuration (*Figure 4-107*).

Axial Location. Moving collets provide axial location for workpieces. *Figure 4-108A* shows a collet chuck, the diameter of which contracts upon axial motion. *Figure 4-108B* shows a collet mandrel which expands upon axial

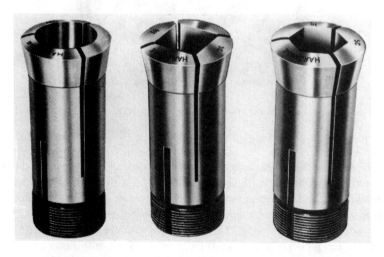

Figure 4-106. Collets are available in a wide range of sizes and standard shapes. (*Courtesy, Hardinge Brothers, Inc.*)

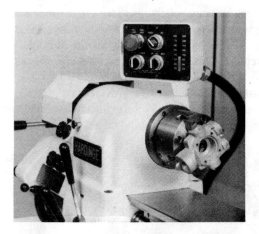

Figure 4-107. Some collets and chucks may be machined to suit part shapes and configurations.

motion. At first, the clearance between the workpiece diameter and the chucking diameter is closed; upon sufficient axial motion, a shrinkfit is created between the workholder and the workpiece. As soon as contact is made at the chucking surface, the workpiece has a tendency to move together with the collet relative to the workpiece holder body. This movement produces an axial shift of the workpiece and prevents precise axial location. *Figure 4-109* shows a method of obtaining precise location by providing a stop surface. The collet moves the workpiece toward the stop and pushes it firmly against the stop. Some slipping

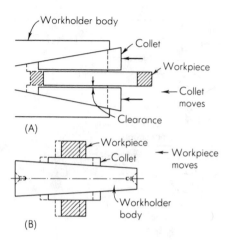

Figure 4-108. Axial location of workpieces as affected by collets.

occurs between the collet and workpiece during the chucking operation. The slippage tends to reduce the driving power of the workholder by absorbing part of the actuating force, and the mechanical efficiency of the chucking operation is thus lowered. To obtain accuracy, the stop surfaces of the workholder and faces of the workpiece must be square with the centerline. Lack of squareness may cause distortion of the arbor by one-sided loading against the stop surface (*Figure 4-110*).

Figure 4-111 shows a collet arrangement designed to eliminate axial shift by not moving the collet. An intermediate bushing interacts with the tapered surfaces of the collet. An additional element with additional mating surfaces between the two cylinders is thus introduced. The need to produce additional surfaces to a high dimensional and relational accuracy will lower the resultant final accuracy of the workholder. The more mating surfaces required in any workholder, the lower the accuracy tends to be.

Some frequently used workholders have sliding elements or inserts actuated by movement on inclined planes (*Figure 4-112*). They are similar to a lathe

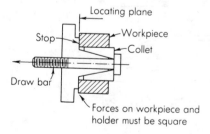

Figure 4-109. Collet with stop plate to ensure correct axial location.

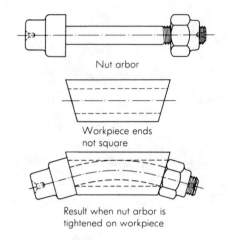

Nut arbor

Workpiece ends
not square

Result when nut arbor is
tightened on workpiece

Figure 4-110. Distortion caused by lack of squareness.

chuck in that their jaws or driving keys are held, guided, or moved in the workholder body. The actuating mechanism consists of a bar with inclined wedge surfaces. The body, jaws, and bar are three different elements working together. Accuracy depends on excellence of workmanship and careful use. A desirable simplification is the combination of the actuating mechanism with

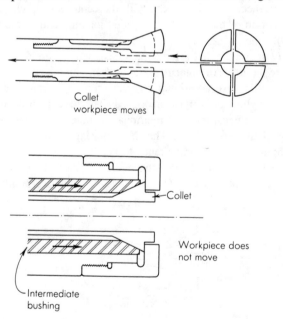

Collet
workpiece moves

Collet

Workpiece does
not move

Intermediate
bushing

Figure 4-111. Collet with intermediate bushing to eliminate axial shift.

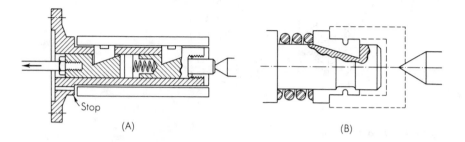

Figure 4-112. Collets for internal chucking.

the workholder body. These workholders are usually built as arbors and mandrels and have good range and gripping power. They are very useful tools for even the largest tubular workpieces.

Self-Actuating Wedge Cam and Wedge Roller Workholders. These workholders are tightened by the tangential work forces of the cutting tools. They actuate themselves once the holding elements are brought even slightly in contact with the workpiece.

Figure 4-113 shows an arbor using the principle of the roller clutch. One or more rollers are nested in cutouts in the body of the workholder, retained by wire clips. Turning the workpiece in the direction shown, relative to the workholder, wedges the rolls between the workpiece and the workholder. An increase in the applied cutting force increases the wedging action almost proportionally.

The wedge-cam chuck (*Figure 4-114*) has two inwardly spring-loaded cams. The jaws are lifted by a ring (not shown), then released until they touch the workpiece. The cam surfaces are usually serrated to obtain a better grip. The tangential cutting forces wedge the cam jaws tightly against the workpiece.

Vises

Vises are perhaps the most widely used and best-known workholders. All vises have in common one fixed and one movable jaw that hold a workpiece between them. In all other respects, such as configuration of the jaws to grip

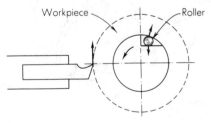

Figure 4-113. Roller-clutch type arbor.

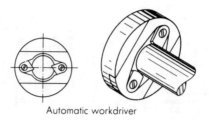

Automatic workdriver

Figure 4-114. Wedge-cam type chuck.

particular workpieces, means of actuation, means of mounting, ability to position the workpiece, and sizes, an endless variety is commercially available. A vise is a good basic workholder element. By reworking the jaws or making special jaws, and adding such details as locating pins, bushings, and plates, vises can be easily converted into efficient, specialized workholders.

Figure 4-115 shows a heavy plain vise for holding flat workpieces. With V-shaped grooves in the jaws, it can locate round workpieces such as bars and tubing. *Figure 4-116* shows a vise mounted on a rotary base permitting angular rotation, positioning, and locking as may be required. *Figure 4-117* shows a universal vise that can position a workpiece at any angle relative to the cutting tool.

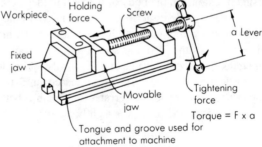

Figure 4-115. Elementary workholder (vise).

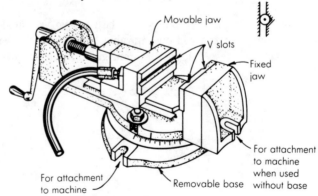

Figure 4-116. Vise with hydraulic clamping.

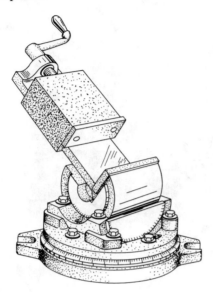

Figure 4-117. Universal-type vise.

Figure 4-118 shows a workholder using a vise with a quick-acting jaw that is moved rapidly into contact with the workpiece. A handle-operated cam then locks the movable jaw to produce great holding force. A special plate attached to the vise precisely positions three guide bushings in relation to the fixed (locator) jaw of the vise.

Another variation of this type of vise is shown in *Figure 4-119*. The solid jaw is made with drill bushing mounted in movable arms. An end stop is also incorporated into the solid jaw. This type of vise is adaptable for many different types and shapes of workpieces.

Figure 4-120 shows a vise that is particularly well-suited for jig and fixture work. As shown, the three gripping areas of this vise are completely movable and can be locked to accommodate almost any part shape within the range of the vise.

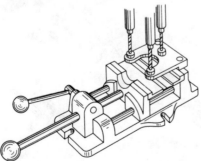

Figure 4-118. Drill jig with quick-acting vise as component.

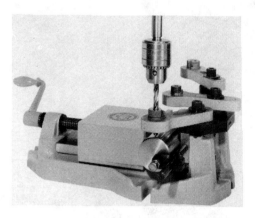

Figure 4-119. A special vise with drill bushings in movable arms. (*Courtesy, Universal Vise and Tool Co.*)

Another form of special purpose vise is the vertical vise (*Figure 4-121*). This vise is very useful for applications where a standard milling machine vise cannot provide adequate support. The height of the workpiece can be regulated by the position of the stop bar (*Figure 4-122*). These bars are movable in half-inch increments for the complete width of the vise jaw and are made in sets to match the range of openings for the vise.

Another style of vise ideally suited for a variety of different size workpieces is the Bi-Lok Machine Vise™ (*Figure 4-123*). This vise is designed to hold either identical parts or two completely different sized parts (*Figure 4-124*). The vise mechanism is totally self-compensating, and the same pressure is applied on both gripping positions regardless of the part sizes.

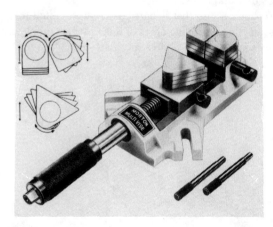

Figure 4-120. Vise with three movable gripping areas. (*Courtesy, James Morton*)

Figure 4-121. Vertical vise. (*Courtesy, Mid-State Machine Products, Inc.*)

Figure 4-122. Regulating workpiece height using a stop bar. (*Courtesy, Mid-State Machine Products, Inc.*)

The same operating principle used for the Bi-Lok vise is also applied to the Multi-Lok Vise™ (*Figure 4-125*). A group of vises are mounted together in units of 2, 4, 8, or more to hold a series of identical parts, or a variety of different parts. This unit may be used for either horizontal or vertical machining operations. The individual jaws on the Multi-Lok are also removable, and may be replaced with soft jaws machined to suit specific part shapes.

Despite their many applications most milling machine vises cannot hold odd shapes or multiple workpieces. These parts often require expensive dedicated

Figure 4-123. Bi-Lok™ machine vise. (*Courtesy, Chick Machine Co.*)

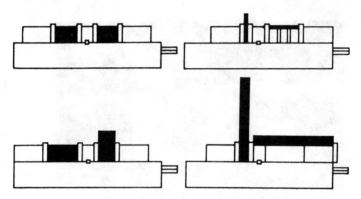

Figure 4-124. Vise mechanism self-compensates to apply equal pressure regardless of part size. (*Courtesy, Chick Machine Co.*)

fixturing. The TWIN-LOCK™ Workholding System (*Figure 4-126*) corrects the problems of conventional vise designs and reduces the need for dedicated fixtures.

As shown, this unit consists of two moving blocks that replace the movable jaw on any standard angle locking-type vise. These moving blocks can hold either single or multiple parts and have the unique ability of applying the holding forces in both horizontal and vertical directions. *Figure 4-127* shows several of the setup possibilities of this jaw unit. The setup shown (1) illustrates how this system holds a rectangular part. Two parts can also be held (2). A removable stop may also be positioned on the solid jaw to act as a reference point for mounting both parts. If the part has two different clamping surfaces (3), the jaws

Figure 4-125. Multi-Lok™ vise. (*Courtesy, Chick Machine Co.*)

will accommodate the variations. The jaws may also be used to hold round parts (4). Each of the jaw elements is keyed to the moving block, and the side of one jaw element may be used to establish a perpendicular mounting plane for smaller parts (5).

When the clamping elements are mounted to the pistons, a completely new set of clamping options are available. Mounting a set of swivel jaws to the piston allows this unit to hold both tapered and larger round parts (6 and 7). Odd shapes and contours may also be held by simply using a set of soft swivel jaws made to conform to the part shape (8). The setup shown (9) is very useful for machining

Figure 4-126. Twin Lock™ workholding system. (*Courtesy, Twin Lock Tool Co.*)

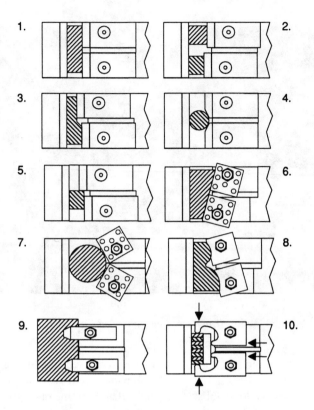

Figure 4-127. Possible setups. (*Courtesy, Twin Lock Tool Co.*)

three sides of a part in one setup. A set of strap clamps are mounted in the pistons to hold the part. Even setups involving multiple small parts (10) are easily performed using another set of swivel jaws. Clamping action is obtained by inserting an anvil plate against the sides of the parts. As the vise is tightened, the swivel jaws contact the anvil plate and push it against the parts. As the pressure increases, the swivel jaws pivot and apply the necessary holding force to both ends of the stacked parts.

Special Vise Jaws. The usefulness of vise-like workholders can be enhanced by making special cast and molded jaws, which adapt to the holding and locating of complex workpieces. There are two methods for making such special jaws.

Figure 4-128 shows the method of casting jaws for holding a nipple by pouring a low-temperature alloy around the workpiece, which is used as the pattern. Two wooden spacers locate the pattern and separate the cast jaw halves. The pattern is coated with a releasing agent for easy removal from the cast jaws. These jaws are then attached to the vise jaws.

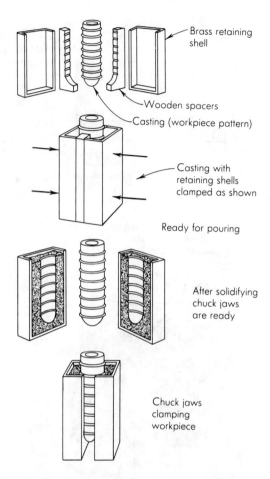

Figure 4-128. Casting the vise jaws with low-temperature alloy.

In the second method (*Figure 4-129*), a plastic material is used. The material has a metal filler to give it more wear resistance and strength. The plastic is of putty-like consistency, and is placed on each jaw of the vise. The workpiece, a *T* fitting in the example shown, again acts as the pattern. Coated with a releasing agent, it is located and pressed into the plastic material on the two sides of the vise by closing the vise jaws. The plastic hardens within two hours to form two precise half impressions of the workpiece, which make an excellent locating and holding arrangement. The plastic material may also be used to locate and secure pins, bushings, and other details used in workholders.

Independent Vise Jaws. Occasionally, independent vise jaws may be used as workholders for large or odd-shaped parts. As shown in *Figure 4-130*, these workholders are available in two general styles. The bigger of the two, at *A*, is very useful for larger parts, while those shown at *B* are well-suited for edge

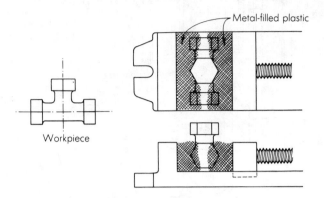

Figure 4-129. Plastic vise jaws.

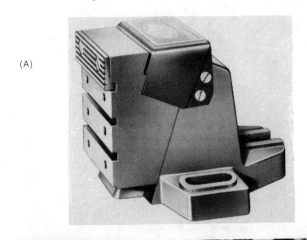

(A)

(B)

Figure 4-130. Independent vise jaws. (*Courtesy, De-Sta-Co Products*)

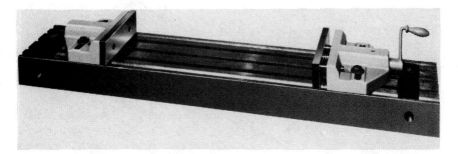

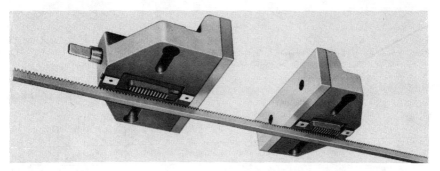

Figure 4-131. Adjustable-range vise. (*Courtesy, Universal Vise and Tool Co.*)

gripping on thinner parts. A third style of independent vise jaws is shown in *Figure 4-131*. This style of vise is actually an adjustable range vise. Its jaws may be set at any convenient distance and held in place with the rack gear. This style of vise is very useful for machines with T-slots in the table since the rack must be located below the working surface of the vise.

Nonmechanical Clamping

Sometimes it is impractical to hold a workpiece by direct clamping pressure because of possible distortion, or because of the size of the workpiece. Magnetic, vacuum, and electrostatic workholders may be of value in such cases.

Magnetic Chucks. Magnetic chucks are available in a variety of shapes. They can hold only ferrous workpieces unless intermediate mechanical workholders permit the holding of workpieces made of nonmagnetic material. Magnetic chucks are suitable for light machining operations such as grinding. Strongly magnetic materials and better utilization of magnetic force permit their use for heavier operations such as light milling and turning. Magnetic chucks can be operated by permanent magnets or by electromagnets powered by direct current. The gripping power attainable depends on the strength of the magnets and the amount of magnetic flux that can be directed through the workpiece. *Figures 4-132* and *4-133* illustrate various magnetic chucks.

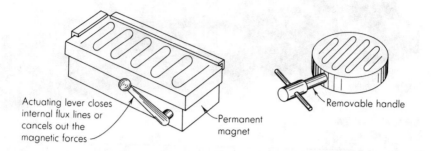

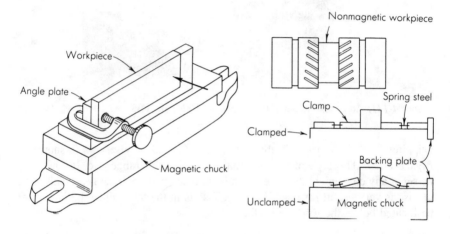

Figure 4-132. Magnetic chucks.

Figure 4-133. Magnetic chuck and angle plate used as workholder.

A magnetic chuck is fast acting and, by holding a large surface of the workpiece, causes a minimum of distortion. Magnetic chucks are available in rectangular shape, in circular shape as a rotary chuck, and as a V block. Magnetic chucks impart some residual magnetism to workpieces. This must be removed by demagnetizing if it would interfere with proper functioning of the workpiece.

Vacuum Chucking. Quite often workpieces of nonmagnetic materials or of special shapes and dimensions must be securely held flat without any mechanical clamping. An example might be a large flat plate. In such cases, vacuum chucking may be the only practical holding method.

The basic principle of vacuum chucking is very simple. On each square inch of a surface, a pressure of approximately 14.7 lb./in. (101.4 kPa) is exerted by the atmosphere. This represents approximately 144 x 14.7 = 2116.8 lb./ft.

(101.4 kPa). Part of this pressure is utilized by creating a vacuum in a closed chamber made up of the locating surface of the workpiece and the mating surface of the workholder. At first, the pressures on the outside and inside of the chamber are equal. As a vacuum is produced inside the chamber, the outside pressure holds the workpiece against the locating surface (*Figure 4-134*). *Figure 4-135* shows a typical vacuum chuck. An O ring seal laid in a groove around the chucking area creates the closed chamber to be evacuated. Holes and a grid of small connecting channels in the chucking surface assist in the speedy creation of the needed vacuum.

Electrostatic Chucks. The attraction of electrically opposite charged parts can hold flat or flat-sided workpieces that cannot be magnetized. Electrostatic chucks (*Figure 4-136*) can hold any electrically conductive material. Glass, ceramics, and plastics also may be held by flash metal plating on one flat side to provide a suitable electrical contact.

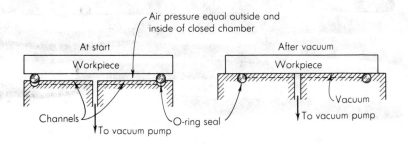

Figure 4-134. Vacuum-chucking principle.

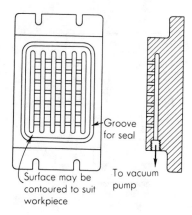

Figure 4-135. Vacuum chuck.

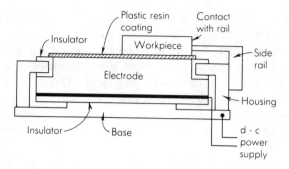

Figure 4-136. Electrostatic chuck.

Power Clamping

Power clamping devices are frequently used for applications where speed and uniform clamping pressures are important considerations. *Figure 4-137* shows how a typical power clamping system is constructed. Power clamps are normally operated by hydraulic pressure, pneumatic pressure, or a combination of both. *Figure 4-138* shows several typical setups for power clamping.

One problem in using any power clamping system is the basic law of "no hydraulic pressure—no clamping pressure." If any clamping element leaks, or

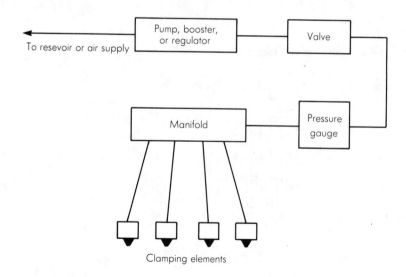

Figure 4-137. Power clamping.

Figure 4-138. Typical setups for power clamping. (*Courtesy, Jergens, Inc.*)

is disconnected, pressure is lost and the clamps loosen. One system of components that uses a mechanical locking principle in combination with a pressurized hydraulic system is the Stay Lock™ clamping system (*Figure 4-139*). These clamps are built using a mechanical lock that is activated by hydraulic pressure. Once locked, the hydraulic pressure and hoses may be removed and the clamps stay locked firmly against the part.

Figure 4-139. Stay Lock™ clamping system. (*Courtesy, Jergens, Inc.*)

The locking action of this clamping system is similar to a wedge action clamp. Working on an inclined plane principle, this locking action operates in much the same way as does a tapered shank drill. The basic operation of this clamp is shown in *Figure 4-140*. This drawing shows the basic wedge lock arrangement and principle behind this clamping concept. In use, the piston is driven to one side of the clamp body by hydraulic pressure. The locking angles on the piston and locking pin engage and lock together. This provides a

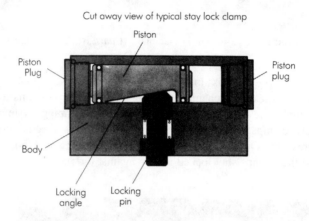

Figure 4-140. Locking action. (*Courtesy, Jergens, Inc.*)

mechanical lock. Thus, once the clamp is activated, the hydraulic connections can be removed and the clamp will not loosen. The clamp is released by switching the hydraulic hoses to the release port. The pressure then drives the piston back off the locking pin.

Multiple Part Clamping

When designing tools for large production runs, it is often desirable to machine more than one part at a time. In these cases, jigs and fixtures should be designed to permit multiple part clamping. The two principal points to remember when designing workholders for multiple clamping are: (1) the clamping pressure must be equal on all parts, and (2) the tool should have a minimum number of operating points. As shown in *Figure 4-141*, a little forethought and ingenuity make it possible to clamp almost any part in a multiple part workholder.

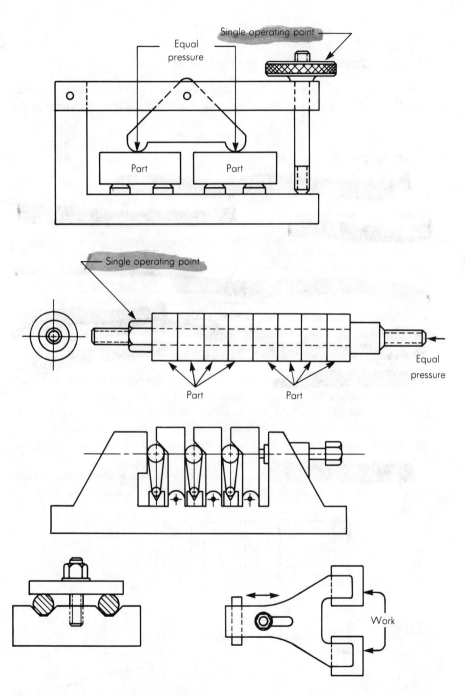

Figure 4-141. Multiple part clamping.

Basic Construction Principles

Every jig and fixture must be constructed properly to function as intended. The degree of accuracy and durability of the workholder is directly related to the way the tool is constructed.

Tool Bodies

Tool bodies are the major element of jigs and fixtures. They form the general size and shape of the tool as well as providing the mounting surfaces for locators, supports, and workholding devices. The three principal types of tool bodies are cast, welded, and built-up (*Figure 4-142*).

Cast Tool Bodies. Cast tool bodies are normally made from cast aluminum, cast iron, or cast epoxy resin. These tool bodies are well suited for permanent, high-volume production workholders. The principal advantages in using cast tool bodies include: stability and vibration dampening, good material distribution, and reductions in machining time. But cast tool bodies normally cost more to make since a pattern must be made for each tool. Another disadvantage of cast tool bodies is the long lead time between design and fabrication of the tool.

Welded Tool Bodies. Welded tool bodies are normally made from any weldable materials, such as steel, aluminum, magnesium, etc. This type of tool body can be easily fabricated with minimum lead time. Welded tool bodies are strong, rigid, and easily modified when required. The major disadvantage of this type of tool body is the additional machining time required to repair and remove the heat-distorted areas of the tool.

Built-Up Tool Bodies. Built-up tool bodies are made from a wide variety of different materials. Steel, aluminum, cast iron, wood, and epoxy resins are

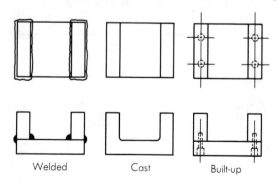

Welded   Cast   Built-up

Figure 4-142. Three types of tool bodies.

commonly used for built-up tool bodies. This is the most versatile type, and the most frequently used for general jig and fixture work. Built-up tool bodies are inexpensive, easily modified, and require less machining after assembly than do cast or welded tool bodies. As shown, this type of tool body normally uses dowel pins and socket head cap screws to align and fasten each tool member.

Tooling Materials

In addition to fabricating tool bodies, many different tooling materials are used for locators, supports, or other jig and fixture elements. To keep the cost of a tool to a minimum, the tool designer should be familiar with the wide variety of preformed materials available.

Precision Ground Materials. Precision ground materials are available in many shapes and sizes. The two primary variations are ground flat stock and drill rod. These materials are commercially available in either standard sized or oversized, in lengths of 18-36″ (450-900 mm). Precision ground materials are normally available in low-carbon steel, high-carbon steel, oil-hardening tool steel, and air-hardening tool steel. The tolerance values of these materials vary with the size and shape of the burr, but they are normally ground to within a few thousandths of an inch accuracy.

Aluminum Tooling Plate. Aluminum tooling plate, like precision ground materials, is available in many sizes and shapes. The major advantage to using this material is the close tolerance of the finished stock. Many times, tools assembled from these materials require no finish machining prior to use.

Precast Bracket Material. Precast bracket materials are commercially available in several shapes and sizes as shown in *Figure 4-143*. These materials are generally made from either cast iron or cast aluminum and are available in

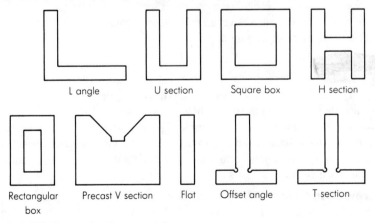

Figure 4-143. Precast bracket materials.

lengths of 18-36″ (450-900 mm). Precast bracket materials are frequently used for locators, supports, or complete tool bodies. The principal advantage to using these sections is the ease with which tools can be made using the convenient forms and shapes available.

Structural Sections. Structural sections are another type of material sometimes used for workholder components. These sections are available in several different shapes and in lengths up to 16 feet (4.5 m) (*Figure 4-144*).

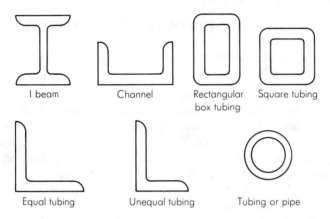

Figure 4-144. Structural sections.

This makes structural sections very useful for large or oversized workholders. The most common material used for these structural sections is low-carbon steel, but sections made from aluminum or magnesium are also available. Structural sections are normally made by rolling and are not made to close tolerance specifications. For this reason, these sections must be machined to suit close tolerance applications. Nevertheless, these sections are very versatile and will suit a wide range of tooling applications.

Fasteners

Almost every jig and fixture uses some type of fastener to align and hold the various members and components in their proper position and relationship. The specific type of fastener used to assemble a particular workholder is normally determined by the joint and the required tolerance of the assembled parts. The following are a few fasteners commonly used for jig and fixture work, and the applications where each is best suited.

Screws and Bolts. Screws and bolts are the most common types of fastening devices used to construct assembled workholders. The principal difference between these fasteners is the method used to hold the assembled parts together. Screws are normally driven by their heads and are turned into threaded holes. Bolts, on the other hand, are installed through holes and are held in place with a nut.

The most common type of screw used for jig and fixture work is the socket head cap screw (*Figure 4-145*). In addition, variations of the socket head cap screw (*Figure 4-146*) are also frequently used to assemble workholders. The principal type of bolt used with jigs and fixtures is the T-bolt (*Figure 4-147*).

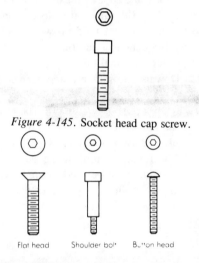

Figure 4-145. Socket head cap screw.

Flat head   Shoulder bolt   Button head

Figure 4-146. Variations of the socket head cap screw.

Figure 4-147. T-bolt. (*Courtesy, Jergens, Inc.*)

These bolts primarily secure the workholder to the machine table. Other types of bolts that are also used occasionally are hex heads, square heads, and studs.

Nuts, Washers, and Inserts. Nuts, washers, and inserts are also used for some workholder designs. The most commonly used nuts are hardened hex nuts, check nuts, T-slot nuts, acorn nuts, flange nuts, and coupling nuts (*Figure 4-148*). The most frequently used washers are plain flat washers, self-aligning washer sets, C-washers, and swinging C-washers (*Figure 4-149*). Thread inserts (*Figure 4-150*) are also used for some applications. These inserts are available in several styles, and serve basically the same purpose—to provide a replaceable thread in an area where threads will not hold up well, such as plastic or soft metals, or in places where frequent installation and removal of the threaded fastener would rapidly wear out a normal thread.

Dowels and Pins. Dowels and pins are another form of fastener often found in workholders. These fasteners, unlike screws and bolts, are intended to align assembled components rather than secure them. The most common type of pin used for alignment is the hardened steel dowel pin (*Figure 4-151*). Other variations include pull dowels, for use in blind holes, and roll pins. Roll pins are frequently used in place of dowel pins to eliminate the need to ream the mounting holes. However, these pins should only be used in areas where the assembled tolerance will permit some variation in the position of the assembled members.

(A)

(B)

(C)

(D)

Figure 4-148. (A) hardened hex nut (B) T-slot nut (C) acorn nut (D) spherical flange nut. (*Courtesy, Jergens, Inc.*)

(A)

(B)

(C)

Figure 4-149. (A) swinging "C" washer (B) flat washer (C) "C" washer. (*Courtesy, Jergens, Inc.*)

Figure 4-150. Thread insert. (*Courtesy, Jergens, Inc.*)

Figure 4-151. Hardened steel dowel pin. (*Courtesy, Jergens, Inc.*)

Retaining Rings. Retaining rings (*Figure 4-152*) are normally used for fastening in areas where there is a minimum amount of thrust against the fastener. These fasteners are normally faster and easier to install than standard threaded fasteners and will work equally well for both internal and external applications.

Fixture Keys. Fixture keys, while not actually a fastening device, are an integral part of many workholders. The two principal styles of fixture keys used for jigs and fixtures are the plain, or standard, type and the interchangeable type (*Figure 4-153*). Generally, the interchangeable style fixture key is easier to use and permits a single set of keys to be used for several workholders. However, the standard keys are very well suited for permanent installations.

Ball Lock™ Mounting System. Many different methods are used in mounting workholders to machine tools during machining. The main problem has always been finding a way to reduce the needed setup time. One method of reducing this time is use of the Ball Lock Mounting System. These units accurately position and securely hold the workholder on the machine table. As shown in *Figure 4-154*, this mounting system has three major parts, a flanged shank, a liner bushing, and a Ball Lock bushing. In use, the bushings are installed in both the workholder base and the machine base or mounting subplate. The flanged shank is then installed through the liner bushing into the bushing and is tightened down using the socket head cap screw (*Figure 4-155*).

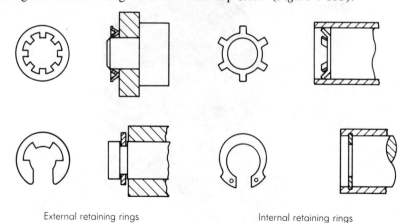

External retaining rings   Internal retaining rings

Figure 4-152. A variety of retaining rings.

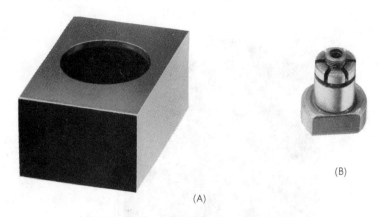

Figure 4-153. (*A*) square fixture key (*B*) single fixture key. (*Courtesy, Jergens, Inc.*)

The operation of the Ball Lock unit is quite simple. As the socket head cap screw in the head of the flanged shank is tightened, a large ball is forced against three smaller balls. This action moves the three locking balls outward in their sockets against the conical sides of the Ball Lock bushing. When the three balls contact the conical area of the Ball Lock bushing they both center the flanged shank and apply a downward force against the shank, locking it against the fixture base. Rather than locating on the cylindrical diameter of the shank, the actual location is achieved by the locking balls against the conical area of the Ball Lock bushing.

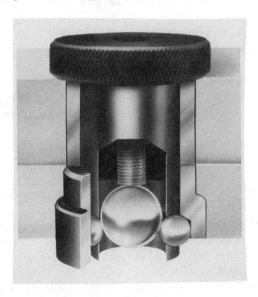

Figure 4-154. Ball Lock™ mounting system. (*Courtesy, Jergens, Inc.*)

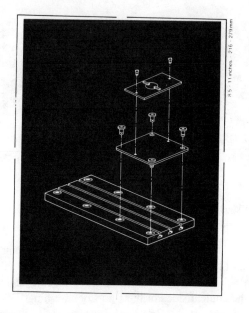

Figure 4-155. Installation of bushings and flanged shank. (*Courtesy, Jergens, Inc.*)

WORKHOLDING PRINCIPLES

Review Questions

1. What is the purpose and function of the workholder?
2. Describe the degrees of freedom of a workpiece located in space.
3. Design a 3-2-1 locating system to locate and support the part shown in *Figure A*.
4. What is the advantage of using assembled locators over integral locators?
5. What are the major disadvantages of a nest type locator?
6. What is the main function of a relieved locator?
7. When designing a workholder, what areas must be designed to resist the major thrust of the cutting force?
8. Design a two-pin locator for the part shown in *Figure B*. Specify the pin sizes using the dimensions shown.
9. Design a set of vise jaws to hold the part shown in *Figure C* for the drilling operation indicated.
10. Design a workholder to turn the outside diameter of six of the parts shown in *Figure D*.

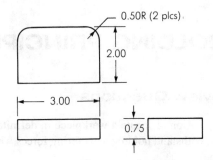

Figure A.

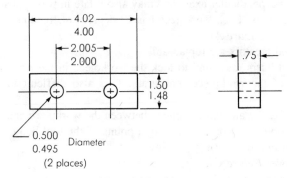

Material: Naval brass

Figure B.

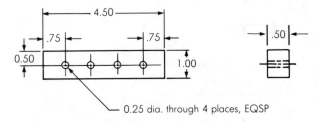

Material: 7075-T6 Alum.

Figure C.

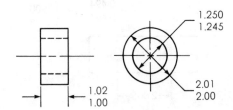

Figure D.

WORKHOLDING PRINCIPLES

Answers to Review Questions

1. A workholder must position or locate a workpiece in definite relation to the cutting tool and must withstand holding and cutting forces while maintaining that precise location.
2. A workpiece can move in two opposite directions along each of three mutually perpendicular axes, and may also rotate in two opposite directions about each axis. Each workpiece has 12 degrees of freedom.
3. (See *Answer Figure A*).
4. Assembled locators are replaceable.
5. Chips and burrs may tend to lock the workpiece in the nest and make the location of the workpiece inaccurate. It is also difficult to remove nest locators.
6. To minimize the amount of contact between the workpiece and locating pin.
7. Locators and the strongest nest rigid points of the tool body.
8. (See *Answer Figure B*).
9. (See *Answer Figure C*).
10. (See *Answer Figure D*).

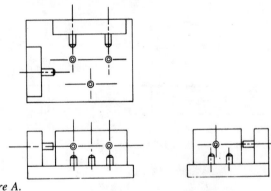

Answer Figure A.

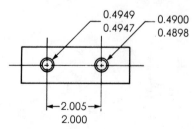

Answer Figure B.

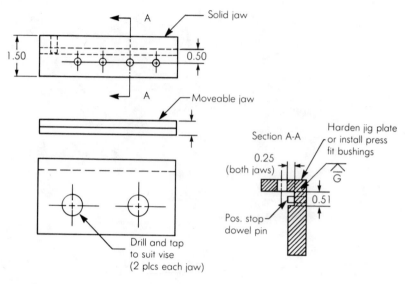

A — Solid jaw

1.50

0.50

A — Moveable jaw

Section A-A

Harden jig plate or install press fit bushings

0.25 (both jaws)

G

0.51

Pos. stop dowel pin

Drill and tap to suit vise (2 plcs each jaw)

Answer Figure C.

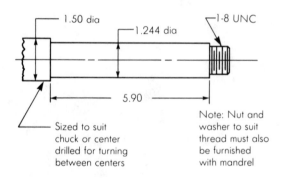

1.50 dia

1.244 dia

1-8 UNC

5.90

Sized to suit chuck or center drilled for turning between centers

Note: Nut and washer to suit thread must also be furnished with mandrel

Answer Figure D.

5

JIG DESIGN

Jigs are workholders which are designed to hold, locate, and support a workpiece while guiding the cutting tool throughout its cutting cycle. Jigs can be divided into two general classifications: drill jigs and boring jigs. Of these, drill jigs are, by far, the most common. Drill jigs are generally used for drilling, tapping, and reaming, but may also be used for countersinking, counterboring, chamfering, and spotfacing. Boring jigs, on the other hand, are normally used exclusively for boring holes to a precise, predetermined size. The basic design of both classes of jigs is essentially the same. The only major difference is that boring jigs are normally fitted with a pilot bushing or bearing to support the outer end of the boring bar during the machining operation.

In designing any jig, there are numerous considerations that must be addressed. Although several of these points, such as locating, supporting, and clamping, have already been covered, they are included in this section because they apply to jig design. Since all jigs have a similar construction, the points covered for one type of jig normally apply to the other types as well. Jig design and selection begins with an analysis of the workpiece and the manufacturing operation to be performed.

General Considerations

One of the first considerations in the design of any workholder is the relative balance between the cost of the tool and the expected benefits of using the tool for production. All workholders should save more in production costs than the tool costs to design and construct. In many instances, tool designers may have to complete detailed estimates to justify the cost of special tooling. This involves a close look at the part drawing, process specifications, and other related documents.

Typically, the complexity of the part, location and number of holes, required accuracy, and the number of parts to be made are all points which must be considered to determine if the cost of a particular jig is warranted. Once the tool designer is satisfied that the cost of special tooling is justified, the remaining data required to produce a suitable workholder is compiled and analyzed.

Machine Considerations

The size, type, and capabilities of the machine tool specified for a particular operation have a direct bearing on how a particular jig should be made. For example, a jig intended for small holes drilled on a sensitive drill press will not normally have the same features as a jig designed for use with a larger radial drill press. While both may share many details, the machines they are used with, as well as the specific tooling required, will generally dictate the overall construction of the workholder.

Often, the tool designer is not the individual who specifies the particular machine tool for a machining operation. In most instances the process engineer is charged with this responsibility. However, the tool designer should check the specified machines to insure they are the best possible choices. The following are a few machine considerations to keep in mind when designing any jig.

1. The machine should be large enough and rigid enough to perform the desired operations.
2. The production capabilities and accuracy of the machine tool should be suitable for the operation.
3. The machine tool must be able to safely accommodate the workholding device.
4. The machine selected should be close to the machines which are to be used for subsequent machining operations. Here, the shop layout is important to minimize the distance between machining stations and to prevent backtracking across the shop.
5. The specified cutting tools must be compatible with the machine tool. A lathe with a #2 Morse taper in the tailstock spindle cannot hold #3 Morse taper tools without an adapter.
6. Whenever possible, standard size cutting tools should be specified to reduce costs and simplify procurement of drills, reamers, and other cutting tools.

When possible, each tool designer should have all relevant specifications for each type of machine tool used for part production. This information should include table sizes, T-slot sizes, machine travel in all axes, and similar data. Such information is normally available in the maintenance manuals that are furnished with each machine. The designer could make the required measurements directly on the machine tool. A similar specification sheet should be maintained for each standard cutter normally used in the shop. This information is available from the cutter manufacturer and should include the length of the tool, length of effective cutting area, shank size, and similar data.

Process Considerations

Process considerations deal with the actual processes used to produce the required workpieces. Here the type of jig, number of jigs needed, and the specific step-by-step processing of the workpiece must be determined. While the proposed processing of a part is normally a function of process engineering, the tool designer should always double check the processing to insure the proposed tooling will be compatible with the process selected.

For workpieces without prior machining or any reliable reference surface, first-operation jigs are normally used. Here the first holes are put into the part and act as a reference point for any subsequent machining. First-operation jigs generally use adjustable supports, adjustable locators, or sight locators to set the initial position of the part to machine the first holes.

When several different operations must be performed on the same part, more than one jig may be used. The most important point to consider is the repeatability of the location in each of the jigs. In these cases, the same location point should be used for all machining operations on the part.

Other process considerations which should be analyzed before a final design is determined include the actual processing methods and chip control and disposal. Several parts may, in some cases, be stacked or aligned so more than one part can be drilled at a time. Likewise, when large drills must be used to produce the required size hole, smaller step drills can be used to lessen the torque required to drill the holes. Here a smaller or lighter-duty drill press must be used if a drill press of the proper size is unavailable. Chip control and disposal are considerations which must be remembered when designing a workholder. Chips and coolant are normally removed with a brush or air flow, so adequate slots or open areas should be provided to permit easier removal (*Figure 5-1*).

Developing the Preliminary Jig Design

The following represents a systematic approach to the design of a jig for the part shown in *Figure 5-2*. These considerations and their effects on the jig design are intended to illustrate, in a tabular form, the step-by-step process of

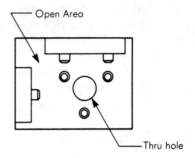

Figure 5-1. Areas to permit easy coolant
and chip expulsion.

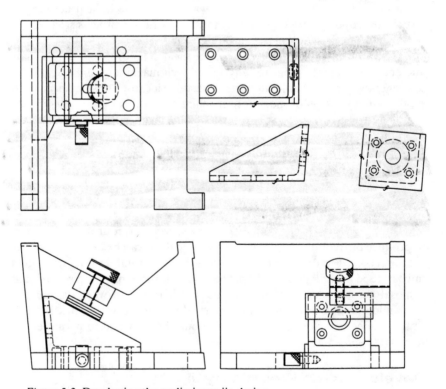

Figure 5-2. Developing the preliminary jig design.

evaluating each of the considerations before deciding on a final jig design. The list in *Figure 5-3* is only meant as an example and not as a rule for all jigs. Just as each part has its peculiarities, each tool also has its own specific characteristics which must be addressed.

Product analysis considerations	Effect on jig design
Size, weight	Relatively small; lightweight construction.
Wall thickness and general shape provide regidity	No special supporting and clamping methods are necessary to prevent part distortion.
Surface finish	Clamps, locators, or other details must not incorporate sharp points or edges to mar flat surfaces of part.
Machinability index	Permits metal removal with moderate machining forces.
Angular surface relation (95° ± 15′)	Clamps and locators must not change close angular tolerance.
Surface flatness tolerance, ±0.010″ (0.25 mm)	Clamping forces must not alter flatness tolerance.
Normality of hole axes	Clamping and locating must maintain axes normal to respective flat surfaces.
Location (±0.005″ [0.13 mm]) and diameter (±0.0156″ [0.396 mm]) tolerances on 0.1405″ (3.57 mm) holes	Standard drill bushings satisfactory.
Diameter tolerance (+0.001, -0.000″) [+0.03] on 0.500″ (12.7 mm) hole.	Tolerances can be held with drill and ream jig, suitably bushed.

Location tolerance (1.000 ± 0.001″ [25.4 ± 0.03 mm]) on 0.500″ (12.7 mm) hole	Tolerance
	    Cannot be held with single jig. Can be held with a drill jig and a separate jig with a jig ground bushing for reaming.
	    Can be held with a drill jig for rough-drilling all holes and a separate simple holding fixture for finish boring of the 0.500″ (12.7 mm) hole.

Maximum possible mislocations, in. (mm)
1. From true position
  of liner in jig   0.0005 (0.013 mm)
2. Due to fit of bushing
  and liner   0.0004 (0.01 mm)
3. Due to fit of reamer
  and bushing   0.0002 (0.005 mm)
    Total   0.0011 (0.028 mm)
	Design decision: most accurate results will be obtained with a tumble jig for drilling all holes and a separate holding fixture for boring the 0.500″ (12.7 mm) hole.

Operation considerations	Effect on jig design
Operation 1, grinding 2 x 3, 2 x 1¾″ (51 x 76, 51 x 44 mm) surfaces, with horizontal disk grinder, according to process sheet	No fixture required. Grinder table provides adequate holding, positioning, and locating facilities.
Operation 2, drilling (10) 3.569 mm-diam holes and (1) 12 mm-diam pilot hole for 12.7 mm-diam hole on Avey number 2 two-spindle drill press, according to process sheet	Torque and feed force allows hand-held tumble-jig design. Clamps must clear drills. Adequate three-point supporting for stability of jig on feet and for normality of hole axes.
Operation 3, bore 12.7 mm-diam hole on Heald Borematic, according to process sheet	Simple holding fixture or vise, not shown; normality of hole axis provided by shims or pins.

Figure 5-3. Considerations for jig design.

Machine considerations	Effect on jig design
Option 2, Avey number 2 two-spindle drill press, according to process sheet	Tooling area, bed, size, and kind of chuck will not limit tumble-jig design.
Operation 3, Heald Borematic, according to process sheet	Simple clamping, positioning, and supporting to ensure 51 x 44 mm face at 90° to boring bar (this fixture not shown).

Operator considerations*	Effect on jig design
Operation 2, loading, unloading, and fixture handling	Operator loads, unloads, and clamps with lower thumbscrew with left hand; tightens hand knob with right hand. Small tumble jig easily turned for drilling second set of holes.
Operation 3, loading, unloading, and fixture handling	This simple fixture (not shown) will not be handled; unloading, loading, locating, and clamping are simple, nontiring operator motions.

Production considerations	Effect on jig design
200 parts per month, 2,400 per year; possible future production of 8,000 per year	Air clamping, indexing, etc., and various automated designs not justified by production rates and quantity.

Economic evaluation	Effect on jig design
Jig cost, $360 design and make; $0.15 per part, at 2,400 annual rate Boring fixture cost, $146.40 design and make; $0.061 per part, at 2,400 annual rate. Total fixture costs are $506.40; $0.21 per part Operations 2 and 3 costs total $100.65; $0.5032 per part, at 2,400 annual rate	For 2,400 annual production rate, reducing the time to operate the jig and fixture by faster clamping, etc., would not be justified because of increased fixture cost. Setup and run time for both operations is 5.49 min per piece; cost studies show that the cost of timesaving fixture details, such as air cylinder clamping, can be absorbed by the reduction possible just in labor cost to operate the fixtures, when the setup and run time are considerably longer and production is 5 to 10, or more, times the present annual production.

* There are no particular problems of operator and machine safety or operator fatigue.

Figure 5-3. Continued

Drill Jigs

The workpiece, production rates, and machine availability normally determine the size, shape, and construction details of any jig. However, all jigs must conform to certain design principles which will provide for the efficient and productive manufacture of quality workpieces by providing a method to:

1. Correctly locate the workpiece with respect to the tool.
2. Securely clamp and rigidly support the workpiece during the operation.
3. Guide the tool.
4. Position and/or fasten the jig on a machine.

These features will ensure interchangeability and accuracy of parts, plus provide the following advantages:

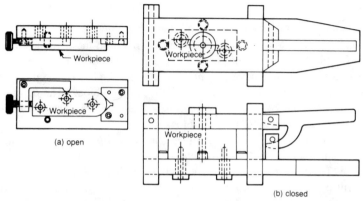

Figure 5-4. Open and closed types of drill jigs.

1. Minimize tool breakage.
2. Minimize the possibility of human error.
3. Permit the use of less skilled labor.
4. Reduce manufacturing time.
5. Eliminate retooling for repeat orders.

Jigs are often divided into two broad categories, open and closed. Open jigs are generally used when machining a single surface of a workpiece, whereas closed jigs are used when machining multiple surfaces. Examples of open and closed jigs are shown in *Figure 5-4*. More often, jig types are identified by the method used to construct the jig (for example: template, plate, leaf, channel, etc.). The main types are discussed in the following sections.

Template Drill Jigs. Template drill jigs are not actually true jigs because they do not incorporate a clamping device. However, they can be used on a wide variety of parts and are among the simplest and least expensive drill jigs to build. Template drill jigs are simply plates containing holes or bushings to guide a drill. They are usually placed directly on a feature of the part itself to permit the

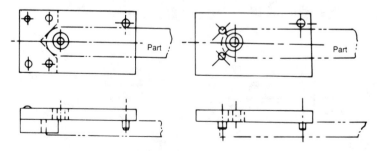

Figure 5-5. Flat plate template drill jigs.

drilling of holes at the desired location. When this is impractical, they are located on the part by measurement or by sight lines scribed on the template.

Two flat plate template drill jigs are shown in *Figure 5-5*. Both are designed to drill a hole through the center of the rounded end of a lever. The jig shown in *a* consists mainly of a drill guide plate and a locating V-block. The jig in *b* does the same job, but has been further simplified by using dowel pins to accomplish the same centralizing action as the V-block.

Another flat plate template drill jig is shown in *Figure 5-6*. This jig was designed to drill a three-hole pattern into either a left- or right-hand version of a workpiece. This was made possible by having the pins protrude from both sides of the drill plate, thereby permitting it to be flip-flopped to suit the workpiece being drilled. A common practice with template drill jigs is to place a pin into the first hole drilled to prevent excessive movement of the jig while drilling the remaining holes.

Three circular-type template drill jigs are shown in *Figure 5-7*. All are designed to locate from the maximum material condition (MMC) of the part diameter. Jig *a* was designed to locate from the OD of a shaft, jig *b* from the ID of a part, and jig *c* from a boss diameter. In all cases, a pin of the proper size was placed into the first hole drilled to properly position the jig to drill the second hole.

Figure 5-8 illustrates two nesting type template drill jigs. Jig *a* is designed to locate a small sheet metal workpiece in a cavity to permit drilling two holes, which are located from the periphery of the workpiece. Jig *b* was designed to perform the same operation by using five dowel pins press-fitted into the jig in lieu of the cavity to locate the workpiece, reducing the cost to build the jig. A template drill jig is often used to drill holes in one portion of a large workpiece

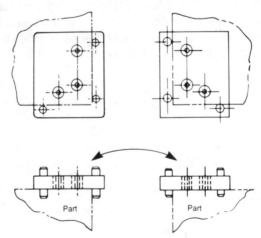

Figure 5-6. Flat plate template drill jig—L/R hand.

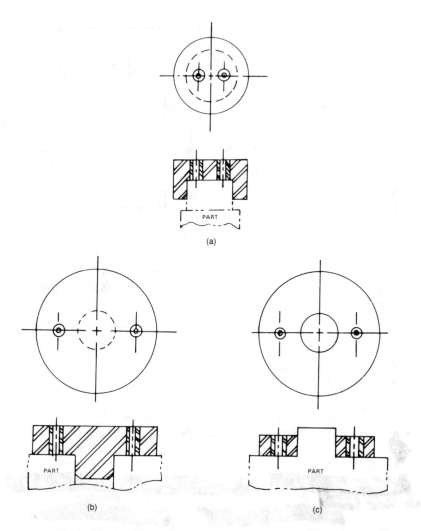

Figure 5-7. Circular plate template drill jigs.

where a conventional jig large enough to hold the entire part would be impractical and costly. Template jigs usually cost much less than conventional jigs, often making the use of two or three template drill jigs more economical than using one large conventional jig.

Some of the main disadvantages of template jigs are:

1. They are not as foolproof as most other types, which may result in inaccurate machining by a careless operator.
2. Orientation of the hole pattern to workpiece datums may not be as accurate

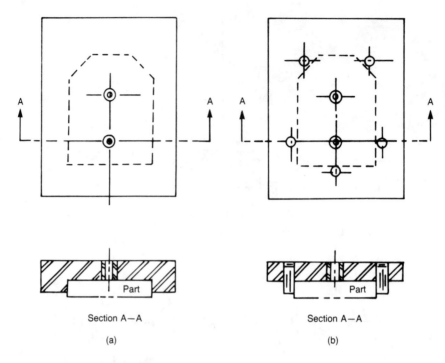

Section A—A   Section A—A

(a)   (b)

Figure 5-8. Nesting template drill jigs.

as with other types. However, the accuracy of the hole pattern within the template jig itself is comparable to that of any conventional jig.

3. They are usually not practical when locating datums are dimensioned, regardless of feature size (RFS).

Plate Jigs. Plate jigs are basically template jigs equipped with a workpiece clamping system. Initial construction costs are greater for plate jigs than for template jigs, but plate jigs are generally more accurate and last longer.

A plate jig incorporates a plate, which is generally the main structural member, that carries the drill or liner bushings. Slip bushings of various sizes can be used with liner bushings, allowing a series of drilling and related operations without the need to relocate or reclamp the workpiece. The plate jig's open construction makes it easy to load and unload a workpiece and to get rid of chips.

Three different types of plate jigs are shown in *Figure 5-9.* The open plate jig, *a*, is basically a template jig equipped with a means to clamp the workpiece, with work being supported by the drill press table. The table plate jig, *b*, consists of a drill plate, locating stud, and clamping screw, with standard screws used as jig feet. This type of jig is usually hand-held on the drill press table, rather than clamped to the table, so it may be easily inverted for loading and unloading.

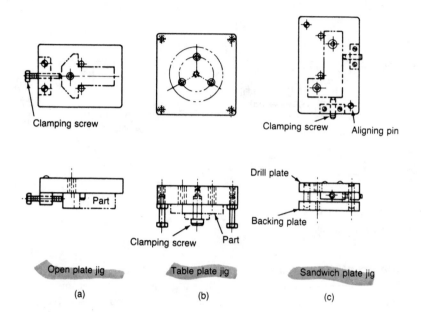

Figure 5-9. Plate jigs.

Consequently, table plate jigs are generally used for small parts. Note in particular that tool thrust in this type of jig is directed toward the clamps rather than the rigid portions of this jig. Therefore, it is imperative that the clamping method be strong enough to resist the thrust of the drill.

For obvious reasons, the jig shown in *Figure 5-9c* is called a sandwich plate jig. With this type of jig, the workpiece is positioned between two plates; a drill plate containing the drill jig bushings, locators, and clamps and a backup plate to provide support only. The backup plate has clearance holes for the drill and is aligned with the drill plate by two pins. The backup plate makes it ideal for drilling thin parts which would otherwise buckle from the thrust of the drill.

The angle plate jigs shown in *Figure 5-10* are primarily used to drill workpieces at an angle to the part locators. The plain angle plate jig, *a*, is designed to drill holes perpendicular to the locating surface, while the modified angle plate jig, *b*, is designed to drill holes at angles other than 90° to the locating surface.

Plate type jigs are usually moved around the table by hand. Therefore, special safety precautions should be provided to prevent the jig from whirling around the spindle whenever a cutting tool jams. The best way to prevent this is to build the jig with an extension handle long enough for the machine operator to overcome the torque of the jammed tool. When a plate jig is used with a radial drill, provision can often be made to clamp the drill jig or the workpiece to the machine table.

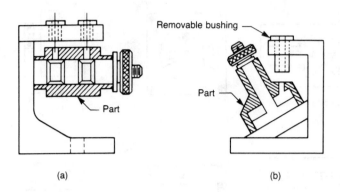

Figure 5-10. Angle plate jigs.

Universal Jigs.   Universal jigs (sometimes called pump jigs) utilize a handle connected to a cam or rack and pinion to move either a bushing plate or a nest plate, often vertically, to clamp the workpiece. Parts held in universal jigs have surfaces adaptable to fitting against the surfaces of the bushing plate and nest.

Universal jigs are readily available in a wide variety of styles and sizes similar to those shown in *Figure 5-11*. Universal jigs are well designed, ruggedly built, and can be easily prepared to drill a specific part. Usually all that is required is to add part locators and drill jig bushings.

Some typical applications are shown in *Figure 5-12*. Jig *a* was set up to drill a hole through the center (RFS) of a small rod. It features a self-locating V-bushing liner, an adjustable end stop, and a riser block. This same operation could be accomplished with another type of universal jig called a cross-hole jig, *b*. In this type of jig, the drill is guided by a standard slip-fixed renewable bushing, which is fitted into the jig clamping plate and precisely centered above a hardened and ground V-block.

Either of these jigs could easily be adapted to drill a wide variety of parts, each having a different diameter or hole size, merely by changing the renewable slip-fixed jig bushing and adjusting the end stop.

Compare these jigs with the plate jig shown in *Figure 5-13*, which was specifically designed and built to perform a similar operation. Needless to say, the universal type jig could be put on the job in a fraction of the time and at a lower cost. By utilizing the wide variety of standard drill bushings, liners, and tooling components, universal jigs can be adapted to drill a wide variety of parts. They are also reusable for other jobs, although it may require replacement of the drill plate, which is interchangeable and available separately from the manufacturer. Because of this versatility, universal jigs are ideal for limited production manufacturing. One manufacturer states that tooling costs can be reduced by

Figure 5-11. Universal jigs. *(Courtesy, Swartz Fixture Div., Universal Vise & Tool Co.)*

one-third. With few exceptions, universal jigs can be designed and adapted to do a job in a fraction of the time required to design and build a conventional jig that performs the same function.

Leaf Jigs. A leaf jig is generally small, and incorporates a hinged leaf that carries the bushings, and through which clamping pressure is applied. Although the leaf jig can be used for large and cumbersome workpieces, most designs are limited in size and weight for easy handling. A leaf jig can be box-like in shape, with four or more sides for drilling holes perpendicular to each side. Leaf jigs with additional feet are often called tumble jigs, and permit operations from more than one side.

(a)

(b)

Figure 5-12. (a) Universal jig ready for production, (b) adjustable cross-hole drill jig. *(Courtesy, (a) Acme Industrial Company, (b) Heinrich Tools, Inc.)*

Off-the-shelf, tumble-type leaf jigs, such as those shown in *Figure 5-14*, are available in a variety of sizes from many manufacturers. Construction consists of a drill plate (leaf) attached to the jig body with a precise fitting hinge at one end and a positive positioning clamp at the other end. As with universal jigs, all that is required to prepare it for use is to provide a means to locate the part and to add

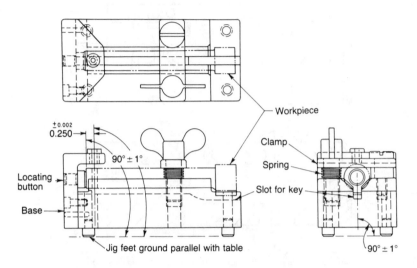

Figure 5-13. Plate jig for cross-hole drilling. *(Courtesy, SME Chapter 100)*

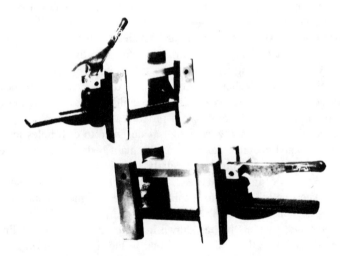

Figure 5-14. Leaf jigs. *(Courtesy, Carr Lane Manufacturing Co.)*

drill bushings. Lids swing up to provide easy loading from three sides. Occasionally, a side plate is attached to the lid or base to permit cross-drilling.

The leaf jig shown in *Figure 5-15* was specifically designed and built to drill two holes in a small connecting rod. The hinged drill plate contains the drill bushings and is precisely located at both ends by the slots in the body of the jig. The workpiece is located and clamped between two V-blocks, one fixed and the

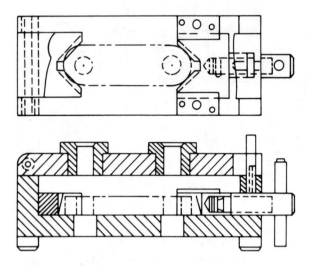

Figure 5-15. Leaf jigs for drilling two holes in a small connecting rod.

other movable. The V-blocks are tapered to force the workpiece down against the base of the jig body.

Channel and Tumble Box Jigs. Channel and tumble box jigs (see *Figure 5-16* and *Figure 5-17*) permit drilling into more than one surface of a workpiece without relocating the workpiece in the jig. This results in greater accuracy with less handling than required using several separate jigs. These jigs can be quite complicated and more expensive to build than several simpler types, but they can still be very cost-effective if properly designed.

The channel jig shown in *Figure 5-16* was designed to drill holes into three surfaces. A U-shaped channel was used as the main body along with press fit drill bushings, locators, and clamping details. The U-shaped channels used to construct this type of jig can be cast, built-up, or of welded construction. However, to keep building cost to a minimum, the designer should seriously consider using standard U-shaped sections discussed previously.

Tumble box jigs, such as the one shown in *Figure 5-17*, permit drilling and similar operations from all six sides. The one in the figure is shown with the hinged top open to permit loading and unloading. Tumble box jigs are commercially available in a variety of sizes which can be prepared to machine a particular part by adding drill bushings and a means to locate and clamp the part. Because of this off-the-shelf availability, designers often choose a tumble box jig over a channel jig when machining only two or three sides of a workpiece.

Indexing Jigs. Indexing jigs are used to drill holes in a pattern, usually radial. Location for the holes is generally taken from the first hole drilled, a datum hole in the part, or from registry with an indexing device incorporated in the jig.

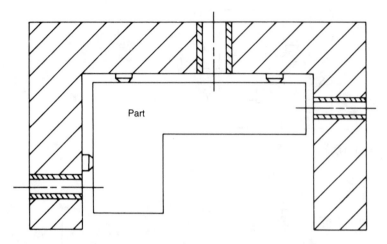

Figure 5-16. Cross section of a typical channel jig (clamping details not shown).

Figure 5-17. Tumble box jig. *(Courtesy, Carr Lane Manufacturing Co.)*

The simple jig shown in *Figure 5-18* features a base made from a standard angle iron section into which a locating stud (2) has been placed to position a bored cylindrical workpiece (1) which is clamped on the stud with a C-washer (4) and a hex nut (3). A drill bushing (5) is press fit into the bushing plate (8). In use, the hex nut is loosened after the first hole is drilled, the workpiece is revolved, the index pin (6), which is held in place with a flat spring (7), is pushed into the hole, and the second of four holes 90° apart, is drilled after the nut is tightened. Indexing is repeated until all four holes have been drilled.

Another indexing jig, shown in *Figure 5-19*, utilizes an angle plate of welded construction. A spindle pressed into the jig's pivot point is threaded on one end for the locking wheel. Clamping for the workpiece is not shown. The workpiece,

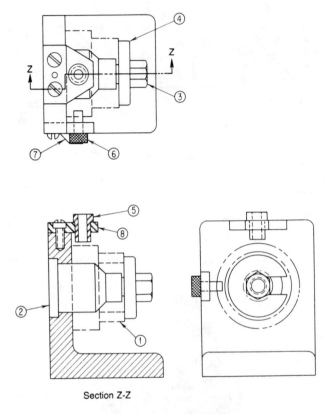

Section Z-Z

Figure 5-18. Simple indexing jig with a base of standard angle iron.[1]

bushing plate, and bushings are rotated for drilling holes at various angles which are determined by the location of the index holes in the jig and angle plate.

Miscellaneous Drill Jigs. Drill jigs can be made from many standard commercial items such as premachined section forms and bases; angle irons, parallels, and V-blocks; section components; and standard structural forms. Following are some examples of drill jigs constructed of these materials:

1. Premachined sections were used to construct the drill jig shown in *Figure 5-20*. The base was made from a standard commercial T- section to which a bushing plate made from a flat section was attached with screws and dowels. The addition of standard jig feet, drill bushings, and clamps completed the assembly.

2. The drill jig shown in *Figure 5-21* was built from a U-shaped, premachined cast iron section and a flat section. It was built in 9.5 hours with a savings of 30% in build time.

3. The drill jig shown in *Figure 5-22* was built from standard structural form

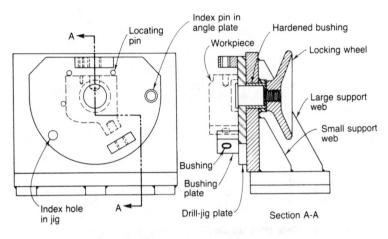

Figure 5-19. Indexing jig of welded construction.[2]

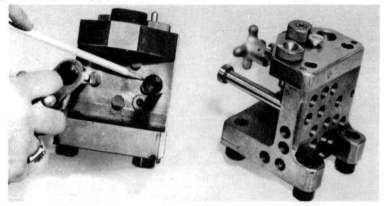

Figure 5-20. Drill jig made from standard T-section. *(Courtesy, Standard Parts Co.)*

materials. It consists of two pieces of angle iron welded to a channel iron to form a V-section. A larger angle iron was welded over the V-section to serve as a bushing plate and clamping mount. The addition of a standard thumb screw, a drill bushing, and a section of rod completed the assembly.

Wooden Drill Jigs. Wooden tooling is often overlooked. An example of its use for drilling is shown in *Figure 5-23*. The jig consists of a plywood base, two wood side spacer rails, a plywood bushing plate joined together with wood screws, and four bolts which also serve as jig feet. To reduce wear and provide additional support, a sheet metal plate is bonded to the top of the base. The part is located against a pressed-in roll pin and two more bolts thru the side rail. The hand knob clamp, thumb screw, threaded socket, and serrated type bushings are all standard tooling components.

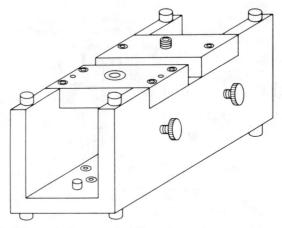

Figure 5-21. Drill jig made from standard U-section tooling material.

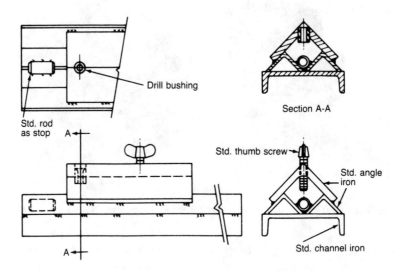

Figure 5-22. Drill jig made from standard structural form material (channel and angle iron).

Plastic Drill Jigs. Some of the things that require special consideration when using these materials for drill jigs include the following:

1. Conventional drill bushings should not be used in this type of tooling. Castable drill bushings are designed for this purpose and should be used.
2. Drill bushings must be positioned in the exact location before pouring the liquid epoxy, thermoplastic compound, or low-melt alloy.

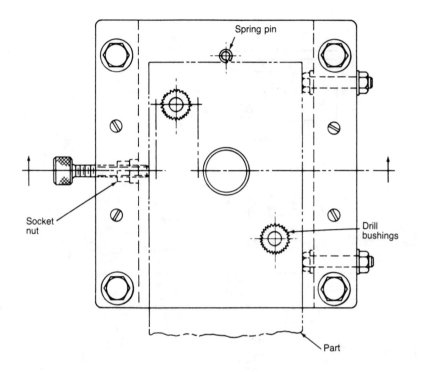

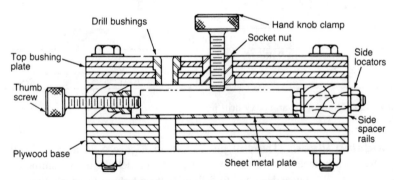

Figure 5-23. Wooden drill jig incorporating standard tooling hardware.

3. When designing epoxy tooling, the tool designer should understand the various construction methods available along with the advantages and disadvantages of each method. He or she should also be familar with the many grades of epoxies, or seek help from the epoxy formulators, especially when designing complex tooling.

General construction of a cast epoxy drill jig is shown in *Figure 5-24*. A drill jig constructed by the laminating method is shown in *Figure 5-25*.

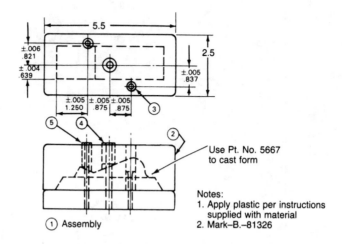

Use Pt. No. 5667
to cast form

Notes:
1. Apply plastic per instructions
   supplied with material
2. Mark–B.–81326

5	1	Drill Bushing (Plastic Type)	American Type DGV. 0.156 ID x 0.406 x 1 LG.	STD.
4	1	Drill Bushing (Plastic Type)	American Type DGV. 0.25 ID x 0.5 ID x 0.875 LG.	STD.
3	1	Drill Bushing (Plastic Type)	American Type DGV. 0.125 ID x 0.312 x 0.75 LG.	STD.
2	1	Body	REN Plastics Inc. #RP–3203–2	
1	1	Assembly		
Det. No.	No. Req'd	Name	Stock Size	Mat.
Stock List				
Drill Jig				
Drawn By – E. C. Key Date – 5-1-85			App. By – CD Date – 5-6-85	
			B–81236	Rev 1

Figure 5-24. Drawing of an epoxy drill jig.

Modified Vises. The drill jigs shown in *Figure 5-26* along with their respective workpieces, were all made by adding details to a commercially available vise. Jig *a* was designed to drill a locating hole in a sprocket shaft. Jig *b* was designed to both drill one 0.277'' (7.04 mm) diameter hole and also to drill, countersink, and tap a 0.3125-24 hole in a pivot block. Jig *c* was designed to drill (1) No. 38 and (1) 0.094'' (2.38 mm) diameter hole in a pivot pin. For positioning on the machine table, the vises are bolted to precision ground sub-bases and located against a rail and stop pins as shown in *d*.

Collet Fixtures. As shown in *Figure 5-27*, 5C collet blocks and collet fixtures mounted vertically can be used to hold workpieces for drilling, tapping, and related operations. The collet fixture shown in the figure has been mounted on a sub-base to permit long workpieces to pass through the fixture.

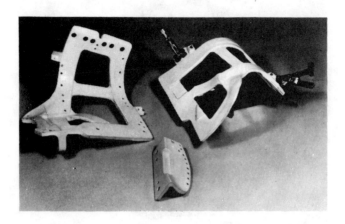

Figure 5-25. Plastic drill jigs and scribe templates. *(Courtesy, Ciba-Geigy Tooling Systems)*

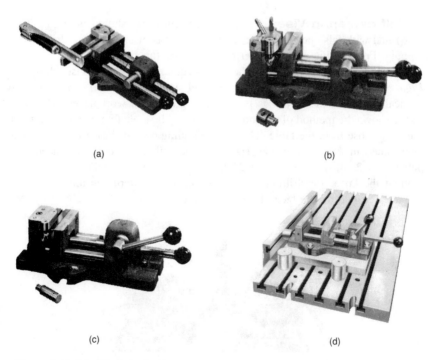

(a)   (b)

(c)   (d)

Figure 5-26. Modified vises. *(Courtesy, Heinrich Tools, Inc.)*

Figure 5-27. Air collet fixture. *(Courtesy, Heinrich Tools, Inc.)*

Self-centering Vises. The self-centering vise shown in *Figure 5-28* was fitted with false jaws to center and clamp a cast iron elbow for drilling and tapping. Form-fitting cast jaws perform a similar function.

Drilling Accessories. The safety quick-acting drill vise shown in *Figure 5-29* features built-in recessed parallel jaw inserts to hold work square. A vertical V-groove in the jaw is provided to grip round workpieces. The same figure shows the method of mounting the vise to the table to prevent spinning or breaking loose from the table. The quick dividing device shown mounted to an angle knee in *Figure 5-30* can speed up the drilling of equally spaced hole patterns of 2, 3, 4, 6, 8, 12, or 24 holes on a workpiece OD, or when used without the knee, for drilling equally spaced hole patterns on any bolt circle diameter desired on the face of the workpiece.

Figure 5-28. Self-centering vise with false jaws. *(Courtesy, Heinrich Tools, Inc.)*

Figure 5-29. Safety drill vise. *(Courtesy, Heinrich Tools, Inc.)*

Figure 5-30. Quick divider. *(Courtesy, Willis Machinery & Tools Co.)*

The universal drill jig shown in *Figure 5-31* can be readily adapted for drilling a wide variety of workpieces. It features a solid jaw with an adjustable end stop and several movable bushing carriers, which can be oriented for the different workpieces.

Drilling applications using a multi-purpose vise are shown in *Figure 5-32*. The mounting table (see *Figure 5-33*) can be used for light drilling as well as electrical discharge machining (EDM), milling, boring, and jig grinding. The workpiece is located and held on the square ledges of the rails and a movable support block, which permits the fixture to be used for a wide range of workpiece sizes. The application shown is a setup for an NC drill press.

Figure 5-31. Universal drill jig. *(Courtesy, Vise & Tool Co.)*

Figure 5-32. Multipurpose vise. *(Courtesy, James Morton, Inc.)*

A drill press clamp (two Hol-Down products are shown in *Figure 5-34a*) can be very useful for clamping a workpiece directly on the drill press table. Standard clamping attachments are shown in *Figure 5-34b*. The unit can also be used to guide a drill by replacing the clamp screw with a suitable drill bushing.

Drill Jig Bushings

Drill jig bushings position and guide the tools that do the cutting. The basic styles are shown in *Figure 5-35*. They are available in hardened steel or carbide. Bushings made of other materials such as bronze, stainless steel, etc., are also available from the bushing manufacturers by special order. Most bushings and liners can be supplied with the OD ground to industry standards, or left unground (type U) for custom grinding.

Figure 5-33. Edge mounting table. *(Courtesy, Harig Manufacturing Co.)*

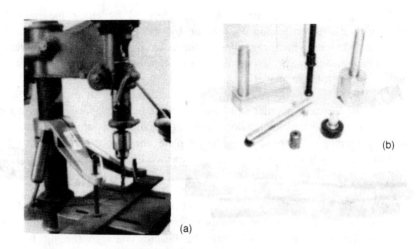

Figure 5-34. Drill press clamps. *(Hol-down drill press clamp, Courtesy, James Morton, Inc.)*

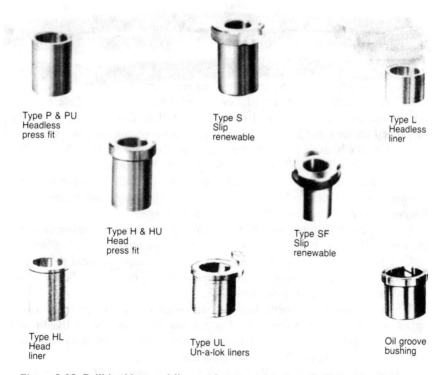

Type P & PU
Headless
press fit

Type S
Slip
renewable

Type L
Headless
liner

Type H & HU
Head
press fit

Type SF
Slip
renewable

Type HL
Head
liner

Type UL
Un-a-lok liners

Oil groove
bushing

Figure 5-35. Drill bushings and liners. *(Courtesy, American Drill Bushing Co.)*

Drill jig bushings and liners may be specified by either the individual manufacturer's identification system or by the universally accepted ANSI designation system which, regardless of the manufacturer, will insure delivery of the proper bushing. Specifying the type of cutting tool and its diameter will assure correct tool clearance.

Headless Press Fit Bushings. Headless press fit bushings, type P and PU, are the most popular and least expensive bushings. They are used for a single size cutting tool application where light axial loads are expected. Since they are permanently pressed into the jig plate, they are generally used where replacement is not anticipated during the expected life of the tool. They are ideal where the top of the bushing must be flush with the jig plate or where hole spacing is too close to use headed bushings.

Head Press Fit Bushings. Head press fit bushings, type H and HU, are also used for permanent installations requiring greater bearing area or where heavy axial loads that could force the bushing through the jig hole are anticipated. Since the bearing area extends beyond the jig plate, the thickness of the jig plate can often be reduced, thereby lightening the overall weight of the jig. Cases where the head must be flush with the jig plate require counterboring of the mounting hole in the jig plates.

Although designed for permanent installations, press fit bushings are easily replaced, but at the expense of losing some mounting hole accuracy with each replacement.

Slip Renewable and Slip-Fixed Renewable Bushings. Slip renewable bushings, type S, and slip-fixed renewable bushings, type SF, are used with a headless liner, type L, or a head liner, type HL, where multiple operations, such as drilling and reaming or drilling and tapping, are to be performed on the same hole; or, where long production runs require occasional changing of the bushing to maintain jig integrity. They are available for use with lock screws, type LS and TW; round clamps, type CL; round end clamps, type RE; or flat clamps, type FC, as shown in *Figure 5-36*. Both diameters of the bushings and the ID of the liner are finish ground to industry standards, while the liners are available with the OD finish ground or unground.

When used for multiple operations, the first operation bushing is installed and the hole drilled. The bushing is then replaced with the second operation bushing and the second operation performed. This process is repeated until the hole is completely machined. The first operation bushing is then reinstalled and the procedure repeated on the next workpiece.

When slip-fixed renewable bushings are used on jigs to perform single operations on long production runs, they are usually secured in the fixed mode by a lock screw. When used to perform more than one operation on the same hole, they are secured in the slip mode for easy changing.

Type UL liners (such as UN-A-LOC) are used with ANSI slip renewable

Lock screws

Type LS and TW

Used for locking renewable bushings in liners. Available for all styles of renewable bushings.

Round clamps

Type CL

Used instead of lock screws for locking ANSI and Extended Range fixed renewable bushings in liners.

Round end clamps

Type RE

Used instead of lock screws for locking ANSI and Extended Range fixed or slip renewable bushings in liner. Available for applications requiring either flush or projected mounting of liners.

Flat clamps

Type FC

Used for locking ANSI and Extended Range Type FX fixed renewable bushings with heads milled for flat clamp mounting. Available for applications requiring either flush or projected mounting of liners.

Figure 5-36. Accessories. *(Courtesy, American Drill Bushing Co.)*

bushings. These special liners eliminate the need for a bushing locking device. In use, the bushing is locked tight in the liner by the torque of the drill bit. UN-A-LOC liners must be installed with an arbor press.

Oil Groove Bushings. Oil groove bushings are designed to provide complete lubrication between the cutting tool and the bushing when maximum cooling is required during a machining operation. Over two dozen separate groove patterns are available.

Drill Bushings for Special Applications. Some of the more common drill bushings for special applications are shown in *Figure 5-37.* When installing any of these bushings, the individual manufacturer's recommendations should be followed closely.

Bushings and Liners for Plastic, Castable, and Soft Material Tooling. Bushing types HGV, HGP, and DGV have serrated or

Type HGV   Type HGP   Type DGV

Type SGP   Type ULD

Type TB template bushing   Type TB Lock ring

Figure 5-37. Special drill bushings. *(Courtesy, American Drill Bushing Co.)*

grooved ODs for casting in-place with epoxy resins, thermoplastic tooling compounds, or low-melt alloys. Bushing type SGP is for press-in installation in soft materials such as aluminum, magnesium, wood, or masonite. The liner type ULD serves the same function as the type UL, except that it is knurled for use in plastic tooling.

Template Bushings. Type TB bushings are used with thin template materials ranging from 0.0625-0.375" (1.6 to 9.5 mm) thick to provide low cost tooling. Installation is shown in *Figure 5-38*.

Rotary Bushings. Rotary bushings (not shown) feature precision tapered roller or ball bearings capable of handling high thrust and/or radial loads encountered in some jig applications, such as supporting a piloted cutting tool for extremely close machining.

Drill Bushing Tips and Accessories. Drill bushing tips (see *Figure 5-39*) are used with automatic self-feed drill motors. Most are constructed by a two-piece method consisting of a collar which contains the screw threads, alignment diameter, and the lock flange. It also contains a pressed-in shank which is a plain bushing with ID ground to guide and support the cutting tool, and the OD ground to suit the liner bushing mounted on the jig.

Mounting of Bushings. Jig mounting with a lock liner bushing and lock nut is the conventional method of mounting the drilling unit to a jig or fixture. To install, a hole is bored into the jig or fixture to suit the lock liner

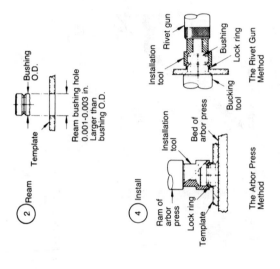

① Lay out holes

Lay out holes observing hole spacing and edge distance minimums

Bushing OD	A Min. Hole Spacing	B Min. Edge Distance
0.375	0.60	0.250
0.500	0.73	0.312
0.750	0.98	0.438

② Ream

Ream bushing hole 0.001-0.003 in. Larger than bushing O.D.

Bushing O.D.

Template

③ Countersink

Lock ring to be flush with or a max. of 0.015 above bushing groove

Flush to 0.015

Aluminum lock ring

Bushing

Bushing chamfer must seat here

Reamed hole   Template

90°

④ Install

Ram of arbor press

Installation tool

Bed of arbor press

Lock ring

Template

The Arbor Press Method

Installation tool   Rivet gun

Bushing

Lock ring

Bucking tool

The Rivet Gun Method

Countersink reamed hole with 90° included angle countersink tool to permit bushing to seat flush to 0.015 inside surface of template. The countersink must be normal and concentric to the reamed hole and free of chatter. For best results, use micro stop countersink tool with piloted countersink cutter.

The arbor press method of installation is recommended whenever template size permits. The rivet gun method is used for large templates. Template Bushing Tool is threaded for adaptation to rivet gun. Use lowest impact pressure adjustment that will properly form the aluminum lock ring. Note that impact pressure will vary for bushings of different outside diameters. Excessive impact pressure will flatten lock ring causing insecure mounting of bushing.

Figure 5-38. Installing template bushings. (Courtesy, American Drill Bushing Co.)

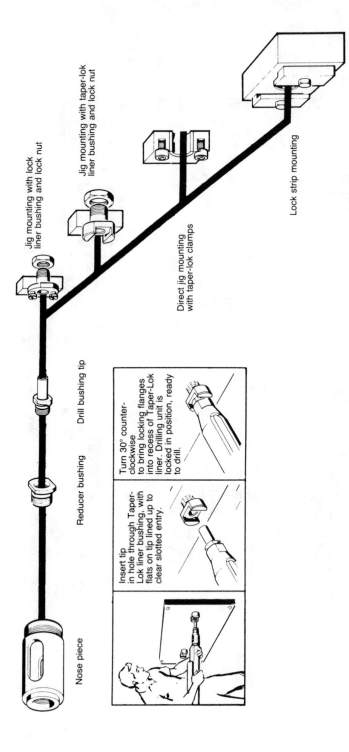

Figure 5-39. Drill bushing tips. *(Courtesy, American Drill Bushing Co.)*

Jig mounting with lock liner bushing and lock nut

Jig mounting with taper-lok liner bushing and lock nut

Direct jig mounting with taper-lok clamps

Lock strip mounting

Drill bushing tip

Reducer bushing

Nose piece

Insert tip in hole through Taper-Lok liner bushing, with flats on tip lined up to clear slotted entry.

Turn 30° counter-clockwise to bring locking flanges into recess of Taper-Lok liner. Drilling unit is locked in position, ready to drill.

bushing, which is then assembled to the jig or fixture and then held in place with a lock nut. (The Taper-Loc liner bushing is a compact, one-piece version of the lock liner bushing.)

Direct jig mounting is an alternate method used when holes are so closely spaced that lock liner bushings cannot be used. In this case, lock screws are mounted directly on the jig and a hardened, headless liner is pressed into the jig to accept the drill bushing tip shank. Lock strip mounting is another mounting method for holes that are closely spaced. It features a lock strip along each side of a row of holes in the jig. A hardened, headless liner to accept the shank of the drill bushing tip is also pressed into the jig with this method.

Installation of Drill Bushings

To ensure accuracy in the workpiece, drill bushings must be properly located and installed using the following rules:

1. Mounting holes should be round and properly sized to prevent bushing closure and jig plate distortion. For this reason it is recommended that the mounting holes be jig bored or reamed to size. Headless press fit and liner bushings are generally installed with a diametral interference of 0.0005-0.0008" (0.0127-0.020 mm), while headed press fit bushings are generally installed with a diametral interference of 0.0003-0.0005" (0.0076-0.0127 mm). Interference greater than this may reduce the diameter of the bushing ID to the point where the tool may seize, or, in the case of liners, prevent insertion of a renewable bushing. On the other hand, too little interference will result in a loosely installed bushing which may spin or be forced out of place. Drill bushings for special applications should be installed in accordance with the individual manufacturer's recommendations.

2. Sufficient chip clearance, as illustrated in *Figure 5-40*, should be provided between the bushing and the workpiece to allow for chip removal, except in cases where extreme accuracy is required. In this case, the bushing should be in direct contact with the workpiece. In most cases, a clearance of 1-1.5 times the bushing ID should be used when machining materials such as cold-rolled steel which produces long stringy chips, while a clearance of one-half the bushing ID is recommended when machining materials such as cast iron, which produces small chips. Excessive chip clearance should be avoided because (1) most cutting tools are slightly larger at the cutting end due to back taper, and (2) excessive clearance will reduce the guiding effect of the bushing, resulting in less accurate drilling.

Chip control should be closely monitored after the jig is put into production. As a general rule, if the chips have a tendency to lift the bushing, more clearance is needed. If the cutting tool wanders or bends, less clearance is needed.

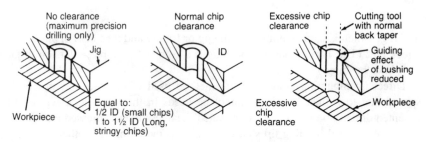

Figure 5-40. Recommended clearance between workpiece and bushing. *(Courtesy, American Drill Bushing Co.)*

When performing multiple operations such as drilling and reaming, slip renewable bushings of different lengths should be used to provide optimum chip clearance for each operation as shown in *Figure 5-41*.

Clearance between the bushing and the workpiece must also be provided when drilling wiry metals such as copper, which tends to produce secondary, or minor, burrs around the top of the drilled hole, as shown in *Figure 5-42*. This, in turn, causes the jig to lift from the workpiece or makes it difficult to remove the workpiece from side-loaded jigs. A burr clearance of one-half the bushing ID is recommended. Provisions must also be made to the jig itself to provide clearance for the primary, or major, burrs that form as the tool exits the workpiece.

In order to properly support and guide the cutting tool, the length of the drill bushing under normal circumstances should range between 1.5-2.5 times the diameter of the bushing ID and the jig plate supporting the bushing should be thick enough to sustain the bushing. Usually a thickness between one and two times the cutting tool will be sufficient.

When drilling irregular work surfaces, the ends of the bushings should be formed to the contour of the workpieces as shown in *Figure 5-43*. This will prevent the tool from pushing off center. The distance between the workpiece and the bushing should also be held to a minimum under these circumstances.

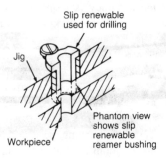

Figure 5-41. Chip clearance for multiple operations. *(Courtesy, American Drill Bushing Co.)*

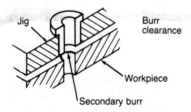

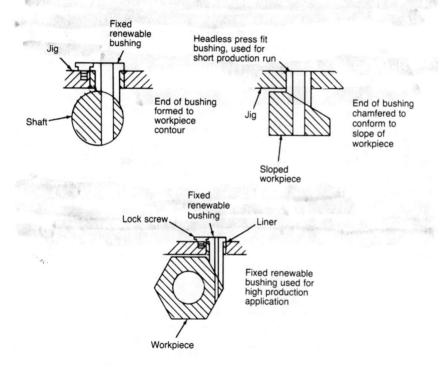

Figure 5-42. Burr clearance. *(Courtesy, American Drill Bushing Co.)*

Figure 5-43. Drilling irregular work surfaces. *(Courtesy, American Drill Bushing Co.)*

Sometimes hole center distances are so close that there is not enough room for the bushings. In these cases, extra thinwall bushings or standard headed or headless bushings which have flats ground on them might work. Sometimes a single, hardened steel insert with two or more holes must be used.

Designing A Jig

The most efficient method of designing any jig is to follow a systematic approach. Each individual element of the jig must first be thought out and sketched. The elements are then brought together in a sketch of the complete tool. The following represents one method to systematically design a jig around the proposed workpiece.

The simplest and easiest way to begin a jig design is to sketch the part in three or more views (*Figure 5-44*). This sketch should be made to a scale size of the actual part, and the jig must be sketched to the same scale size. Make sure to allow sufficient room between each view to permit the remaining elements to be sketched without crowding. It is sometimes better to sketch the part in a different color, such as red, to allow it to stand out in the sketch.

Once the part is sketched the next step is to add the locators and supports (*Figure 5-45*). Again, these details must be drawn to the proper scale and the their proper size. If commercial locating devices are specified, check a catalog to find the exact size of each area of the locator or support. Each locating element must be drawn as it would appear in each view.

Next, the tool body should be added to the sketch. Since this element is an integral part of the entire tool, it should also be drawn to the proper proportion and size. Once the part, locators, supports and tool body are sketched, a thorough inspection must be made to insure that there is no interference among these elements. Look for areas where the elements are positioned close together. When constructed, these points could interfere with each other. Make sure the locator design will permit adequate space for chip and coolant removal. As a preliminary check, make sure there are no redundant locators and that the part can only be loaded correctly. Any location problem must be resolved before the clamping element is added to the jig.

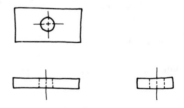

Figure 5-44. Sketch of the part.

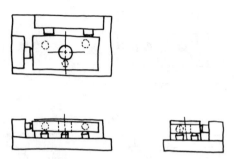

Figure 5-45. Sketch in locators and supports.

Once the jig design has been checked and no conflict between the elements exists, the clamping device can be added to the design (*Figure 5-46*). When designing the clamping device remember: this is the area of the tool that can either save or add time in production. Always specify clamping devices that are simple to use and fast operating. Spending an additional hour in construction time to make a better clamp could easily save thousands of dollars in production costs. Once a clamp design is determined, it should be sketched into the existing drawing of the jig. Remember to include the clamping device in each view. With the clamp sketched in place, once again begin a thorough inspection of the design to insure there is no conflict or interference between the various elements.

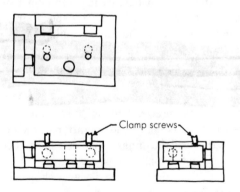

Figure 5-46. Add the clamping device to the design.

The next step in the initial design of the jig should be to add the drill bushings (*Figure 5-47*). The location of these bushings is critical to the accuracy of the jig and should be placed carefully. The tolerance values

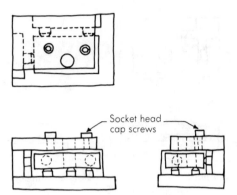

Figure 5-47. Add the drill bushings.

specified for the bushing placement should reflect a positional relationship of about 50% of the part tolerance. So, if a part dimension is specified as 1.000" ± 0.010" (25.4 mm ± 0.25 mm), the bushing location should be specified as 1.000" ± 0.005" (25.4 mm ± 0.13 mm). A general rule of 50% of the part tolerance applied to the tool will not only provide parts within the required accuracy, but will also reduce the cost of constructing the jig since the tolerance value allows the toolmaker more space to place the bushings. In cases where extreme accuracy or high volume production tools must be made, a tight tolerance could easily nullify the cost effectiveness of the jig and result in a loss of money rather than an increase in profit from using the tool. When selecting the bushings to use in the jig, always specify those bushings which will permit adequate support for the cutting tool. If space is not a problem, use head type bushings. If, however, space is limited, headless bushings should be used. In those cases where the bushings will be replaced frequently, either fixed renewable or slip renewable bushings should be specified. If these bushings are used, remember to include liner bushings. Once the bushings are added to the sketch, check the complete tool once again to insure the elements will not interfere with each other or the operation of the jig. At this stage the clearance between the drill bushings must be determined. If clearance is desired, provisions should be made to insure adequate space exists in the tool. If, however, no clearance is desired, the designer should insure the jig place and bushings will fit flush against the part.

The initial sketch of the basic jig is now complete; however, the relationship between the jig and the machine tool it is to be used with must now be checked. Using the size specifications for the machine tool selected for the operation, sketch the relationship between the jig and the machine table, spindle, and frame. Look for areas where any interference might occur.

Sketch the proposed mounting device on the table to make sure the T-slots are the proper size, or in cases where strap clamps are used, make sure they will not interfere with the operation of the jig. Determine if the cutting tool specified will perform the desired operation without interfering with the jig body or other elements. Check the clearance between the cutting tool and the machine tool members. Always specify standard, off-the-shelf, cutting tools whenever possible. Do not modify a standard tool to accommodate a flaw in the jig design. It is simpler to alter the jig design than to justify the additional costs of modifying standard tools. All tooling problems must be resolved on the drawing board, not in the shop during production.

Once the initial sketches are complete and the designer is convinced that the design is sound and workable, the final tool drawings are initiated. Here again the designer should look for any problem areas where the jig elements interfere with each other or with the operation of the machine tool. In cases where the jig is to be used with numerically controlled or other automatic machinery, the designer should check the program to insure the jig will not interfere with the operation of the machine tool. Look for points such as the height of the tool during the return cycle, or the height when moving from hole to hole. Make sure the actual height of the highest element of the jig is considered, not just the height of the jig plate. When manual machines are specified to use the jig, make sure the length of the spindle travel is sufficient to permit the easy use of the tool. Here again, make sure the actual height of the tool will permit the jig to be easily moved under the spindle. Little would be saved if the table of the machine tool had to be lowered and raised to clear an obstruction or to move the jig from hole to hole.

Designing A Typical Plate Jig

Using the methods previously outlined, the process of designing a plate jig for the part shown in *Figure 5-48* is as follows: The first step is to inspect the part to determine the basic requirements and condition of the workpiece material. In this case, the requirement is for a jig to drill two, 0.188" (4.78 mm) diameter holes in the locations shown in the figure. The part is 4.00 x 2.00 x 0.75" (101.6 x 50.8 x 19.1 mm) cold drawn AISI 1030 bar stock which weighs 1.71 lbs. (0.77 kg). The production plan calls for a one-time production run of 600 pieces, so the jig must be made as inexpensively as possible, yet the production rate must be rapid.

The first step is to sketch the part in three views as shown in *Figure 5-49*. If the part is small enough, the sketch should be made full size. This permits the designer to see the full-size tool and eliminates any error which might occur in converting to a scale size. Remember, it is often a good idea to sketch the part in a different color to allow it to stand out in the sketch and to minimize the chance of misinterpreting the meaning of the lines in the sketch.

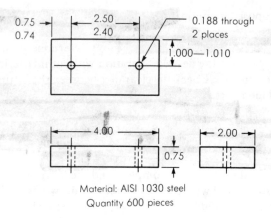

Material: AISI 1030 steel
Quantity 600 pieces

Figure 5-48. Part to be designed.

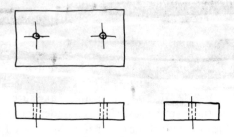

Figure 5-49. Preliminary sketch.

Once the part is sketched, the locators should be positioned in the sketch (*Figure 5-50*). Here a simple 3-2-1, or six-point, locational arrangement is used, since the part does not present any spacial locational problems. The three locators on the primary datum surface, in addition to establishing the location of this plane, also serve as risers to provide the necessary clearance between the drill bushings and the part. If a zero clearance is desired, the three locators on this surface can be eliminated and the part located against the flat jig plate. The secondary locators, positioned on the long side of the part, should be spaced about three inches (76 mm) apart to gain the maximum practical distance between these locators. The tertiary locator should be positioned approximately 1.50″ (38 mm) from the secondary locators to locate the end position of the part. The length of the secondary and tertiary locators is specified as 1.375″ (34.9 mm). This will permit accurate location of the workpiece and will prevent the ends of the locators from interfering with the machine table while drilling the part. With the locators in place, the remainder of the tool body, or in this case, the jig plate, is sketched in (*Figure*

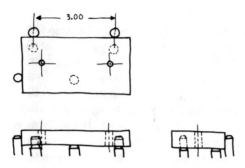

Figure 5-50. Position the locators in the sketch.

5-51). The thickness specified for this jig plate is 0.50″ (12.7 mm). This thickness will accommodate the headless press fit bushings which are specified, and will provide adequate support for the drill. If head press fit bushings are specified, a thinner jig plate could be used.

The next step in the design of this jig is to add the clamping device. This type of plate jig can use several different clamping devices; however, the simplest and fastest acting clamp would be an L-shaped clamp, as shown in *Figure 5-52*. Here the clamp is mounted as shown and is activated by the screw in the long leg of the L. As the screw is tightened, the part is pressed against the secondary locators. At the same time, the short leg of the L presses against the part, forcing it against the tertiary locator. This type of locator may take a little longer to make than a simple screw clamp positioned at the side and end, but the faster operation of this clamp will more than offset the additional

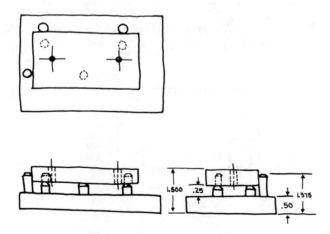

Figure 5-51. Sketch in the jig plate.

costs of its manufacturing.

To add drill bushings to this design, the first step is to locate the bushings over the holes sketched in the part (*Figure 5-53*). Next, dimensions and tolerances of the bushing location must be established. Using the 50% rule, the dimensions and tolerances should be shown.

The initial sketch of the jig is now complete. Since the holes are only 0.188″ (4.78 mm) diameter, the jig can be hand held and does not require an additional holding device. Likewise, since the jig is so small, there is little chance of its interfering with the operation of the machine tool, so the machine elements may or may not be sketched, depending on the desire of the designer. The final step in the design of this jig is to complete the final tool drawings. Remember, look for any interferences that may have been overlooked in the sketch.

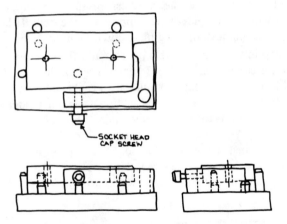

Figure 5-52. The simplest and fastest clamp for this part is the L-shaped clamp.

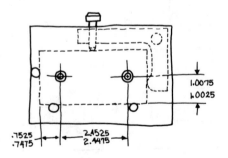

Figure 5-53. Add drill bushings.

Reaming

Jig design for reaming is basically the same as for drilling, which has been discussed throughout the chapter. The main difference is the need to hold closer tolerances on the jigs and bushings, and provide additional support to guide the reamer. For long holes, it is essential to guide the reamer at both ends, as shown in *Figure 5-54a*, using special piloted reamers designed for this purpose. Jigs should be designed so that the pilot enters the bushing before the reamer enters the workpiece, and remains piloted until the reaming operation is completed. For short holes, the reamer is usually guided at one end only, as seen in *Figure 5-54b*, with the bushing sized to fit the OD of the reamer. Additionally, bushings for reaming are generally longer than for drilling, usually three or four times the reamer diameter. Chip clearance is also generally less for reaming than for drilling, varying from one-fourth to one-half the tool diameter down to a maximum of 0.125-0.24'' regardless of the reamer diameter.

Bushing bores must be closely controlled. Bushings that are too small can cause tool seizure and breakage. Bushings that are too large will result in bellmouthed or out-of-round holes. Data in the *Handbook of Jig and Fixture Design*[3] can be used as a guide.

Carbide bushings should be considered for long production runs or where abrasive conditions are present. Roller or ball bearing, rotary-type bushings also provide maximum wear while maintaining close tolerances under high loads.

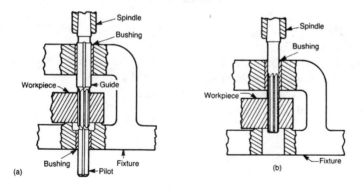

Figure 5-54. Fixtures for guiding reamers.

Boring Jigs. Boring jigs are quite often constructed similar to box jigs. The principal difference between boring jigs and box drill jigs is the bearing point on the side opposite the surface to be machined. As shown in *Figure 5-55*, the workpiece is located and clamped inside the box. The boring bar is

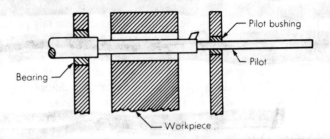

Figure 5-55. Boring jig.

then passed through the part and the pilot bearing is aligned in the bore on the opposite side of the jig. The part is then bored by feeding the rotating boring bar into the part. In some designs both the outer end of the boring bar and the driven end of the bar are supported during the boring cycle. This design is normally used where extreme accuracy is of prime importance.

REFERENCES

1. F.L. Rush, "Drill Jig for Uniformly Spaced Radial Holes," *Machinery* (September 1957).
2. H.G. Frommer, "Rotating Fixture on Angle Plate Indexes Parts for Drilling," *American Machinist* (December 1, 1949).
3. William E. Boyes, *Handbook of Jig and Fixture Design*, 2nd ed. (Dearborn, MI: Society of Manufacturing Engineers, 1989).

JIG DESIGN

Review Questions

1. What is a jig?
2. What is a box jig?
3. What is an advantage to using a sandwich jig?
4. What are the three general styles of drill bushings?
5. What is the recommended thickness of a jig plate?
6. What should be done to the drill bushing when drilling an angular surface?
7. Design a plate jig for the workpiece shown in *Figure A*. This is a prototype jig for 50 pieces.

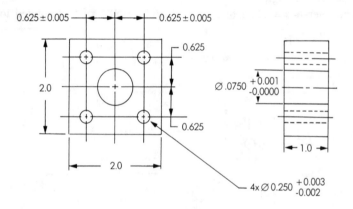

Figure A

8. What relationship should exist between the workpiece tolerance and the tolerance applied to the jig?
9. When making the initial sketch of the jig, what is the purpose of using a second color for the sketch of the workpiece?
10. List four criteria a jig must meet to provide for the efficient and productive manufacture of quality workpieces.

JIG DESIGN

Answers to Review Questions

1. A tool or workholder designed to hold, locate, and support a workpiece while guiding the cutting tool throughout its cutting cycle.
2. A jig that is constructed to completely enclose a workpiece.
3. The workpiece is supported on both sides. It can be drilled from one side, and a secondary operation such as reaming can be performed from the opposite side.
4. Press fit, renewable, and liner.
5. One to two times the diameter of the cutting tool.
6. Modify the end to suit the angular surface.
7. (See *Answer Figure A.*)
8. The tolerance applied to the jig should be 50% of that applied to the workpiece.

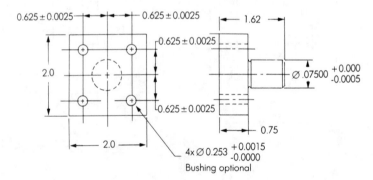

Square drill jig with top of part and drill first hole.
Pin and drill remaining holes.

Answer Figure A.

9. To make the workpiece standout from the workholder.
10. a. Correctly locate the workpiece with respect to the tool.
    b. Securely clamp and rigidly support the workpiece during the operation.
    c. Guide the tool.
    d. Position and/or fasten the jig on a machine.

6

FIXTURE DESIGN

Fixtures are workholders which are designed to hold, locate, and support the workpiece during the machining cycle. Unlike jigs, fixtures do not guide the cutting tool, but rather, provide a means to reference and align the cutting tool to the workpiece. Fixtures are normally classified by the machine with which they are designed to be used. A sub-classification is sometimes added to further specify the fixture classification. This sub-classification identifies the specific type of machining operation the fixture is intended to perform. For example: a fixture used with a milling machine is called a milling fixture; however, if the operation it is to perform is gang milling, it may also be called a gang-milling fixture. Likewise, a bandsawing fixture designed for slotting operations may also be referred to as a bandsaw-slotting fixture.

The similarity between jigs and fixtures normally ends with the design of the tool body. For the most part, fixtures are designed to withstand much greater stresses and tool forces than jigs, and are always securely clamped to the machine. For these reasons, the designer must always be aware of proper locating, supporting, and clamping methods when fixturing any part.

In designing any fixture, there are several considerations in addition to the part which must be addressed to complete a successful design. Cost, production capabilities, production processing, and tool longevity are some of the points which must share attention with the workpiece when a fixture is designed. The following is an explanation of the basic considerations which must be addressed before a design is completed and a fixture built. As with other special tools, these should be modified to suit the particular workpiece or production situation.

General Considerations

As with all tooling, the first consideration in fixture design is the cost versus the benefit. The production quantity, rate, or accuracy must warrant the added expense of special tooling. In addition, the fixture must pay for itself with savings derived from its use in as short a time as possible.

Once the decision is made that a fixture is required, the part print, process specifications, and other production documents are studied to determine the best and least expensive type of tool. Details such as locating and supporting the part, clamping, and tool referencing must all be considered in the initial design.

The tool designer should know enough of economics to determine the relative cost/benefit relationship of any tool design. Decisions such as whether temporary or permanent tooling should be used or just how much money should be requested for a workholder are decisions the designer must make to present management with a realistic picture of production costs. Not only is a tool designer expected to prepare accurate cost estimates, but these estimates must be completely defensible.

Machine Considerations

The machine tool specified for the machining operation should also be examined. Every detail which might affect the mounting and operation of the fixture must be addressed. Table sizes, table travel in all axes, spindle size and movement, spindle swing, distance between centers, and workholder mounting methods are typical examples of points the designer must consider before beginning a fixture design.

A major consideration is the condition of the workpiece to be held by the proposed workholder for the next operation to be performed. This includes the physical characteristics of the workpiece, i.e., whether it is round, irregular, large, heavy, of weak or strong sections, and the like. The operation to be performed and the physical characteristics will dictate whether the workpiece has to remain stationary or be moved along a definite path relative to the cutting tool. Different machine tools have means to provide the needed motions for the workpiece and the cutting tools. Most of these motions are straight-line or rotary, or combinations of both. There are operations that to all appearances seem to be producing and requiring an irregular path of the cutting tool on the workpiece. This seemingly irregular path is often the result of the combination of several straight-line and rotary motions. Except for speed consideration and the degree of simplicity obtainable, only the relative motion of the cutting tool to the workpiece is of importance. For instance, the turning of flanges on a valve body or a flanged-T pipe fitting could be performed with either the workpiece or the tool revolving (*Figure 6-1*).

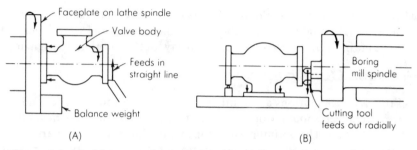

Figure 6-1. Machining a valve body by either (*A*) workpiece rotation or (*B*) tool rotation.

Whether the workpiece or cutting tool moves in a straight line, revolves, or moves in some combination of both, design requires careful coordination of workholder to the workpiece, and the workholder to the machine tool. Operations in which the workpiece revolves require great care in the attachment of the workholder to the machine tool and the means of actuation of the workholder. Unbalanced masses in the workholder and the workpiece must be minimized by proper balancing. This is particularly true in high-speed applications such as turning with tungsten carbide, diamond, and ceramic cutting tools.

Workpiece weight and size influence what type and size of machine tool can or should be used for a particular operation. The combined weight of the workholder and the part must be carefully matched to the capacity of the ways, bed, table, and spindles of the machine tools. Excessive weight may cause distortion in the machine tool and produce inaccurate work.

In cases where the machine specified is not the best choice, or where the machine cannot perform the required operation, the designer should consult the process engineer to see if another machine tool could be used. While it is generally good practice to fully utilize a machine's capabilities, overloading a small machine tool is not efficient. Many times money can be saved by using a larger machine tool. Running a small machine to its maximum limits or capabilities over a long production run will not only shorten the life expectancy of the machine tool, but may also fail to give the desired degree of precision. If there is any question as to the estimated power required for a job and the available power in the machine tool, the designer or process engineer should select the machine with more power available than that required for the operation. Before any fixture design is finalized, the designer must make sure the machine tool will accommodate the fixture in every respect.

When designing a fixture, another factor which must be carefully evaluated is the cutting tool. As with jigs, cutters specified for fixtures should, whenever possible, be standard, off-the-shelf sizes. Designing a special cutter when a standard cutter would work is very costly. Even if the standard cutter requires a slight modification of the fixture, the benefits and savings derived over the life of the fixture would more than pay for the additional design time.

Process Considerations

Process considerations are another area where the tool designer and the process engineer should work together to determine the best method to produce the part. The optimum situation in fixturing is where the part can be clamped one time and not removed from the fixture until the part is completely machined. This reduces costly part handling and also minimizes the chance of errors in switching the part from one fixture to another. In cases where the part must be removed from the fixture, the tool designer and process engineer should find the methods of processing the part which require the least amount of fixture changes. When more than one machine is required, consider moving the fixture and part from machine to machine.

To achieve the lowest cost per part in production, the fixture must be fast operating, easy to load and unload, and have a positive, foolproof method of locating the part. In addition, when the volume of production permits more money to be spent, semi-automatic or automatic tooling should be considered. As the volume of production increases, the opportunity to save more per part increases dramatically. Multi-part machining and power clamping are also areas which could be explored to reduce the cost per part in production.

On all operations, the magnitude and direction of the forces produced by the material-removing operation determine the necessary holding forces. Cutting forces must be held within limits, so that the part itself cannot be distorted to an amount that would affect the required accuracy. Rigidity and strength of the workpiece limit the applicable holding forces and the speed and amount of metal removal per unit of time. A thin-walled part may not be able to sustain heavy cutting and holding forces without distortion or damage. The mounting or attachment of the workholder or the workpiece to the machine tool should be so arranged that the forces produced in the material-removing operation are absorbed by the strongest and most rigid parts of the machine tool. The cutting forces should tend to hold the workholder down against the bed of the machine rather than lift it away. Projection between the point at which the cutting force is applied and the nearest support should be minimized. Cutting forces that act parallel to the bed, table, and face plate should be applied as close to the bed as possible. Cutting forces should never be permitted the advantage of a large lever arm which could increase the tendency to loosen or pry away the workpiece and the workholder from their attachment. This is in contrast to the holding forces, where the effect of a large lever arm or mechanical advantage is always desirable.

Figure 6-2 shows a lathe chuck and a short workpiece to be turned. The cut is relatively close to the supporting spindle bearing and no difficulty is expected. Increasing the workpiece length, however, will give the cutting tool greater leverage; higher side thrust will be produced and may cause difficulties in the spindle bearing. The workpiece will deflect much more if the same size

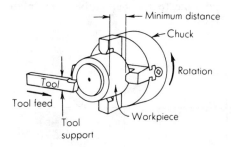

Figure 6-2. Minimizing cutting force by applying holding force as near as possible to point of tool application.

of the chip is removed as on the short piece. The results may be inefficient material removal, low accuracy, and poor tool life. To reduce the effect of the large lever arm, a center support can be used. If this does not improve performance, then a steady rest must be used. The center and the steady rest may be called workholders. They hold, locate, and support. *Figure 6-3* shows a turning operation with the workpiece held by a lathe chuck. The steady rest

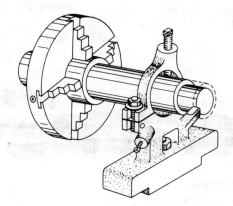

Figure 6-3. Steady rest used to support workpiece in area of cutting-force application.

supports the bar to be turned while three adjustable shoes center and support the bar. This support absorbs most of the bending forces. *Figure 6-4* shows a turning operation with a steady rest and center support. The cutting tool is directed against the workpiece near the center support. Cutting forces should be absorbed as near as possible to the point where the force is created. Place the supports for the workpiece near the cutting forces. Apply the cutting forces as close as possible to the bed, table, face plate, and the spindle bearings.

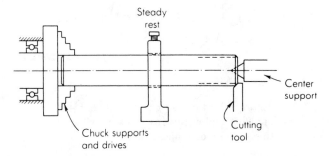

Figure 6-4. Steady rest and center used to support workpiece in area of cutting-force application.

Product Considerations

In conjunction with the other areas, the part must also be a major consideration in the design of a fixture. All relevant details such as location surfaces, clamping surfaces, clamping methods, areas requiring machining, the amount of machining required, and the degree of surface finish desired are all elements which must be considered in the initial phase of fixture design. While each of these design considerations has been treated separately in practice, all are usually considered simultaneously. So, when designing a fixture, constantly consider all effects a design decision may have on the complete, overall function and operation of the tool. Remember, changes in a fixture design are much easier and less expensive to make while the tool is still on paper. Changes or corrections in a fixture design are very expensive when they must be made after the tool is built, or in production.

Types of Fixtures

Fixtures are classified either by the machine they are used on, or by the process they perform on a particular machine tool. However, fixtures also may be identified by their basic construction features. For example, a lathe fixture made to turn radii is classified as a lathe radius turning fixture. But if this same fixture were a simple plate with a variety of locators and clamps mounted on a faceplate, it is also a plate fixture. Like jigs, fixtures are made in a variety of different forms. While many fixtures use a combination of different features, almost all can be divided into five distinct groups. These include plate fixtures, angle plate fixtures, vise jaw fixtures, indexing fixtures, and multipart or multistation fixtures.

Plate Fixtures

Plate fixtures, as their name implies, are constructed from a plate with a variety of locators, supports, and clamps (*Figure 6-5*). Plate fixtures are the most common type of fixture. Their versatility makes them adaptable for a wide range of different machine tools.

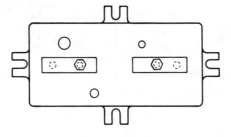

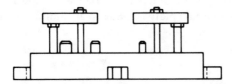

Figure 6-5. Plate fixtures.

Plate fixtures may be made from any number of different materials, depending on the application of the fixture. For example, if a very large fixture is needed, and it is only required to make a few parts, aluminum or magnesium plate may be selected to keep the weight of the fixture to a minimum. If, however, weight is not a factor and a very large number of parts must be made, another material such as tool steel may be selected. Likewise, if the part or process requires a material which is resistant to corrosion, a nickel-based alloy might be selected. But, as is usually the case, a combination of different materials may also be used to construct the plate fixture. The part being machined and the process being performed are the only guides the tool designer has when selecting the material to use for any fixture.

Angle Plate Fixtures

The angle plate fixture (*Figure 6-6*) is a modified form of plate fixture. Here, rather than having a reference surface parallel to the mounting surface, the angle plate fixture has a reference surface perpendicular to its mounting surface. This construction is very useful for those machining operations which are performed perpendicular to the primary reference surface of the fixture.

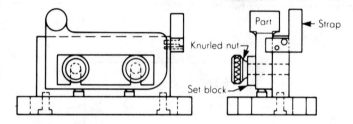

Figure 6-6. Angle plate fixture.

Another variation of the basic angle plate fixture is the modified angle plate fixture (*Figure 6-7*). This design differs from the basic angle plate in that where the angle plate fixture is designed to be at a 90° angle to its mounting surface, the modified angle plate is made to accommodate angles other than 90°, as shown.

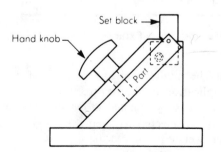

Figure 6-7. Modified angle plate fixture.

Vise Jaw Fixtures

Vise jaw fixtures are basically modified vise jaw inserts which are machined to suit a particular workpiece. In use, these modified vise jaws are installed in place of the standard, hardened jaws normally furnished with milling machine vises. Vise jaw fixtures are the least expensive type of fixture to produce, and since there are so few parts involved, they are also the simplest to modify. *Figure 6-8* shows several examples of parts which could easily be fixtured with this type of workholder.

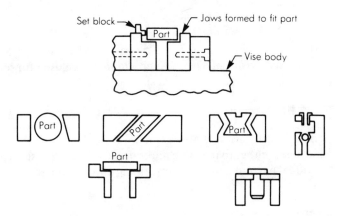

Figure 6-8. Vise jaw fixture.

The principal advantage of using this type of fixture is that only the locating elements need to be constructed to suit each part. The milling machine vise contains the clamping elements and the means to attach the fixture to the machine table. In addition, if simple or compound angles need to be machined, a vise with these capabilities may also be used. The only limitations to using this type of fixture are the size of the part and the capacities of the available vises.

Indexing Fixtures

Indexing fixtures (*Figure 6-9*), like indexing jigs, are used to reference workpieces which must have machine details located at prescribed spacings. The typical indexing fixture will normally divide a part into any number of equal spacings, such as those used for geometric shapes or gears. However, some indexing fixtures may also be used to locate and reference a workpiece for unequal spacings. Regardless of the configuration of the workpiece,

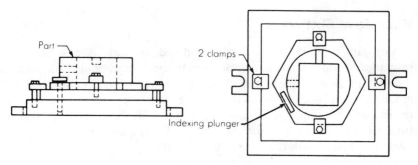

Figure 6-9. Indexing fixture.

indexing fixtures must have a positive means to accurately locate and maintain the indexed position. The most common device used for location and indexing is a simple indexing pin, as shown.

Multi-Part Fixtures

Multi-part or multi-station fixtures (*Figure 6-10*) are normally used for one of two purposes: either to machine multiple parts in a single setting, or to machine individual parts in sequence, performing different operations at each station.

Fixture Classifications

As mentioned before, fixtures are normally classified by the machine tool with which they are designed to be used. The following is a brief decription of the major fixture classifications and their basic design characteristics.

Milling Fixtures

Milling fixtures are the most common type of fixture in general use today. The simplest type of milling fixture is a milling vise mounted on the machine table. However, as the workpiece size, shape, or complexity becomes more sophisticated, so too must the fixture. The following are a few points which should be considered in designing fixtures for milling operations:

1. The design should permit as many surfaces of the part to be machined as possible, without removing the part.
2. Whenever possible, the tool should be changed to suit the part. Moving the part to accommodate one cutter for several operations is not as accurate or as efficient as changing cutters.
3. Locators must be designed to resist all tool forces and thrusts. Clamps should not be used to resist tool forces.

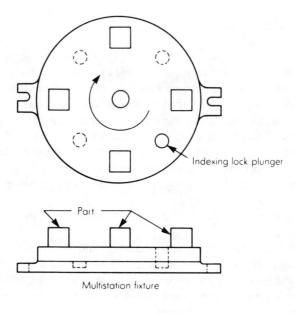

Indexing lock plunger

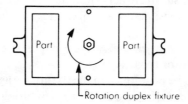

Part

Multistation fixture

Part   Part

Rotation duplex fixture

Figure 6-10. Multi-part or multi-station fixture.

4. Clearance space or sufficient room must be allotted to provide adequate space to change cutters or to load and unload the part.
5. Milling fixtures should be designed and built with a low profile to prevent unnecessary twisting or springing while in operation.
6. The entire workpiece must be located within the area of support of the fixture. In those cases where this is either impossible or impractical, additional supports, or jacks, must be provided.
7. Chip removal and coolant drainage must be considered in the design of the fixture. Sufficient space should be permitted to allow the chips to be easily removed with a brush.
8. Set blocks or cutter setting gages must be provided in the fixture design to aid the operator in properly setting up the tool in production.

Lathe Fixtures

The same basic principles that apply to the design of milling fixtures also apply when designing turning fixtures. The only major difference between the two is the relationship between the workpiece and the cutting tool. In milling, the workpiece is stationary and the cutting tool revolves. However, with turning operations, the workpiece revolves and the cutting tool is stationary. This situation creates another condition which the tool designer must deal with—centrifugal force. The complete fixture must be designed and constructed to resist the effects of the rotational, or centrifugal, forces present in the turning. The following are a few basic design principles which apply directly to lathe fixtures:

1. Since lathe fixtures are designed to rotate, they should be as lightweight as possible.
2. Lathe fixtures must be balanced. While perfect balance is not normally required for slow-speed turning operations, high rotational speeds require the fixture to be well-balanced. In most fixtures, balance is achieved by using counterweights positioned opposite the heaviest part, or area, of the workpiece.
3. Projections and sharp corners should be avoided since these areas will become almost invisible as the tool rotates and they could cause serious injury.
4. Parts to be fixtured should, whenever possible, be gripped by their largest diameter, or cross section.
5. The part should be positioned in the fixture so that most of the machine operation can be performed in the first fixturing.
6. Clamps should be positioned on surfaces, or areas, which are rigid before and after machining. Clamping over an area which is to be bored to a very thin wall thickness could cause the part to warp or deform, thus causing the hole to be bored incorrectly.
7. As with other fixtures, some means of cutter setting should also be incorporated into the design. However, since the workholder will be rotating, this setting device should be removed.
8. Whenever possible, standard lathe accessories should be adapted in the design of turning fixtures. Lathe faceplates are an ideal method to mount large fixtures. Likewise, a standard lathe chuck, or collets, can and should be modified for many fixturing applications.

Grinding Fixtures

Grinding fixtures are actually a family of fixtures rather than a single classification. The major types of fixtures used for grinding are surface grinding fixtures and cylindrical grinding fixtures. The following is a brief discussion of the design characteristics peculiar to grinding fixtures.

Surface Grinding Fixtures

1. Surface grinding fixtures are similar in design to milling fixtures, but made to much closer tolerances.
2. Whenever practical, use magnetic chucks to hold the workpiece. In these cases, the fixture is simply a device to contain the workpiece and prevent any lateral or transverse movement of the part.
3. Provide adequate room or slots to permit the escape of coolant and to allow easy removal of built-up grinding sludge.
4. Provide coolant containment devices or splash guards to keep the fixture from spilling coolant on the floor around the machine.
5. Fixture elements which are in contact with the magnetic chuck should be made from ferrous materials if they are intended to be held on the chuck. If, however, they are not to be held to the chuck, then a nonferrous metal should be specified.
6. Include provisions for rapid wheel dressing and truing in the design of the fixture, if not built into the machine.
7. All locators must be accurately and positively positioned.

Cylindrical Grinding Fixtures

1. Cylindrical grinding fixtures are often very similar to lathe fixtures.
2. Since cylindrical grinding is normally a secondary operation, performed after turning, it is often desirable to use the same center holes for grinding as were used for turning the part.
3. Coolant build-up is seldom a problem with cylindrical grinding; however, sludge removal is still one area that must always be considered.
4. Cylindrical grinding fixtures should always be perfectly balanced to achieve the desired results.
5. When possible, use standard accessories and attachments. These include grinding collets, chucks, and drive plates with grinding dogs.
6. Incorporate provisions for wheel dressing and truing into the design.

Boring Fixtures

Boring fixtures are designed to hold the workpiece while the part is bored. These fixtures differ from boring jigs in that a boring fixture does not have any provision for guiding or supporting the boring bar. Boring fixtures are normally used for large parts with large holes where the boring bar is rigid enough to provide additional support. A pilot bushing is not needed.

Boring fixtures, like milling fixtures, should have some provision for setting the position of the cutting tool relative to the part. In cases where a boring fixture is to be used on a very large machine, such as a boring mill, or vertical turret lathe, it is also good practice to include areas on the fixture to insure proper alignment with the machine.

Broaching Fixtures

Broaching fixtures are normally designed to simply hold and locate a part relative to either an internal or external broach. Since there is a great deal of cutting force exerted during broaching, the complete fixture must be built more substantially than those for other processes.

Internal broaching fixtures need only locate and hold the part in proper position relative to the hole in the broaching machine. Most broaching performed today is of the pull type and tends to keep the part firmly seated on the fixture. However, clamping devices are necessary to establish the proper relationship and maintain the position of the part until the broaching pressure pulls the part against the table.

External, or surface broaching requires a different approach to fixturing. Since this type of broaching is performed on the outside of a part, the fixture must be designed to resist both pulling thrust and perpendicular thrust which tends to try to push the part away from the broach. In either case, the principal purpose for a broaching fixture is to maintain the proper relationship between the part and the cutting tool and to prevent the part from moving or tilting.

Sawing Fixtures

The two primary machines commonly used for production sawing operations are the vertical bandsaw and the horizontal bandsaw. With both types of machines, the main intent is to accurately position and hold the workpiece so it can either be sawed into pieces or slotted with the saw blade. The following are a few design characteristics which are peculiar to these saws and the sawing process in general:

1. When possible, standard saw accessories and attachments should be used in conjunction with fixturing elements.
2. Clamps, locators, supports, or similar fixture details must be kept clear of the blade path. Since the area occupied by the saw blade on both types of bandsaws extends well above and below the actual working area, any fixture elements which overhang could interfere with the normal operation of the saw.

3. Provisions for coolant drainage and chip disposal must be planned into the fixture design. While most bandsaws have an internal chip disposal system, a significant amount of chips will also collect in the fixture unless some means for their elimination is planned.
4. When practical, the table slots should be used to reference the fixture to the saw blade.
5. Use power feed whenever possible. This may require designing a means to secure the power feed chain to the fixture.

Standard Mounting of Fixtures

Quite often, only standard accessories such as clamps, straps, T-slot bolts, T-slot nuts, and jacks are needed to hold a workpiece. This is especially true where only a few pieces have to be produced and where economic considerations do not justify more elaborate workholders. Most machine tools have provisions to receive such equipment. The beds and tables of machine tools such as drill presses, boring mills, and jig-bores have T-slots milled into their work tables. The spindles of machine tools such as lathes and grinders have spindles to which, directly or by means of suitable adapters, the various types of workholders, chucks, arbors, and collets may be attached. There are standards for the sizes and the spacing of T-slots of tables, beds, and other equipment to which workholders are to be attached. Standards for spindles are also established. Adherence to standards can result in economic interchangeability and multiple sources of tooling. Tool costs will be lower because the supplier can produce standard components in larger quantities at lower cost.

Always be sure the fixture can be solidly located on the machine. A fixture that rocks or cannot be kept from twisting under cutting forces will not provide acceptable accuracy.

Relationship Between Fixture and Cutting Tool

The direction and magnitude of the forces created at the cutting tool-workpiece interface must be known. Use this knowledge to minimize the size of force moments (force times distance of force application) by reducing the moment arm. Avoid excessive projection. Provide additional supports where needed.

The relative motions between the workpiece and the cutting tool may change the tool geometry during the cutting cycle. Rake and clearance angles may change from a selected optimum condition to a bad condition.

Every material removing operation has as its final objective the removal of a certain amount of material per unit of time, to a certain depth, thickness, diameter, contour, and other related specifications. These quantities can be obtained by control of machine motions with stops, gages, or tape controls. Workholders may be equipped with stop surfaces such as the top of a drill bushing against which a drill stop abuts. Setting gages may be placed between a milling cutter and a reference plane on a workholder to obtain the right thickness of a workpiece after milling (*Figure 6-11*). Swing stops on lathe fixtures and arbors may be used to locate workpieces for removing stock evenly from both sides (*Figure 6-12*).

It is necessary to check for and control interference between any cutting edge of the cutting tool and any part of the workholder during any possible

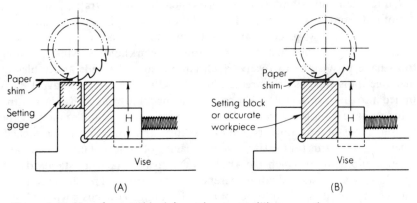

Figure 6-11. Use of a gage block in setting up a milling operation.

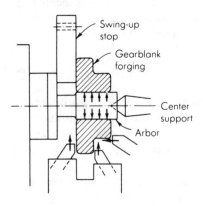

Figure 6-12. Lathe fixture with swing stop.

contact of workholder and cutting tool. It is advisable to simultaneously check for, minimize, and control excessive nonproductive approaches of cutting tools to the workpiece. Planning should be done to avoid "cutting air."

Space must be provided to remove and load workpieces easily without danger to the operator or damage to the workpiece and equipment. Space is also required for the insertion and removal of the cutting tools. This includes space for the application of any wrenches, keys, or other tools used for change of cutting tools. Such change should be possible without removing the workpiece.

Tool Positioning

Tool positioning refers specifically to locating the tool with respect to the work, or vice versa.

Prior to setting up the workpiece, the blueprint and workpiece are studied to determine primary and secondary locating points or surfaces. Once these are determined, it is then necessary to visualize how these points or surfaces may be accurately located in relation to the locating means.

Relationship to Locators

The locating means are the alignment or gaging surfaces of any angle, plate, bar, V-block, vise, or the like that is secured to or is a part of the work table or fixture for properly positioning the work relative to the tool (*Figure 6-13*). A T-slot may be considered a locating means, but generally such slots are only accurate enough for rough machining.

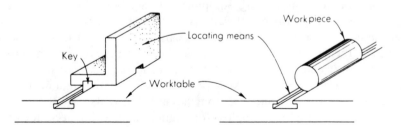

Figure 6-13. Positioning a workpiece relative to locating means.

Keys are used under the base of an angle plate, vise, or workholding fixture. They provide an easy and accurate method of aligning a workholding fixture to a T-slot to the same degree of accuracy as the T-slot itself. Before using the T-slots as a basic locating reference, their accuracy should be established in relation to table or cutter movement. Removable keys are used extensively, especially in vises with slots at right angles to each other for lengthwise or crosswise mounting. Both the key and the T-slots should be periodically inspected for wear to ensure proper dimensions and accuracy between the key and locating means.

The most practical procedure for establishing the relationship of the tool to the work will be governed by the type and size of the machine, type and size of work, production rate, and specified dimensions and tolerances of cut. The results of correct tool positioning are proper depth and location of the finished cut. Regardless of the type of machine involved, there are several different techniques for locating the work in relation to the tool, the exact technique being determined by the specific job requirement. Mass production may dictate great expenditure of money to minimize the time required for locating the part, compared with less expenditure and more time for location of a single part.

Cutter Setting Devices

Gage or setup blocks are common means of reference for cutter setting (*Figure 6-14*). In many cases, the reference may be a designated surface on a locator (locating means). In its correct position, the cutter should clear the setting surface by at least 0.031″ (0.8 mm). Usual shop practice employs only one feeler thickness to avoid use of the wrong size on any particular operation. The thickness of the feeler to be used should be stamped on the fixture base near the setup block.

Optimal methods are also used extensively for gaging the accuracy of tool position in relation to the work and also to the locating means. The multiplicity of optical tooling equipment design makes possible broad applications in tool-to-work locations. One typical application is the establishment of an exact drill center location. An optical instrument is inserted in the chuck. Through high magnification, the eyepiece and cross hairs are used to determine the exact center. The work is clamped securely to the locating means and rechecked for proper tool center alignment. The optical instrument is then removed from the chuck and replaced with the drill. The same principle may be applied for gaging the indexing accuracy of a rotary fixture or index plate onto which work is positioned for machining.

Fixture Design Process

The following step-by-step approach will prove of value for the design of all fixtures:

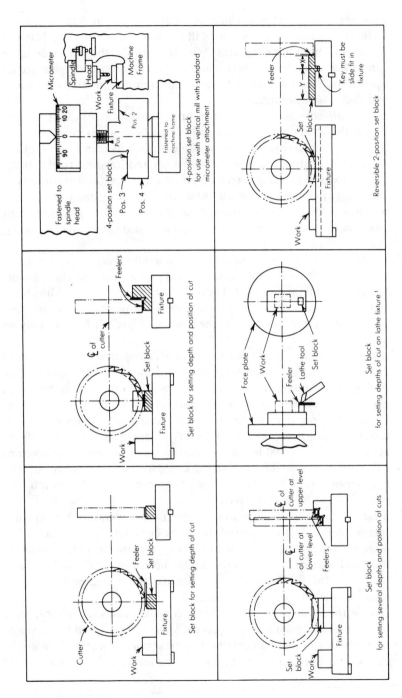

Figure 6-14. Cutter setting with set blocks and feelers.

1. *View the workpiece as a whole.* A study of the complete workpiece, including its intended function, will disclose the relationship between the various workpiece features. With this insight, the fixture designer can either establish or better understand a proposed sequence of operations. Such knowledge may enable him to combine operations and thus, minimize fixturing.

2. *Gather all necessary data.* All information that may affect jig and fixture design should be readily available. The designer must know the physical characteristics of the workpiece, such as composition of the material, condition (hardness), and rough and finished weight. If the workpiece is to be made directly from raw material (mill extrusions or sheet stock), the designer must know the shape, size, and tolerances of the mill stock. All production data, including total pieces, rate of production, tooling budget allotted, and proposed production sequence, must be available.

3. *Consider standard workholders.* Many operations can be performed by using available commercial workholders such as machine vises, T-slots and bolts, jacks, and clamps. The design of a special workholding jig or fixture must be economically justifiable. If the planned operations are similar to present operations, the rework of present fixtures may be considered.

4. *Determine what special workholders will be required.* Every proposed operation should be carefully examined to see whether it can be performed economically with commercial workholders (machine vises and so on), or whether available fixtures can be used. The designer should also consider whether available fixtures with minor alterations can be used. After assigning as many operations as possible to commercial or available workholders, a small number of operations will remain for which special workholders must be provided. The number of special workholders may be further reduced by combining operations within a single fixture.

5. *Study fixtures for similar operations.* Every operation for which a special workholder is required should be considered individually. The designer should seek out—within his own plant, in other plants, in technical journals, and so on—similar operations for which fixtures were provided. By examining a number of existing fixtures, the designer can combine the best features of each.

6. *Review the fixturing plan.* The designer should consider, in turn, every operation of the production sequence and review his fixturing decisions. At every step he should confirm that the proposed workholder will be structurally adequate to withstand cutting forces and will be as precise

as required (location aspects). The preliminary design of the required special workholders should be completed.

7. *Execute the fixturing plan*. The designer should remain available during the execution of the fixturing plan. No plan can be regarded as final because of possible changes in workpiece dimensions and the typical cut-and-try aspects of much fixture planning. After the line has been used for production, further process improvement may suggest fixturing changes.

A well-designed fixture may be used on several different machines. *Figure 6-15* shows a workpiece and its fixture as it might be processed in several different ways. The workpiece is a magnesium casting; the base and primary reference plane are established at the first milling operation. For this, the

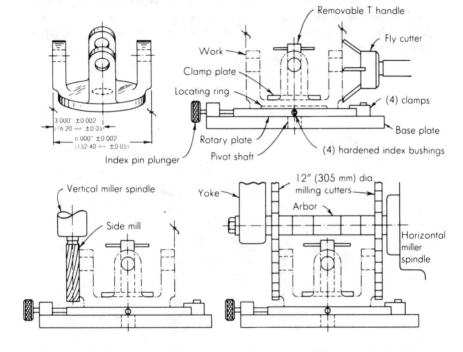

Figure 6-15. Fixtured workpiece processed by several different methods.

second milling operation, the workpiece is clamped to the fixture and is repeatedly indexed to present four surfaces to the milling cutters. The rotating plate of the fixture has hardened bushings at 90° intervals into which an index pin plunger can nest.

Designing a Fixture

Fixtures, like jigs, should be designed using a systematic approach. This will ensure each element in the design is considered and that nothing is forgotten or overlooked. Using the process outlined in Chapter 5, JIG DESIGN, the following method may be used to design a simple plate fixture.

The part to be machined is the stop block in *Figure 6-16*. The first step is to study the part and determine the basic requirements and relevant information about the part. In this case, the requirement is for a simple fixture to hold the part for gang milling a stepped shoulder. The part is 3" x 2" x 1" (76 x 51 x 25 mm) 6061-T6 aluminum bar stock. The production plan calls for the part to be cut from 1 x 2" (25 x 51 mm) bar to a tolerance of three inches ±0.005" (76± 0.13 mm). The production run is specified as 750 parts per month. The fixture should be designed as a permanent tool, yet the cost must be kept to a minimum. Since gang milling is specified, a horizontal milling machine will be used to machine these parts.

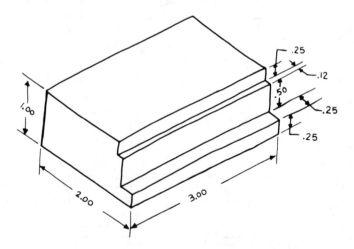

Figure 6-16. Rough sketch of a stop block.

The first step is to sketch the part in three views (*Figure 6-17*). If possible, the part should be sketched full size. However, if the part is too large, use half or quarter scale. Remember, make the sketch as accurate and true to scale as possible. Sketching the part in a different color will also help reduce confusion as the design develops and more lines are added to the initial design sketch.

Once the part is sketched, the base plate and part locators should be added to the sketch. As shown in *Figure 6-18*, a simple six-point locating method is

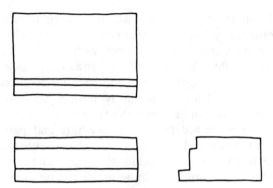

Figure 6-17. Sketch the part in three views.

used, since the part presents no special locational problems. As shown, the base plate acts as the primary reference surface. Since the primary reference surface on the part is flat, no special locators, other than the base plate itself, are installed. The secondary reference surface is located with two dowel pins positioned as shown, toward the rear of the plate. The final locator is positioned on the short side of the part, toward the end where the tool thrust

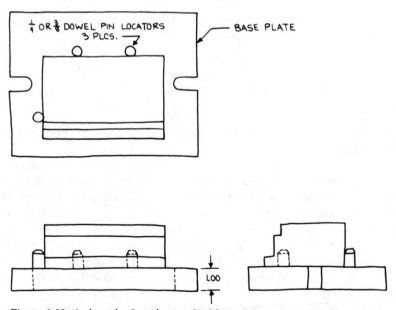

Figure 6-18. A six-point locating method is used.

will be directed. This locational method will be sufficient to accurately locate the part and resist the cutting forces that can be expected with this part. If, however, a different material, such as steel, were to be machined, a more substantial locator might be better suited. In such a case, a block, screwed and doweled to the base plate, might be desired to resist the additional tool thrust

The base plate selected for this fixture is one-inch (25.4 mm) steel plate This material was selected for its low cost and durability. The top and bottom surfaces should be machined to provide accurate and parallel locating surfaces for both the part and the fixture. The slots in either end provide a means to secure the fixture to the machine table. Drill and ream fixture keys are used to maintain the proper position of the fixture in the table T-slot (*Figure 6-19*). These were selected because of the relatively short time

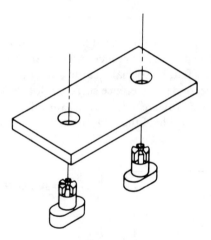

Figure 6-19. Fixture keys.

required for installation. If desired, other standard fixture keys could also be used (*Figure 6-20*).

The set block is now sketched into the design. The block selected is made 0.0625″ (1.588 mm) smaller than the actual part on both the referencing surfaces to accommodate the feeler gage. A feeler gage made from 0.0625″ (1.588 mm) stock will be used to position the cutters in the initial setup of the fixture. The location of the set block is determined by the dimensions on the part print. As shown, the first step is dimensioned as 1.625″ (4.28 mm) from the secondary reference surface and 0.75″ (19.1 mm) from the primary reference surface. The set block is then positioned to suit these dimensions (*Figure 6-21*).

The clamping device selected for this fixture is a toggle action clamp. While

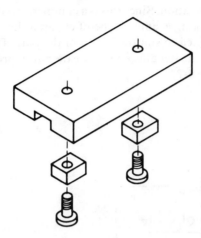

Figure 6-20. Standard fixture keys.

any of several different types of clamps could be used, a toggle action clamp offers the advantages of being fast acting, capable of being moved completely clear of the part, and easily modified to suit the part shape. The cost of this

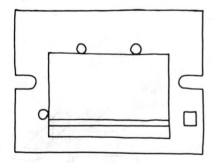

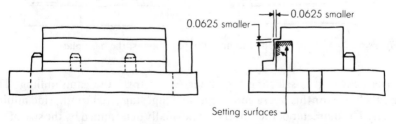

Figure 6-21. The set block is positioned.

clamp is also a consideration. Since this is a commercial component, the cost is much less than making a similar type of clamp in-house. The end of the toggle clamp is modified to suit the length of the part. This will spread the clamping force over a greater area and insure the part is properly held against the base plate (*Figure 6-22*).

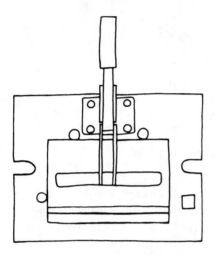

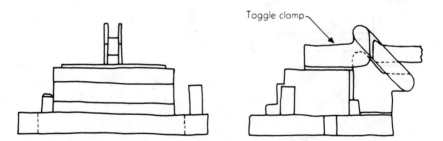

Figure 6-22. A clamp is added to hold the part against the base plate.

The next step is selecting the cutters to perform the gang milling. The logical choice for this operation is interlocking, staggered tooth, side milling cutters. The diameter of the cutters are normally determined by the size of the milling machine specified for the machining. However, for this example, the cutters specified have diameters of five inches and six inches (127 and 152 mm). Both cutter widths are 0.25″ (6.35 mm). The cutters are now sketched into the initial design sketch (*Figure 6-23*). The entire design must now be checked to make sure there is no interference between members or between the fixture and the machine tool. Once this inspection is complete, the final tool drawings are prepared.

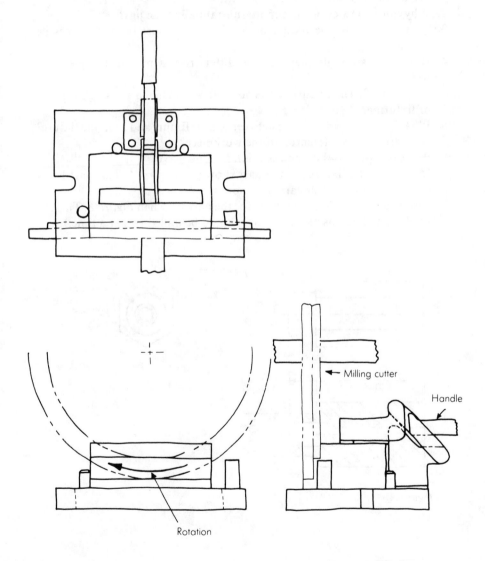

Figure 6-23. Cutters sketched into the initial design.

FIXTURE DESIGN

Review Questions

1. What are the four functions of a fixture?
2. Why must fixtures be built more substantially than jigs?
3. To what part of the fixture or machine tool must the tool forces be directed?
4. How does an angle plate fixture differ from a modified angle plate fixture?
5. Why should as many surfaces as possible be machined at a single setting, or fixturing?
6. What is the major difference between a lathe fixture and a milling fixture?
7. How are cutters referenced to the workpiece?
8. What two purposes do multistation fixtures serve?
9. Design a lathe fixture for the part shown in *Figure A*. This part has a production rate of 500 parts per month.
10. Design a straddle milling fixture for the part shown in *Figure B*. This is a single run of 750 parts. (The parts should be milled two at a time.)

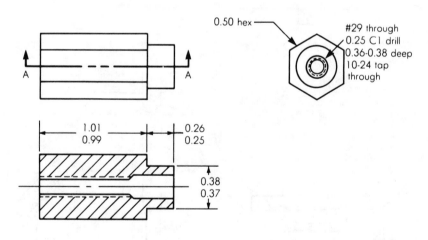

Figure A.

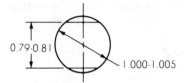

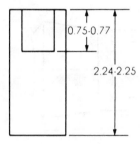

Figure B.

FIXTURE DESIGN

Answers to Review Questions

1. Hold, support, and locate the workpiece while referencing the cutting tool.
2. Because of the increased cutting forces.
3. Toward the most rigid part of the fixture or machine tool and toward the locators.
4. Angle plate fixtures are used for right angles (90°), while modified angle plate fixtures are used for all other angles.
5. To reduce the change of error in changing positions with the part and to reduce the cost of part handling.
6. The relationship between the cutting tool and the workpiece. With a lathe, the workpiece rotates and the tool is stationary. With a milling machine the workpiece is stationary and the tool rotates.
7. By set blocks, optics, or setting gages.
8. They permit the machining of multiple parts in a single setting, or they

allow machining of individual parts in a sequence where different operations are performed at each station.

9. (See *Answer Figure A*.)
10. (See *Answer Figure B*.)

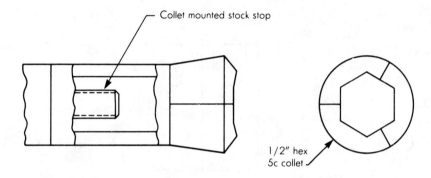

Collet mounted stock stop

1/2" hex
5c collet

Answer Figure A.

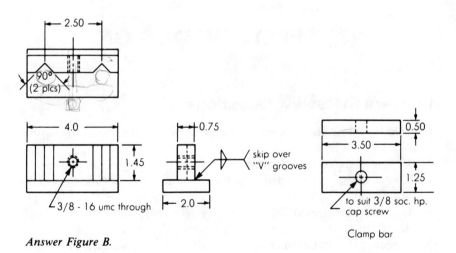

2.50

90°
(2 plcs)

4.0

1.45

3/8 - 16 umc through

0.75

skip over
"V" grooves

2.0

0.50

3.50

1.25

to suit 3/8 soc. hp.
cap screw

Clamp bar

Answer Figure B.

7

DESIGN OF
PRESSWORKING TOOLS

Characteristic of the pressworking process is the application of large forces by press tools for a short time interval, which results in the cutting (shearing) or deformation of the work material. A pressworking operation, generally completed by a single application of pressure, often results in the production of a finished part in less than one second. Pressworking forces are set up, guided, and controlled in a machine referred to as a press.

Power Presses

Essentially, a press is comprised of a frame, a bed or bolster plate, and a reciprocating member called a ram or slide, which exerts force upon work material through special tools mounted on the ram and bed. Energy stored in the rotating flywheel of a mechanical press (or supplied by a hydraulic system in a hydraulic press) is transferred to the ram for its linear movements.

Press Types

An open-back inclinable (OBI) press (*Figure 7-1*), also called a gap-frame press, has a C-shaped frame that allows access to its working space (between the bed and the ram). The frame can be inclined at an angle to the base, allowing for the disposal of finished parts and scrap by gravity. The open back allows the feeding and unloading of stock, workpieces, and finished parts through it from front to back.

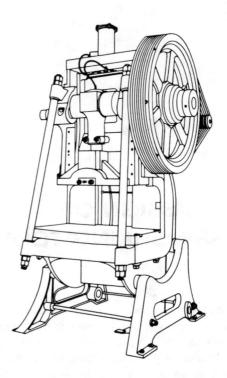

Figure 7-1. Open-back inclinable (OBI) press.

Major components of a press are as follows:

1. *Press Bed*. A rectangular part of the frame, often open in its center, which supports a bolster plate.
2. *Bolster Plate*. A flat steel or cast-iron plate, upon which press tools and accessories are mounted. Bolsters having standard dimensions and openings are available from press manufacturers. The Joint Industrial Council (JIC) standard dimensions for press bolsters are listed in reference 1.
3. *Ram* or *Slide*. The upper press member that moves through a stroke a certain distance depending upon the size and design of the press. The position of the ram, but not its stroke, can be adjusted. The distance from the top of the bed (or bolster) to the bottom of the slide, with its stroke down and adjustment up, is the maximum shut height of a press.
4. *Knockout*. A mechanism operating on the upstroke of a press, which ejects workpieces or blanks from the upper half of a press tool.
5. *Cushion*. A press accessory located beneath or within a bolster for producing upward motion and force; it is actuated by air, oil, rubber, high-pressure nitrogen, or springs.

A straight side press of conventional design has columns (uprights) at the ends of the bed, usually with windows (square or rectangular openings) to allow the feeding and unloading of stock, workpieces, and finished parts. This type of press can also be used for feeding from front to back (*Figure 7-2*).

A press brake is illustrated in *Figure 7-3*. It is essentially the same as a gap-frame press except for its long bed, generally from six to 20 feet (1.8-6.1 m) or more. A press brake is primarily used for bending operations on large sheet metal parts. It can also be used with a series of separate sets of press tools to do piercing, notching, and forming. This allows parts of a complex design to be

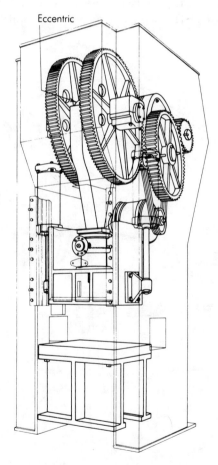

Figure 7-2. Single-action, straight-side, eccentric-shaft mechanical press. (*Courtesy, Verson Allsteel Press Co.*)

accurately produced without a high-cost press tool by simply breaking the complex part down into several simple operations. This type of operation is used on low-production or prototype parts. The tooling cost is usually very low, but labor cost is high as the parts are manually transferred and located in each station.

A double-action press is used for drawing operations on sheet metal parts. This type of press has an outer ram (blank holder) and a second inner ram to actuate the drawing punch. During the operating cycle, the blank holder contacts the material first and applies pressure to allow the punch to properly draw the part.

A triple-action press has the same inner and outer ram as the double-action press, but a third ram in the press bed moves up, allowing a reverse draw to be made in one press cycle. The triple-action press is not widely used.

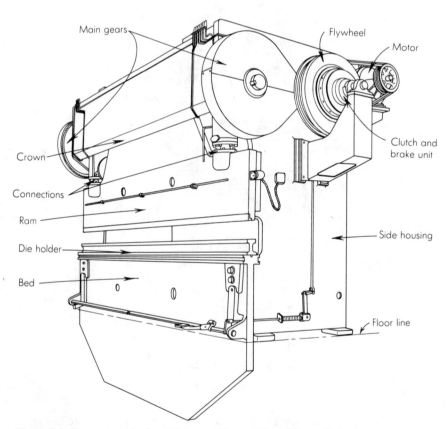

Figure 7-3. Power-press brake. (*Courtesy, Verson Allsteel Press Co.*)

A knuckle press is used for coining operations. The design of the drive allows for very high pressures at the bottom of the ram stroke. This type of press uses a crank, which moves a joint consisting of two levers that oscillate to and from dead center (*Figure 7-4*), and results in a short, powerful movement of the slide with slow travel near the bottom of the stroke.

A hydraulic press has a slower operating cycle time than most mechanical presses. Some advantages of hydraulic presses are that the working pressure, stroke, and speed of the ram are adjustable. A further advantage is that full tonnage is available at any point in the stroke (*Figure 7-5*).

These are the basic press types used in industry, although there are many more types with special applications.

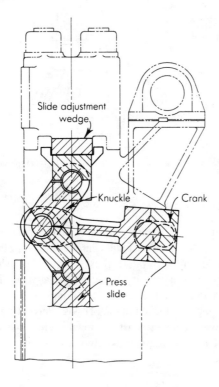

Figure 7-4. Knuckle-joint press.

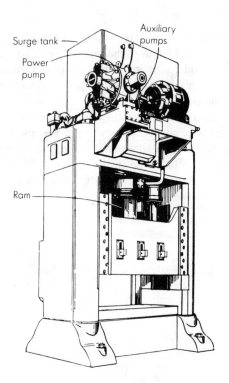

Figure 7-5. Typical hydraulic press. (*Courtesy, Hydraulic Press Manufacturing Co.*)

Safety Precautions

When performing any work on power presses, safety rules must be followed. In the United States, the Occupational Safety and Health Administration (OSHA) promulgates and enforces power press safety rules, which are subject to change. Press safety must be a primary concern for everyone. The operator must follow good safety practices at all times to avoid injury. While working under the ram, the press control must be locked in the off position and safety blocks placed under the ram to prevent it from coasting down. While the press is running, proper guards and safety procedures must always be used.

Cutting (Shearing) Operations

In the following discussion, specific die terminology will be used frequently. *Figure 7-6* displays the most commonly used terms and illustrates a simple die of the type operated in the press shown in *Figure 7-1*.

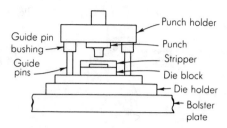

Figure 7-6. Common components of a simple die.

Shear Action in Die Cutting Operations

The cutting of metal between die components is a shearing process in which metal is stressed in shear between two cutting edges to the point of fracture, or beyond its ultimate strength. The metal is subjected to both tensile and compressive stresses (*Figure 7-7*). Stretching beyond the elastic limit occurs, then plastic deformation, reduction in area, and finally, fracturing starts through cleavage planes in the reduced area and becomes complete.

The fundamental steps in shearing or cutting are shown in *Figure 7-8*. The pressure applied to the metal by the punch tends to deform it into the die opening. When the elastic limit is exceeded by further loading, a portion of the metal will be forced into the die opening in the form of an embossed pad on the lower face of the material. A corresponding depression results on the upper face, as indicated at *A*. As the load is further increased, the punch will penetrate the metal to a certain depth and force an equal portion of metal thickness into the die, as indicated at *B*. This penetration occurs before fracturing starts and reduces the cross-sectional area of metal through which the cut is being made. Fracture will start in the reduced area at both upper and lower cutting edges, as indicated at *C*. If the clearance is suitable for the material being cut, these fractures will spread toward each other and eventually meet, causing complete separation. Further travel of the punch will carry the cut portion through the stock and into the die opening.

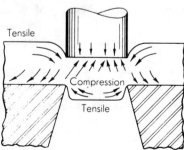

Figure 7-7. Stresses in die cutting.

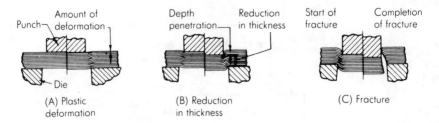

Figure 7-8. Steps in shearing metal.

Center of Pressure

If the contour to be blanked is irregularly shaped, the summation of shearing forces on one side of the center of the ram may greatly exceed the forces on the other side. Such irregularity results in a bending moment in the press ram, and undesirable deflections and misalignment. It is therefore necessary to find a point about which the summation of shearing forces will by symmetrical. This point is called the center of pressure, and is the center of gravity of the line that is the perimeter of the blank. It is not the center of gravity of the area. The press tool will be designed so that the center of pressure will be on the central axis of the press ram when the tool is mounted in the press. If this is not possible, it may be feasible to offset the tool in the press to achieve the same goal.

Mathematical Calculation of Center of Pressure. The center of pressure may be precisely determined by the following procedure:

1. Draw an outline of the actual cutting edges, as indicated in *Figure 7-9.*
2. Draw axes *X-X* and *Y-Y* at right angles in a convenient position. If the figure is symmetrical about a line, let this line be one of the axes. The center of pressure will, in this case, be somewhere on the latter axis.
3. Divide the cutting edges into line elements, straight lines, arcs, etc., numbering each, 1, 2, 3, etc.
4. Find the length L_1, L_2, L_3, etc., of these elements.
5. Find the center of gravity of these elements. Do not confuse the center of gravity of the lines with the center of gravity of the area enclosed by the lines.
6. Find the distance x_1 of the center of gravity of the first element from the axis *Y-Y*, x_2 of the second, etc.
7. Find the distance x_1 of the center of gravity of the first element from the axis *X-X*, y_2 of the second, etc.
8. Calculate the distance X of the center of pressure C from the axis *Y-Y* by the formula:

$$X = \frac{L_1 x_1 + L_2 x_2 + L_3 x_3 + L_4 x_4 + L_5 x_5 + L_6 x_6}{L_1 + L_2 + L_3 + L_4 + L_5 + L_6}$$

9. Calculate the distance Y of the center of pressure from the axis X-X by the formula:

$$Y=\frac{L_1y_1+L_2y_2+L_3y_3+L_4y_4+L_5y_5+L_6y_6}{L_1+L_2+L_3+L_4+L_5+L_6}$$

In the accompanying *Figure 7-9*, the elements are shown, numbered 1, 2, 3, etc. The length of 1 is obtained directly from the dimensions. It has a value of four. The center of gravity is evidently at the geometrical center of the line. Therefore:

$$x_1=0$$

and

$$y_1= 4\ 1/4+4/2$$

For the second element, x_2 is 1.5. The value of y_2 is found from the equation $CG=2r/\pi$, where r is the radius of the element. To find the requirements for line 3, it is necessary to solve the right triangle of which it is the hypotenuse.

The requirements of the other elements are found in a similar manner, all values being entered in a table.

Element	L	x	y	Lx	Ly
1	4.00	0.00	6.25	0.00	25.00
2	4.71	1.50	9.20	7.05	43.33
3	3.20	4.00	7.00	12.80	22.40
4	2.50	4.00	5.00	10.00	12.50
5	3.00	1.50	4.25	4.50	12.75
6	1.57	1.00	0.00	1.57	00.00
	18.98			35.92	115.98

These values are then substituted in the preceding formulas.

$$X=\frac{35.92}{18.98}=1.89''\ (48\ mm); \quad Y=\frac{115.98}{18.98}=6.10''\ (155\ mm)$$

The center of pressure C is therefore located as indicated in *Figure 7-9*.

Wire Method of Locating the Center of Pressure. The center of pressure of a blank contour may be located mathematically as shown above, but it is a tedious computation. Location of the center of pressure within 0.5" (12.7 mm) of true mathematical location is normally sufficient. A simple procedure accurate within such limits is to bend a soft wire to the blank contour. By balancing this frame across a pencil, in two coordinates, the intersection of the two axes of balance will locate the desired point. As an example of the marked influence this factor may have on tool design, a rather unusual blank is shown in *Figure 7-10*. In this illustration, the center of pressure is near one end of the

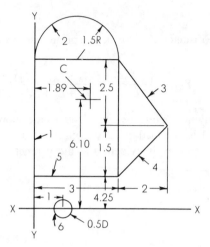

Figure 7-9. Example of center-of-pressure calculation.

blank, and will require the indicated imbalance in the press tool design. The use of computer aided design (CAD) provides many methods for locating the center of an irregular area. If CAD is utilized in designing the part, the center of pressure (not center of area) can also be located by the designer.

Clearances

Clearance is the space between the mating members of a die set. Proper clearances between cutting edges enable the fractures to meet. The fractured portion of the sheared edge will have a clean appearance. For optimum finish of a cut edge, proper clearance is necessary and is a function of the type, thickness, and temper of the work material. Clearance, penetration, and fracture are shown schematically in *Figure 7-11*. In *Figure 7-12*, characteristics of the cut edge on stock and blank, with normal clearance, are shown schematically. The upper corner of the cut edge of the stock (indicated by A') and the lower corner of the blank (indicated by A'-1) will have a radius where the punch and die edges, respectively, make contact with the material. This radiusing is due to the plastic

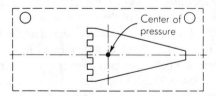

Figure 7-10. Effect of center-of-pressure location on tool design.

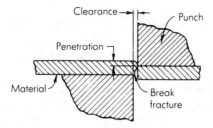

Figure 7-11. Punch-and-die clearance; punch penetration into and fracture of die-cut metal.

deformation taking place, and will be more pronounced when cutting soft metals. Excessive clearance will cause a large radius at these corners, as well as a burr on opposite corners.

In ideal cutting operations, the punch penetrates the material to a depth equal to about one-third of its thickness before fracture occurs, and forces an equal portion of the material into the die opening. That portion of the thickness so penetrated will be highly burnished, appearing on the cut edge as a bright band around the entire contour of the cut adjacent to the edge radius indicated at B' and B'-1 in *Figure 7-12*. When the cutting clearance is not sufficient, additional

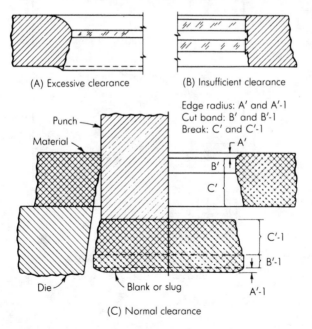

Figure 7-12. Cut-edge characteristics of die-cut metal; effect of excessive and insufficient clearances.

bands of metal must be cut before completion separation is accomplished, as shown at *B'* in *Figure 7-12*. When correct cutting clearance is used, the material below the cut will be rough on both the stock and the slug. With correct clearance, the angle of fracture will permit a clean break below the cut band because the upper and lower fractures extend toward one another. Excessive clearance will result in a tapered cut edge since, for any cutting operation, the opposite side of the material that the punch enters will, after cutting, be the same size as the die opening.

The width of the cut band is an indication of the hardness of the material, provided that the die clearance and material thickness are constant. The wider the cut band, the softer the material. Harder metals require larger clearances and permit less penetration by the punch than the ductile metals. Dull tools create too small a clearance, and will produce a burr on the die side of the stock. The effects of various amounts of clearance are shown in *Figures 7-11* and *7-12*. Defective or nonhomogeneous material cut with the proper amount of clearance will produce nonuniform edges.

The edge conditions *C* and the hypothetical load curve *B* (*Figure 7-13*, cases 1, 2, 3, and 4) are shown, as well as the amount of deformation and extent of

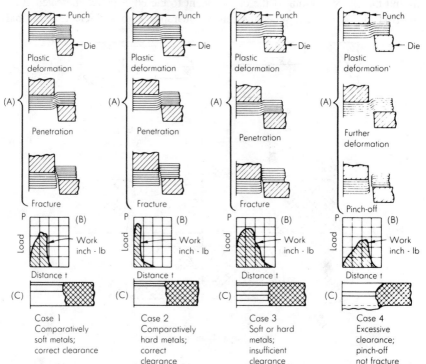

Figure 7-13. Effect of different clearances on soft and hard metals.

punch penetration. Location of the proper clearance (*Figure 7-14*) determines either hole or blank size; punch size controls hole size, and die size controls blank size.

At *A* (*Figure 7-14*), which shows clearance *C* for blanks of a given size, make die to size and punch smaller by total clearance 2*C*. At *B*, which shows clearance for holes of a given size, make the punch to size and die larger by the amount of the total clearance 2*C*. The application of clearances for holes of irregular shape is diagrammed in *Figure 7-15*; at *B* the hole will be of punch size, while at *A* the blank will be of the same dimensions as the die. One manufacturer charts clearance per side for groups of materials up to and including thicknesses of 0.125″ (3.18 mm) (*Figure 7-16*). These clearances are intended to maximize die life by permitting as many sharpenings as possible before the clearance between the punch and die becomes excessive. Many manufacturers specify greater clearances, typically 10%, to reduce cutting pressure and the required precision of alignment, provided that tooling life is not a critical factor.

The die clearance chart (*Figure 7-16*) may be used to find the recommended die clearance allowed and to be provided for in designing a die for service as determined by the materials groups listed below. The chart may also be used for the preestablished percentage of material thickness of the original part which the die is designed to produce.

Group 1. 1100S and 5052S aluminum alloys, all tempers. An average clearance of 4.5% of material thickness is recommended for normal piercing and blanking.

Group 2. 2024ST and 6061ST aluminum alloys; brass, all tempers; cold-rolled steel, dead soft; stainless steel, soft. An average clearance of 6% of material thickness is recommended for normal piercing and blanking.

Group 3. Cold-rolled steel, half hard; stainless steel, half hard and full hard. An average clearance of 7.5% is recommended for normal piercing and blanking. *Example:* In *Figure 7-17* it is seen that, for a nominal stock thickness of 0.080″ (2.03 mm), the die clearance for any Group 1 materials would be

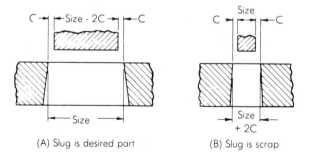

Figure 7-14. Control of hole and blank sizes by clearance location.

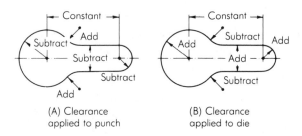

(A) Clearance
applied to punch

(B) Clearance
applied to die

Figure 7-15. How to apply clearances.

0.0036″ (0.09 mm); for any Group 2 materials, 0.0048″ (0.12 mm); for any Group 3 materials, 0.006″ (0.15 mm).

Angular Clearance. Angular clearance is defined as the clearance below the straight portion of a die surface introduced to enable the blank or slug (piercing operation) to clear the die (*Figure 7-15*). Angular clearance between 0.25° to 1.5° per side is usually provided. The amount of angular clearance is occasionally higher, depending mainly on stock thickness and the frequency of sharpening. High-production dies usually have a 0.25° clearance angle.

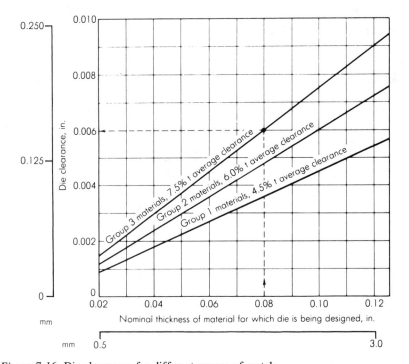

Figure 7-16. Die clearances for different groups of metals.

Cutting Forces

The force (in tons) required to cut (shear) work material can be calculated using the following formulas:

$$P = SLT \text{ or } TN = \frac{SLT}{2000} \text{ (for contours)}$$

$$P = \pi DST \text{ or } TN = \frac{\pi DST}{2000} \text{ (for round holes)}$$

The shear strengths of many common materials in PSI (MPa) are listed in *Figure 7-17*.

SHEAR STRENGTH OF COMMON FERROUS MATERIALS

0.10 carbon steel annealed	35,000	(241.3)
0.20 carbon steel annealed	42,000	(289.5)
0.30 carbon steel annealed	52,000	(358.5)
0.40 carbon steel annealed	80,000	(551.5)
1.00 carbon steel annealed	110,000	(758.3)

SHEAR STRENGTH OF COMMON NONFERROUS MATERIALS

Aluminum 1100-O	9,000	(62)
Aluminum 6061-T6, -T651	30,000	(206.8)
Brass, yellow annealed	33,000	(227.5)
Bronze, annealed commercial	28,000	(193)
Copper, annealed	22,000	(151.7)

SHEAR STRENGTH OF COMMON NONMETALLIC MATERIALS

Cellulose acetate	10,000	(69)
Cloth	8,000	(55.2)
Fiber, hard	18,000	(124.2)
Hard rubber	20,000	(138)
Leather, tanned	7,000	(48.3)
Leather, rawhide	13,000	(89.7)
Mica	10,000	(69)
Paper	6,400	(44.2)

Figure 7-17. Shear strength, PSI (MPa) of various materials.

Stripping Forces

The actual force required to strip the work material off the punches depends on factors such as the type of material, the smoothness of the punches and the type and amount of lubricant. The following empirical equation is used for rough approximations:

$$Ps = 3500 \ LT \qquad TNs = \frac{3500 \ LT}{2000} \text{ (for contours)}$$

$$Ps = 3500 \ \pi DT \qquad TNs = \frac{3500 \ \pi \ DT}{2000} \text{ (for round holes)}$$

The above cutting and stripping force equations utilize the following terms:

P = Cutting force in pounds.
TN = Cutting force in tons.
Ps = Stripping force in pounds.
TNs = Stripping force in tons.
S = Shear strength in pounds per square inch.
L = Length of cut in inches.
T = Thickness of material in inches.
D = Diameter in inches.

Press Tonnage

This is a total of forces required to cut and form the part. Practical experience with presses equipped with tonnage meters shows close agreement between calculated versus measured values. In many cases, you will need to add the stripping force to the cutting force if a spring-loaded stripper is used, because the springs must be compressed while cutting the material. Any spring pressure for forming, draw pads, and the like will have to be added. Using tunnel strippers will keep the press load to a minimum, but not control the stock as well as spring-loaded strippers.

Methods of Reducing Cutting Forces

Since cutting operations are characterized by very high forces exerted over very short periods of time, it is sometimes desirable to reduce the force and spread it over a longer portion of the press stroke. Punch contours of large perimeter or many smaller punches will frequently result in tonnage requirements beyond the capacity of an available press. Also, whenever abnormally high tonnage requirements are concentrated in a small area, the likelihood of design difficulties increases.

Two methods generally reduce cutting forces and smooth the shock impact of heavy loads. Keep in mind that during a piercing operation with proper clearance, complete fracture occurs when the punch has penetrated one-third the material thickness.

1. By stepping punches one-third the material thickness, they can cut individually. Typically using three punches of the same diameter stepped properly, one-third of the total tonnage required is used to do all three simultaneously.

2. Adding shear to the die or punch equal to one-third the material thickness reduces the tonnage required by 50% for that area being cut with shear applied. Note that shear is applied to the die member (punch or die) that contacts the scrap. Therefore, deformation due to the shear angles has little effect on the part *(Figure 7-18)*.

In piercing, the direction of the shear angles generally are such that the cut proceeds from the outer extremities of the contour toward the center. This avoids stretching the material before it is cut free. Limiting factors in how much timing can be used to reduce tonnage requirements is the allowable tonnage as a

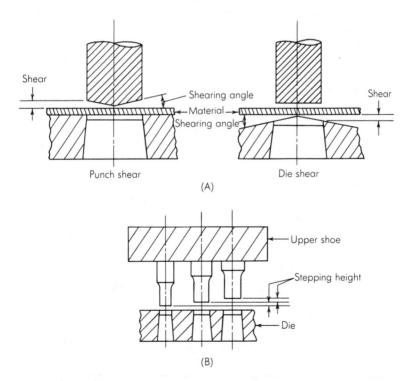

Figure 7-18. Reducing cutting forces.

function of the distance above the bottom of the press stroke and the allowable loss of flywheel energy per stroke. Timing is important in many cases to reduce the snap through that can damage the press.

Analysis of Snap-Through Forces[2]

The loud boom that characterizes a snap-through problem when cutting through metal is the direct result of the sudden release of the potential energy stored in press and die members as strains or deflection. The deflection is a normal result of the pressure required to cut through the material.

In extreme cases, the energy released can damage the press. The press connection (the attachment point of the pitman to the slide) is easily damaged by the reverse load generated by snap-through. As a general rule, presses are not designed to withstand reverse loading of more than 10%, and the shock may result in die components working loose.

In timing punch entry or die shear, care must be taken to provide for a gradual release of developed tonnage. The shock is not normally generated by the impact of the punches on the stock. In fact, when the punches first contact the stock, the initial work is done by the kinetic energy of the slide. To complete the cut, energy must be supplied by the flywheel. As this occurs, the press members deflect. An analysis of the quantity of energy involved will show why a gradual reduction in cutting pressure prior to snap-through is very important.

The magnitude of the actual energy released increases as the square of the actual tonnage developed at the moment of final breakthrough. The actual energy is given by:

$$E = \frac{F \times D}{2}$$

F = Pressure at moment of breakthrough in standard tons.
D = Amount of total deflection in inches.
$E \times 166.7$ = Energy in foot pounds (1 foot pound = 1.356 joules or watt-seconds)

or:

F = Pressure at moment of breakthrough in metric tons (kgf $\times$ 1000).
D = Amount of total deflection in millimeters.
$E \times 9.807$ = Energy in joules or watt seconds (1 joule or watt second = 1.356 foot pounds)

For example, if 400 tons (363 t), which resulted in 0.080" (2.03 mm) total deflection, was required to cut through a thick steel blank, the energy released at snap-through would be 2,667 foot pounds (3,616 J).

If careful timing of the cutting sequence results in a gradual reduction in the amount of tonnage at the moment of snap-through so that only 200 tons (181.5 t)

Employee Comments:

X .85

X .85
Y $250

600 2.185
X .88
601 X .79 F .002
600 X .88

10.9
6 9

X .85

REQUEST

Last Name ______________________

Soc. Sec. # ______________________

Week Beginning Date ______________________

.173648

Dept. ______________________

.3125

Shift ______________________

.0543

Trade ______________________

.1875

____

Mark all boxes you are willing to work; unmarked boxes indicate refusal of voluntary overtime.

.109

3125

	Early	Over	Any Dept. Y–N
Monday	☐	☐	☐
Tuesday	☐	☐	☐
	☐	☐	☐

was released, the reduction in shock and noise would be dramatic. Half the
tonnage would produce half as much deflection or 0.04″ (1.02 mm). The
resultant snap-through energy would only be 667 foot pounds (904 J), or
one-fourth the former value.

An Example of Snap-Through Reduction by Die Timing[3]

Figure 7-19 illustrates a waveform resulting from an operation to punch two
1.625″ (41.275 mm) holes and parting a chain side bar from fine-grained
AISI-SAE 1039 steel. The steel was 0.5″ (12.7 mm) thick by 3″ (76.2 mm) wide.
A 300 ton (2.669 MN) Verson straightside press was used for this operation. The
allowable reverse load is 30 tons (.267 MN). Point *A* on *Figure 7-19* illustrates
a peak load of 191 tons (1.7 MN), which was also displayed on the tonnage
meter. This is well within press capacity.

In this case, the reverse load *B* was 87 tons (.774 MN), which is nearly three
times the allowable amount. The die was immediately taken to the repair bench
and one punch shortened 0.312″ (7.92 mm). Balanced angular shear was ground
on the punches. Balanced shear was also ground on the parting punch.

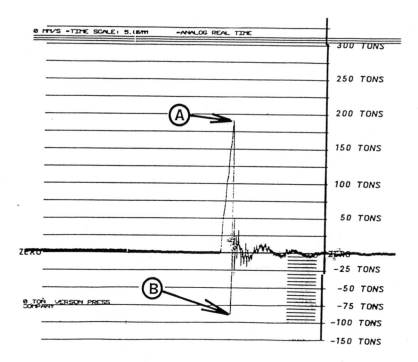

Figure 7-19. The actual waveform signature of a combined piercing and cut-off
operation having excessive snap-through, or reverse load (Ref. 3).

Figure 7-20 illustrates the improvement achieved by modifying the tool. The peak tonnage was reduced to 82.8 tons (0.737 MN), which is less than half the initial value. The reverse load was reduced to 22 tons (.196 MN) or about one-fourth the former value. This is keeping with the square law formula for the amount of energy suddenly released.

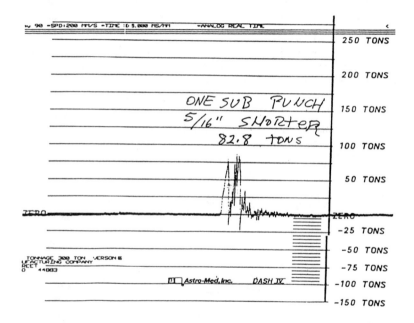

Figure 7-20. Waveform signature of the job illustrated in *Figure 7-19* after adding timing and balanced shear (Ref. 3).

Die Block General Design

Overall dimensions of the die block will be determined by the minimum die wall thickness required for strength and by the space needed for screws and dowels and for mounting the stripper plate. Wall thickness requirements for strength depend upon the thickness of the stock to be cut. Sharp corners in the contour may lead to cracking in heat treatment, and so require greater wall thickness at such points.

Only two dowels should be provided in each block or element that requires accurate and permanent positioning. They should be spaced as far apart as possible for maximum locating effect, usually near diagonally opposite corners.

Two or more screws will be used, depending on the size of the element mounted. Screws and dowels are preferably located about 1.5 times their diameters from the outer edges or blanking contour. Die block thickness (*Figure 7-21*) is governed by the strength necessary to resist the cutting forces, and will depend upon the type and thickness of the material being cut. For very thin materials, 0.5" (12.7 mm) thickness should be sufficient but, except for temporary tools, finished thickness is seldom less than 0.875" (22 mm), which allows for blind screw holes. Another issue is maintaining a constant pass height for quick die change considerations.

Stock thickness, in. (mm)	Die thickness, in. (mm)*	Stock thickness, in. (mm)	Die thickness, in. (mm)*
		0.6 (15.2)	0.15 (3.8)
0.1 (2.5)	0.03 (0.8)	0.7 (17.8)	0.165 (4.19)
0.2 (5.1)	0.06 (1.5)	0.8 (20.3)	0.18 (4.6)
0.3 (7.6)	0.085 (2.2)	0.9 (22.9)	0.19 (4.8)
0.4 (10.2)	0.11 (2.8)	1.00 (25.4)	0.20 (5.1)
0.5 (12.7)	0.13 (3.3)		

* For each ton per sq in. of shear strength.

Figure 7-21. Die thickness per ton of pressure.

Die Block Calculations

Method 1 (Rule of Thumb). Assuming a die block is of tool steel, its thickness should be 0.75" (19 mm) minimum for a blanking perimeter of 3" (76 mm) or less; 1" (25 mm) thick for perimeters between 3 and 10" (76 mm and 254 mm) and 1.25" (32 mm) thick for larger perimeters. There should be a minimum of 1.25" (32 mm) margin around the opening in the die block.

The die opening should be straight for a minimum of 0.125" (3.2 mm); the opening should then angle out at 0.25° to 1.5° to the side (draft). The straight sides provide added die life for resharpening the die and the tapered portion enables the blanks to drop through without jamming. To secure the die to the die plate or die shoe, the following rules provide sound construction:

1. On die blocks up to 7" (178 mm) square, use a minimum of two, 0.375" (M10) cap screws and two, 0.375" (10 mm) dowels.
2. On sections up to 10" (254 mm) square, use three cap screws and two dowels.
3. For blanking heavy stock, use cap screws of 0.5" (M14) diameter and dowels of 0.5" (14 mm) diameter. Counterbore the cap screws 0.125" (3.2 mm) deeper than usual, to allow for die sharpening.

Method 2. This method of calculating the proper size of the die was derived from a series of tests whereby die plates were made increasingly thinner until breakage became excessive. From these data the calculation of die thickness was divided into four steps:

1. Die thickness is provisionally selected from *Figure 7-21*. This table takes into account the thickness of the stock and its ultimate shear strength (see *Figure 7-17*).
2. The following corrections are then made:
   a. The die must never be thinner than 0.3" (7.6 mm).
   b. Data in *Figure 7-21* apply to small dies; i.e., those with a cutting perimeter of 2" (51 mm) or less. For larger dies, the thicknesses listed in *Figure 7-21* must be multiplied by the factors in *Figure 7-22*.
   c. Data in *Figure 7-21* and *Figure 7-22* are for die members of low-cost tool steels, properly machined and heat treated. If a special alloy of steel is selected, die thickness can be decreased.
   d. Dies must be adequately supported on a flat die plate or die shoes. Thickness data above do not apply if the die is placed over a large opening or is not adequately supported. However, if the die is placed into a shoe with a hardened backing plate, the thickness of the member can be decreased up to 50%.
   e. A grinding allowance up to 0.1" to 0.2" (2.5 mm to 5.1 mm) must be added to the calculated die thickness.
3. The critical distance A (*Figure 7-23*), between the cutting edge and the die border must be determined. In small dies, A equals 1.5 to two times the die thickness; in larger dies it is two to three times the die thickness.
4. Finally, the die thickness must be checked against the empirical rule that the cross-sectional area $A \times T$ (*Figure 7-23*) must bear a certain minimum relationship to the impact force for a die put on a flat base. In *Figure 7-25*, impact force equals stock thickness times the perimeter of the cut times ultimate shearing strength. If the die height, as calculated by steps 1 and 2, does not give sufficient area for the critical distance A (*Figures 7-23* and *7-24*), the die thickness must be increased accordingly.

Cutting perimeter, in. (mm)	Expansion factor
2-3 (51-76)	1.25
3-6 (76-152)	1.5
6-12 (152-305)	1.75
12-20 (305-508)	2.0

Figure 7-22. Factors for cutting edges exceeding two inches.

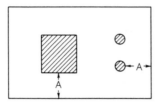

Figure 7-23. Critical distance *A* must not be less than 1.5 to two times die thickness.

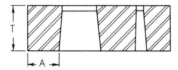

Figure 7-24. The critical area between the die hole and the die border must be checked against minimum values in *Figure 7-25*, and die thickness *T* corrected if necessary.

With the die block size determined, the exact size of the die opening can be determined. Assuming a clearance of approximately 10% of the metal thickness, and by the rule-of-thumb method, for a metal thickness of 0.064″ (1.63 mm), the clearance would be as follows:

0.064″ (1.63 mm) x 10% = 0.0064″ (0.16 mm) clearance per side

If the finished part opening is 1.000″ (25.4 mm) in diameter, add clearance to the die opening 0.006″ (0.16 mm) per side giving 0.012″ + 1.00″ = 1.012″ dia. (0.30 mm + 25.4 mm = 25.7 mm). If the blank is to be the size of the die opening, the clearance must be applied to the punch.

1.000″ – 0.012″ = 0.988″ in dia. (25.4 mm – 0.3 mm = 25.1 mm)

Impact pressure, tons (kN)	Area between die opening border, sq in. (cm^2)
20 (178)	0.5 (3.2)
50 (445)	1.0 (6.5)
75 (668)	1.5 (9.7)
100 (890)	2.0 (12.9)

Figure 7-25. Minimum critical area versus impact pressure.

Punch Dimensioning

The determination of punch dimensions has been generally based on practical experience. When the diameter of a pierced round hole equals stock thickness, the unit compressive stress on the punch is four times the unit shear stress on the cut area of the stock, from the following formula:

$$\frac{4S_s t}{S_c d} = 1$$

S_c = unit compressive stress on the punch, psi.
S_s = unit shear stress on the stock, psi.
t = stock thickness, inch
d = diameter of punches hole, inch.

The diameters of most holes are greater than stock thickness; a value for the ratio d/t of 1.1 is recommended.

The maximum allowable length of a punch can be calculated from the formula:

$$L = \frac{\pi d}{8} \frac{E}{S_s} \frac{d^{1/2}}{t}$$

Where:
    d/t = 1.1 or higher
    E = modulus of elasticity

This is not to say that holes having diameters less than stock thickness cannot be successfully punched. The punching of such holes can be facilitated by:
1. Punch steels of high compressive strengths.
2. Greater than average clearances.
3. Optimum punch alignment, finish, and rigidity.
4. Shear on punches or dies or both.
5. Prevention of stock slippage.
6. Guiding the punch with a quill.
7. Guiding the punch with the stripper.
8. Piercing at high cutting speeds.

Methods of Punch Support

Figure 7-26 presents a number of methods to support punches to meet various production requirements:

View A. When cutting punch *A* is sharpened, the same amount is ground off spacer *B* to maintain the relative distance *C*.

View B. If delicate punches must be grouped closely together, a hardened guide block with the required number of holes should be used.

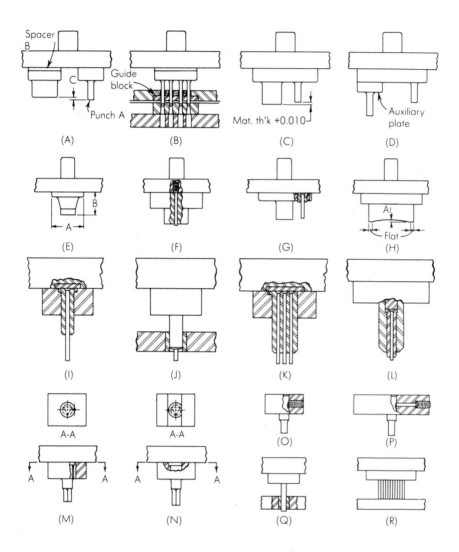

Figure 7-26. Various methods of punch support. (*Courtesy, American Machinist*)

View C. A slender piercing punch (at right) should be made shorter than an adjacent large punch.

View D. If punches must protrude more than 4″ (102 mm) beyond the punch holder, an auxiliary plate may be used to maintain stiffness.

View E. Flange width of the punch should be greater than height *B* to provide stability for unguided punches.

View F. In a large punch, push-off pins can prevent slugs from pulling up and causing trouble.

View G. To avoid cracking a large, hardened punch or a punch plate, do not press a small punch directly into either of these members. Instead, use a soft plug or insert.

View H. Long slotting punches should be hollow ground so that dimension *A* equals the metal thickness, so as to put shear on the punch. The ends should be flat for 0.125″ (3.2 mm) to avoid bending the stock.

View I. A quill is useful for supporting pin punches.

View J. A bushing in the stripper plate can guide the quill for increased punch support.

View K. Quills need not be limited to a single punch. If prevented from turning, they can be used for pin punches on close centers.

View L. Two quills are used for a bit punch—one to support the punch, the other to support the inner quill, when a stripper is not used.

View M. A dowel can prevent rotation of the punches.

View N. For high-speed dies, a flat on the punch head is more positive.

View O. In low-production dies, a setscrew is adequate to hold the punch.

View P. When a deep hole must be drilled, a drill-rod pin can be used to span the distance.

View Q. Light drill-rod punches are guided in the stripper plate to prevent buckling.

View R. Several punches can be set at close center distances.

Ball-Lock Punches

The ball-lock concept of retention and quick punch replacement has been in use for many years. Style changes on a stamping can be made quickly without removing the die from the press simply by pulling or adding punches. In the event that a fragile punch needs replacement during the run, the ball-lock system of punch retention makes this operation quick and easy. *Figure 7-27* illustrates the ball-lock retention system used to positively lock both a punch and die button (matrix) in a die. The ball-lock retention is actually an adaptation of the wedge principle. Once correctly seated, the ball locks the punch in position both vertically and radially, while permitting rapid replacement of the punches in the die assembly.

To install, the punch is inserted into the retainer and twisted until the spring-loaded ball drops into position, locking the punch. Punch removal is accomplished by depressing the ball with a ball-release tool inserted in a hole in the retainer provided for this purpose. This action frees the punch so that it can be pulled out of the retainer. Some retainers utilize a ball release screw as illustrated in *Figure 7-27*. To ensure correct location and certain retention, it is important that ball locking condition be correct. *Figure 7-28* illustrates the

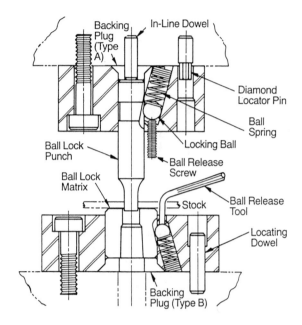

Figure 7-27. The ball-lock retention system used to positively lock both a punch and die button (matrix), while permitting rapid replacement of these components in the die assembly. (Ref. 4, *Courtesy, Dayton Progress Corp.*)

problem of the ball being either too low or too high in the teardrop-shaped ball seat. An important way to avoid this problem is to always use the correct retainer specified for a given punch and to always use the proper size replacement ball.

Stock Stops

In its simplest form, a stock stop may be a pin or small block against which an edge of the previously blanked opening is pushed after each stroke of the press. With sufficient clearance in the stock channel, the stock is momentarily lifted as it clings to the punch, and is thus released from the stop. *Figure 7-29* shows an adjustable type of solid block stop, which can be moved along a support bar in increments up to 1″ (25.4 mm) to allow various stock lengths to be cut off.

A starting stop, used to position stock as it is initially fed to a die, is shown in *Figure 7-30, View A.* Mounted on the stripper plate, the stop incorporates a latch that is pushed inward by the operator until its shoulder (1) contacts the stripper plate. The latch is held in to engage the edge of the incoming stock; the first die operation is completed, and the latch is released. The starting stop shown at *View B,* mounted between the die shoe and die block, upwardly actuates a stop

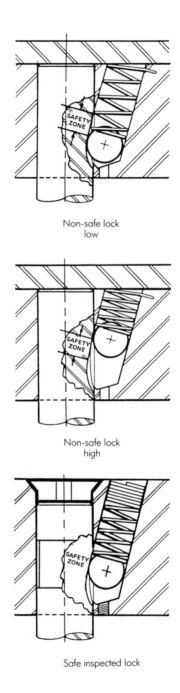

Non-safe lock
low

Non-safe lock
high

Safe inspected lock

Figure 7-28. The ball must be properly seated in the teardrop-shaped pocket in the punch. (Ref. 4, *Courtesy, Dayton Progress Corp.*)

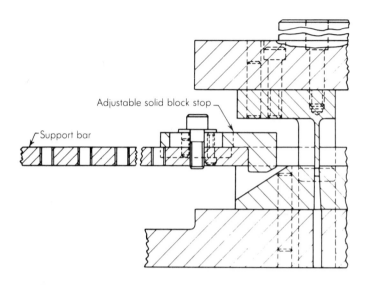

Figure 7-29. Adjustable block stop for a parting die.

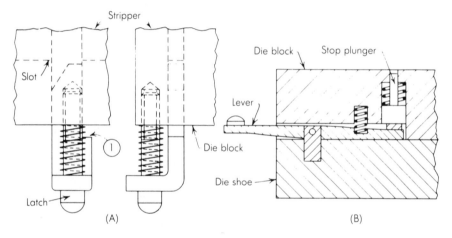

Figure 7-30. Starting stops.

plunger to initially position the incoming stock. Compression springs return the manually operated lever after the first die operation is completed.

Trigger stops incorporate pivoted latches (1, *Figure 7-31, Views A and B*). At the ram's descent, these latches are moved out of the blanked-out stock area by actuating pins, (2). On the ascent of the ram, springs (3) control the lateral movement of the latch (equal to the side relief), which rides on the surface of the advancing stock and drops into the blanked area to rest against the cut edge of the cut-out area.

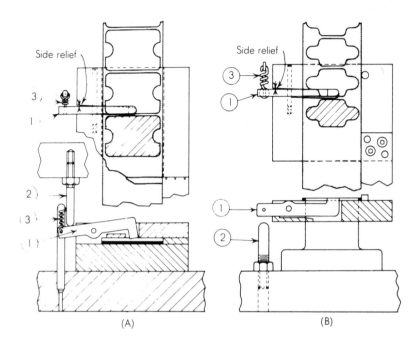

Figure 7-31. Trigger stops: (*A*) top stock engagement; (*B*) bottom stock engagement.

Automatic Stops

When feeding the stock strip from one station to another, some method must be used to correctly locate and stop the strip.

Automatic stops (trigger stops) register the strip at the final die station. They differ from finger stops in that they stop the strip automatically, with the operator having only to keep the strip pushed against the stop during its travel through the die.

Figure 7-32 shows automatic stop designs ranging from simple pin to escapement mechanisms.

View A illustrates a pin stop suitable for low-to-medium-production dies. When the ram ascends, the strip clings to the punch, is stripped, and then is fed until the pin hits the edge of the hole.

View B shows the method of locating the pin stop so that it bears against the blank opening upon an angular edge, so that the strip is crowded against the back stop and accurate piloting is obtained.

View C. If no scrap is left between blanks, as in a combination blank-and-draw die, a bent pin stop is suitable. The sharpened point of the stop faces the incoming strip, thrusting the fins aside and stopping the strip when contact is made at the opposite side of the hole.

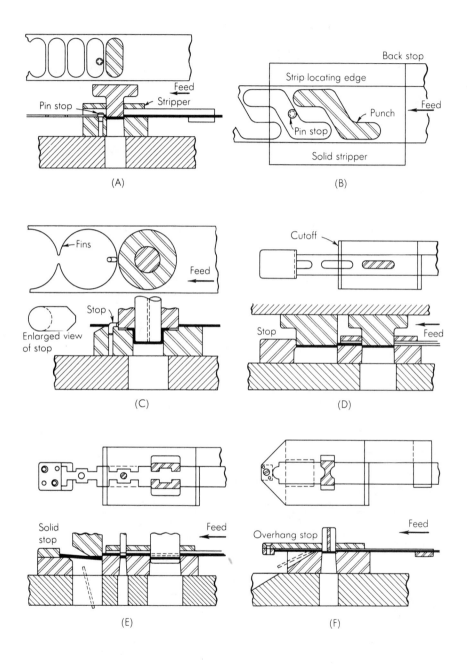

Figure 7-32. Designs of automatic stops.

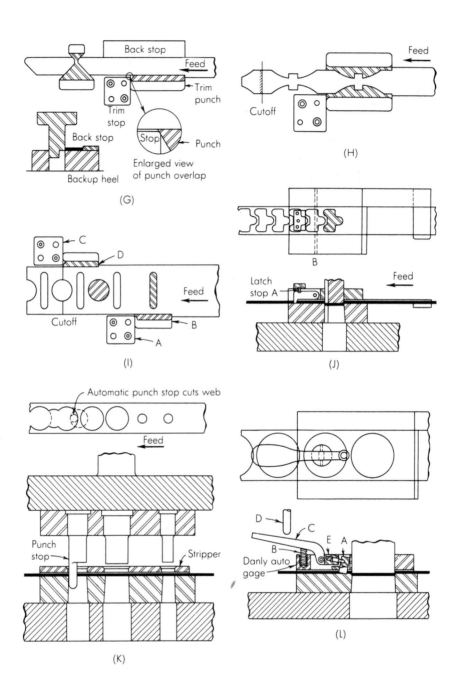

Figure 7-32. (Continued)

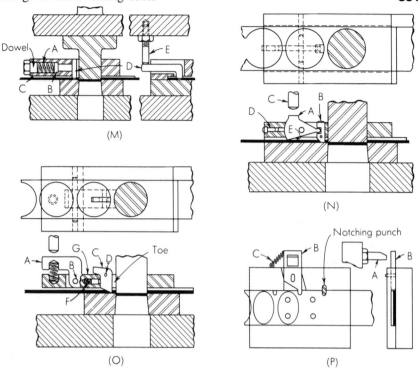

Figure 7-32. (Continued)

View D presents a design in which a combination stop and backup block locates the strip and prevents deflection of the cutoff punch of this two-station die. The part, a flat spring, drops to the punch-holder and slides by gravity to the rear of an inclined press.

View E shows another solid stop. The part, after cut-off, can drop through a hole in the die set into a box directly beneath.

View F. Overhanging stops are useful when the press cannot be inclined, or when the size of the die or press will not allow part removal through the bolster plate.

View G. If the strip must be cut to accurate width, a trimming stop can be used against the shoulder formed by the trimming punch. The length of the trimming punch is made equal to the feed distance for part length. The left edge of the punch overlaps the previous cut to prevent leaving a fin on the strip.

View H. A notching strip is ideal for automatic stopping of the strip when the first operation partially blanks the sides. When the strip is advanced, it is automatically positioned by the shoulder left by the blanking punch. The only extra cost is that for the stop.

View I. Double trimming stops reduce the extent of the waste end. Stop *A* and punch *B* are used during starting and normal running of the strip. Stop *C* and punch *D* cut the last two parts when the end of the strip is reached.

View J. A latch stop, like a pin stop, is ideal for low-production jobs, but should not be used on a high-production job. The latch pivots on a pin and is held down by a spring. In use, the strip is moved until the scrap bridge has lifted the latch and it drops into the hole. Then the strip is pulled back until the bridge is against the latch.

View K. Punch stops are applicable to many types of cutting dies. A round eccentric protrusion on a round or square punch body contacts the scrap bridge. The ram descent causes the punch body to cut out a portion of the scrap strip. Upon ram ascent the strip is advanced, with the eccentric passing through the gap until contact is made with the next scrap bridge.

View L. Commercial trigger stops are often an economical choice. The design shown consists of gage pin *A*, which fits loosely in a hole in the stripper, and spring *B*, which normally holds up lever *C*. The strip advance crowds the gage pin to the slanted position shown. Ram descent causes pin *D* to push the lever down, and the gage pin is lifted above the strip. Spring *E* now pushes the gage pin on top of the scrap bridge. Upon forward movement of the strip, the gage pin falls inside the hole just blanked.

View M. Spring torsion and compression combine to operate a stop. Ends of spring *A* are entered in drilled holes into plunger *B* and cap plug *C*. At assembly, the plug is rotated to apply sufficient torque so that stop *D* is kept firmly in contact with the die block. Strip advance slightly compresses the spring. At ram descent, pin *E* strips the stop and spring *A* pushes it forward to drop on the scrap bridge.

View N. An escapement stop eliminates misfeeds if the stock is fed forward to the stop every time. Rocker arm *A* lifts the square toe stop *B* above the strip when struck by actuator *C* attached to the punch holder. Now spring-operated detent *D* holds the stop up, until the advancing scrap bridge strikes the cam surface *E* on the rocker arm. Thereupon, the toe is dropped in front of the cut edge, just as the next ram descent commences.

View O. Long, high-speed runs on heavy stock often require an automatic toe stop. Lever *A* pivots on pin *B* to rock toe *C*, on pin *D* above the strip surface, when the ram descends. Spring *F*, acting through plunger *G*, pushes the toe until the heel on top rests on the lever. When the ram goes up, the toe falls on the scrap bridge. It eventually contacts the cut edge when the strip is advanced, thereby resetting the stop.

View P. A notch in the strip edge locates the strip in a two-stage piercing and blanking die. At the first station, a small notch is cut in the strip edge. At the second station, actuator *A* enters the cam hole in stop *B* to retract the toe from the notch. Spring *C* swings the toe counter-clockwise so it comes to rest on the strip edge. Upon strip advance, the toe springs into the notch just cut and stops the material in the correct position.

Pilots

Since pilot breakage can result in the production of inaccurate parts and the jamming or breaking of die elements, pilots should be made of good tool-steel, heat-treated for maximum toughness and to a hardness of Rockwell C57 to 60.

Press-fit pilots Press-fit pilots (*Figures 7-33* and *7-34, View C*), which may drop out of the punch holder, are not recommended for high-speed dies but

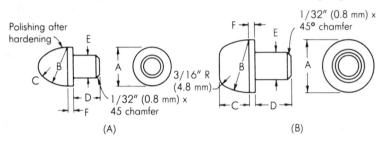

Figure 7-33. Press-fit pilots.

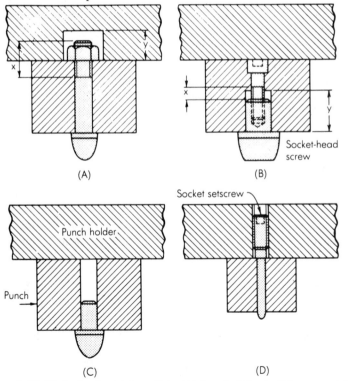

Figure 7-34. Methods of retaining pilots: (*A*) threaded shank; (*B*) screw-retained; (*C*) press-fit; (*D*) socket setscrew.

are used in low-speed dies. A method of establishing the dimensions of pilots is to make the radius *B* equal to the pilot diameter *A*. The spherical nose radius *C* of the acorn type may be 0.25*A*, approximately. Length *C* of a flattened-point-type pilot may be about 0.5*A*.

Pilots may be retained by the methods shown in *Figure 7-34*. A threaded shank, shown at *View A*, is recommended for high-speed dies; thread length *X* and counterbore *Y* must be sufficient to allow for punch sharpening. For holes 0.75″ (19 mm) in diameter or larger, the pilot may be held by a socket head screw, shown at *B*; recommended dimensions *X* and *Y* given for threaded-shank pilots also apply. A typical press-fit type is shown at *C*. Pilots of less than 0.25″ (6.4 mm) diameter may be headed and secured by a socket setscrew, as shown at *D*.

Indirect Pilots. Designs of pilots that enter holes in the scrap skeleton are shown in *Figure 7-35*. A headed design, at *A*, is satisfactory for piloting in holes from 0.188″ to 0.375″ (4.8 mm-9.5 mm) in diameter. A quilled design, at *B*, is suitable for pilots up to 0.188″ (4.8 mm) in diameter.

Spring-loaded pilots should be used for stock exceeding 0.060″ (1.5 mm). A bushed, shouldered design is shown at *C* of *Figure 7-35*. A slender pilot of drill rod shown at *D* is locked in a bushed quill which is countersunk to fit the peened head of the pilot.

Tapered slug-clearance holes through the die and lower shoe should be provided, since indirect pilots generally pierce the strip during a misfeed. Ball-lock retainers also hold pilots. These retainers are the same type illustrated in *Figures 7-27* and *7-28*.

Strippers

There are two types of strippers: fixed and spring-operated. The primary function of either type is to strip the workpiece from a cutting or noncutting punch or die. A stripper that forces a part out of a die may also be called a knockout, an inside stripper, or an ejector. Besides its primary function, a stripper may also hold down or clamp, position, or guide the sheet, strip, or workpiece.

The stripper is usually the same width and length as the die block. In simpler dies, the stripper may be fastened with the same screws and dowels that fasten the die block, and the screwheads will be counterbored into the stripper. In more complex tools and with sectional die blocks, the die block screws will usually be inverted, and the stripper fastener will be independent.

The stripper thickness must be sufficient to withstand the force required to strip the stock from the punch, plus whatever is required for the stock strip channel. Except for very heavy tools or large blank areas, the thickness required for screwhead counterbores, within the range of 0.375″ to 0.625″ (9.5 mm-16 mm), will be sufficient.

The height of the stock strip channel should be at least 1.5 times the stock thickness. This height should be increased if the stock is to be lifted over a fixed pin stop. The channel width should be the width of the stock strip plus adequate clearance to allow for variations in the width of the strip cut, as follows:

Stock thickness	Add to strip width
up to 0.04″ (1.02mm)	0.078″ (1.98mm)
0.04″-0.08″ (1.02mm-2.03mm)	0.094″ (2.38mm)
0.081″-0.12″ (2.06mm-3.05mm)	0.109″ (2.78mm)
over 0.12″ (3.05mm)	0.125″ (3.18mm)

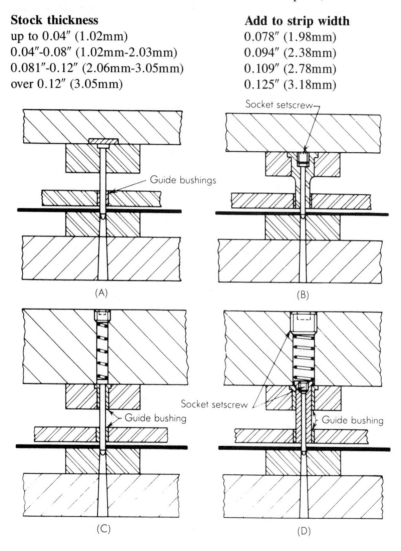

Figure 7-35. Indirect pilots: (A) headed; (B) quilled; (C) spring-backed; (D) spring-loaded quilled.

If the stripper length has been extended on the feed end for better stock guidance, a sheet metal plate or bar should be fastened to the underside of the projecting stripper to support the stock. This plate should extend inward slightly for convenience in inserting the strip. The entry edges of the channel should be beveled for the same reason.

Where spring-operated strippers are used, the force required for stripping in pounds is approximately 3,500 times cut perimeter times the stock thickness in inches. It may be as high as 20% of the blanking force, which will determine the number and type of springs required. The highest of these values should be used.

Die springs are designed to resist fatigue failure under severe service conditions. They are available in medium, medium-heavy, heavy-duty, and extra heavy duty grades, with corresponding maximum permissible deflections ranging from 40% to 25% of free length. The corresponding maximum deflections should be limited to from 25% to 17% if long life is desired. The number of springs for which space is available and the total required force will determine which grade is required. The required travel plus the preload deflection will be the total deflection, and will determine the length of spring required to stay within allowable percentages of defection limits.

As the punch is resharpened, deflections will increase, and allowances should be made. Charts giving spring pressures as a function of linear travel for many types and sizes of die springs are available from distributors of diemakers' supplies. If the manufacturer's recommendations are followed, high-quality die springs can be expected to provide lengthy trouble-free service.

To retain the stripper against the necessary preload of the springs, and to guide the stripper in its travel, a special type of shoulder screw known as a *stripper bolt* is used.

Choice of the method of applying springs to stripper plates depends on the required pressure, space limitations, shape of the die, nature of the work, and production requirements. *Figure 7-36* presents a number of such methods:

View A. The stripper plate has two counterbored holes, usually 0.25" (6.4 mm) deep. The punch holder counterboring is deeper to apply the correct initial spring deflection. Hole edges are chamfered 0.0625" (1.6 mm) x 45°. While machining the counterbored holes, the stripper plate and punch holder are clamped together, and the holes A are drilled. These holes are then used as pilots when counterboring.

View B. In this piercing die, the upper portions of the springs are retained in holes bored through the punch plate. The spring ends bear against the ground underside of the punch holder.

View C. The pilots shown here provide an excellent means of locating the springs.

View D. The springs are placed around stripper bolts. Although this method is widely used, it is not recommended for high-grade dies.

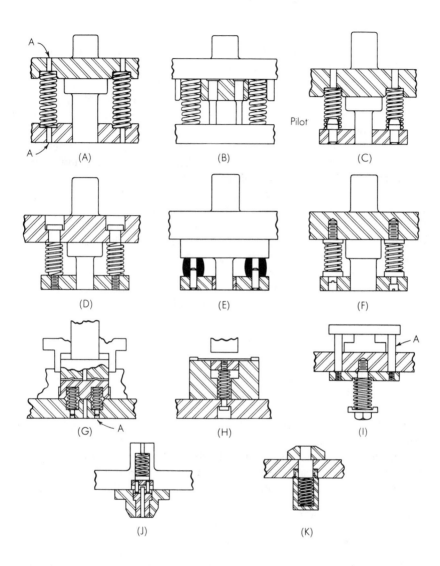

Figure 7-36. Use of die springs with strippers. (*Courtesy, American Machinist*)

View E. Standard rubber springs can provide sufficient stripping pressure, and are inexpensive and easily applied. However, they may be adversely affected by pressworking lubricants and subject to deterioration due to heat buildup.

View F. Standard *Strippit* spring units allow removal of the stripper plate without disturbing the springs. The left-hand Strippit is installed to prevent sideways movement of the lower portion of the unit. The right-hand unit holds the locating plug of the unit to the stripper plate.

View G. This shows application of springs to a blank-and-draw die, where the pressure pad bottoms in such a way that the springs are entirely confined. In this case, the counterbored holes are chamfered at least 0.125" (3.2 mm) x 45°. Safety pins *A* are pressed into the dieholder. If a spring should break, these pins hold the fractured coil and prevent it from getting between the pressure pad and the face of the dieholder.

View H. This shows a bending die used to form a flat part to channel shape. Shoulder bolts limit stripper plate travel, and the springs are placed around them. The stripper bolts will hold the coil if spring breakage occurs.

View I. The stripper plate is actuated by four pins (*A*) through one large die spring. An adjustable spring holder stud threads into the dieholder unit.

View J. An inverted compound die employs a spring-actuated knockout in the upper die. The spring is confined in a counterbored hole in the punch shank, and applies pressure to the knockout through pins.

View K. This shows a method of holding springs below the dieholder, if there is insufficient room for them within the die set. A housing is screwed into a tapped hole, and the confined spring actuates the knockout.

Air and Nitrogen Cylinders

A disadvantage of mechanical springs is the need for preloading for a pad or stripper to exert force upon initial contact. Also, if large forces are required, many springs must be used, which may necessitate the use of large die shoes to accommodate the springs.

Compressed air actuated cylinders provide large forces throughout their stroke, eliminating the need for preloading. Nitrogen actuated cylinders are available that operate at pressures up to 2,000 psi (13,788 KPa). These cylinders, or *gas springs* as they are often called, provide much greater forces in a given space than that obtainable with springs or air cylinders. Their application is detailed in references 5 and 6.

Knockouts

Since the cut blank will be retained in the die block by friction, some means of ejecting on the ram upstroke must be provided. A knockout assembly consists of a plate, a push rod, and a retaining collar. The plate is a loose fit with the die opening contour, and moves upward as the blank is cut. Attached to the plate, usually by rivets, is a heavy pushrod which slides in a hole in the shank of the die set. This rod projects above the shank, and a collar retains and limits the stroke of the assembly. Near the upper limit of the ram stroke, a knockout bar in the press will contact the pushrod and eject the blank. It is essential that the means of retaining the knockout assembly be secure, since serious damage would otherwise occur.

In the ejection of parts, positive knockouts offer the following advantages over spring strippers where the part shape and the die selections allow their use:

1. *Automatic part disposal.* The blank, ejected near the top of the ram stroke, can be blown to the back of the press, or the press may be inclined and the same result obtained.
2. *Lower die cost.* Knockouts are generally of lower cost than spring strippers.
3. *Positive action.* Knockouts do not stick as spring strippers occasionally do.
4. *Lower pressure requirements.* There are no heavy springs to be compressed during the ram descent.

Figure 7-37 shows several good knockout designs.

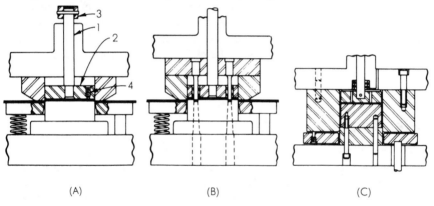

(A)   (B)   (C)

Figure 7-37. Positive knockouts for dies.

View A. This design, applied to a plain inverted compound die, is very simple. It consists of an actuating plunger (1), knockout plate (2), and a stop collar (3) doweled to the plunger (1). The shedder (4) consists of a shouldered pin backed by a spring, which is confined by a setscrew.

View B shows the knockout plate used as a means of guiding slender piercing punches through hardened bushings.

View C shows a design in which the flanged shell, upon completion, is carried upward in the upper die and ejected by a positive knockout.

Type of Die-Cutting Operations

The operations of die cutting (shearing) of work materials are classified as follows:

Piercing (punching) (*Figure 7-38A*) is the operation in which a round punch (or a punch of other contour) cuts a hole in the work material which is supported by a die having an opening corresponding exactly to the contour of the punch. The material (slug) cut from the work material is often scrap.

Blanking (*Figure 7-38A*) differs fundamentally from piercing only in that the part cut from the work material is usable, becoming a blank (workpiece) for subsequent pressworking or other processing.

Lancing (*Figure 7-38B*) combines bending and cutting along a line in the work material. It does not produce a detached slug; it leaves a bent portion, or tab, attached to the work material.

A cut-off (*Figure 7-38C*) operation achieves complete separation of the work material by cutting it along straight or curved lines.

A notching operation cuts out various shapes from the edge of workpiece material (a blank or a part).

Shaving is a secondary shearing or cutting operation in which the surface of a previously cut edge of a workpiece is finished or smoothed. Punch and die clearance for a shaving die, considerably less than that for other cutting dies, allows a thin portion (or shaving) to be cleanly cut from such a surface of the workpiece.

Piercing-Die Design

A complete press tool for cutting two holes in work material at one stroke of the press has been classified and standardized by a large manufacturer as a single-station piercing die and is shown in *Figure 7-39*.

Any complete press tool, consisting of a pair (or a combination of pairs) of mating members for producing pressworked (stamped) parts, including all supporting and actuating elements of the tool, is termed a *die*. Pressworking

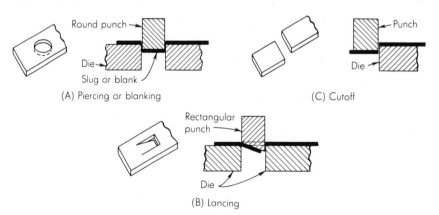

Figure 7-38. Die-cutting (shearing) operations.

terminology also commonly defines the female part of any complete press tool as a die.

The guide pins, or posts, are usually mounted in the lower shoe. The upper shoe contains bushings which slide on the guide pins. The assembly of the lower and upper shoes with guide pins and bushings is a die set. Die sets in many sizes and designs are commercially available. The guide pins guide the stripper in its vertical travel.

A punch holder mounted to the upper shoe holds two round punches (male members of the die), which are guided by bushings inserted in the stripper. A sleeve, or quill, encloses one punch to prevent its buckling under pressure. After

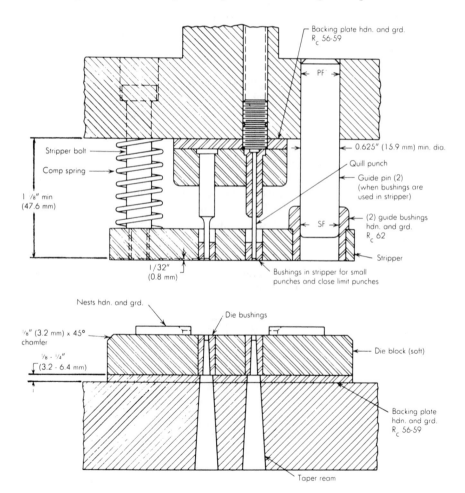

Figure 7-39. Typical single-station die for piercing holes.

penetration of the work material, the two punches enter the die bushings for a sufficient distance to push the slug into them. The female member, or die, consists of two die bushings inserted in the die block. Since this press tool punches holes to the diameters required, the diameters of the die bushings are larger than those of the punches by the amount of clearance. The workpiece is held and located in a nest composed of flat plates shaped to encircle the outside part contours.

Blanking-Die Design

The design of a small blanking die shown in *Figure 7-40* is the same as that of the piercing die of *Figure 7-39* except that a die replaces the die bushings and the two piercing punches are replaced by one blanking punch. A stock stop is incorporated instead of nest plates. This is a drop-through design since the finished blanks drop through the die, the lower shoe, and the press bolster.

Large blanks are often produced by an inverted blanking die (*Figure 7-41*) in which the die is mounted to the upper shoe with the punch secured to the bottom shoe. The passing of a large blank through the bolster is often impractical, but its size may necessitate sectional die design (*Figure 7-42*).

Draft, or angular clearance in an inverted die is unnecessary because the blank does not pass through it. For ease of construction, regrinding, and strength, the cutting edges of each section should not include points and intricate contours. Sections (1) and (2) of the die (*Figure 7-42*) were laid out to include the entire semicircular contour, with straight contours included in the other six sections.

The spring-loaded stripper is mounted on the lower shoe. It travels upward in stripping the stock from the punch fastened to the lower shoe. Stripper bolts hold and guide the stripper in its travel.

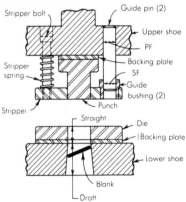

Figure 7-40. A simple blanking die.

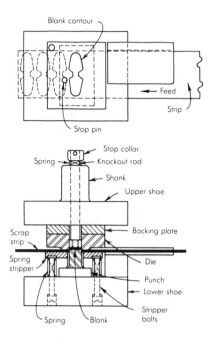

Figure 7-41. An inverted blanking die.

On the upstroke of the ram, the upper end of the knockout rod strikes an arm on the press frame, which forces the lower end of the rod downward, through the die, and ejects the finished blank from the die cavity. A stop collar retains the rods and limits their travel.

Cutting-rule Dies. Instead of a conventional female die, the sharpened edges of steel-rule stock serve as the cutting blades of the dies of *Figures 7-43*

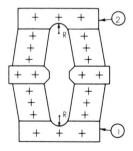

Figure 7-42. Eight-section layout for die shown in *Figure 7-41*.

and *7-44*. The rule, bent to the contour of the blank outline, is used principally for blanking cork, paper, and similar nonmetallic fibrous materials, although the economical blanking of aluminum stock up to 0.04″ (1.02 mm) thick is feasible. The stripper can be of neoprene, cork, or similar resilient sheet material, to force the work material out of the shallow die cavity. This type of die is also known as a *steel-rule* die.

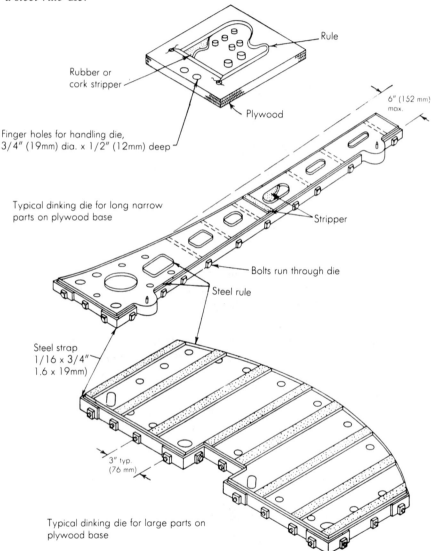

Figure 7-43. Steel-rule blanking dies.

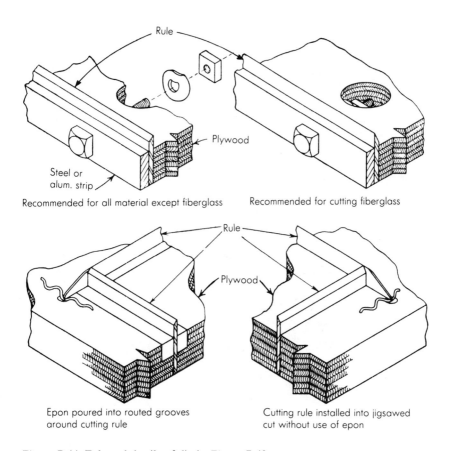

Recommended for all material except fiberglass

Recommended for cutting fiberglass

Epon poured into routed grooves
around cutting rule

Cutting rule installed into jigsawed
cut without use of epon

Figure 7-44. Enlarged details of die in *Figure 7-43*.

Compound-Die Design

A compound die performs only cutting operations (usually blanking and piercing), which are completed during a single press stroke. A compound die can produce pierced blanks to close flatness and dimensional tolerances. A characteristic of compound dies is the inverted position of the blanking punch and blanking die. As shown in *Figure 7-45*, the die is fastened to the upper shoe, and the blanking punch is mounted on the lower shoe. The blanking punch also functions as the piercing die, with a tapered hole in it and in the lower shoe for slug disposal.

On the upstroke of the press slide, the knockout bar of the press strikes the knockout collar, forcing the knockout rod and shedder downward, thus pushing the finished workpiece out of the blanking die. The stock strip is guided by stock

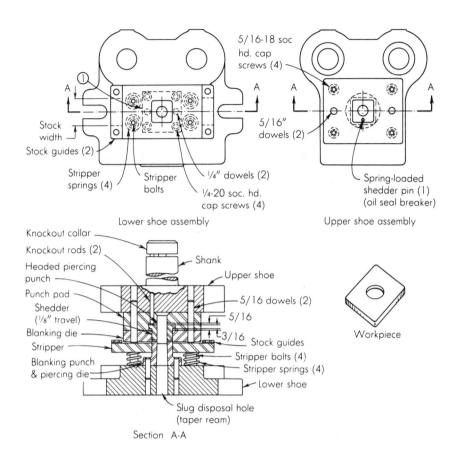

Figure 7-45. A compound die.

guides screwed to the spring-loaded stripper. On the upstroke, the stock is stripped from the blanking die by the upward travel of the stripper. Before the cutting cycle starts, the strip stock is held flat between the stripper and the bottom surface of the blanking die.

Four special shoulder screws (stripper bolts), commercially available, guide the stripper in its travel and retain it against the preload of its springs. The blanking die as well as the punch pad is screwed and doweled to the upper shoe.

A spring-loaded shedder pin (oil-seal breaker) incorporated in the shedder is depressed when the shedder pushes the blanked part from the die. On this upstroke of the ram, the shedder pin breaks the oil seal between the surfaces of the blanked part and shedder, allowing the part to fall out of the blanking (upper) die.

Low-Production and Low-Cost Die Design

When production quantities required for a part are very low, standard design tooling would cause the cost of manufacturing the individual parts to be very high. This cost can be reduced by using a low-cost die design such as a press plate die. This type of die is frequently used in the aircraft industry.

A press plate die can produce complex parts using a compound blank and pierce design. The following is a bill of materials used in a press plate die for the part illustrated in *Figure 7-46*.

Det. 1   1.5 x 4 x 4"   4130 Blanking Punch   1 req.
(12.7 x 102.0 x 102.0 mm)

Det. 2   0.5 x 8 x 12"   CRS B. P. Holder   1 req.
(12.7 x 203.0 x 305.0 mm)

Det. 3   0.5 x 8 x 12"   CRS Die Plate   1 req.
(12.7 x 203.0 x 305.0 mm)

Det. 4   0.5 x 8 x 12"   CRS Retainer Plate   1 req.
(12.7 x 203.0 x 305.0 mm)

Det. 5   0.375 x 2 x 5"   CRS End Riser   2 req.
(9.5 x 51.0 x 127.0 mm)

Det. 6   0.375 x 0.75 x 8"   CRS Slug Riser   3 req.
(9.5 x 19.0 x 203.0 mm)

Det. 7   0.098" Dia.   Lead Set Punches   5 req.
(2.49 mm)

Det. 8   0.187" Dia.   Lead Set Punches   3 req.
(4.75 mm)

Det. 9   0.147" Dia.   Lead Set Punches   1 req.
(3.73 mm)

Det. 10   Guide Pin & Bushing set (Pin Dia. 0.875")   2 req.
(22.23 mm)

Det. 11   0.25 x 1" Dowel   8 req.
(6.4 x 25.4 mm)

Det. 12   1/4-20 x 1"   Flat Head Screw   24 req.
(M6 x 1.5 x 25 mm)

Det. 13   1/4-20 x 0.75"   Flat Head Screw   10 req.
(M6 x 1.5 x 19 mm)

Det. 14   0.5 x 6 x 6"   Cork or Rubber (stripper) 1 req.
(12.7 x 152.0 x 152.0 mm)

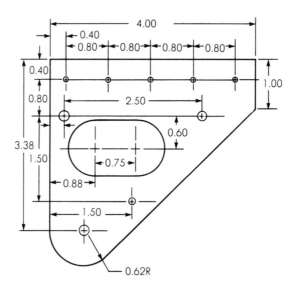

Figure 7-46. Part produced with press plate die.

Typically, a punch is made of 0.5″ (12.7 mm) thick tool steel to the part geometry including internal cutouts and holes (Det. 1). When the punch is complete, it is heat treated or the cutting edges are flame hardened depending on its size and complexity.

The three CRS plates (Dets. 2, 3 and 4) are stacked, and two 0.938″ (23.8 mm) holes are drilled at the guide pin locations. The blank punch holder (Det. 2) is removed, and the two 0.938″ (23.8 mm) holes are redrilled and reamed to 1.250″ (31.75 mm) press fit for the guide bushings in Dets. 3 and 4. Det. 4 is counterbored 1.50″ (38.1 mm) diameter x 0.125″ (3.18 mm) deep for the head of the bushing. The bushings are then pressed in and the three plates restacked. The guide pins are placed through Det. 2 into the bushings and tack welded to the bottom of Det. 2. The punch (Det. 1), end risers (Det. 5) and slug risers (Det. 6) are mounted to the blank punch holder (Det. 2). The die opening and the punch for the oval are cut out of the die plate (Det. 3) at a 15° angle allowing stock for shearing by the punch. The punches (Dets. 7, 8, and 9) are located in the punch (Det. 1) after the die and oval punch are sheared and held in the retainer plate (Det. 4) with a low-temperature eutectic alloy possessing a high shear strength. This alloy, which melts near the boiling point of water, is sometimes called *lead*, and punches so retained are referred to as *lead set punches*.

A 0.313″ (8 mm) hole, counterbored 0.5″ (12.7 mm) diameter x 0.438″ (11.13 mm) deep was located for each of the lead set punches in Det. 4. The strippers are made by placing a piece of cork or rubber (Det. 14) on the punch and cycling the press. This leaves material around the punch and in the die to do the stripping.

Press plate dies are capable of running several thousand parts with little or no maintenance. It should be noted, however, that secondary operations such as manual removal of the part from the stock strip and the oval slug from the part must be done, because the stripping action returns them to the strip after being cut.

Reusable punching and notching die systems are another means of tooling for low cost. The basic system consists of a special die set. Two types are available, one with tee-slots used to mount punch and die retainers, and the second similar to a standard die set with magnets in the punch retainers for mounting. The punch and die retainers are located by a pair of identical steel templates, prebored to conform to the stamping pattern desired (*Figure 7-47*).

The initial cost of this type of system is high compared to the cost of one or two standard dies. The cost savings are realized in the flexibility of the system when several low-production parts are produced on either one-time runs or multiple short runs. The standard items can be used on several parts, and special tooling can be ordered if needed. Templates can be made quickly by stacking them and boring both at the same time. Just a few different hole sizes are required to cover the range of retainers, as the same size hole locates both the punch and die retainer.

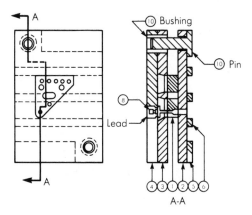

Figure 7-47. Punch and die retainers located by steel templates.

Cam Action Die Operations

In many cases, angular die operations (piercing, lancing, trimming, forming, extruding, etc.) are required. These operations may be performed at any angle off the vertical by use of cams. The basic components in a cam operation are as follows:

1. Cam driver.
2. Cam slide.
3. Slide retainer.
4. Heel block.
5. Slide return.

In *Figure 7-48*, the most common type of cam operation is shown. The driver has a wide base with a 40° cam angle and a vertical heel. The cam slide moves horizontally and is held in a square gib-type retainer. The heel block backs up the driver to eliminate side thrusts during the cam operation. The heel block also holds the spring used to return the cam slide as the driver is retracted. The tooling is mounted on the cam slide's vertical face.

In *Figure 7-49*, a similar cam operation is illustrated except that a gib-type cam return is shown. This positive return device is used when damage to the tooling would result should the cam slide not return.

Figure 7-50 illustrates a positive dogleg cam drive and return. This style drive is useful because many cam motion cycles can be realized by changing the design of the driver. The basic cam action is shown. On the downstroke of the press, the slide is driven forward until the press reaches the bottom of its stroke. On the upstroke, the slide is returned to the dwell area of the dogleg. By making modifications to the driver, the cam cycle can be made to dwell before the ram reaches the bottom of its stroke. There are many more variations possible. The only limitation is the imagination of the designer.

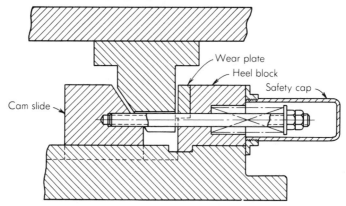

Figure 7-48. Forming die with spring-returned cam slide.

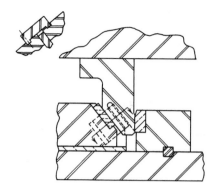

Figure 7-49. Forming die cam slide with positive tabbed-gib drive and return. (*Courtesy, Livernois Automation*)

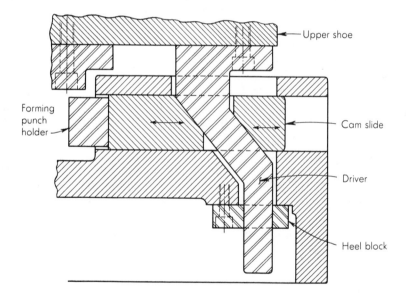

Figure 7-50. Forming die with positive dogleg cam drive and return.

Strip Layout for Blanking

In designing parts to be blanked from strip material, economical stock utilization is very important. The minimum goal should be 75% utilization. A very simple strip layout is shown in *Figure 7-51*.

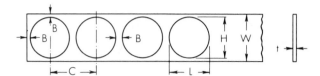

Figure 7-51. Simple progression-strip layout: t is the stock thickness; B is the space between part and edge of strip; C is the progression of the die, i.e., the distance from a point on one part to the corresponding point on the next part; L is the length of the part; H is the part width; W is the width of the stock strip.

Scrap Allowance

A strip layout with insufficient stock between the blank and strip edge, and between blanks, will result in a weakened strip subject to breakage and misfeeds. Such troubles will cause unnecessary die maintenance owing to partial cuts which deflect the punches, resulting in nicked edges. The following formulas are used in calculating scrap-strip dimensions for all strips over 0.032" (0.8 mm) thick:

t = specified thickness of the material;
$B = 1.25t$ when C is less than 2.5" (63.5 mm);
$B = 1.5t$ when C is 2.5" (63.5 mm) or longer;
$C = L + B$, or progression (lead or pitch) of the die.

Example: A rectangular part, to be blanked from 0.060" (1.524 mm) steel. The blank is to be 0.375" x 1.062" (9.525 mm x 26.98 mm) If the progression strip is developed as in *Figure 7-52*, the solution is as follows:

$t = 0.06''$ (1.524 mm);
$B = 1.25 \times 0.0598'' = 0.07475'' = 0.078''$ (1.25 mm x 1.519 mm = 1.823 mm);
$C = 0.375'' + 0.078'' = 0.453'' = (9.53$ mm + 1.98 mm = 11.51 mm);
$W = H + 2B = 1.062'' + 0.156'' = 1.219''$ (27 mm + 4 mm = 31 mm).

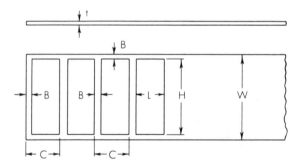

Figure 7-52. Progression strip development.

The nearest commercial stock is 1.125″ (28.6 mm). Therefore, the distance *B* will equal 0.0897″ (2.3 mm). This is acceptable since it exceeds minimum requirements.

Strip Allowance for Thin Materials. If the material to be blanked is 0.025″ (0.64 mm) thick or less, the previously mentioned formulas should not be used. Instead, dimension *B* should be as follows:

Strip width *W*	Dimension *B*
0-3″ (0-76 mm)	0.05″ (1.27 mm)
3-6″ (76-152 mm)	0.093″ (2.36 mm)
6-12″ (152-305 mm)	0.125″ (3.18 mm)
over 12″ (over 305 mm)	0.156″ (3.96 mm)

Other Strip Allowance Applications. *Figure 7-53* illustrates special allowances for one-pass layouts.

View A. For work with curved outlines, $B = 70\%$ of strip thickness t.

View B. For straight-edge blanks: where *C* is less than 2.5″ (63.5 mm), $B = 1t$; where *C* is 2.5″ to 8″ (63.5 mm to 203 mm), $B = 1.25t$; where *C* is over 8″ (203 mm), $B = 1.5t$.

View C. For work with parallel curves, use the same formulas as for View B.

View D. For layouts with sharp corners of blanks adjacent, $B = 1.25t$.

Figure 7-54 illustrates special allowances for two-pass layouts:

View A. Single-row layout for two passes through the die: $B = 1.5t$.

View B. Double-row layout of blanks with curved outlines: $B = 1.25t$.

View C. Double-row layout of parts with straight and curved outlines: $B = 1.25t$.

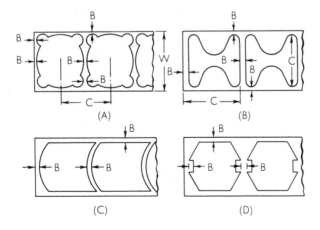

Figure 7-53. Allowances for one-pass layouts.

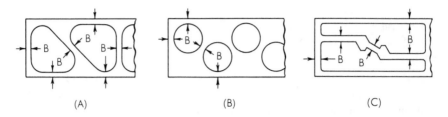

(A)   (B)   (C)

Figure 7-54. Allowances for two-pass layouts.

Percentage of Stock Used. If the area of the part is divided by the area of the scrap strip used, the result will be the percentage of stock used.

If A = total area of strip used to produce a single blanked part, then $A = CW$ (*Figure 7-52*), and a = area of the part = LH.

If $C = 0.453''$ (11.51 mm) and $W = 1.25''$ (31.8 mm), then $A = 0.453''$ x $1.25'' = 0.56625''^2$(11.51 mm x 31.75 mm = 365.44 mm²)

If $L = 0.375''$ (9.53 mm) and $H = 1.0625''$ (26.988 mm), then $a = 0.375''$ x $1.0625'' = 0.3984375''^2$(9.53 mm x 26.99 mm = 257.20 mm²)

Percentage of stock used:

$$\frac{a}{A} = \frac{0.3984375}{0.56625} \frac{(257.20)}{(365.44)} = 70\% \text{ approx.}$$

Commercial Die Sets

Two-post commercial die sets are available in many styles and sizes. Commonly used die sets are the round series diagonal (*Figure 7-55*), and the back-post (*Figure 7-56*). These die sets are covered in ANSI Standard B5.25, "Punch and Die Sets."

In addition to providing enlarged mounting bases for the punch and die elements, the die set is equipped with heavy guide posts which maintain alignment of the two members. The dieholder (lower shoe) should be at least $0.25''$ (6.4 mm) larger all around than the die block. If the bolster opening is excessively large, an oversize dieholder may be needed to bridge the opening.

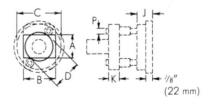

(22 mm)

Figure 7-55. Dimensions of round diagonal-post die sets.

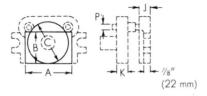

Figure 7-56. Back-post die sets.

Between the rear edge of the die block (or any portion that requires resharpening) and the guide posts there should be a 0.625" (15.9 mm) minimum clearance to allow for the surface grinder grinding-wheel guard.

The punch holder (upper shoe) is usually equipped with an integral round shank, which is gripped by a clamp in the press ram. The shank is located in the center of the die space area, as cataloged by the maker of the die set. The die set must be laid out so the center of pressure of the blank is approximately central with this shank. The shank diameter is determined by the press in which the tool is to be used, and must be specified on the tool drawing.

The determined shut height will establish the length of the guide posts, which must be at least 0.5" (12.7 mm) shorter to allow for the reduction of shut height due to resharpening.

Evolution of a Blanking Die

In the planning of a die, the examination of the part print immediately determines the shape and size of both punch and die as well as the working area of the die set.

Die Set Selection

A commercially available, standard two-post die set (*Figures 7-56 and 7-57*) with 6" (152.4 mm) overall dimensions side-to-side and front-to-back allows the available 3" (76.2 mm) wide stock to be fed through. It is large enough for mounting the blanking punch on the upper shoe (with the die mounted on the lower shoe) for producing the blank shown in *Figure 7-58*, since the guide posts can be supplied in lengths of from 4" to 9" (102 mm to 229 mm). Since the stock in this example was available only in a width of 3" (76.2 mm), the length of the blanked portions extended across the stock left a distance of 0.25" (6.35 mm), or twice the stock thickness, between the edges of the stock and the ends of the blank. This allowance is satisfactory for the 0.125" (3.18 mm) stock.

Die Area			Die holder diam	Thickness		Min guide-post diam
Rectangular Area		Diam		Die holder	Punch holder	
A	B	D	C	J	K	P
1 3/4 (44.4)	3 1/2 (88.9)	2 3/4 (69.8)	5 (127)	1 3/4 (44.4)	1 1/4 (31.8)	1/2 (12.7)
2 1/4 (57.2)	4 1/2 (114.3)	3 1/2 (88.9)	6 (152)	1 3/4 (44.4)	1 1/4 (31.8)	5/8 (15.9)
2 3/4 (69.8)	5 1/2 (139.7)	4 (102)	7 (178)	2 (51)	1 1/2 (38.1)	3/4 (19.0)
3 1/2 (88.9)	7 (178)	5 1/4 (133.4)	9 (229)	2 (51)	1 1/2 (38.1)	1 (25.4)
4 1/2 (114.3)	9 (229)	7 (178)	11 (279)	2 (51)	1 1/2 (38.1)	1 1/8 (28.6)

All dimensions are given in inches and (mm).

Figure 7-57. Dimensions of round diagonal-post die sets.

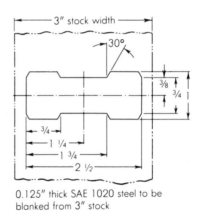

0.125" thick SAE 1020 steel to be
blanked from 3" stock

Figure 7-58. Part to be blanked.

Die Block Design

By the usual "rule-of-thumb" method previously described, die block thickness (of tool steel) should be a minimum of 0.75" (19.1 mm) for a perimeter between 3" (76.2 mm) and 4" (101.6 mm). For longer perimeters, die block thickness should be 1.25" (31.75 mm). Die blocks are seldom thinner than 0.875" (22.2 mm) finished thickness to allow for grinding and for blind screw holes. Since the perimeter of the blank is approximately 7" (177.8 mm), a die block thickness of 1.5" (38.1 mm) was specified, which includes a 0.25" (6.4 mm) grinding allowance.

There should be a margin of 1.25" (31.75 mm) around the opening in the die block; its specified size of 6" by 6" (152.4 mm by 152.4 mm) allows a margin of 1.75" (44.45 mm) in which four 0.375" (9 mm) cap screws and 0.375" (9 mm) dowels are located at the corners 0.75" (19.1 mm) from the edges of the block.

The wall of the die opening is straight for a distance of 0.125" (3.18 mm) (stock thickness); below this portion or the straight, an angular clearance of 1.5° allows the blank to drop through the die block without jamming.

The dimensions of the die opening are the same as that of the blank. Those of the punch are smaller by the clearance, which is 6% of stock thickness, or 0.075" (1.91 mm). This is required to produce the blanks to print (and die) size.

The top of the die was ground off a distance equal to stock thickness (*Figure 7-59*) with the result that shearing of the stock starts at the ends of the die and progresses toward the center of the die. Less blanking pressure is required than if the top of the die were flat.

Punch Design

The shouldered punch, which is 2.25" (57.2 mm) long, is held against a 0.250" (6.4 mm) thick hardened steel backup plate by a punch retainer 0.75" (19.1 mm) thick, which is screwed and doweled to the upper shoe. A die of this size can be accommodated in a 32 ton (285 kN) (JIC Standard) open-back inclinable press.

The following conditions apply in this case study:

Shear strength S = 60,000 psi (413.7 MPa).
Blanked perimeter length L = 7" (178 mm) approximately.
Thickness T = 0.125" (3.18 mm).

From the equation $P = SLT$, the pressure P = 60,000 psi x 7" x 0.125" = 26.25 tons (413.7 MPa x 178 mm x 3.18 mm = 233.9 kN). This value is well below the 32-ton capacity of the selected press and has substantial added safety value due to the angular die shear.

The shut height (*Figure 7-59*) is 7" (177.8 mm) less the 0.063" (1.6 mm) travel of the punch into the die cavity.

Stripper Design

The stripper that was designed is the fixed type with a channel or slot having a height equal to 1.5 times stock thickness and a width of 3.125" (79.38 mm) to allow for variations in the stock width of 3" (76.2 mm). The same screws that hold the die block to the lower shoe fasten the stripper to the top of the die block.

If, instead of 0.125" (3.18 mm) stock, thinner 0.031" (0.79 mm) stock were to be blanked, a spring-loaded stripper such as that shown in *Figures 7-39* and *7-40* would firmly hold the stock down on top of the die block and help to prevent distortion of the stock. A spring-loaded stripper should clamp the stock until the punch is withdrawn.

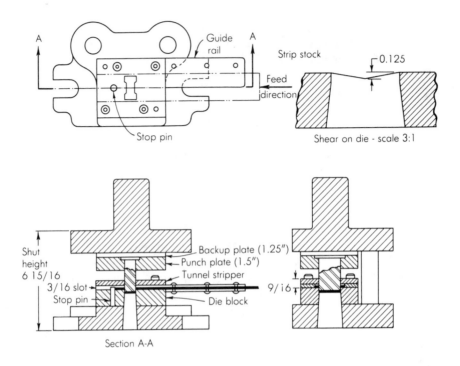

Figure 7-59. Blanking die for part shown in *Figure 7-58.*

Stock Stops

The pin stop pressed in the die block is the simplest method for stopping the hand-fed strip. The right-hand edge of the blanked opening is pushed against the pin before descent of the ram and the blanking of the next blank. The 0.188″ (4.78 mm) depth of the stripper slot allows the edge of the blanked opening to ride over the pin and engage the right-hand edge of every successive opening.

The design of various types of stops adapted for manual and automatic feeding is covered in a preceding discussion.

Evolution of a Progressive Blanking Die

Figure 7-60 gives the blanked dimensions of a linkage case cover of cold rolled steel, stock size 0.125 x 2.375 x 2.375″ (3.2 x 60.3 x 60.3 mm). Production is stated to be 200 parts made at one setup, with the possibility of three or four runs per year.

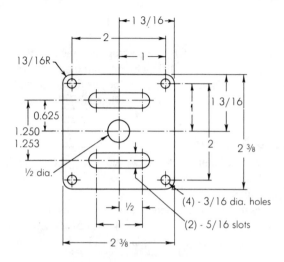

Figure 7-60. Linkage case cover to be blanked.

Step 1: Part Specification

a. The production is low volume; therefore, a second-class die will be used.

b. Tolerances required: No close tolerances are specified. A compound die is not necessary; a two or three-station progressive die will be adequate.

c. Thickness of material: Specified as 0.125" (3.18 mm) standard cold rolled steel.

Step 2: Progression Strip Development

From the production requirements, a single-row strip will suffice. After several trials, the strip development shown in *Figure 7-61* was decided upon. Owing to the closeness of the holes, it was decided to make a four-station die to avoid weak die sections. From discussion related to *Figure 7-61*:

$B = 1.25t = 1.25 \times 0.125" = 0.156"$ (1.25 × 3.18 mm = 3.97 mm)

$W = H + 2B = 2.375" + 0.25" = 2.625"$ (60.3 mm + 6.4 mm = 66.7 mm)

$C = L + B = 2.375" + 0.156" = 2.0531"$ (60.3 mm + 4.0 mm = 64.3 mm)

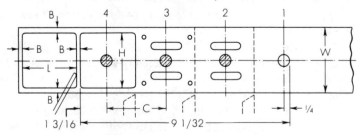

Figure 7-61. Progression strip development for the part shown in *Figure 7-60*.

The strip is first fed into the first finger stop, and the center hole pierced. The strip is then moved into the second finger stop, and the two holes are pierced. At the third stage and third finger stop, a pilot locates the strip, and the four corner holes are pierced. At the fourth stage, a piloted blanking punch cuts out the finished part.

Step 3: Press Tonnage

Now determine the amount of force needed. Only the actual blanking in the fourth stage need be calculated, since the work in the first three stages will be done by stepped punches.

The shear strength S of cold rolled steel is 58,000 psi (400 MPa). The length L of the blanked perimeter equals 2.375" x 4 = 9.5" (60 mm x 4 = 240 mm). The depth of cut (stock thickness t) equals 0.125" (3 mm). From the equation $P = SLt$:

P = 58,000 psi x 9.5" x 0.125" = 68,875 lbs. or 34.4 tons
P = (400 MPa x 240 mm x 3 mm = 307.2 kN)

This force is greater than can be handled by the available presses. To lower the force, shear is ground on the blanking punch to reduce the needed force by one-third. Thus, 1/3 x 34.5 = 11.5; 34.5–11.5 = 23 tons (1/3 x 307 = 102; 307–102 = 205 kN). A 30-ton (267 kN) press with a 7.5" (190.5 mm) shut height and a 2" (50.8 mm) stroke is available and is selected. The bolster plate is 12" (305 mm) deep. There is 5.5" (140 mm) from centerline of ram to back edge of bolster, and the bed is 24" (610 mm) wide. The shank diameter is 2.5" 63.5 mm).

Step 4: Calculation of the Die

a. *The die*. The perimeter of the cut equals 9.5" (241.3 mm), and therefore the thickness of the die must be 1" (25.4 mm).

   The width of the strip opening is 2.375" (60.3 mm). With 1.25" (31.8 mm) extra material on each side of the opening, it will be:
   2.375" + 2.5" = 4.875"
   (60.3 mm + 63.5 mm = 123.8 mm),
   or 5" (127 mm) in width.
   The distance from the left side of the opening in station 4 to the edge of the opening in station 1 equals
   $3C$ + 0.656" + 0.25" = 7.594" + 1.188" + 0.25" = 9.031"
   ($3C$ + 30.2 mm + 6.35 mm = 192.88 mm + 30.2 mm + 6.35 mm = 229.39 mm)
   plus 2.5" (63.5 mm) = 11.531" (292.89 mm). Therefore, the die should be 1" x 5" x 11.625" (25.4 mm x 127 mm x 295.28 mm) long.

b. *The die plate*. As a means of filling in between the die and the die shoe, a die plate of mild steel is used. To secure the die plate to the die shoe, 0.5"

(M14) cap screws and 0.5″ (14 mm) dowels are used. A minimum of twice the size of the cap screw for the distance from the edge of the die to the edge of the die plate is needed, which is 1″ (25.4 mm). Twice this distance equals 2″ (50.8 mm), which, when added to the size of the die will result in a die plate of 1″ x 7″ x 13.625″ (25.4 mm x 177.8 mm x 346.1 mm).

Figure 7-62 illustrates the die and die plate fitted together, including the die openings with a straight portion for sharpening and angular relief.

Step 5: Calculation of Punches

Many experts specify that 10% of the stock thickness should be allowed for cutting clearance. This value is used on the die opening, where holes are to be pierced in the blank. The clearance rule will be applied to the die opening in Stations 1, 2, and 3, and to the punch in Station 4 (see *Figure 7-63*).

The punch and the die opening will have straight sides for at least 0.125″ (3.18 mm) for sharpening, and will have a taper relief of 1.5° to the side. *Figure 7-63* also illustrates a 0.125″ (3.18 mm) angular shear applied to the die at Station 4 and the punches of Station 2. The punches for all stations are stepped.

Step 6: Springs

A solid stripper plate can be used for this job.

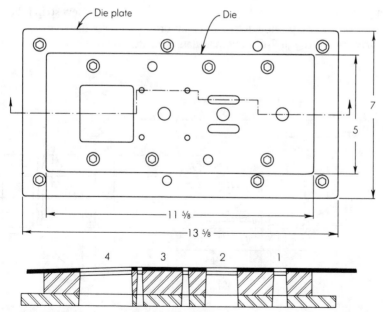

Figure 7-62. Fitting the die and die plate: note shear on Station 4, and straight edge and relief at the die opening (dimensions in inches).

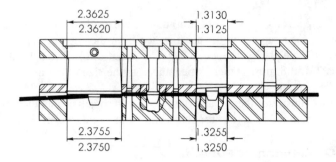

Figure 7-63. Calculations of clearance and shear on punches and die, and stepped arrangement of punches to reduce cutting pressure.

Step 7: Piloting

Figures 7-61 and *7-63* illustrate the arrangement for piloting. In this case, direct piloting is used. However, if the part did not have a center hole, and the slots and other holes were too small, indirect piloting would have to be provided.

Step 8: Automatic Stops

Finger stops, illustrated in *Figure 7-64*, act as stops when a new strip is being inserted, but afterward, an automatic spring drop stop halts the strip.

Figures 7-64 through *7-67* illustrate details of the completed drawing of the die.

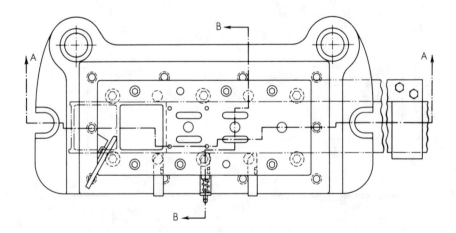

Figure 7-64. Top view of die with punch holder removed.

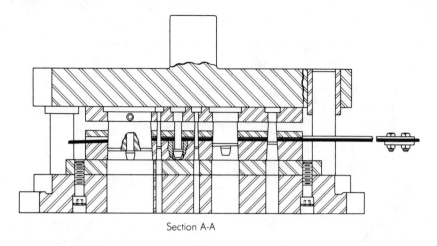

Section A-A

Figure 7-65. Front sectional view of complete die.

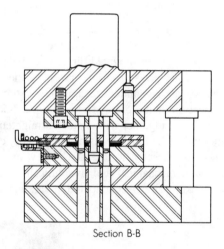

Section B-B

Figure 7-66. Side sectional view of complete die.

Mechanical Feeders

All the die designs discussed incorporate stock guides and stops intended to facilitate an operator in hand-feeding strips of stock through the die. These strips are either sheared from larger rectangular sheets of metal or cut from coiled stock.

For low to moderate production volumes, hand feeding cut strips is satisfactory. The dies illustrated are simple to set up and operate. If a large variety of small parts are needed in small lot sizes, this may be the most economical method of production.

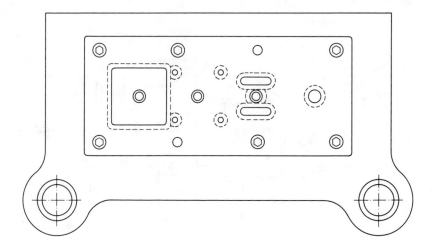

Figure 7-67. Bottom view of punch assembly.

Coil Feeders. The feeder can be a part of the press equipment, a separate accessory outside of the press, or mounted directly to the die.

Roll feeds are part of the press, and are driven by the press crankshaft. A ratchet mechanism allows an intermittent feeding of the stock to provide a dwell during the working portion of the press stroke. *Figure 7-68* illustrates a press equipped with such a feeder together with a stock straightener and an uncoiler.

Slide feeds (sometimes called grip feeds or hitch feeds) can be freestanding, attached to the press or directly to the die. They are usually driven by compressed air, although mechanical and hydraulically driven types are also used. On the feed stroke, grippers clamp the stock, moving it forward to a

Figure 7-68. Roll feed attached to the press with a powered stock straightener and uncoiler. (*Courtesy, F. J. Littell Machine Co.*)

positive stop. The stock is clamped by a second gripper while the feed slide returns to the start position. One type of air-powered slide feed is illustrated in *Figure 7-69*. Another type of slide feed is operated by a cam that returns the slide to the start position. It has a dwell at the start position during the working cycle of the press. On the upstroke, the cam allows the slide to move forward under spring pressure to a positive stop (*Figure 7-70*). This type of feed can be mounted on the press or directly on the die.

Figure 7-69. Air-driven slide feeder. (*Courtesy, Rapid-Air Corp.*)

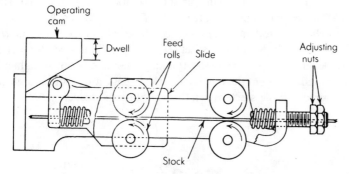

Figure 7-70. A die or press mounted cam-driven grip feeder. (*Courtesy, H. E. Dickerman Manufacturing Co.*)

Stock Uncoilers and Straighteners

Coil cradles (*Figure 7-71*) and stock reels (*Figure 7-72*) uncoil stock as it is needed while the press is running. Uncoilers are often motorized for the heavier materials, but are often unpowered when used for light materials.

When the stock is unwound by the decoiler, there remains a normal curvature or *coil set* in the stock. The coil set is often removed by a stock straightener for

Figure 7-71. A cradle. (*Courtesy, Cooper-Weymouth, Peterson*)

smooth feeding in the stock guides of dies and as an aid to producing uniform parts. This is done by subjecting the stock to a series of up and down bends as it passes through a series of rollers. The bending action must be severe enough to exceed the yield point of the stock as the outer fibers of the metal are alternately stretched and compressed.

Figure 7-73 illustrates the principle of operation of a powered stock straightener. The first pair of powered feed rolls (*D1*) feeds the stock into a series of seven straightening rollers (*D2*). A second set of powered rollers (*D3*) operating in synchronism with the first set (*D1*) act to pull the stock through the straightener. Depending upon the application, a greater number of straightening rollers may be used. For light-duty and noncritical applications, the straightener may be unpowered and the stock drawn through it by the feeder.

Figure 7-72. A stock reel. (*Courtesy, Cooper-Weymouth, Peterson*)

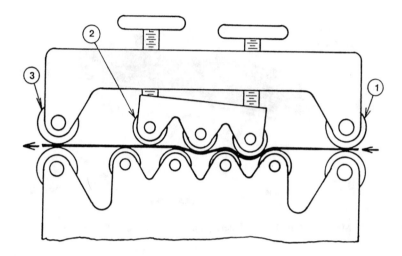

Figure 7-73. The principle of operation of a powered stock straightener: the first pair of powered feed rolls (*D1*) feeds the stock into a series of seven straightening rollers (*D2*); a second set of powered rollers (*D3*) operating in synchronism with the first set (*D1*) acts to pull the stock through the straightener.

Parts Feeders

Parts feeders can also be used to dispense individual parts to the press. These can be index wheels, magazine feeds, hopper feeds, or mechanical arms. An index feed is usually a round dial with a number of pockets to hold and carry the parts into the die, and in some cases, out of the die. They can be loaded manually or by a parts hopper.

A magazine feed has a magazine that holds parts vertically or slightly off the vertical so the edges of the parts are staggered slightly. A slide with a step equal to two-thirds the part thickness peels the bottom part from the stack and deposits it in position. Then, the slide returns to the start position, and the stack drops so the next part can be fed. This type of feeder usually delivers blanks into a single-station form die or a multiple-station transfer die. Generally, the magazine is loaded manually (*Figure 7-74*).

A hopper feed delivers parts to a die for secondary operations, such as coining or staking. A hopper feed may also be used on parts to be installed into the stock strip of a progressive die, such as eyelets or nut-plates. They also feed parts into an assembly machine. The two basic types of hoppers are a rotating drum and vibrating bowl. The function of a hopper is to orient the parts the same way and load them into a track for use by tooling, whether in a die or assembly machine.

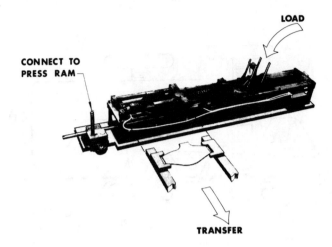

Figure 7-74. Magazine-type blank feeder. (*Courtesy, Livernois Automation Co.*)

The subject of mechanical feeders is treated in detail in reference 7.

A mechanical arm simply takes the place of an operator's hand and arm to load and unload parts. The mechanical function may be as simple as extend-and-grip, retract-and-release for an unloader. Loading could entail grip, raise, rotate, extend, lower, locate with pitch and yaw (wrist actions), release, raise, retract, rotate back to home position, lower, and grip next part. All these functions can be mechanical, fixed functions, or controlled by a computer to tailor the motion to a number of applications. The grip function may be fingers that close to hold the part or vacuum cups. *Figure 7-75* illustrates a simple mechanical arm.

Figure 7-75. A mechanical arm part loader. (*Courtesy, Livernois Automation Co.*)

A robot is basically a mechanical arm with programmable functions and a great range of motion. It can be used for a simple pick-and-place operation, or can be so complex that it performs every movement of the human hand and arm. The operating functions are controlled by a computer, and when the robot has completed one job, it can be programmed to do another (*Figure 7-76*).

Figure 7-76. Computer-controlled robot. (*Courtesy, Cincinnati Milacron*)

DESIGN OF
PRESSWORKING TOOLS

Review Questions

1. List the six types of presses discussed.
2. When the contour to be blanked is irregularly shaped, why should the center of pressure be calculated? What is the center of pressure?

3. Compute the die clearance in inches of a Group 1 material, 0.125" thick; Group 2, 0.250" thick; and Group 3, 0.065" thick.
4. What are the four advantages of positive knockouts over spring strippers?
5. In die-cutting operations, define lancing.
6. In the design of a compound banking die, why is angular clearance not required?
7. List the basic components in CAM action die operations.
8. What is a realistic goal in stock utilization?
9. In die block design, is the die opening or the punch the same size as the blank? By how much?
10. Why are mechanical feeds used on presses?

DESIGN OF PRESSWORKING TOOLS

Answers to Review Questions

1. Open back inclinable (OBI) or gap frame press, straightside press, double-action press, triple-action press, knuckle press, and hydraulic press.
2. The summation of shearing forces on one side of the center of the ram may greatly exceed the forces on the other side. This results in a bending moment in the press ram, and undesirable deflections and misalignment. The center of pressure is a point about which the summation of shearing forces will be symmetrical.
3. Group 1 = 0.0056", Group 2 = 0.015", Group 3 = 0.0049".
4. Automatic part disposal, lower die cost, positive action, and lower pressure requirements.
5. Lancing combines bending and cutting along a line in the work material.
6. Angular clearance is not necessary because the blank does not pass through the die.
7. Cam driver, cam slide, slide retainer, heel block, and slide return.
8. 75% utilization.
9. The die opening. Depending on the type of material, from 4.5% to 10% of stock thickness.
10. Increased productivity, the automation of operations, and for increased safety.

References

1. David A. Smith, *Die Design Handbook*, 3rd Edition, Chapter 27, "Press Data," Society of Manufacturing Engineers, Dearborn, Michigan, 1990.
2. David A. Smith, *Die Design Handbook*, 3rd Edition, Chapter 4, "Shear Action in Metal Cutting," Society of Manufacturing Engineers, Dearborn, Michigan, 1990.
3. David A. Smith, *Quick Die Change*, 1st Edition, Chapter 29, "Control the Process with Waveform Signature Analysis," Society of Manufacturing Engineers, Dearborn, Michigan, 1991.
4. *The Tooling*, Dayton Progress Corporation.
5. David A. Smith, *Die Design Handbook*, 3rd Edition, Chapter 22, "Die Sets and Components," Society of Manufacturing Engineers, Dearborn, Michigan, 1990.
6. David A. Smith, *Quick Die Change*, 1st Edition, Chapter 5, "Advanced Diesetting Topics," Society of Manufacturing Engineers, Dearborn, Michigan, 1991.
7. David A. Smith, *Die Design Handbook*, 3rd Edition, Chapter 23, "Designing Dies for Automation," Society of Manufacturing Engineers, Dearborn, Michigan, 1990.

8

BENDING, FORMING, DRAWING, AND FORGING DIES

Bending Dies

Bending is the uniform straining of material, usually flat sheet or strip metal, around a straight axis which lies in the neutral plane and normal to the lengthwise direction of the sheet or strip. Metal flow takes place within the plastic range of the metal, so that the bend retains a permanent set after removal of the applied stress. The inner surface of a bend is in compression; the outer surface is in tension. A pure bending action does not reproduce the exact shape of the punch and die in the metal; such a reproduction is one of forming.

Terms used in bending are defined and illustrated in *Figure 8-1*. The neutral axis is the plane area in bent metal where all strains are zero.

Bend Radii. Minimum bend radii vary for different metals; generally, different annealed metals can be bent to a radius equal to the thickness of the metal without cracking or weakening.

Bend Allowances. Since bent metal is longer after bending, its increased length, generally of concern to the product designer, may also have to be considered by the die designer if the length tolerance of the bent part is critical. The length of bent metal may be calculated from the equation:

$$B = \frac{A}{360} \times 2\pi \ (R_i + Kt) \qquad (8\text{-}1)$$

Where:

B = bend allowance, in. (mm) (along neutral axis)
A = bend angle, deg.
R_i = inside radius of bend, in. (mm)
t = metal thickness, in. (mm)
K = 0.33 when R_i is less than $2t$ for edge bending and 0.43 for rotary bending

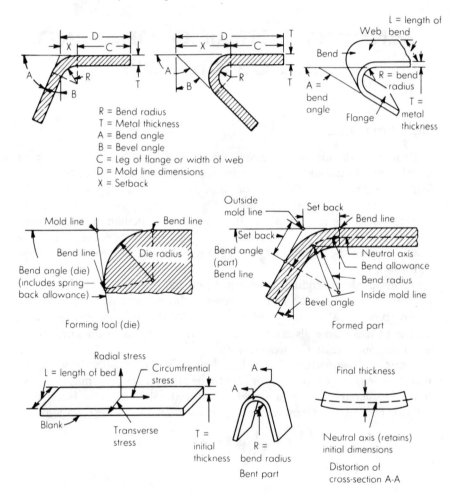

Figure 8-1. Bending terms.

Bending Methods. There are three bending methods commonly used in press tools: (1) vee bending, (2) edge or wipe bending, and (3) rotary bending.

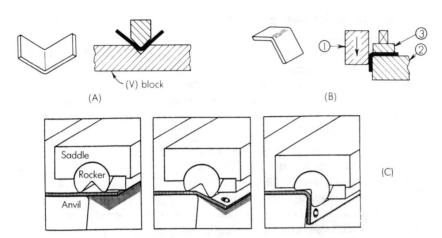

Figure 8-2. Bending methods: (*A*) V bending; (*B*) edge bending; (*C*) rotary READY®
Bender method. (*Courtesy, Ready Tools, Inc.*)

In vee bending, commonly used in press brakes, a metal sheet or strip supported by a V block (*Figure 8-2A*) is forced into the block by a wedge-shaped punch. This method produces a bend having an included angle which may be acute, obtuse, or of 90°.

Edge or wipe bending (*Figure 8-2B*) is cantilever loading of a beam. There are three separate bending pressures to be considered: (1) bending force, (2) hold-down force, and (3) bottoming force.

Rotary bending transfers the vertical motion of the press into a rotary bending action, which overforms the material around the anvil radius to counteract the material springback.

Bending Force. The force required for V bending is as follows:

$$P = \frac{KLSt^2}{W} \tag{8-2}$$

Where:

P = bending force, tons (for metric usage, multiply number of tons by 8.896 to obtain kilonewtons)

K = die opening factor: 1.20 for a die opening of 16 times metal thickness, 1.33 for an opening of eight times metal thickness

L = length of part, in.

S = ultimate tensile strength, tons per sq in.

W = width of V or U die, in.

t = metal thickness, in.

For U bending (channel bending) pressures will be approximately twice

those required for V bending; edge bending requires about one-half those
needed for V bending.

Springback. After bending pressure on metal is released, the elastic
stresses also are released, which causes metal movement resulting in a
decrease in the bend angle (as well as an increase in the included angle between
the bent portions). Such a metal movement, termed springback, varies in steel
from $1/2°$ to $5°$, depending upon its hardness; phosphor bronze may spring
back from $10°$ to $15°$.

V-bending and rotary bending dies customarily compensate for springback
with V blocks and wedge-shaped punches having included angles of somewhat
less than the that required in the part. The part is bent through a greater angle
than required, but it springs back to the desired angle.

Parts produced with cams and toggles are also overbent through an angle equal
to the spring-back angle with an undercut or relieved punch.

Evolution of a Bending Die

The production of a workpiece of *Figure 8-3* in the die of *Figure 8-4*
required blank development before die design began.

The straight length of the vertical leg is 1.000"—0.060" (25.4 mm—1.52
mm) or 0.940" (23.88 mm); the straight length of the horizontal leg is 6.000"—
0.060" (152.4 mm—1.52 mm) or 5.940" (150.88 mm).

The bend length (since R_i is less than twice metal thickness) is,

$$B = \frac{90}{360} \, 2\pi \left(0.0625 + \frac{0.060}{3} \right) \quad \text{or} \quad B = \frac{90}{360} \, 2\pi \left(1.588 + \frac{1.52}{3} \right)$$

$$= 0.1296'' \qquad\qquad\qquad\qquad = 3.291 \text{ mm}$$

The developed length is,

$$0.940 + 5.940 + 0.1296 = 7.0096''$$

$$23.88 + 150.88 + 3.291 = 178.051 \text{ mm}$$

To hold the tolerance of $\pm 1/2°$ allowed for the $90°$ bend, the designer
decided that an edge-bending die, with a slight ironing action on the stock,
be used.

Based upon Eq. (8-2), the bending pressure needed without ironing is:

$$P = \frac{1.33 \times 10 \times 30 \times (0.060)^2}{2 \times 1.414} = 0.51 \text{ ton} \times 8.9 = 4.5 \text{ kN}$$

The total spring force required of six springs in the pressure pad is 1020 lb.;
each spring will supply a pressure of 170 lb. Commercial 1" (25 mm) diam die

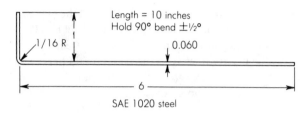

Figure 8-3. Part bent in the die of *Figure 8-4.*

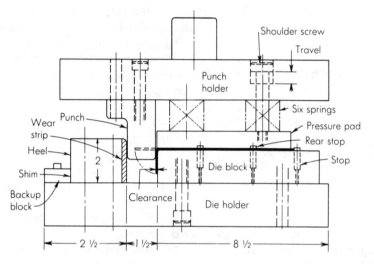

Figure 8-4. Die design for part of *Figure 8-3.*

springs, 2" (51 mm) long, will easily supply this pressure. Almost any small OBI press will supply these pressures with an ample allowance for slight ironing of the blank, and has a bed area large enough to accommodate a die set. There are no formulae for determining ironing pressure; it can be approximated by multiplying the yield strength of the metal by the thickness of the metal after reduction times its length.

Since the size of the blank to be bent is 10 x 7" (254 mm x 178 mm), the area of a die set 14" (356 mm) (right-to-left) by 10"(254 mm) (front-to-back) permits for mounting the punch and pressure pad on the upper shoe, and mounting the die block and heel to the lower shoe.

The blank is located on the die block against an end-stop pin and two rear-stop pins. On the down stroke of the press, the pressure pad clamps the blank in this location.

The descent of the punch forces the end of the blank against the end of the die block. Its wiping action results in some ironing of the blank, the amount of which is determined by the clearance between the heel block and the punch. To establish optimum clearance and to allow for wear on punch and heel block, shims can be inserted between the backup and heel blocks.

The surface of the heel block against which the punch rubs can be hardened or can have a bronze wear strip as shown.

Forming Dies

Forming dies, often considered in the same class with bending dies, are classified as tools that form or bend the blank along a curved axis instead of a straight axis. There is very little stretching or compressing of the material. The internal movement or the plastic flow of the material is localized and has little or no effect on the total area or thickness of the material. The operations classified as forming are bending, drawing, embossing, curling, beading, twisting, spinning, and hole flanging.

A large percentage of stampings used in the manufacturing of products require some forming operations. Some are simple forms that require tools of low cost and conventional design. Others may have complicated forms, which require dies that produce multiple forms in one stroke of the press. Some stampings may be of such nature that several dies must be used to produce the shapes and forms required.

A first consideration in analyzing a stamping is to select the class of die to perform the work. Next to be considered is the number of stampings required, and this will govern the amount of money that should be spent in the design and building of the tools. Stampings of simple channels in limited production can be made on a die classed as a solid form die. It would be classified under channel forming dies. Others—the block and pad type—are also channel forming dies. Such operations as curling, flanging, and embossing as well as channeling employ pressure pads.

A forming die may be designed in many ways and produce the same results; at this point the cost of the tool, safety of operation, and also the repairing and reworking must be considered. The tool that is cheapest and of the simplest design may not always be best because it may not produce the stamping to the drawing specifications. Where limited production is required, and a liberal tolerance is allowed in a stamping, a solid form die can be used.

Solid Form Dies

The solid form die is usually of simple construction and design. Stampings produced in these dies are usually of the flanged clamp type, such as pipe straps, etc. They are made of metal which is of a soft grade, mostly strip stock, and the grain runs parallel to the form. Some distortion is encountered; this can be compensated for in the design of the templates. A male and a female template are usually made. The male template is made to the contour that will be shaped on the punch of the die—the same as the inside contour of the stamping. The female template is made to the outside contour of the stamping and will be shaped on the die halves. By so doing, allowance has been made for the thickness of the metal, when both die halves are set in place. *Figure 8-5* illustrates both templates in place.

A forming die of this kind need not be mounted on a die set. A die set should be considered because of the amount of time that can be saved in setting the die for production; also it eliminates the chances for misalignment in setting the die. A die that is not properly set could cause some pinching of the metal, thereby causing the stamping to break. Considering that a die shoe and punch holder are required in each case, the cost of adding the leader pins to complete the die set is nominal, and should be saved in a short time.

A great deal of side pressure is exerted on the die blocks, and must be considered in the designing. The die block should be made of more than one piece of tool-steel. This is necessary to eliminate the possibility of cracking the die block if the operator should feed a double blank or if metal of a thicker gage is used. Tie the die blocks together by means of cross pieces. Sink the blocks into the die shoe to obtain some support from the edge of the sinking. The blocks should be constructed wider than they are high, a proportion of at least 1:1 1/2, with large sturdy dowel pins. The form edge of the blocks must be of proper radius to prevent digging on the side of the stamping. The radius should not be less than twice material thickness, and for best results, the edge should have a high polish. A smaller radius could cause fatigue in the material when formed.

The punch is made of tool-steel, hardened according to the severeness of

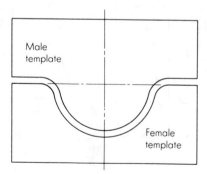

Figure 8-5. Templates for forming.

the operation. Design the punch to be long enough to allow complete forming of the part without interference with the punch holder. The width and shape are governed by the part to be formed, and at all times it must be twice material thickness smaller than the width or span of the die blocks.

The screws and dowel pins should be spaced properly and located so that they will not mar the stamping. Consideration must be given to stripping the formed part from the punch. This can be accomplished by means of a knockout, stripper hooks, or stripper-pin (spring) construction. It is important to consider and plan for the removing of the formed part. *Figure 8-6* shows these details and also illustrates the gages necessary for locating the stamping.

The shoe (1) should be thick enough to withstand the pressure required for the forming operation, and in selecting its thickness, the size and shape of the hole in the bolster plate of the press must also be considered. The *A* and *B* dimensions of the shoe when placed over the hole of the bolster plate should also be long and wide enough to allow ample space for clamping it to the bolster plate. Consider the depth of the sunk-in *d* section when selecting shoe thickness. The die shoes can be cast iron, semisteel, or steel. Select the proper material for the die shoe; the shoe must withstand a good share of the force applied during the forming operation. A die shoe made of steel will give good service, and costs only a little more than a cast or semicast shoe. Select the diameter of the leader pins (3) according to the working area of the shoe, or by consulting a die set manufacturers catalog. The length of the pins should be at least 1/4" (6.4 mm) shorter than the shut height of the die, as listed on the drawing.

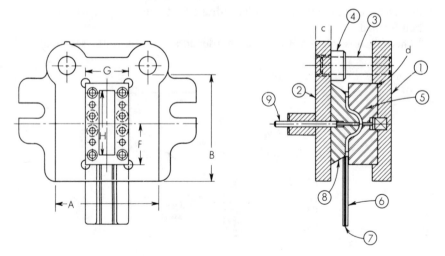

Figure 8-6. A solid form die.

The guide bushings (4) should be of the shoulder type, and for a die of this kind can be the regular-length type. (Guide bushings are made in three lengths—regular, long, and extra long.) Die sets are made in two types—precision and commercial. Precision die sets are best for stamping work that requires great accuracy in alignment, such as between punch and die parts of cutting dies. For secondary operations, such as bending, forming, or other noncutting operations, specify commercial die sets.

The punch holder (2) is the same as the shoe in the a and b dimensions. The thickness c should be 1 1/4″ (31.75 mm) and have a 2″ (51 mm) diam shank. Always place the shank on the centerline of the punch holder of a regular stock die set. If it is necessary to locate it off center, the set becomes a special, therefore costing more. The shanks of die sets come in various diameters—1 1/2″, 1 9/16″, 2″, 2 1/2″, and 3″ (38, 40, 51, 64, and 76 mm) —also can be had in special diameters when specified, at extra cost. The shank is used to align the centerlines of the die with the centerlines of the press. The shank is clamped securely in the press ram, and must help lift the punch and stamping from the die. Punch shank diameters are selected according to the hole in the press ram. The 2″ (51 mm) diam shank is the most popular, because it is strong and eliminates the use of collars. The shanks 1 1/2 and 1 9/16″ in diameter are 2 1/8″ long (38 and 40 mm, 50 mm long); the others are 2 7/8″ (73 mm) long.

Die blocks (5) are of a two-piece construction. They should be made of tool-steel that will withstand excessive wear and galling. Consider toughness and shock resistance when selecting the tool-steel. It should be an oil or air-hardening steel, and in many applications, a double draw will produce some added toughness and abrasion (galling) resistance in the steel. The die blocks take most of the wear, so design them properly. The length F of the die blocks must always be longer than the height G, at least one and one-half times as long. The width should be governed by the width of the stamping to be formed, taking into account also the space required to fasten gages, to locate the stamping.

The screws and dowels should be spaced to help withstand some of the forces that will be encountered in the forming operation. Locate the screw and dowel holes so that they will not mar the stamping. The dowel pins should be large enough in diameter to help withstand some of the spreading force that will be encountered. The screws should be large enough to compensate for these forces and should be located to help control them. The screw and dowel hole locations should also be considered for locating the gage plate (6). Designing the gage plate shape when designing the die blocks, and then placing the screw and dowel holes accordingly, accomplishes a dual purpose. It helps eliminate some of the holes required in the die blocks, as well as the work of drilling and tapping the holes and the cost of the extra screws.

Study the shape of the stamping to be formed. If small radius corners are required, design for a pad-type rather than a solid form die (*Figure 8-7*). The radius in the corners should be at least five times metal thickness in the solid form die design.

Study the gaging of the stamping, and whenever possible, the blank should be fed the long way. Locate the die on the die set to permit the blank to feed the long way. When blanks are fed the short way, they have a chance to twist, and the operator loses control over them. This causes a loss of time in locating the blank in the die, and can cause mislocation of the blank.

Make the gage plate so the blank can slide into the proper location for forming. Provide a slide with sides (7) whose height is at least one and one-half times metal thickness. The pocket for locating the blank in the length should stop it on both ends and the clearance should be governed by the stamping drawing tolerances. The width of the slide pocket should be sufficient to include the mill tolerance of the strip. When the blanks are wide and flat, provide ribs or wires to help slide the blank into the die and reduce some of the friction caused by the oil on the blank surfaces. Design the slide long enough to prevent the operator from sliding his fingers under the punch. The safety factor most commonly used is to have it long enough for the operator to have the web of his thumb rest at the front edge of the gage, and with his hand spread forward, have the center finger clear the punch or die opening (approximately 6 to 6 1/2" [152.4 mm to 165.1 mm]).

The punch (8) should be of tool-steel; it is usually made of the same kind as the die blocks. The width of the punch is usually governed by the width of the stamping being formed. When narrow-width stock, 1/2" (12.7 mm) or less, is used, add flanges to the sides to make it wide and strong enough at the base.

Place the screw and dowel holes to eliminate the possibility of marring the stamping. The screws and dowels should be large enough to take care of the forces encountered by the punch in the forming operation. Locate the screws and dowels so they do not cut into the punch shank. The only hole we should consider putting into the punch shank at any time is for a knockout rod, and this is usually in the center of the shank diameter. When possible, employ a knockout arrangement using a knockout rod (9) for releasing the finished

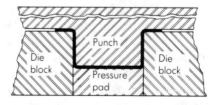

Figure 8-7. A pad-type form die.

stamping from the punch. This usually is simple to design, and making it is most economical. Also, if the press bar for pushing the knockout rod is properly adjusted, removing the stamping is easy and should increase production.

All the comments made in this section (on the solid form die) relating to die sets and their selection, screws and dowels, die blocks, punches, etc., apply to the design of other forming dies. They are necessary details to be considered for most die construction, and will not be referred to again.

Forming Dies with Pressure Pads

When the forming of stampings requires accuracy dies employing pressure pads are often designed. The pressure pad helps to hold the stock securely during the forming and eliminates shifting of the blank. The pressure can be applied to the pad by springs or by the use of an air cushion (*Figure 8-8A* and *B*). When springs are used, they can be located directly under the pad and confined in the die shoe (*Figure 8-8A*). They may also be located in or under the press bolster plate; and by the use of pressure pins, which are located under the pad, and through the die shoe, pressure is applied to the pressure plate.

Pressure pins are also used with an air cushion. The construction of the pressure plate and pressure pins would be the same as shown in *Figure 8-8B* except that an air cushion is substituted for the springs.

When springs are used to apply pressure to a pressure pad, spring pressure increases with the pad travel. Each fraction of an inch of travel increases the pressure on the pad. This could cause some trouble in stampings of light-gage material, because too much pressure may cause the metal to stretch. When springs are used, a certain amount of pressure is lost owing to the springs' setting (losing height after being worked). When an air cushion is used, the proper amount of pressure on the pressure pad is assured as long as the air supply is set properly. It is important to have a set amount of pressure on the pressure pad to control the quality of the stampings.

The pressure pad, the moving member of the die, must always be controlled in its travel between the die blocks. This can be done by means of retaining shoulders, or by shoulder screws. When using the retainer shoulder construction, a recess is machined into the form blocks, and a corresponding shoulder is machined on the pad. The retainer shoulder should always be made strong enough to withstand the pressure applied by either springs or air cushion. The size of the shoulder to be used varies according to the size and metal thickness of the stamping. A good rule to employ is to have the height of the shoulder one and one-half times the width (*Figure 8-9*).

Always design the shoulders of the pad with a radius in the corner. When

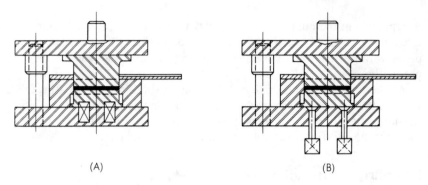

(A)   (B)

Figure 8-8. Pressure-pad-type form dies.

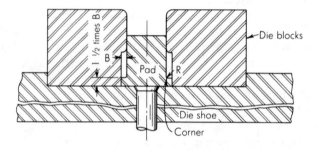

Figure 8-9. Pressure pad design.

the pad is made of hardened tool-steel, heat treatment should specify a double draw of the shoulder section.

When using shoulder screws to control the travel of the pad, the die shoe must be thick enough to permit sufficient travel.

The pressure pad should always travel so that it extends slightly above the die blocks. This will insure uniform parts, because there will be pressure to lock the part between the punch and pad faces, before the actual forming takes place.

The amount of travel the pad should have depends upon the height of the form die. It is not always necessary to travel the full height; in many cases half the die's form height is sufficient. When a blank is distorted, or has a tendency to curl, which may cause the completed blank to be out of square, it may be necessary for the pad to travel the full length. It is necessary for the pad to bottom on the die shoe, to allow the punch to give the part a definite set at the bottom of the stroke. When a stamping must have sides that are square with the bottom, after forming, the corner radius should be set. This is done by designing the die blocks with the correct radius *A*, *Figure 8-10*. The pressure

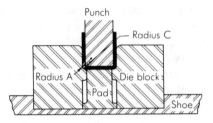

Figure 8-10. Radii considerations for form die design.

pad is made to match the height of the die blocks' radius edge. The punch radius *C* is made slightly smaller, approximately 10% less than the die block radius.

It may be necessary to machine a slight angle on the side of the punch to allow a slight overbending of the side being formed. This ensures that the sides of the formed part will be square with the base after forming.

Single and multiple pressure pads are used in die construction. The single pressure pad is used when the forming is done in one direction. It is most commonly used for forming short flanges, tabs, lugs, or ears at right angles to the base of the part. The pressure pad is used to support the base of the part accurately, either by pilot pins or other gages, and by its securing the part properly, the part is formed with great accuracy. The side or tab to be formed may be bent downward as well as upward. When the side or tab is being bent downward, the length may vary slightly, because the metal is stretched or drawn more. Some of this may be overcome by having an angle and radius on the punch as shown in *Figure 8-11*. The greater the angle and radius, the less bending pressure is required. When a side is bent down, a heel block is required to help support the punch before it starts to do any forming. It should be at least two metal thicknesses higher than the die block. The pressure pad must travel at least 1/8″ (3.2 mm) beyond the edge of the form

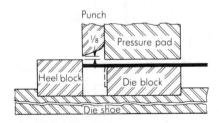

Figure 8-11. Die with a heel block and relieved punch.

punch. This is done to assure holding pressure before any forming work is done. The punch should travel far enough beyond the corner radius to smooth out the formed side.

Multiple pressure pads are used when a series of forms are necessary; they are used mostly in progressive dies, when several bends are required on small precision parts. A combination of stationary form blocks, supplied by pressure pads, helps lift the strip so it can be advanced from one station to the next.

Using Rubber and Urethane for Bending and Forming

A rubber or urethane forming die is often used in press brake operations. Tooling costs are reduced as only the male forming punch has to be made. This method is well adapted to forming angles, channels, and radii. Another advantage is that the rubber die leaves no visible die marks on the part. This makes a part with the best visual appearance for visible covers, guards, and panels (*Figure 8-12*).

The Guerin process is another method of forming parts with a rubber pad. A rubber pad is attached to the ram of the press and a form block is placed on

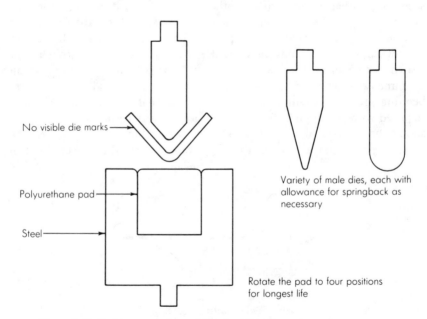

No visible die marks

Polyurethane pad

Steel

Variety of male dies, each with allowance for springback as necessary

Rotate the pad to four positions for longest life

Figure 8-12. Rubber or polyurethane forming die and punches.

the lower platen. When the ram is lowered the rubber forms the part around the form block. The block has the proper bend radii and over-bend allowance for spring-back machined on it. This process is limited to forming of relatively shallow parts in light materials, normally not over 1 1/4" (31.8 mm) to 1 1/2" (38.1 mm) deep. Parts with straight flanges, stretch flanges, and beads formed from developed blanks are most suitable for this process.

The Wheelon process applies hydraulic pressure directly to the rubber forming pad. This process can do about the same operations as the Guerin process, using the same type of form block. The advantage is that with the higher pressures available, practically all wrinkling is eliminated.

Embossing Operations

In an embossing operation, a shallow surface detail is formed by displacement of the metal between two opposing mated tool surfaces. In one surface we have the depressed detail, on the other the relief detail. The metal is stretched into the detail rather than being compressed. Embossing is used for various purposes, the most common being the stiffening of the bottom of a pan or container; the embossing is designed to follow the outside profile of the part. A round can may have an embossed circle or raised grooves of various widths or panels. When the can or box is square or rectangular, such embossments follow the contours. Embossings often are ribs or crosses stamped in the metal to help make a section of a blank stronger by stiffening. An embossing die can be a male and female set of lettering dies, or a profile of one of various shapes.

The method of constructing the die blocks for an embossing operation depends on the size and shape of the form, also the accuracy and flatness required. When embossing simple shapes such as stiffening ribs, it is not necessary to fit up the die to strike the bottom. The metal stretches over the punch and across the two radius edges of the die hole (*Figure 8-13*).

The die opening has the same width of the rib or embossing *a*, and a slight radius *b* is added to the edges of the opening to allow the metal to flow freely. The punch is made slightly smaller than the required metal thickness per side,

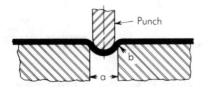

Figure 8-13. An embossing die.

so that it does not strike along this area. By constructing it in this manner the pressure required to stamp the embossing is reduced.

When embossings are of the lettering type, such as depressed or raised letters for name plates, care must be taken to see that both dies are properly located and doweled in place (*Figure 8-14*).

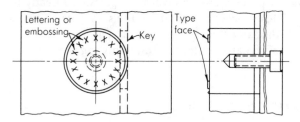

Figure 8-14. Lettering (embossing) die detail.

The male die is located in a pocket or recess, and keyed in place. By lining up the female die profile to correspond with the male die profile, and keying it in place, a good stamping or embossing can be made. Stamping operations of this kind require precision work by the toolmaker; the dies are easily damaged by misalignment.

A small embossing is often used as a weld projection nib. These nibs are used to weld piece-parts together. There are two kinds: a button type, which we use for light-gage metal, 3/32″ (2.4 mm) or less thick, and a cone type for heavier-gage metal (*Figure 8-15*). Care must be taken in their design, because if the projection is too weak the nib will collapse before a good weld can be made. A nib that is too thick or heavy through the section will require too much pressure to produce good welds. The piece-parts are heated electrically, and this causes the projections to melt, so if the projections are too heavy or too light, the heat and pressure required can cause trouble. *Figure 8-16* lists the correct design for both kinds of projections.

One or more projections can be used on a part for welding purposes. The design of the part and its use govern the number and size of the projections.

When embossing ribs or shapes in blanks, it is best to have the die blocks

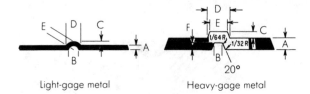

Light-gage metal   Heavy-gage metal

Figure 8-15. Die for embossing weld-projection nibs.

Type	U.S.S. Ga.	A	B	C	D	E	F	G
Button-type projection	24	0.025 (0.64)	0.050 (1.27)	0.025 (0.64)	0.109 (2.77)	0.025 (0.64)	---	---
	23	0.0281 (0.714)	0.050 (1.27)	0.025 (0.64)	0.109 (2.77)	0.025 (0.64)	---	---
	22	0.0312 (0.792)	0.050 (1.27)	0.030 (0.76)	0.125 (3.18)	0.030 (0.76)	---	---
	21	0.0344 (0.874)	0.050 (1.27)	0.030 (0.76)	0.125 (3.18)	0.030 (0.76)	---	---
	20	0.0375 (0.953)	0.050 (1.27)	0.035 (0.89)	0.125 (3.18)	0.035 (0.89)	---	---
	19	0.0437 (1.110)	0.050 (1.27)	0.040 (1.02)	0.125 (3.18)	0.035 (0.89)	---	---
	18	0.050 (1.27)	0.050 (1.27)	0.040 (1.02)	0.156 (3.96)	0.040 (1.02)	---	---
	17	0.0562 (1.427)	0.055 (1.40)	0.040 (1.02)	0.156 (3.96)	0.045 (1.14)	---	---
	16	0.0625 (1.588)	0.060 (1.52)	0.045 (1.14)	0.172 (4.37)	0.050 (1.27)	---	---
	15	0.0703 (1.786)	0.075 (1.90)	0.045 (1.14)	0.172 (4.37)	0.055 (1.40)	---	---
	14	0.0781 (1.984)	0.075 (1.90)	0.050 (1.27)	0.180 (4.57)	0.065 (1.65)	---	---
	13	0.0937 (2.380)	0.075 (1.90)	0.050 (1.27)	0.180 (4.57)	0.065 (1.65)	---	---
Cone-type projection	12	0.1093 (2.776)	0.080 (2.03)	0.055 (1.40)	0.172 (4.37)	0.080 (2.03)	0.090 (2.29)	0.131 (3.33)
	11	0.125 (3.18)	0.080 (2.03)	0.055 (1.40)	0.172 (4.37)	0.080 (2.03)	0.100 (2.54)	0.138 (3.50)
	10	0.1406 (3.571)	0.080 (2.03)	0.060 (1.52)	0.172 (4.37)	0.080 (2.03)	0.110 (2.79)	0.145 (3.68)
	9	0.1562 (3.967)	0.080 (2.03)	0.060 (1.52)	0.172 (4.37)	0.080 (2.03)	0.122 (3.10)	0.155 (3.94)
	8	0.1718 (4.364)	0.080 (2.03)	0.060 (1.52)	0.190 (4.83)	0.080 (2.03)	0.138 (3.50)	0.166 (4.22)
	7	0.1875 (4.762)	0.080 (2.03)	0.070 (1.78)	0.203 (5.16)	0.094 (2.39)	0.166 (4.22)	0.185 (4.70)
	6	0.2031 (5.159)	0.080 (2.03)	0.070 (1.78)	0.203 (5.16)	0.094 (2.39)	0.182 (4.62)	0.206 (5.23)
	5	0.218 (5.537)	0.080 (2.03)	0.075 (1.90)	0.210 (5.33)	0.100 (2.54)	0.200 (5.08)	0.220 (5.59)

All dimensions are given in inches, (mm).

Figure 8-16. Design guides for projections (24- to 5-gauge steel).

large enough to cover the whole blank. A die block that is too small holds the metal between the die and pressure pad. The portion of the blank not held will distort and cause the metal to twist, wrinkle, and pucker.

When designing a die with embossings, regardless of whether they are formed up or down, a way must be provided to lift the form out of the die pocket. This is usually done by pressure pads or ejector pins.

Beading and Curling Dies

In beading and curling operations, the edges of the metal are formed into a roll or curl. This is done to strengthen the part or to produce a better-looking product with a protective edge. Curls are used in the manufacturing of hinges, pots, pans, and other items. The size of the curl should be governed by the thickness of the metal; it should not have a radius less than twice metal thickness. To make good curls and beads, the material must be ductile, otherwise it will not roll and will cause flaws in the metal. If the metal is too hard the curls will become flat instead of round. If possible, the burr edge of the blank should be the inside edge of the curl. This location facilitates metal flow and also helps keep the die radius from wearing or galling. In making curls and beads a starting radius is always helpful and should be provided if possible (*Figure 8-17*).

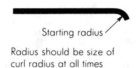

Starting radius

Radius should be size of
curl radius at all times

Figure 8-17. Starting curl
radius.

The curling radius of the die must always be smoothly polished and free of tool marks. Any groove or roughness will tend to back up the metal while it is rolling and cause defective curls. The inside surface of the blank must be held positively in line with the inside curling radius of the punch (*Figure 8-18*).

When curling or beading pots, pans, cans, or pails, wires are often rolled inside the curls to make them stronger. The wire is made to the contour of the pan and placed on a spring pad. When the curling die descends, the edge of the pan is forced to curl around the wire as shown in *Figure 8-19*.

Bulging Operations

In a bulging operation, a die forms or stretches the metal into the desired contour. By using rubber, heavy grease, water, or oil the metal is forced, under pressure, to take the shape of the die. To facilitate the removal of the part from the die after bulging, the die is of a split construction. Dies that are split in halves or sections must have strong hinges and latches. The sections

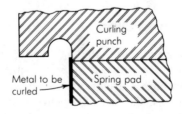

Curling
punch

Metal to be
curled

Spring pad

Figure 8-18. Curling punch design.

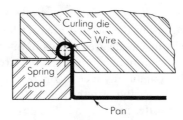

Figure 8-19. Curling die design.

when operated must work freely and rapidly and, when under pressure, be tight enough so that the sections will not mark the stamping. The pressures required for these operations are usually found by experimenting during the first setup of the die. This is necessary because other variables besides metal thickness and the annealed condition of the stamping must be considered. Most stampings that are to be bulged have been work-hardened by previous drawing operations and therefore need annealing. Most metal that is not soft enough will rupture in the bulging operation.

Rubber is the cleanest and easiest material to use for bulging operations (*Figure 8-20*). Once it is designed and tried out, the shape and hardness can be duplicated for replacement. Neoprene rubber of medium hardness is the most suitable.

The rubber should be confined between two punch sections. The upper section slides on the arbor on which the rubber pad is assembled. The lower section is fastened to the arbor. It acts as a stop to locate the rubber in the correct place for the bulging operation. As the punch descends, the rubber forces the metal into the exact shape of the die. Once the press is set to the required height, all the stampings should be of equal quality.

The use of grease or tallow results in a handling problem. First of all, the same amount must be used for each stamping. Secondly, these substances are messy to handle both before and after the bulging operation. They also require extra work in the cleaning of the finished stamping because all traces

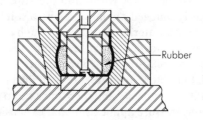

Figure 8-20. A bulging die.

of grease or oil must be removed. The die and punch must be designed to confine the grease securely in the desired area. A loss of grease owing to escaping or squirting out of the die area results in inferior stampings.

Water or oil requires a pump to produce the required pressure. This also causes other undesirable conditions such as operators' clothing getting wet, rusting of the tools and press, and also the necessity for pump maintenance. The die for bulging using water pressure must be designed with care. Its open end or the ends of the stamping should be sealed securely. For tube forming, enough extra length must be allowed on each end for the end-plugs (*Figure 8-21*).

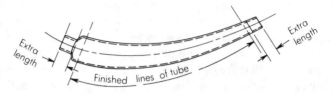

Figure 8-21. Tube design.

One end-plug must be designed so that water and pressure can be fed into the tube. The other end-plug can be designed without any feed lines, but for each end-plug a seal-off gasket must be designed. The die is usually split along the centerline of the contour to be formed, so that it can be lifted out of the die without any trouble. The press or press brake that is used to hold the die halves down during the forming operations should have a bed or bolster area that is longer than the die for proper die support.

The press must have enough force to keep the die halves securely locked during the pumping cycle of the bulging operation; otherwise a large flashing line will be produced on the stamping along the parting line of the die. This is not desirable because it requires extra labor to remove the line; also it tends to break down the edges of the die along the profile of the form. Bulging dies are costly, and extra care should be taken to keep them in good shape.

When using a die to bulge a tube with water pressure, the first step is to see that the tube is properly plugged on one end with a solid plug (no feed line). Next the tube is filled with water, leaving only enough room for the end-plug with the feed line or hose attached. After fastening the end-plug in place, the tube is laid in the die. The next step is to lower the press ram and close the die. The pump is started, but before any pressure is applied to the water, the tube must be filled with water to capacity and there must be no air pocket. The tube is now ready for water pressure application and expansion.

The water pressure necessary to expand a stamping varies according to the thickness and kind of metal to be formed. The pump must be large to do a variety of work.

In most work the pumping cycle is of short duration, from 15 to 30 seconds to complete the form. It takes a much longer time to get the tube ready for the operation; therefore, care must be taken to have everything in order at all times. For uniform stampings in bulging operations using water, the metal must be controlled for thickness of the wall and hardness, and the press and water pressures must be the same at all times.

Twisting Operations

A twisting operation usually is done on flat strip blanks. The demand for this operation is limited; therefore, the cost of the die and the amount of production required must be considered. On simple strip blanks of light-gage stock that require a twist on one end, a hand fixture may be most economical. A good operator can produce almost as many stampings with a hand fixture as with a press die. Most stampings, after being twisted, are difficult to remove from the tools. The twisted stamping usually lies in the die, and, if the blank is short, it generally falls between the form blocks. A twist on flat strip blanks is usually 90°, but the angle can vary to suit the stamping design. The metal is usually hot rolled SAE 1010 steel, but must be one-quarter hard or softer in temper. Too hard a strip will show fractures along the edges after twisting. It is best to use rolled strip stock, with the mill edge, because a sheared edge could produce fractures in the twisting of the metal.

Twisted strips or brackets make sturdy braces or brackets at low cost. Some items, such as fan blades, also require a twisting operation with dies that are made with a great deal of care. Each blade must be bent the same amount, and the spring-back of the metal must be taken into consideration. In designing a die of this kind, the feeding of the blank into the die, its location in the gages, and the removing of the finished stamping must be considered. Uniform location is the first requirement, and next is the ease of removing the blade without distorting it. Blanks for fan blades usually have a hole in the center, which is used for assembling the finished blade to the drive shaft. This hole should always be used to locate the blank centrally, and other gages should be provided along the outside profile as shown in *Figure 8-22*.

A pilot can be used for the center hole locator. Spring pins which disappear when the punch comes down can also be used to locate the outside contour of the blade. For stationary pins, clearance must be provided in the punch. Gages should always hold the blank in the same position. In twisting blades, the blank must be prevented from rolling while being formed. The dies for twisting blades must be made with care, so that each blade is made with a uniform and equal amount of form. A slight difference could cause the blade

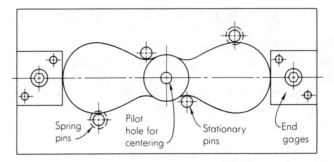

Figure 8-22. Twisting die design.

to vibrate. This would mean a costly balancing operation, which should be eliminated if possible.

The die blocks for twisting flat strip blanks are simple in design. In fact the upper and lower dies can be made or shaped in one piece and then cut to the desired length. When locating them on the die shoe and punch shoe, care must be taken to provide the proper amount of opening for the metal thickness (*Figure 8-23*). The angle on the die block faces (b) can be made to the desired angle of the part, but an allowance must be made for springback of the metal.

A die of this kind can be made for a blank with a single twist, or it can be made to make any number of twists in a part. The designer must be sure that the blank area that is to remain flat is held securely in the die before the twisting operation is performed. This is usually accomplished by having a stationary block on the bottom or die shoe, and a block with spring pressure on the upper die shoe. The spring pressure should be strong enough to keep the part from tipping or moving while being twisted.

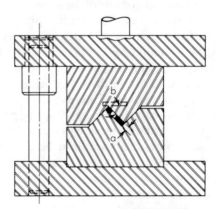

Figure 8-23. Twisting die block considerations.

For flat blanks that require a twist, and where accuracy is not a requirement, a fork lever can be made. The part can be placed and gaged in a bench vise for length location, and the operator slips the forked (slot in the lever) lever over the area to be twisted. The amount of twist is governed by the operator and can be controlled by setting a gage to show him where to stop. The slot width, *A*, *Figure 8-24*, in the fork lever is made to suit the metal thickness. The depth *B* of the slot should always be made 1/4″ (6.35) mm or more than the width of the part. This will give the operator control over the full face width of the part being twisted. The fork ends must be made strong enough to withstand the pressure that is required to twist the blank.

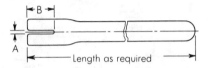

Figure 8-24. A fork lever for twisting.

Coining Dies

Coining dies are used to emboss on the part the detail engraved on the faces of the punch and die. By compressing the metal between the punch and die, metal flows into the detail or embossing. Coining dies produce coins, medals, medallions, jewelry parts, ornamental hardware, plates, and escutcheons.

Hydraulic or knuckle-joint presses are usually used to perform coining operations. To force the metal to flow into the detail requires very high force. One of the reasons for the high force is that when metal is compressed, it hardens and toughens. This is called work-hardening of the metal, and the more complex a part, the greater its resistance and the force required.

During the coining operation the part must be properly confined. The thickness of the part is controlled by the die block surfaces. The closed height of the die is checked by the surfaces of the blocks. The die block surfaces, on which the detail is inscribed, must be highly polished and free of scratches or mars. That is necessary because the slightest scratch or mark will be embossed into the part. The sides of the die that control the outside contour of the part must be slightly tapered to allow removal of the part from the dies.

When designing dies for a coining operation, extra caution must be taken to build the dies strong enough to withstand severe pressures. The pressures of the metal being coined, regardless of the kind, are severe, and heavy die sections are necessary. Polish and the finish of the die surfaces must be the best. The slightest mark could cause the die to split. The matching sections of the dies must be made to very close tolerances, especially if the die has moving parts. The slightest opening will allow the metal to squirt, or be forced into the opening, and could cause the die to burst. The die shoe and punch holder are designed thicker, or heavier, than for most dies because of the heavy force required. Tool-steel backup plates hardened to a good tough hardness of R_C 50-52 are used to back up the dies on the shoe and punch holder to prevent the dies from sagging, or bending, under pressure (*Figure 8-25*).

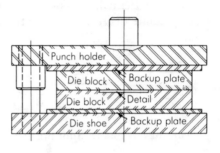

Figure 8-25. A coining die.

The plates should be wider and longer than the die blocks, so that they will support the full face of the die blocks. The bolster plate of the press should be solid, or have as small a hole as possible, to help support the die.

The screws and dowels must be located away from the coining detail, so that they will not mar the part, nor weaken the dies. The screws and dowels should be large enough to hold the dies firmly and securely. Dies for coining receive rough wear, and they wash out owing to the pressure of the metal as it flows into the detail. This makes for short die life, adding to the cost of the operation.

Swaging Dies

A swaging operation is similar to a coining operation, except that the part contours are not as precise. The metal is forced to flow into depressions in the tool faces, but the remaining metal is unconfined, and it flows generally at an angle to the direction of the applied force. The flow of the metal is restricted somewhat by the tool faces, but an overflow flash is usually encountered, and must be removed in a subsequent operation. The upsetting of heads of bolts, rivets, pins, and many cold- and hot-forging operations are classified as swaging operations. The sizing of faces or areas on castings, which is often referred to as planishing, is also classified as swaging. The planishing of faces, especially around hubs or bosses of casting or forging, is considered desirable, because it increases the wear resistance of the area as much as 80%, as compared to a similar machined surface. The faces of connecting rods and piston rings are cold sized, in order to increase the hardness of the wearing surfaces and to make them smooth. A surface of this kind cannot be accomplished by milling or machining; moreover, the squeezing operation can be done with a die about ten times as fast as by milling. Copper electrical terminals are also made by swaging dies.

The presses used for swaging operations must be selected according to the size of the work and the interval of time necessary to complete the operation. Knuckle-joint presses and hydraulic presses, of extra-heavy force capacities, are usually used. Hydraulic presses have a decided advantage over knuckle-joint presses because of the extra dwell at the bottom of the stroke, which puts a definite "set" in the work. Knuckle-joint presses have short powerful strokes and can compress the metal but, lacking the extra-dwell feature at the bottom of the stroke, slight variations of the swaged parts result.

Pressures up to 100 tons per square inch (1379 MPa) are applied to metals in swagging operations. Presses ranging from 25 to 2500 tons (222-22,240 kN) are used.

The most practical way of determining the size of the press for swaging a part is to squeeze the first parts with the dies placed in a hydraulic press provided with force gages which will eliminate any guesswork. If a mechanical or knuckle-joint press is used, one should be selected with a safety factor of three or more times the maximum work pressure. Swaged and cold-sized parts are highly compressed; the metal becomes harder (work-hardened) and more dense. The more metal to be moved, the greater its resistance. All these factors should be considered when selecting a press for swaging to avoid the possibility of broken machinery. When computing the force required for swaging or cold sizing, the correct ultimate compression strength of the metal must be used. The equation used to compute is,

$$P = \frac{A \times S}{2000} \qquad (8\text{-}3)$$

Where:

P = force required, tons (for metric unit, multiply tons by 8.9 to obtain kilonewtons)

A = area to be sized, in.^{2}

S = ultimate compressive strength of the metal, psi

Tool steel used for swaging and sizing operations must be of high strength. Chrome-tungsten oil-hardening steel, which combines high hardness with maximum toughness, is used for swaging dies. High-carbon, high-chrome steels are also used, but they are more difficult to machine, and their resistance to shock is not as high as that of chrome-tungsten oil-hardening steels.

The die body sizes must be extra heavy. The shut height of the die is checked by the surfaces of the blocks. The area of the stop and the top and bottom surfaces of the block should be large enough to allow the block to withstand three times the yield strength of the workpiece metal. The contour or profile of the part to be sized is usually machined into the surface of the die blocks. These surfaces are in a plane coinciding with the longitudinal center plane of the workpiece. To avoid using excessive pressures, the dies are relieved where no pressing is done. A draft is provided along all edges of the work that are not squeezed. There is also a 45° draft around the bosses to be sized to facilitate removal of the finished work. In some cases, the parts are machined about 0.031″ (0.79 mm) oversize before being placed in the sizing dies, which eliminates the chances of overtaxing the dies and also improves the dimensional quality of the parts.

Die shoes and punch holders are similar to the ones used for coining dies. They must be heavy, with wide and long base surfaces. *Figure 8-26* illustrates general principles for designing a swaging or sizing die.

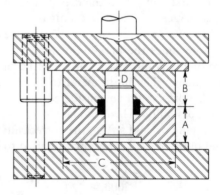

Figure 8-26. A swaging die.

The die blocks have identical thicknesses, *A* and *B*, and lengths, *C*. The center of the part is located in exactly the same position, *D*, in each block. The screw and dowel holes must not be located near or in the area of the part to avoid marking it.

Hole Flanging or Extruding Dies

The forming or stretching of a flange around a hole in sheet metal is termed hole flanging or extruding. The shape of the flange can vary according to the part requirements. Flanges are made as countersunk, burred, or dimpled holes.

When countersunk shaped extruded holes are made in steel, it is necessary to coin the metal around the upper face and beveled sides to set the material. The holes are also made about 0.005" (0.13 mm) deeper than the required height of the rivet or screw head, which allows bunching that occurs when squeezing the rivet in place. A section of a die for this purpose is shown in *Figure 8-27*.

The hole can be pierced before it is placed in the countersinking die, or it can be formed and pierced in a single stroke of the press.

As shown in *Figure 8-27*, the sheet is placed over the pilot diam *A* which locates it centrally in the die. The die body, 1, descends and forces the metal down around the flange surface of the punch. Spring pressure strips the part from the punch and releases the formed part from the die.

Figure 8-28 shows a two-step punch, 1, which first punches the hole in the part and then forces the metal around to the countersunk shape of the die block, 2. The hole punched by this method is always somewhat smaller than the size of the hole in the finished part. Spring pressure is used to strip the finished part from the punch. A shedder pin should be provided in the piercing point of the punch to remove the slug.

The size of the pierced hole for a 90° hole flange can be calculated, but should never be used until it has been proved correct by using the same tools that will be employed in the die. To calculate the hole size the same principles are employed when finding a 90° bend.

Dimensional details of *Figure 8-28* are identified as follows:

T = thickness of metal to be flanged
A = diam of calculated hole
B = diam of hole inside of flange (punch body size)
G = diam of hole in die (outside of flange)
R = radius on edge of die; can be specified from $\dfrac{T}{3}$ to $\dfrac{T}{4}$
H = height of flanged hub

When the flanges are stretched more than 2 1/2 times metal thickness in height, the wall can split. This can be prevented to some extent by burring the edge around the hole before the extruding operation.

90° Hole Flanging. Forming a flange around a previously pierced hole at a bend angle of 90° (the most common operation) is nothing more than the formation of a stretch flange at that angle.

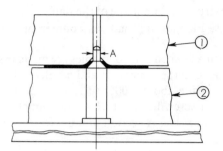

Figure 8-27. A forming die for countersunk holes.

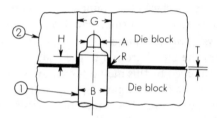

Figure 8-28. A die for punching and countersinking a hole.

One manufacturer has standardized flange widths (*Figure 8-29*, dimension *H*) for holes to be tapped in low-carbon steel stamping stock, as follows:

$$B = A + \frac{5T}{4} \text{ when } T \text{ is less than } 0.045'' \text{ (1.14 mm)} \tag{8-4}$$

$$B = A + T \text{ when } T \text{ is more than } 0.045'' \text{ (1.14 mm)} \tag{8-5}$$

$$H = T \text{ when } T \text{ is less than } 0.035'' \text{ (0.89 mm)} \tag{8-6}$$

$$H = \frac{4T}{5} \text{ when } T \text{ is } 0.035'' \text{ (0.89 mm) to } 0.050'' \text{ (1.27 mm)} \tag{8-7}$$

$$H = \frac{3T}{5} \text{ when } T \text{ is more than } 0.050'' \text{ (1.27 mm)} \tag{8-8}$$

$$R = \frac{T}{4} \text{ when } T \text{ is less than } 0.045'' \text{ (1.14 mm)} \tag{8-9}$$

$$R = \frac{T}{3} \text{ when } T \text{ is more than } 0.045'' \text{ (1.14 mm)} \tag{8-10}$$

$$J = \sqrt{\frac{TB^2 + 4TA^2 + 4HA^2 - 4HB^2}{9T}} \tag{8-11}$$

The radius *P* on the nose of the punch should be blended into the body diameter, eliminating any sharpness which could cause the metal to score as it passes over it. The radius on the body *B* or hole-sizing portion of the punch must be as large as possible, and smooth. The portion between the *A* and *B* diameters of the punch should have a radius *C* which should be as large as possible (*Figure 8-30A*). When using the single-station method (*Figure 8-30B*), controlling the length of the flange is more difficult (*Figure 8-29, H*).

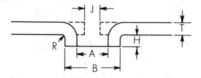

Figure 8-29. Hole flange design.

Drawing Dies

Drawing is a process of changing a flat, precut metal blank into a hollow vessel without excessive wrinkling, thinning, or fracturing. The various forms produced may be cylindrical or box-shaped with straight or tapered sides or a combination of straight, tapered, or curved sides. The size of the parts may vary from 0.250″ (6.35 mm) diameter or smaller, to aircraft or automotive parts large enough to require the use of mechanical handling equipment.

Metal Flow

When a metal blank is drawn into a die, a change in its shape is brought about by forcing the metal to flow on a plane parallel to the die face, with the

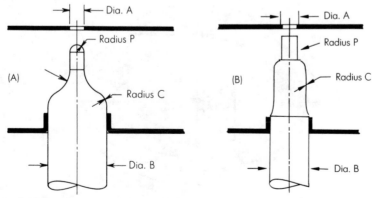

Figure 8-30. (A) Two-station flanging punch design; (B) Single-station flanging punch design.

result that its thickness and surface area remain about the same as the blank. *Figure 8-31* shows schematically the flow of metal in circular shells. The units within one pair of radial boundaries have been numbered and each unit moved progressively toward the center in three steps. If the shell were drawn in this manner, and a certain unit area examined after each depth shown, it would show (1) a size change only as the metal moves toward the die radius; (2) a shape change only as the metal moves over the die radius. Observe that no change takes place in area 1, and the maximum change is noted in area 5.

The relative amount of movement in one unit or in groups of units is shown in *Figure 8-32A* and *B*, in which two methods of marking the blanks are used to illustrate size, shape, and position of the units of area, before and after drawing. The blank in view *A* is marked with radial lines and concentric circles, and in view *B* with squares. If, after these blanks are marked and drawn, sections are cut out of the shell, flattened, and compared with the original triangular portions, a change in shape of the triangular pieces will be found. The illustration shows that the inner portion of the triangle, which becomes the base of the shell, remains unchanged throughout the operation. The portion which becomes the side wall of the shell is changed from an angular figure to a longer, parallel-sided one as it is drawn over the die radius from which point no further change takes place. The particular areas observed have been enlarged and superimposed upon each other, respectively, to show more clearly their size, shape, and position before and after drawing.

The general change in circular draws, due to flow, may be summarized as follows:

1. Little or no change in the bottom area because no cold work was done in this area.

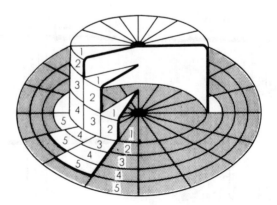

Figure 8-31. A step-by-step flow of metal.

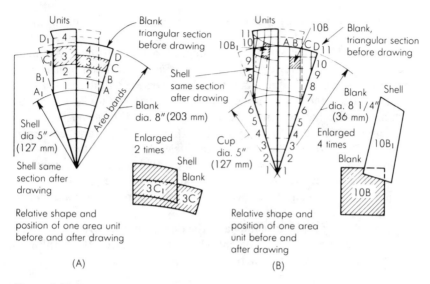

Figure 8-32. Two methods of marking blanks to illustrate size, shape, and position of the units of area, before and after drawing.

2. All radial boundaries of the units of area remain radial in the bottom area. The units in the top flange area remain radial until they move over the die radius; then they become parallel and assume dimensions equal to their dimensions at the point where they move over the die radius.
3. There is a slight decrease in surface area and increase of thickness in the units involving maximum flow. The increase in thickness is limited to the space between the punch and die.
4. The flow lines on a circular shell indicate that the metal movement is uniform.

Flow in Rectangular Shells. The drawing of a rectangular shell involves varying degrees of flow severity. Some parts of the shell may require severe cold working and others, simple bending. In contrast to circular shells, in which pressure is uniform on all diameters, some areas of rectangular and irregular shells may require more pressure than others. True drawing occurs at the corners only; at the sides and ends metal movement is more closely allied to bending. The stresses at the corner of the shell are compressive on the metal moving toward the die radius and are tensile on the metal that has already moved over the radius. The metal between the corners is in tension only on both the side wall and flange areas.

The variation in flow in different parts of the rectangular shell divides the blank into two areas. The corners are the drawing area, which includes all the metal in the corners of the blank necessary to make a full corner on the drawn shell. The sides and ends are the forming area, which includes all the metal necessary to make the sides and ends full depth. To illustrate the flow of metal in a rectangular draw, the developed blank in *Figure 8-33B* has been divided into unit areas by two different methods. In *Figure 8-33A* the corners of the shell drawn from the blank in view (*B*) are shown. The upper view is the corner area which has been marked with squares, and the lower view is the corner area which has been marked with radial lines and concentric circles. The severe flow in the corner areas is clearly shown in the lower view by the radial lines of the blank being moved parallel and close together, and the lines of the concentric circles becoming farther apart the nearer they are to the center of the corner and the edge of the blank. The relatively parallel lines of the sides and ends show that little or no flow occurred in these areas. The upward bending of these lines indicates the flow from the corner area to the sides and ends to equalize the height where these areas on the blank were blended to eliminate sharp corners.

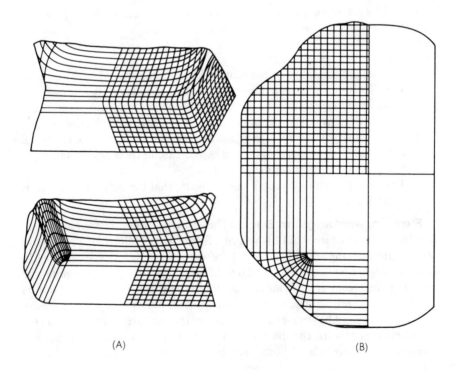

(A)   (B)

Figure 8-33. Metal flow in rectangular draws: (*A*) blank marked before drawing; (*B*) corner areas after drawing.

Single-Action Dies

The simplest type of draw die is one with only a punch and die. Each component may be designed in one piece without a shoe by incorporating features for attaching them to the ram and bolster plate of the press. *Figure 8-34A* shows a simple type of draw die in which the precut blank is placed in the recess on top of the die, and the punch descends, pushing the cup through the die. As the punch ascends, the cup is stripped from the punch by the counterbore in the bottom of the die. The top edge of the shell expands slightly to make this possible. The punch has an air vent to eliminate suction which would hold the cup on the punch and damage the cup when it is stripped from the punch.

The method by which the blank is held in position is important, because successful drawing is somewhat dependent upon the proper control of blankholder pressure. A simple form of drawing die with a rigid flat blankholder for use with 13-gage and heavier stock is shown in *Figure 8-34(B)*. When the punch comes in contact with the stock, it will be drawn into the die without allowing wrinkles to form.

Another type of drawing die for use in a single-action press is shown in *Figure 8-35*. This die is a plain single-action type where the punch pushes the metal blank into the die, using a spring-loaded pressure pad to control the metal flow. The cup either drops through the die or is stripped off the punch

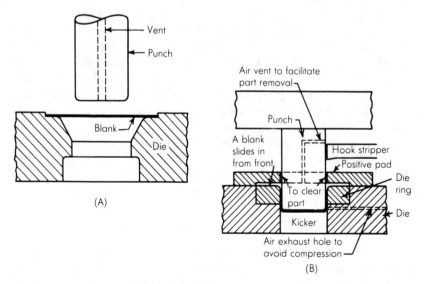

Figure 8-34. Draw die types: (*A*) simple type; (*B*) simple draw die for heavy stock.

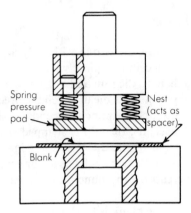

Figure 8-35. Draw die with spring
pressure pad.

by the pressure pad. The sketch shows the pressure pad extending over the
nest, which acts as a spacer and is ground to such a thickness that an even and
proper pressure is exerted on the blank at all times. If the spring pressure pad
is used without the spacer, the more the springs are depressed the greater the
pressure exerted on the blank, thereby limiting the depth of draw. Because of
limited pressures obtainable, this type of die should be used with light-gage
stock and shallow depths.

A single-action die for drawing flanged parts, having a spring-loaded
pressure pad and stripper, is shown in *Figure 8-36*. The stripper may also be
used to form slight indentations or re-entrant curves in the bottom of a cup,
with or without a flange. Draw tools in which the pressure pad is attached to
the punch are suitable only for shallow draws. The pressure cannot be easily
adjusted, and the short springs tend to build up pressure too quickly for deep
draws. This type of die is often constructed in an inverted position with the
punch fastened to the lower portion of the die.

Deep draws may be made on single-action dies, where the pressure on the
blankholder is more evenly controlled by a die cushion or pad attached to the
bed of the press. The typical construction of such a die is shown in *Figure
8-37*. This is an inverted die with the punch on the die's lower portion.

Double-Action Dies

In dies designed for use in a double-action press, the blankholder is
fastened to the outer ram which descends first and grips the blank; then the
punch, which is fastened to the inner ram, descends, forming the part. These
dies may be a push-through type, or the parts may be ejected from the die with

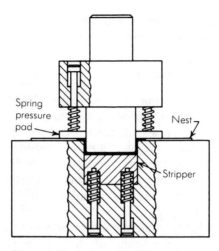

Figure 8-36. A draw die with spring
pressure pad and stripper.

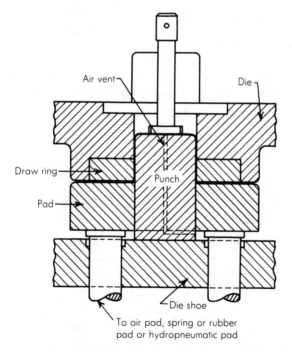

Figure 8-37. Cross section of an inverted draw die
for a single-action press; die is attached to the ram;
punch and pressure pad are on the lower shoe.

a knockout attached to the die cushion or by means of a delayed action kicker. *Figure 8-38* shows a cross-section of a typical double-action draw die.

Development of Blanks

The development of the approximate blank size should be done first (1) to determine the size of a blank to produce the shell to the required depth and (2) to determine how many draws will be necessary to produce the shell. This is determined by the ratio of the blank size to the shell size. Various methods have been developed to determine the size of blanks for drawn shells. These methods are based on (1) mathematics alone; (2) the use of graphic layouts; (3) a combination of graphic layouts and mathematics. The majority of these methods are for use on symmetrical shells.

It is rarely possible to compute any blank size to close accuracy or to maintain perfectly uniform height of shells in production, because the thickening and thinning of the wall vary with the completeness of annealing. The height of ironed shells varies with commercial variations in sheet thickness, and the top edge varies from square to irregular, usually with four more or less pronounced high spots resulting from the effect of the direction of the crystalline structure of the metal. Thorough annealing should largely remove the directional effect.

For all these reasons, it is ordinarily necessary to figure the blank sufficiently large to permit a trimming operation. The drawing tools should be made first; then the blank size should be determined by trial before the

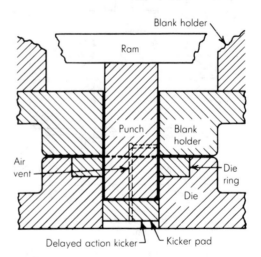

Figure 8-38. A typical double-action cylindrical draw die.

blanking die is made. There are times, however, when the metal required to produce the product is not immediately available from stock and must be ordered at the same time as the tools are ordered. This situation makes it necessary to estimate the blank size as closely as possible by formula or graphically in order to know what sizes to order.

Blank Diameters

The following equations may be used to calculate the blank size for cylindrical shells of relatively thin metal. The ratio of the shell diameter to the corner radius (d/r) can affect the blank diameter and should be taken into consideration. When d/r is 20 or more,

$$D = \sqrt{d^2 + 4dh} \qquad (8\text{-}12)$$

When d/r is between 15 and 20,

$$D = \sqrt{d^2 + 4dh - 0.5r} \qquad (8\text{-}13)$$

When d/r is between 10 and 15,

$$D = \sqrt{d^2 + 4dh - r} \qquad (8\text{-}14)$$

When d/r is below 10,

$$D = \sqrt{(d-2r)^2 + 4d(h-r) + 2\pi r(d-0.7r)} \qquad (8\text{-}15)$$

Where:

D = blank diameter
d = shell diameter
h = shell height
r = corner radius

The above equations are based on the assumption that the surface area of the blank is equal to the surface area of the finished shell.

In cases where the shell wall is to be ironed thinner than the shell bottom, the volume of metal in the blank must equal the volume of the metal in the finished shell. Where the wall-thickness reduction is considerable, as in brass shell cases, the final blank size is developed by trial. A tentative blank size for an ironed shell can be obtained from the equation

$$D = \sqrt{d^2 + 4dh\frac{t}{T}} \qquad (8\text{-}16)$$

Where:

t = wall thickness
T = bottom thickness

Reduction Factors

After the approximate blank size has been determined, the next step is to estimate the number of draws that will be required to produce the shell and the best reduction rate per draw. As regards diameter reduction, the area of metal held between the blankholding faces must be reasonably proportional to the area on which the punch is pressing, since there is a limit to the amount of metal which can be made to flow in one operation. The greater the difference between blank and shell diameters, the greater the area that must be made to flow, and therefore the higher the stress required to make it flow. General practice has established that, for the first draw, the area of the blank should not be more than three and one-half to four times the cross-sectional area of the punch.

One of the important factors in the success or failure of a drawing operation is the thickness ratio, or the relation of the metal thickness to the blank or previous shell diameter; this ratio is expressed as t/D. As this ratio decreases, the tendency to wrinkling increases, requiring more blankholding pressure to control the flow properly and prevent wrinkles from starting. The top limit of about 48% seems to be substantiated by practice for single-action first draws. The 30% limit for double-action redraws is dictated by practice and is modified by corner radii, friction, and the angle of the blankholding faces with respect to the shell wall. Because of strain-hardening stresses set up in the metal, third and subsequent draws should not exceed 20% reduction without an annealing operation.

Force

Drawing Force. The force applied to the punch, necessary to draw a shell, is equal to the product of the cross-sectional area and the yield strength s of the metal. Taking into consideration the relation between the blank and shell diameters and a constant C of 0.6 and 0.7 to cover friction and bending, the force P for a cylindrical shell may be expressed by the empirical equation,

$$P = \pi \, dts \left(\frac{D}{d} - C \right) \tag{8-17}$$

Blankholder Force. The amount of blankholder force required to prevent wrinkles and puckers is largely determined by trial and error. The force required to hold a blank flat for a cylindrical draw varies from very little to a maximum of about one-third or more of the drawing force. On cylindrical draws, the force is uniform and balanced at all points around the periphery because the amount of flow at all points is the same. On rectangular and irregularly shaped shells, the amount of flow around the periphery is not uniform; hence the force required varies also.

Evolution of a Draw Die

The first step in planning a die (*Figure 8-40*) for the part of *Figure 8-39* is not the development of the die but that of the blank for this drawn shell or cup.

Since the ratio d/r is more than 20, the diameter of the blank D is, from Eq. (8-12):

$$D = \sqrt{2.75 + (4 \times 2.75 \times 1.5)}$$
$$= 4.9'' \ (124.5 \ mm)$$

The area of this blank is slightly less than four times the cross-sectional area of the punch (approximately 4.9 in.^{2} 3161.3 mm^2. This blank can be drawn into a shell of the dimensions specified in one draw because the ratio of its area to that of the punch nose is approximately 4:1.

The length of the draw radius (radius of the toroidal zone at the entrance to the die) must be determined before any other die design details are considered.

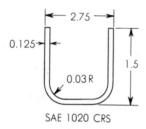

SAE 1020 CRS

Figure 8-39. A drawn shell, the basis for blank development and draw die design (*Figure 8-40*).

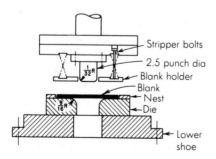

Figure 8-40. Draw die for producing the shell of *Figure 8-39*.

The radius of the draw die should be kept as large as possible to aid in the flow, but if it is too large, the metal will be released by the blankholder before the draw is completed and wrinkling will result. When the radius is too small, the material will rupture as it goes over the radius, or against the face of the punch. *Figure 8-41* gives the practical drawing radii for certain stock thicknesses. The values in this table are based on a radius of approximately four times the stock thickness. In some cases the radius may vary from four to six times the stock thickness. The length of the draw radius from *Figure 8-41* is 0.563" (14.28 mm).

The required drawing force is determined; Eq. (8-17) is used:

P = 20 tons (182 kN)

Thickness of stock, in. (mm)	Drawing radius, in. (mm)
1/64 (0.4)	1/16 (1.6)
1/32 (0.8)	1/8 (3.2)
3/64 (1.2)	3/16 (4.8)
1/16 (1.6)	1/4 (6.4)
5/64 (2.0)	3/8 (9.5)
3/32 (2.4)	7/16 (11.1)
1/8 (3.2)	9/16 (14.3)

Figure 8-41. Practical drawing radii for certain thicknesses of stock.

Blankholder force is determined largely by trial and error; an allowable range is from a few pounds up to one-third of the drawing force; approximately six tons (53.4 kN) is more than adequate. An ordinary 30-ton (267 kN) OBI press has enough capacity for the total possible maximum force of 26 tons (231 kN) required for the operation.

The blank is nested in a semicircular plate. Its thickness is established by the optimum blankholder pressure on the blank during drawing of the shell to specifications. The unit pressure on the blank becomes greater as it is pushed into the die.

Shimming the nest plate will vary the pressure on both nest and blank, and results in a critical pressure on the latter without the production of defective shells.

The blankholder is a 0.625" (15.88 mm) thick circular plate suspended from the punchholder by six stripper bolts equally spaced around its circumference.

A like number of die springs fit around the bolts and are retained by pockets counterbored in the punchholder and blankholder.

In addition to its function of holding the blank, the blankholder strips the shell from the punch.

The draw clearance (space between the punch and die) from *Figure 8-42* should be from 0.1375″ (3.49 mm) to 0.1400″ (3.56 mm). The ID of the die is

Blank thickness, in. (mm)	First draws	Redraws	Sizing-draw*
Up to 0.015 (0.38)	$1.07t$ to $1.09t$	$1.08t$ to $1.1t$	$1.04t$ to $1.05t$
0.016 to 0.050 (0.41 to 1.27)	$1.08t$ to $1.1t$	$1.09t$ to $1.12t$	$1.05t$ to $1.06t$
0.051 to 0.125 (1.30 to 3.18)	$1.1t$ to $1.12t$	$1.12t$ to $1.14t$	$1.07t$ to $1.09t$
0.136 (3.45) and up	$1.12t$ to $1.14t$	$1.15t$ to $1.2t$	$1.08t$ to $1.1t$

* Used for straight-sided shells where diameter or wall thickness is important, or where it is necessary to improve the surface finish in order to reduce finishing costs.

t = thickness of the original blank.

Figure 8-42. Draw clearance.

1.6375″ (41.59 mm) for tryout of the die. If there is too much ironing of the shell, this ID can be increased up to approximately 1.640″ (41.66 mm). The OD of the draw punch is the same as the ID of the drawn shell.

The methods of determining die dimensions (and hold-down size) discussed in Chapter 7 apply to the same elements in a drawing die, but may generally be increased approximately in proportion to the press forces applied to and the stresses in die blocks and blankholders.

Draw Rings. The draw ring of many draw dies is the die itself, made of tool-steel with the edge of the cavity forming a spherical zone, over which the metal is drawn. In many large drawing dies, the die is not made in one piece, but has an inserted draw ring for tool-steel economy and lower replacement cost.

Draw Die Materials. The selection of material for a draw punch and a draw die is determined the same as for any other tool. It is determined largely by the number of parts to be drawn and their ultimate unit cost. For low production, material cost is a major factor. If less than 100 parts are to be drawn, a plastic or zinc alloy material is satisfactory.

For production of approximately 1000 parts, a plain cast iron ring is suitable. Cast iron punches and dies can be chrome plated for medium-production runs. A plating thickness of 0.003″ (0.08 mm) is satisfactory.

The best tool steels should be used for runs of 10,000 or more parts. The choice of the type of tool steel is based upon factors of wear, strength, etc., considered in the selection for any other tool.

Carbide punches and dies or carbide inserts for these tools have proven to be the most economical material for extremely large runs of 1,000, 000 parts or more.

Material selection for blankholders, knockouts, and other parts of a draw die is based upon these same factors.

Lubricants. The functional requirements of lubricants for drawing operations are much more severe than for shafts and bearings.

Possible corrosive action of some compounds on some metals and the ease of their removal from drawn metal parts must be considered.

Handbooks and manufacturers of various metals should be consulted for lubrication recommended for shallow and deep drawing operations.

The following lubrication data for drawing mild steel are included as a guide:

Mild Operations

1. Mineral oil of medium-heavy to heavy viscosity
2. Soap solutions (0.03 to 2%, high-titer soap)
3. Fat, fatty-oil, or fatty and mineral-oil emulsions in soap-base emulsions
4. Lard-oil or other fatty-oil blends (10 to 30% fatty oil).

Medium Operations

1. Fat or oil in soap-base emulsions containing finely divided fillers such as whiting or lithopone.
2. Fat or oil in soap-base emulsions containing sulfurized oils.
3. Fat or oil in soap-base emulsions with fillers and sulfurized oils.
4. Dissimilar metals deposited on steel plus emulsion lubricant or scrap solution.
5. Rust or phosphate deposits plus emulsion lubricants or soap solution.
6. Dried soap film.

Severe Operations

1. Dried soap or wax film, with light rust, phosphate, or dissimilar metal coatings.
2. Sulfide or phosphate coatings plus emulsions with finely divided fillers and sometimes sulfurized oils.
3. Emulsions or lubricants containing sulfur as combination filler and sulfide former.
4. Oil-base sulfurized blends containing finely divided fillers.

Progressive Dies

A progressive die performs a series of fundamental sheet metal operations at two or more stations during each press stroke in order to develop a workpiece as the strip stock moves through the die. This type of die is sometimes called cut-and-carry, follow, or gang die. Each working station performs one or more distinct die operations, but the strip must move from the first through each succeeding station to produce a complete part. One or more idle stations may be incorporated in the die, not to perform work on the metal, but to locate the strip, to facilitate interstation strip travel, to provide maximum-size die sections, or to simplify their construction.

The linear travel of the strip stock at each press stroke is called the progression, advance, or pitch and is equal to the interstation distance.

The unwanted parts of the strip are cut out as it advances through the die, and one or more ribbons or tabs are left connected to each partially completed part to carry it through the stations of the die. Sometimes parts are made from individual blanks, neither a part of, nor connected to, a strip; in such cases, mechanical fingers or other devices are employed for the station-to-station movement of the workpiece.

The operations performed in a progressive die could be done in individual dies as separate operations, but would require individual feeding and positioning. In a progressive die, the part remains connected to the stock strip which is fed through the die with automatic feeds and positioned by pilots with speed and accuracy.

Selection of Progressive Dies

The selection of any multioperation tool, such as a progressive die, is justified by the principle that the number of operations achieved with one handling of the stock and the produced part is more economical than production by a series of single-operation dies and a number of handlings for each single die.

Where total production requirements are high, particularly if production rates are large, total handling costs (man-hours) saved by progressive fabrication compared with a series of single operations are frequently greater than the costs of the progressive die.

The fabrication of parts with a progressive die under the above-mentioned production conditions is further indicated when:

1. Stock material is not so thin that it cannot be piloted or so thick that there are stock-straightening problems.
2. Overall size of die (functions of part size and strip length) is not too large for available presses.
3. Total press capacity required is available.

Strip Development for Progressive Dies

Individual operations performed in a progressive die are often relatively simple, but when they are combined in several stations, the most practical and economical strip design for optimum operation of the die often becomes difficult to devise.

The sequence of operations on a strip and the details of each operation must be carefully developed to assist in the design of a die to produce good parts. A tentative sequence of operations should be established and the following items considered as the final sequence of operations is developed:

1. Pierce piloting holes and piloting notches in the first station. Other holes may be pierced that will not be affected by subsequent noncutting operations.
2. Develop blank for drawing or forming operations for free movement of metal.
3. Distribute pierced areas over several stations if they are close together or are close to the edge of die opening.
4. Analyze the shape of blanked areas in the strip for division into simple shapes so that punches of simple contours may partially cut an area at one station and cut out remaining areas in later stations. This may suggest the use of commercially available punch shapes.
5. Use idle stations to strengthen die blocks, stripper plates, and punch retainers and to facilitate strip movement.
6. Determine whether strip grain direction will hinder or facilitate an operation.
7. Plan the forming or drawing operations either in an upward or a downward direction, whichever will assure the best die design and strip movement.
8. The shape of the finished part may dictate that the cutoff operation should precede the last noncutting operation.
9. Design adequate carrier strips or tabs.
10. Check strip layout for minimum scrap; use a multiple layout if feasible.
11. Locate cutting and forming areas to provide uniform loading of the press slide.
12. Design the strip so that scrap and part can be ejected without interference.

Figure 8-43 illustrates the use of a three-station die to avoid weak die blocks. At *A*, the pierced hole is near the edge of the part where it is cut off, thereby weakening the die block at this point. If an idle station is added so that the piercing operation is moved ahead one station, the die block is stronger and there is less chance of cracking in operation of fabrication. At *B*, the pierced holes are centered on the strip but close together. In this case the holes should be pierced in two stations to avoid thin sections in the die block

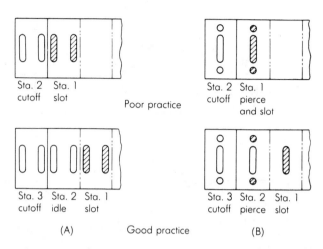

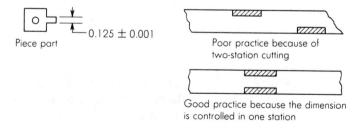

Figure 8-43. Use of three-stage die to avoid weak die blocks: (*A*) pierced hole close to edge of part: (*B*) pierced holes close together.

between the holes. The adding of stations also provides better support for the piercing punches.

Figure 8-44 shows the use of one die station instead of two stations to maintain a close-toleranced dimension. If two stations were used, the variation in the location of the stock guides and cutting punches could make it difficult to hold the ±0.001″ (±0.03 mm) tolerance.

The strip development for shallow and deep drawing in progressive dies must allow for movement of the metal without affecting the positioning of the part in each successive station. *Figure 8-45* shows various types of cutouts and typical distortions to the carrier strips as the cup-shaped parts are formed and then blanked out of the strip. Piercing and lancing of the strip around the periphery of the part as shown at *A*, leaving one or two tabs connected to the carrier strip, is a commonly used method. The semicircular lancing as shown

Figure 8-44. Use of one station versus two to hold a close tolerance.

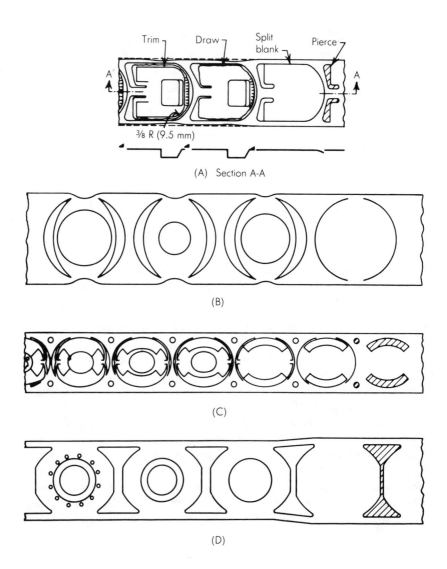

Figure 8-45. Cutout reliefs for progressive draws: (*A*) lanced outline; (*B*) *circular lance;* (*C*) double lance suspension; (*D*) hourglass cutout.

at *B* is used for shallow draws. The use of this type of relief for deeper draws places an extra strain on the metal in the tab and causes it to tear. The carrier strip is distorted to provide stock for the draw. A popular cutout for fairly deep draws is shown at *C*. This double-lanced relief suspends the blank on narrow ribbons, and no distortion takes place in the carrier strips. Two sets of

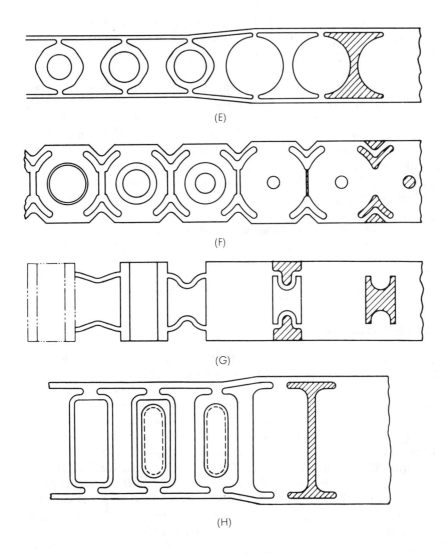

(E)

(F)

(G)

(H)

Figure 8-45.(Continued)(E) cutout providing expansion-type carrier ribbon for circular draws; (*G*) cutout providing expansion-type carrier ribbon for rectangular draws; (*H*) I-shaped relief for rectangular draws.

single rounded lanced reliefs of slightly different diameters are placed diametrically opposite each other to produce the ribbon suspension. The hourglass cutout in *D* is an economical method of making the blank for shallow draws. The connection to the carrier strips is wide, and a deep draw would cause considerable distortion. An hourglass cutout for deep draws is shown in *E*, which provides a narrow tab connecting the carrier strip to the blank. The cupping operations narrow the width of the strip as the metal is drawn into the cup shape.

The hourglass cutout may be made in two stations by piercing two separated triangular cutouts in one station, and lancing or notching the material between them in a second station. The cutouts shown at *F* and *G* provide an expansion-type carrier ribbon that tends to straighten out when the draw is performed. These cutouts are made in two stations to allow for stronger die construction. Satisfactory multiple layouts may be designed using most of the reliefs by using a longitudinal lance or slitting station to divide the wide strip into narrower strips as the stock advances. The I-shaped relief cutout in *H* is a modified hourglass cutout used for relatively wide strips from which rectangular or oblong shapes are produced.

Straight slots or lances crosswise of the stock are sometimes used on very shallow draws or where the forming is in the central portion of the blank. On the deeper draws, this type of relief tends to tear out the carrier strips or cause excessive distortion in the blank and is not satisfactory.

Stock Positioning

Of prime importance in the strip development is the positioning of the stock in each station. The stock must be positioned accurately in each station so that the operation can be done in the proper location. A commonly used method of stock-positioning is the incorporation of pilots in the die.

There are two methods of piloting in dies: direct and indirect. Direct piloting consists of piloting in holes punched in the part at a previous station. Indirect piloting consists of piercing holes in the scrap-strip and locating these holes with pilots at later operations. Direct piloting is the ideal method for locating the part in subsequent die operations. Unfortunately, ideal conditions may not exist, and in such cases, indirect piloting must be used to achieve the desired results of part accuracy and high production speeds. The advantages of locating pilots in the scrap material area are as follows:

1. Not readily affected by workpiece change.
2. Size and location not as limited.

Disadvantages of locating pilots in the scrap section are as follows:

1. Material width and lead may increase.
2. Scrap-strip carriers distort on certain types of operations and make subsequent station use impossible.

How to pilot is an arbitrary decision that the tool designer must make. It is impossible to give definite rules and formulas, because the material and the hardness of the stock influence the decision. However, in the indirect piloting method, it is possible to use pilots of greater diameter than if holes in the part are used for piloting such as in *Figure 8-46A*. The greater the diameter of the pilot, the less chance there is of distortion of either the strip or the pilot. Also, small-diameter pilots introduce the possibility of broken pilots.

When holes in the part are held to close tolerances (*Figure 8-46B*), it is possible for the pilots to affect the hole size in their effort to move the strip to proper location.

When holes in the part are too close to the edges (*Figure 8-46C*), the weak outer portions of the part are likely to distort upon contact with the pilots, instead of the strip's moving to the correct location. This possibility is often overlooked in planning a progressive die, and leads to subsequent runs of scrap parts and expensive die alterations.

Just what constitutes a condition where the edge of the hole is too close to the edge of the part is, like many aspects of design, a matter of personal judgement. Many designers use the rule-of-thumb: The distance between the two must be at least twice the stock thickness.

A similar problem exists when the part holes are located in a weak portion of the inside area of the part (*Figure 8-46D*). Here, there is a possibility of the part's buckling before the pilots can position the stock strip. In this case it is advisable to pilot in the scrap strip.

To achieve accurate part location, the pilots must be placed as far apart as possible. When the holes in the workpiece are close together, as in *Figure 8-46E*, holes in the scrap strip should be used for piloting. A second method would be to place a pilot in one hole in one station and in the same hole in a succeeding station. The feasibility of the second method depends upon the availability of an additional die station.

When slots are punched in the blank parallel to the stock movement (*Figure 8-46F*), the slots are not suitable piloting holes. Therefore, indirect pilots must be used.

Disposition of Scrap Strip

A strip development is illustrated in *Figure 8-47B* utilizing pierce, trim, form, and blank-through operations and carriers on both sides of the strip. The workpiece is dropped through the die, while the carrier bars continue to the scrap cutters to be cut into short lengths. The dropping of the workpiece through the die is the most desirable method of part ejection, but cannot always be obtained. Cutting the scrap into small sections simplifies the material handling problems and produces a greater dollar return when sold as scrap metal. *Figure 8-47C* shows an alternate strip development with one side carrier. The workpiece is pierced, trimmed, cut off, and formed on a pad with air or gravity ejection, and the carrier bar is cut into short pieces by the scrap

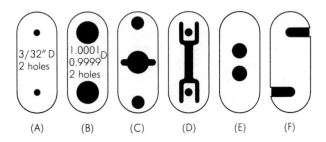

Figure 8-46. Part conditions that require indirect pilots: (*A*) small holes; (*B*) close-tolerance holes; (*C*) holes too near edge; (*D*) hole in fragile areas of the part; (*E*) holes too close together; (*F*) slots in parts.

cutter. Remember that if a part is to be ejected as this one is, the double carrier bar design in *Figure 8-47B* should be avoided, because the part may become trapped in these bars and cause die damage.

The design of the part in *Figure 8-47A* requires that the carrier be outside the part configuration. This necessitates the use of stock wider than the part width plus the normal trimming allowance. The part shown in *Figure 8-48A* can be made of stock the same width as the part.

The strip development of *Figure 8-48D* illustrates how the strip is pierced,

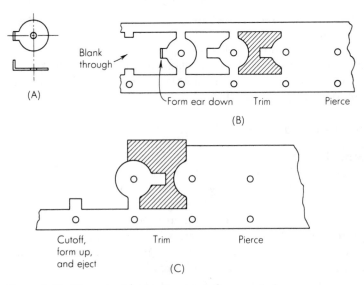

Figure 8-47. Alternate strip developments for a workpiece.

trimmed, and the part cut off and formed. A slug-type cutoff punch is used and the flange is formed downward. The part is then ejected by an air jet or by gravity. This arrangement is often referred to as a scrapless development since no carrier strips remain after the part is cut from the strip.

Figure 8-48E shows a strip development for the same part using a shear-type cutoff. The flange is formed upward as the combination cutoff and form punch descends. A spring-loaded pad supports the workpiece during forming and assists in ejecting the part from the die. The progression of this type of development is shortened by the width of the cutoff slug.

Figure 8-49 shows another development in which the stock is the same as the developed width of the workpiece. The strip is pierced in station 1; piloted and notched in station 2; piloted, pierced, and formed in station 3.

Progressive Die Elements

The die elements used in progressive dies such as punches, stops, pilots, strippers, die buttons, punch guide bushings, die sets guide posts, and guide post bushings are of similar design to those used in other types of dies. Refer to Chapter 7 and previous text in this section for their design.

General Die Design

A progressive die should be heavily constructed to withstand the repeated shock and continuous runs to which it is subject. Precision or antifriction guide posts and bushings should be used to maintain accuracy. The stripper plates (if spring-loaded and movable), when also serving as guides for the punches, should engage guide pins before contacting the strip stock. Lifters should be provided in die cavities to lift up or eject the formed parts, and carrier rails or pins should be provided to support and guide the strip when it is being moved to the next station. A positive ejector should be provided at the last station. Where practical, punches should contain shedder or oil-seal-breaker pins to aid in the disposal of the slug. Adequate piloting should be provided to ensure proper location of the strip as it advances through the die. For more die-design consideration, see Chapter 7.

Evolution of a Progressive Die

Figure 8-48A shows a workpiece to be blanked and formed at high production volume in a progressive die.

The first step in the design of the die is the development of the blank (*Figure 8-48B*). The grain of coiled metal is normally parallel to the length of the strip. Forming operations are normally performed perpendicular to the

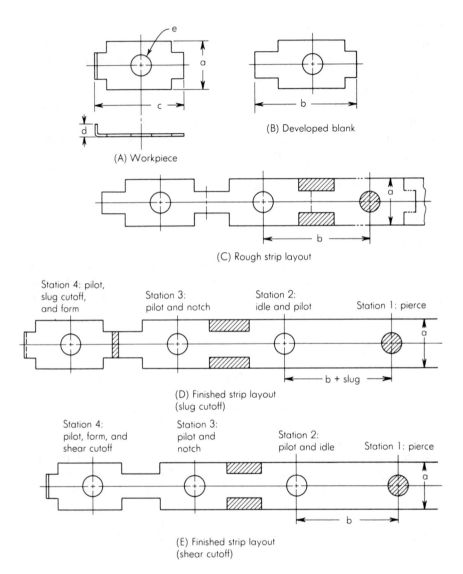

Figure 8-48. Scrapless strip development.

grain to forestall any tendency to fracture. The workpiece is therefore sketched with the bend line perpendicular to the strip.

The *a* dimension appears to coincide with stock width. A comparison is made between the exact dimension with tolerance, and available purchased

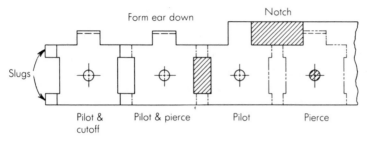

Figure 8-49. Four-station strip development.

coil stock. If the *a* dimension is not within the slitting tolerance of the purchased stock, the part must be blanked from a wider strip and perhaps shaved to the finished dimension. In this case, the *a* dimension is within the slitting tolerance, the slit edge will be satisfactory, and no work need be performed to obtain the dimension.

The *b* dimension is developed by adding the *c* dimension, the *d* dimension, and a bend allowance, minus twice the outside radius. Tolerances must be carefully considered in developing the *b* dimension, and will determine whether the dimension can be achieved with a single shear or whether shaving will be required. In this case the *b* dimension is not critical and is, for the moment, accepted as the station-to-station distance.

With the *a* and *b* dimensions established, a centerline is drawn, and the template of the blank is traced along the centerline (*Figure 8-48C*). The areas from which stock must be removed are clearly indicated. The straight edges of the stock appear acceptable for guidance. The absence of scrap stock on either side of the workpiece determines that the strip must be piloted by registry of workpiece features. Hole *e* appears most suitable for piloting purposes. Hole dimensions are examined to determine whether the hole can be pierced conventionally (hole diameter not less than 1 1/2 times stock thickness), or whether a secondary operation may be needed. The hole dimensions on this case are not critical, so the hole can be pierced in the first station and can be used as a pilot hole in subsequent stations. The first station can now be drawn and will include a guidance method for stock direction and elevation, the punch, and the die. The punch location can be transposed to the other stations where it will indicate the presence of a bullet-nosed pilot.

As the strip enters the second die station, it will register on the pilot. The next operation required on the workpiece is the notching of the edges. The workpiece dimensions are such that the notching should be performed by one stroke in one station. A preliminary layout discloses, however, that a notching operation in the second die station would be close to the piercing operation of the first station. The second station is tentatively designated as an idle station.

Idle stations are of great value. They enable the designer to distribute the work load uniformly throughout the length of the die. Individual die details can consequently be less complex, and can be easier to build and maintain. Idle stations also permit later changes in workpiece design by providing space for added operations.

The third die station can now be drawn to include a pilot and the notching punches.

The stock strip will enter the fourth die station and register on the pilot. Two operations are required to finish the workpiece: the leading edge must be formed 90°, and the workpiece must be separated from the strip. Once the workpiece has been separated, no further operations are normally possible, but the workpiece must be separated for forming. The two operations can, however, be performed simultaneously in one station, just as the two sides were notched in one station. Two methods may be considered. If the stock is to be separated by removing a slug, the width of the slug must be added to the *b* dimension and the station-to-station distance of the entire layout must be correspondingly amended. This method, however, would permit a second punch to simply wipe down (form) and thereby complete the part without changing the elevation of the stock. If the separation is to be performed by direct shear without a slug, the workpiece must change elevation. The die station would be spring-loaded to permit the shear action, and the forming action would take place at the depressed level.

Figure 8-50 shows the proposed die including stock elevation, stripper pads, and the ejection method. *Figure 8-50(A-A)* shows the die construction for a shear-type cutoff.

Force Determination. The force required for each station is computed by the methods detailed in Chapter 7. If the work load has been properly distributed throughout the length of the die, no rocking action should occur. If the required force is concentrated in any one station or area of the die, the planned operations should be shifted to achieve balance. When reasonable balance is assured, the forces may be added to determine the required press capacity.

Press Selection. The press selected must be of adequate capacity for the planned operation and must also be able to withstand the overloads that may accidentally be encountered. It must be equipped with a precise stock feed mechanism. The bolster plate must not deflect under the planned load even after it has been weakened by perforation for slug disposal. The sets are commercially available in a wide variety of sizes. An inexpensive die set would suffice. If precise shaving operations are necessitated by close workpiece tolerances, an expensive precise die set will be required.

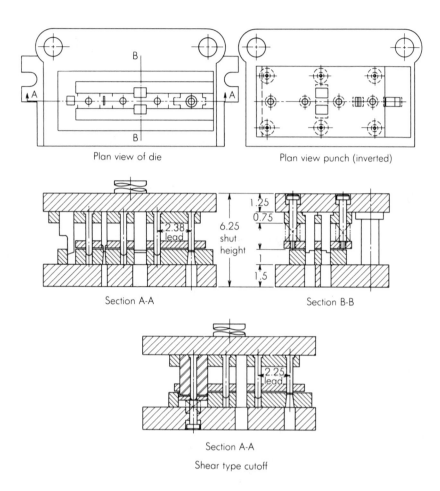

Plan view of die   Plan view punch (inverted)

Section A-A   Section B-B

Section A-A

Shear type cutoff

Figure 8-50. Die for part of *Figure 8-48.*

Examples of Progressive Dies

The part to be made is a small clip, *Figure 8-51*, made of 0.008″ thick phosphor bronze, eight numbers hard. Because of the delivery date for the parts and the anticipated total production, it was decided to pierce and blank the part in a progressive die, then form in a separate die.

The product designer indicated that the grain direction for optimum part performance is to be perpendicular to the bends. The grain direction favors the bends shown on the part and will decrease the tendency to crack at the bend lines.

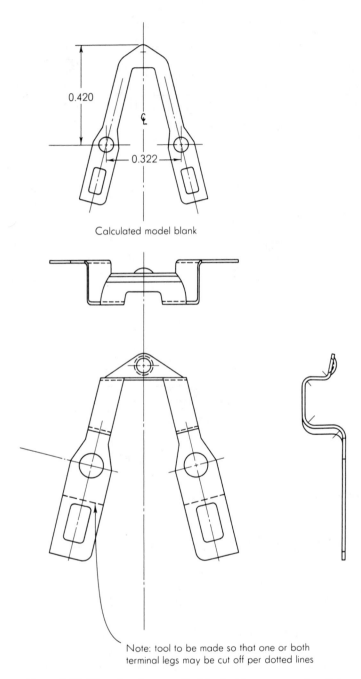

0.420

0.322

Calculated model blank

Note: tool to be made so that one or both
terminal legs may be cut off per dotted lines

Figure 8-51. Phosphor bronze clip blanked in a progressive die.

The pilots were located in the scrap area of the strip because of possible design changes, thin walls on the blanking punch if pilots were inserted in the holes of the part, and the opportunity to use larger-diameter pilots since the holes are only 0.0655" ±0.0015" diameter.

The perimeter of the blank and holes to be pierced is about 2.54". Multiplying the length of cut by the stock thickness of 0.008" and a shear stress of 80,000 psi gives a shear load of approximately 1600 lb. Therefore, the size of press will be more dependent on the die area than the capacity.

Figure 8-52 shows the strip development for the part. In the first station, the two 0.067/0.064" diam holes are pierced, the tip of the blank dimpled, and a 0.094" diam pilot hole is pierced. When the full-length legs are required, station 2 is a piloting station only. When one or both of the legs are to be cut off, punches are inserted in this station.

In station 3, the strip is piloted and the two rectangular holes are pierced. In station 4, the strip is piloted while the part is blanked through the die.

There are clearance spots on the die blocks for the dimple, thus permitting the strip to be flat on the die blocks. The die blocks are made in three sections for ease of machining as shown in *Figure 8-53*. A channel-type stripper, 1, was selected for this die because of its more reasonable cost, and less design and build time. The second-operation forming die would remove some of the curvature caused by the blanking operation. Hardened guide bushings, 2, were used for the pilot hole piercing punch, 3, and the punches, 4, for the 0.067/0.064" diam holes. Guide bushings were used with these punches since they were small in diameter and were for the most important holes in the part.

The punches, 5, for cutting off the legs were backed up with a socket set screw, 6. When the punches were not in use, the screws were backed off a few

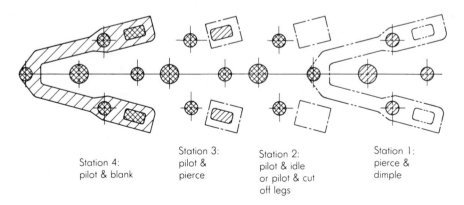

Station 4:
pilot & blank

Station 3:
pilot &
pierce

Station 2:
pilot & idle
or pilot & cut
off legs

Station 1:
pierce &
dimple

Figure 8-52. Strip development for part of *Figure 8-51*.

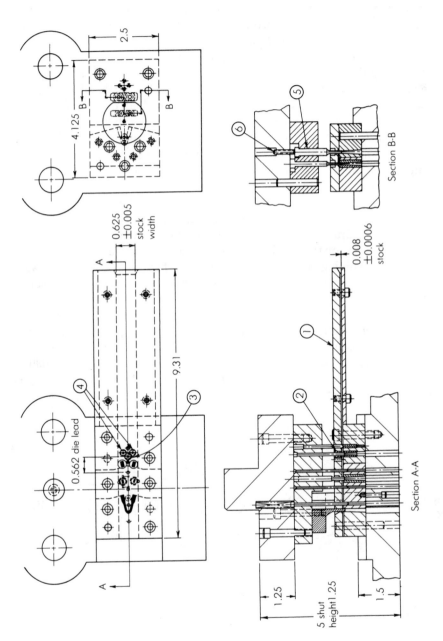

Figure 8-53. Progressive die for part of *Figure 8-51.*

turns and the tight fit between the punch and punch holder held them away from the stock.

A 5 x 5″ two-post precision die set was selected for this die. The lower shoe was 1 1/2″ thick and the upper shoe was 1 1/4″ thick with a 1 9/16″ diam by 2 1/8″ long shank.

The stripper was extended beyond the die set close to the push-type roll feed. This was to support the stock and avoid buckling and subsequent damage to the die.

Figure 8-54 illustrates the die layout for a small, irregularly-shaped stamping having eight small holes and one large hole. For accurate positioning of the strip, pilots are provided in each station. There are six stations in the die with progression of 1.718″ between stations.

In station 1, the eight holes, the 0.1875″ diam pilot hole, and a 0.302 x 0.259″ hole are pierced. In station 2, a 0.500″ diam hole, two 0.062 x 0.068″ holes, and a 0.095 x 0.379″ slot are pierced.

In station 3, a slot is pierced symmetrically about the centerline. The strip is notched and slotted in the next station. Station 5 is an idle station because of the length of the notching punch in the previous station. The last station, 6, incorporates a slug-type cutoff punch. This punch trims the left-hand side of the blank piloted in station 5 and the right-hand side of the blank in station 6.

For ease of grinding, the die was made in five segments as shown in *Figure 8-55*. The die set was specially made with diagonally-placed guide posts. One

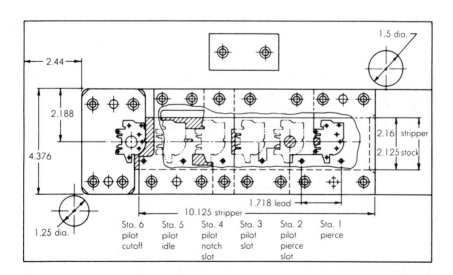

Figure 8-54. Die layout for a small, irregularly-shaped stamping.

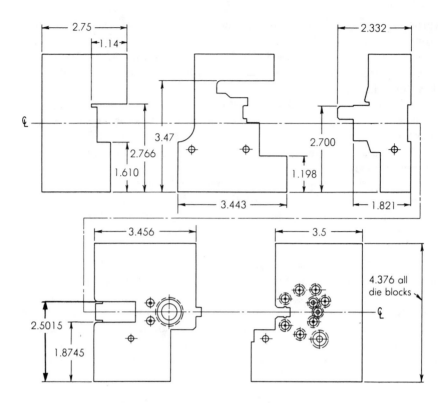

Figure 8-55. Details of die blocks for die of *Figure 8-54.*

post is 1 1/2″ diam while the other is 1 1/4″ diameter. Guide posts of two different diameters were used to prevent placing the upper shoe on the lower shoe incorrectly and damaging the punches.

Die buttons were used for each of the piercing punches; the die also incorporated guide bushings in the channel stripper for the piercing punches.

Transfer Die Systems

Transfer die systems increase production rates over manual line die operations (where parts are run through several dies in more than one press and moved manually) and progressive dies with secondary forming operations that are done manually.

In many cases, there can be large material cost savings by the elimination of carrier strips and webs between blanks or using offal from larger parts to make blanks for smaller parts.

Transfer systems may reduce direct manufacturing costs. In many cases, one press and operator using a transfer system can do the work of many.

Parts can be made on a transfer system that couldn't be made in progressive die because the carrier strip would not allow the parts to be formed properly.

They improve safety for the operator and equipment by running in a hands-off condition. The operator only loads blanks into the feeder or, in combination progressive transfer system, starts new coils of stock. They can incorporate fail-safe systems that detect a miscycle in the system or parts that are not in their proper place and automatically stop the press to avoid damage.

Coil stock or blanks can be fed from any direction. Feeders can be part of the automation and designed for special applications which conventional press feeds will not do.

A good example of material and labor savings is shown in *Figure 8-56.*

Extrusion Dies

Impact extrusion, also known as cold extrusion or cold forging, is closely allied to coining, sizing, and forging operations. The operations are generally performed in hydraulic or mechanical presses. The press applies sufficient pressure to cause plastic flow of the workpiece material (metal) and to form the metal to a desired shape. A metal slug is placed in a stationary die cavity into which a punch is driven by the press action. The metal is extruded upward around the punch, downward through an orifice, or in any direction to fill the cavity between the punch and die. The shape of the finished part is determined by the shape of the punch and the die.

Product design is influenced by methods of manufacturing such as machining from the solid, drawing, spinning, stamping, or casting. Impact extrusion may combine more manufacturing processes into a single operation than any other metalworking method. Cold extrusion can result in excellent surface finish and accurate size, and can improve the mechanical and physical properties of the workpiece material. Cold extruded parts may have smooth surfaces ranging between 30 and 100 μ in. (0.76 μ m and 2.54 μ m). Very close tolerances can be achieved by cold extrusion. Owing to severe cold working of the metal, a fine, dense grain structure is developed parallel to the direction of metal flow. These continuous flow lines increase the fatigue resistance of the material.

Figure 8-57 lists the pressures required to extrude common metals. The pressures required depend upon the alloy, its microstructure, the restriction to flow, the severity of work hardening, and the lubricant used. With high speed and restricted flow, the press load may suddenly increase to three times

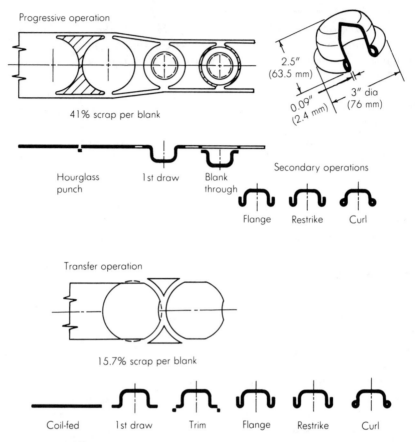

Figure 8-56. Material scrap loss of a progressive operation compared with a transfer operation for the same part.

the anticipated tonnage requirement. Pressures have been recorded up to 165 tons/in.^{2} (2275 MPa). Owing to the high pressures required, presses must be carefully selected.

The pressures required for extrusion also depend upon the percentage of reduction in area. When the percentage of reduction of area, for parts made of various aluminum alloys, is known, the required pressure may be taken from *Figure 8-58*. Although the alloys mentioned are usually extruded at room temperature, a reduction in press pressure can be achieved by extruding at elevated temperatures. Closely related to, and a factor in establishing reduction of area, is part wall thickness. The increasing effect on punch pressure required as the wall becomes thinner is illustrated in *Figure 8-59*.

Material	Pressure, tons/in.^{2}/(MPa)
Pure aluminum "extrusion grade"	40.70 (561)
Brass (soft)	30.50 (421)
Copper (soft)	25.70 (354)
Steel C1010 "extrusion grade"	50.165 (692)
Steel C1020 (spheroidized)	60.200 (830)

Figure 8-57. Extrusion pressures for common metals.

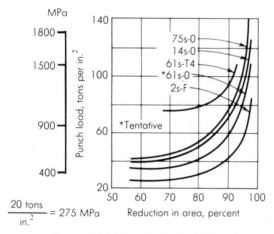

$$\frac{20 \text{ tons}}{\text{in.}^2} = 275 \text{ MPa}$$

Figure 8-58. The effect of reduction in area to the punch load in extruding.

The relationship between reduction of area and extrusion pressures for a series of plain carbon steels is shown in *Figure 8-60*. The steels referred to in the different curves have carbon content in the range of 0.05 to 0.50%, and less than 0.03% each of sulfur and phosphorus. Steel 11 contained 0.58% chromium, 0.11% carbon, and 0.36% manganese, and 0.03% each sulfur and phosphorus.

Correct lubrication reduces considerably the pressures required and makes possible the cold extrusion of steel. Ordinary die lubricants break down because of the high pressures and the excessive surface heat generated by the plastic flow of material. A bonded steel-to-phosphate layer, and a bonded phosphate-to-lubricant layer is a satisfactory lubricant to prevent metal-to-metal welding or pickup. The workpiece is usually phosphatized before extrusion. If the workpiece is subjected to several extrusion operations, it is usually annealed and phosphatized between operations.

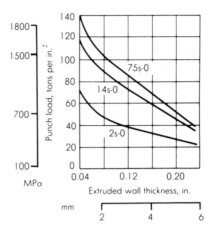

Figure 8-59. The effect of extruded wall thickness on the punch load.

Because the slugs used for impact extrusion are cut from commercial rod or bar, it is desirable to coin them to a desired size and shape for better die-fitting characteristics prior to extrusion. Using coiled wire stock, headers can be used to cut and coin or upset as a continuous high-speed operation. *Figure 8-61* illustrates slug coining and upsetting. *Figure 8-62* illustrates how a profiled slug fits the die cavity and how the material flows when pressure is applied to cause plastic flow. Voids in the die cavity are filled as the slug collapses under initial pressure. Profiled slugs thus cushion impact of the punch, and allow higher ram velocity. Plastic flow then continues through small orifices between the diepot and the punch.

Basic methods of impact extrusion are backward, forward, and combination methods. In backward extrusion (*Figure 8-63*) metal flows in a direction opposite to the direction of punch movement. The punch speed can be from seven to 14 ips (178 to 356 mm/sec). As the punch strikes the slug, the heavy pressure causes the metal to flow through the orifice created by the punch and the die, and forms the side wall of the part by extrusion. In forward extrusion (*Figure 8-64*) plastic flow of metal takes place in the same direction as punch travel. The orifice in this case is formed between the extension on the punch and the opening through the die. To prevent reverse flow of metal, the body of the punch seals off the top of the die. The finished part is ejected backward after the punch is restricted.

The design of a punch and die with double orifices to permit the plastic flow of metal in forward and reverse extrusion simultaneously will produce parts by combination impact extrusion. *Figure 8-65* illustrates examples of combination extrusion.

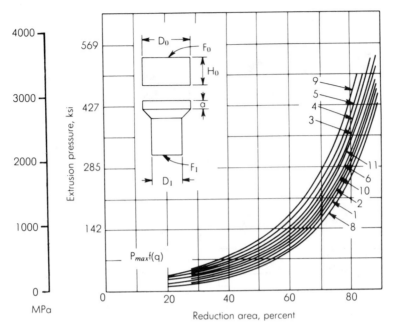

Figure 8-60. Extrusion pressures: reduction relationship for the forward extrusion of a series of steels with carbon contents in the range 0.005 to 0.50%.

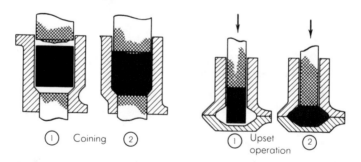

Figure 8-61. Slug coining and upsetting. (*Courtesy, American Machinist*)

Impact extrusion dies are also classified as open or closed. In open dies, the metal does not completely fill the die cavity. In closed dies, the material fills out the details of the die. To compensate for variations in slug volume, it is desirable to relieve closed dies and thus minimize excessive pressures.

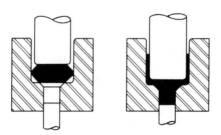

Figure 8-62. Profiled slugs. (*Courtesy, American Machinist*)

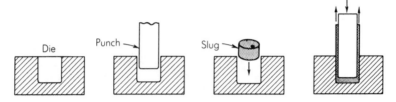

Figure 8-63. Backward extrusion. (*Courtesy, Design Engineering*)

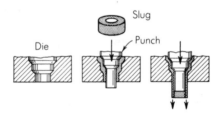

Figure 8-64. Forward extrusion. (*Courtesy, Design Engineering*)

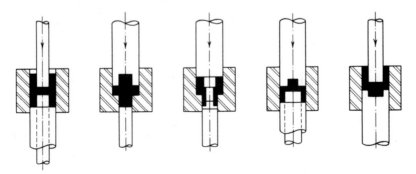

Figure 8-65. Combination extrusion. (*Courtesy, American Machinist*)

Die Design Principles

Coining Dies. In backward extruding dies the punch is always smaller in diameter than the die cavity in order to give the clearance between punch and die equalling the desired wall thickness of the part to be produced. The punch is loaded as a column. To minimize punch failure it is desirable to coin the slugs to a close fit in diameter to assure concentricity. *Figure 8-66* illustrates a coining die to prepare a slug for backward extrusion. Coining the slug to fit the diepot and coining the upper end to fit and guide the free end of the punch will minimize punch breakage of the extruding die.

Backward Extrusion Dies. A typical backward extrusion die is shown in *Figure 8-67*. The use of a carbide die cavity will minimize wear due to excessive pressures. The carbide insert is shrunk into a tapered holder. The holder has a 1° side taper that prestresses the carbide insert to minimize expansion and fatigue failure. The inserts are well supported on hardened blocks. The extruding punch is guided by a spring-loaded guide plate which in turn is positioned by a tapered piloting ring on the lower die. Ejection of the finished part from the die is by cushion or pressure cylinder. *Figure 8-68* illustrates a backward extrusion die with an unusual punch penetration ratio of 5:1 made possible with a modified flat-end punch profile.

Forward Extrusion Dies. *Figure 8-69* is an example of a typical forward extrusion die in which the metal flows in the same direction as the punch, but at a greater rate owing to change in the cross-sectional area. The lower carbide guide ring is added to maintain straightness. The nest above the

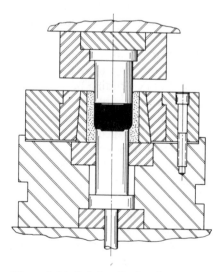

Figure 8-66. Coining die for slug preparation.

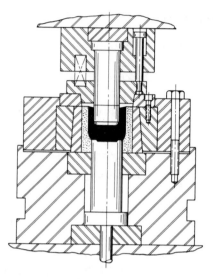

Figure 8-67. Backward extrusion die.
Note that the centering ring for the
punch and the carbide die cavity are
preloaded in shrink rings and supported
on toughened load-distributing steels.
(*Courtesy, E.W. Bliss Company*.)

upper carbide guide ring serves as a guide for the punch during the operation.
Figure 8-70 illustrates another forward extrusion die in which the punch
creates the orifice through which the metal flows. The extruding pressure is
applied through the punch guide sleeve.

Combination Extruding Dies. A typical combination forward and
backward extrusion die is shown in *Figure 8-71*. In this die, the two-piece
pressure anvil acts as a bottom extruding punch and a shedder. The upper
extruding punch is guided by a spring-loaded guide plate into which the guide
sleeve is mounted. To maintain concentricity between the punch and die, the
punch guide sleeve is centered into the die insert.

Punch Design. The most important feature of punch design is end
profile. A punch with a flat end face and a corner radius not over 0.020″ (0.51
mm) can penetrate three times its diameter in steel, four to six times its
diameter in aluminum. A punch with a bullet-shaped nose or with a steep
angle will cut through the phosphate coat lubricant quicker than a flat-end
punch. When the lubricant is displaced in extrusion, severe galling and wear
of the punch will take place. The punch must be free of grinding marks and
requires a 4 μ in. (0.10 μ m) finish, lapped in the direction of metal flow. The
punch should be made of hardened tool steel or carbide. In some backward

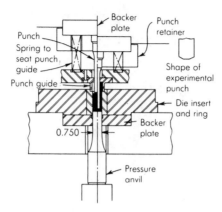

Figure 8-68. Backward extrusion die.
(*Courtesy, American Machinist*)

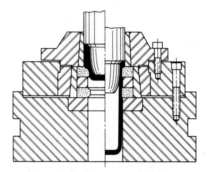

Figure 8-69. Forward extrusion die.

extrusion dies a shoulder is provided on the punch to square up the metal as it meets this shoulder.

Pressure Anvil Design. The function of the pressure anvil is to form the base of the diepot, to act as a bottom extruding punch, and to act as a shedder unit to eject the finished part. Heat treatment and surface finish requirements are the same for pressure anvils and for punches.

Diepot Design. To resist diepot bursting pressure, the tool steel or carbide die ring is shrunk into the shrink ring or die shoe. The die shoe is normally in compression. A shrink fit of 0.004″ per in. (0.004 mm per mm) of diameter of the insert is desirable. Material, heat treatment, and finish requirements of the diepot are the same as for the punch. The recommended material for shrink rings is a hot-worked alloy tool steel which is hardened to

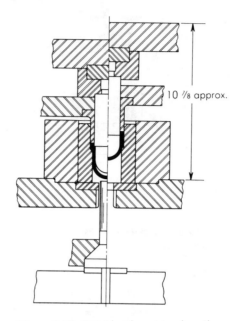

Figure 8-70. Combination extrusion die.

R_C 50 —R_C 52. A two-piece diepot insert is sometimes used for complex workpiece shapes.

Punch Guide Design. The guide ring minimizes the column loading on the punch by guiding the punch above the diepot. The spring-loaded guide sleeve pilots the punch into the diepot and maintains concentricity between them. The guide ring can also act as a stripper. The proper use of guide sleeves permits higher penetration ratios.

Four-Slide Dies

Multislide high-speed presses, *Figure 8-72*, are used to mass-produce various parts such as terminal clips, bobby pins, and various bent and formed wire and sheet metal parts (*Figure 8-73*). Coiled wire or sheet metal (usually 1″ [25 mm] or less in width) can be fed directly into the machine, or precut blanks can be chute-fed. Cycle times run from 40 strokes per minute on 1/2″ (12.7 mm) wire or sheet stock to 250 strokes per minute or more on 1/32″ (0.79 mm) wire.

Typically, wire is fed in and cut off, transferred, and gripped. Then as many of the four slides as are required successively form the part. Tooling is moderate in cost and the machine is suited for high-volume production.

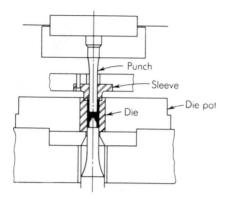

Figure 8-71. Combination extrusion die with two-piece pressure anvil. (*Courtesy American Machinist.*)

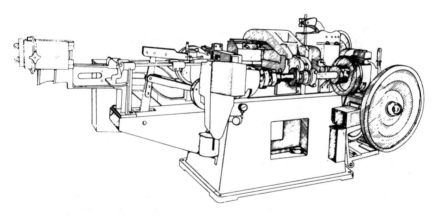

Figure 8-72. Multislide machine. (*Courtesy, U.S. Tool Co., Inc.*)

Forge Plant Equipment

Forgings in general are seldom used in the as-forged condition but are finished by various machining processes. A typical forging is the engine connecting rod (*Figure 8-74*). As can be seen by a comparison of the illustrations, very little machining is required to bring this vital component to its usable form. What then is forging? It is the plastic deformation of a metal caused by the stresses imposed upon it through hammering or squeezing operations. All forgings are not made as close to their final configuration as in the case of the connecting rod. The shape of the forging is governed by the

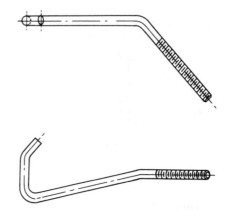

Figure 8-73. Part made of 0.325-in. wire on
a multislide (four-slide) machine

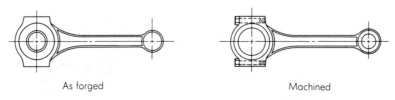

As forged   Machined

Figure 8-74. Designs of engine connecting rod.

equipment and dies used in its production. Forging equipment, which will be considered first, may be divided into three broad classifications. First is the hammer, which imparts stress on the material by impact. Second is the forging machine or upsetter. This unit delivers its stress to the material by pressure or squeeze; it operates in a horizontal position. The third category of forge equipment is the forging press, which operates in a vertical position like the hammer but imparts pressure to the workpiece instead of impact. These three pieces of equipment constitute the main forging units; they do not cover the various pieces of equipment required to bring a particular forging to its final form. These will be discussed under "Auxiliary Forging and Finishing Machinery".

Hammers. The basic unit of the forging shop is the drop hammer (*Figure 8-75*), the function of which is to form metal heated to a plastic state into a desired shape by the blows of the falling weight of the ram (a). There are three principal designs of drop hammers: board, air-lift-gravity-drop, and steam piston. The board-drop hammer is used for illustration in this discussion.

Forging dies into which impressions have been cut are keyed into the movable ram (a) and the stationary anvil cap (b). The dies must be precisely aligned, or the forgings produced will not be in match; i.e., the impressions in the upper and lower dies will not be precisely opposed and will produce an off-center forging. Ways in the columns and in the sides of the ram assure that with each blow, the dies strike in the same place.

The operation of securing the dies in the hammer is known as the setup. Setting up dies and tools in a hammer and its auxiliary equipment may take from several hours to a day, depending upon the complexity of the forging to be made and the size of the hammer to be used.

When the dies have been set up, they are preheated to prevent breaking under the impact of forging. A steel bar or billet which has been cut to the proper length is brought to the forging temperature, usually between 2100° F (1149°C) and 2400° F (1316°C), and placed over the first impression in the die. The hammerman then steps on the treadle (c) which allows the ram to drop. After the blow has been struck, the rolls (d) engage the boards (e) keyed to the ram and raise it for the next blow.

A drop hammer is classified by the weight of the ram; a hammer having a ram weighing 2500 pounds is known as a 2500-pound hammer. A well-equipped gravity-drop hammer shop has hammers ranging in size from 1000 to 5000 lbs. (540 to 2270 kg). A set of dies for producing connecting rods is shown in place in the hammer illustration.

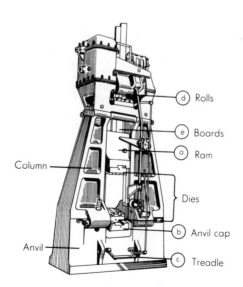

Figure 8-75. Board-type drop hammer.

Forging Machines or Upsetters. Forgings that require a symmetrical gathering of stock on the end of a shaft are usually forged in an upsetter (*Figure 8-76*). This machine is fitted with two gripping dies which come together horizontally to hold the forging bar in a rigid position, while the third die, called the plunger, upsets the stock on the end of the bar into an impression. This equipment is occasionally used to supplement hammer forging but generally it is used as a finished forging unit.

Forging Presses. These forging units are capable of varied forging operations which can be accomplished by means of exerting high pressures upon a heated material. Their greatest capability lies in the type of forging that is symmetrical in design. The construction compares closely with that of the crank-type presses used in stamping and drawing. In the very high tonnage ratings of over 2000 tons (18 MN), forging presses are generally of the hydraulic type, whereby a high-pressure fluid acts on a piston and transmits power to the dies.

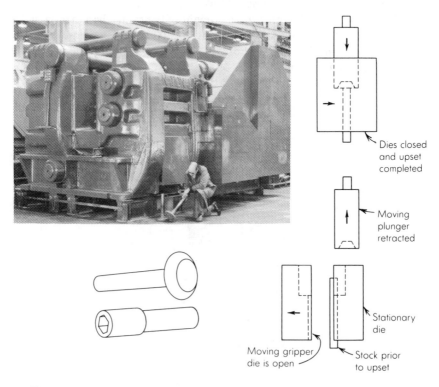

Figure 8-76. Upsetter in operation. Two typical upset parts are shown, also a schematic of steps in the operation. (*Courtesy, National Machinery*)

Auxiliary Forging and Finishing Machinery

The machinery discussed in the preceding section is generally used to move and shape the metal into a die cavity which contains an impression of the part to be forged, but since the initial volume of stock is seldom distributed evenly throughout the cavity, it is necessary to perform other operations to obtain a finished piece. For a clear understanding of these auxiliary units, a brief explanation of the equipment will be given, followed by an example of the forging produced as well as the basic design of the dies.

General Utility Hammers. These single frame hammers employ steam or air as the source of pressure. They are available in sizes ranging from 100 to 500 lb. (45 to 225 kg).

Utility hammers, as the name implies, can handle a variety of jobs, from tool dressing and general light blacksmithing to emergency forging or maintenance work.

Stock is worked by a series of sharp, rapid blows between a ram and an anvil.

A general utility hammer is shown in *Figure 8-77*. *Figure 8-78* depicts forging practice from rough stock to finished part.

Presses. Presses are used for trimming, punching, bending, and straightening operations which can often be performed at the forging heat.

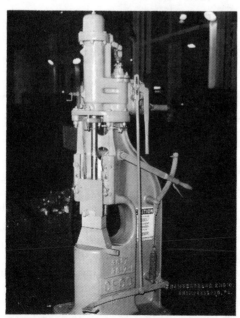

Figure 8-77. The general utility hammer
(*Courtesy, Chambersburg Engineering Co.*)

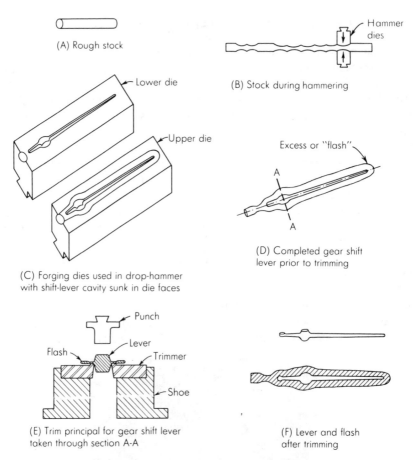

Figure 8-78. Forging practice from rough stock to finished part.

The presses are usually the crank type (*Figure 8-79*). As can be seen from the illustration of the helve-hammered stock, a material saving in stock has been accomplished by drawing a short bar out to a rough shape which approximates the die cavity. In addition to the saving of a considerable amount of material, a further benefit is gained in finishing the forging since fewer blows are required in the hammer. This is a result of throwing out less flash around the die cavity. Since the stock cannot be proportioned to fill the die cavity exactly, the excess flows out between the die faces and results in a thin flash. The flash cools very rapidly, and if it is not kept to a minimum, excessive pounding will be necessary to complete the forging. The result therefore is low production and decreased die life. Thus, in this forging operation as well as most others, the die designer must consider various operations to get the stock proportioned as close as possible to the ultimate shape of the piece before placing it into the final die cavity. Because stock cannot be provided with an exact volume to fill the die cavity, it is generally understood that the die cavity will throw a flash outside of its contour. Therefore, the majority of forgings must be trimmed by a press of the proper capacity. Material at the forging temperature of 2100-2400° F (1149-1316° C) will require a smaller press than if the same part were trimmed cold. Conservatively speaking, it can have a capacity one-third that of a cold trimming press. Thus, if the periphery of a mild steel part was 20″ (508 mm) and the flash was 1/8″ (3.18 mm) thick, a force of approximately 75 tons (667.2 kN) would be required to trim the forging.

$$F = \frac{PTS}{2000} \tag{8-18}$$

Where:

F = force requirement in tons
P = periphery (inches) of forging
T = thickness of flash
S = shear strength of material in psi.

Low carbon steel has a shear strength of 60,000 psi.

$$F = \frac{PTS}{2000} = \frac{20 \times 0.125 \times 60,000}{2000} = \frac{150,000}{2000} = 75 \text{ tons (667.2 kN)}$$

If the trimming is done at forging temperature, the shear strength S should be divided by three and the same formula applied.

$$F = \frac{PTS}{2000} = \frac{20 \times 0.125 \times 20,000}{2000} = \frac{50,000}{2000} = 25 \text{ tons (222.4 kN)}$$

In addition to the trimming operations, the crank-type press is used for a variety of the other operations as previously mentioned. For example, the forged connecting rod in *Figure 8-74* will have a web of material in the hole after forging and, since it is desirable from a finish machining standpoint to have this web removed, it must be punched out. *Figure 8-80* illustrates a punching die.

Often a forging is of such an intricate nature that in normal processing by the hammer and through the trimming and piercing operation, the forging will become distorted. This can be rectified by mounting a set of restriking dies in the press and sizing the forging all over. A set of restrike dies is shown in *Figure 8-81*. Often presses are used for bending operations such as illustrated in *Figure 8-82*.

Coining Presses. It is possible, by using a heavy duty coin press (*Figure 8-83*), to obtain very close tolerances on the thickness dimensions of

Figure 8-79. Heavy-duty forging press.
(*Courtesy, National Machinery*)

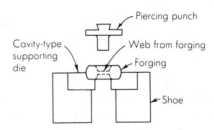

Figure 8-80. Punching die.

some forgings. A tolerance of ±0.005″ (±0.13 mm) is not uncommon, and some dimensions can be held as close as ±0.002″ (±0.05 mm). Coining dies (*Figure 8-84*) are designed to allow a flow of metal so that the pressure applied will force any excess metal on the surfaces being coined to move to adjacent surfaces on which no pressure is exerted. The dimensions which are coined, therefore, must be those between opposed surfaces, and there must be adjacent surfaces to which the excess metal may move. Coining has effected savings to many forging users by replacing expensive machining.

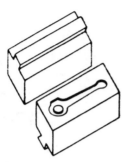

Figure 8-81. Restrike die.

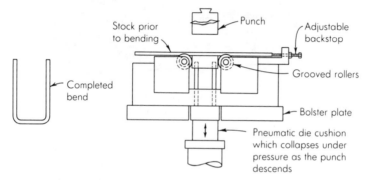

Figure 8-82. Bending die.

Forging Rolls. A forging roll (*Figure 8-85*) is used to draw stock from a large diameter down to a straight or tapered rod of smaller diameter and greater length. Its function is similar to that of the helve hammer; it is also used for finishing operations after a portion of the forging has been processed in a hammer. The rolls are semicylindrical to permit clearance for the preforged section. An excellent example of the work that can be accomplished on this machine is the tapered shaft of a gear shift lever. The ball end of this lever is forged in a hammer, and the shaft is later drawn out to its finished dimensions on a forging roll.

The Forging Process

The elasticity of the metal or its size usually governs whether or not the operation of forging is to be accomplished in the hot or cold state. A good example of cold forging is the common nail, in which a small rod is gripped between moving dies and a small amount of the stock is upset on the end,

Figure 8-83. Screwpress (900-ton) for coining. (*Courtesy, National Machinery*)

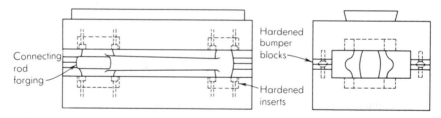

Figure 8-84. Set of cold coining dies with hardened inserts at points of greatest wear.

forming a head. Most often, metals are forged hot because the material will shape more readily in this condition and the necessary fiberlike flow line structure is imparted. In the grain flow lies the inherent strength and toughness of forgings; the die designer must make the dies in such manner as to take full advantage of this highly desirable quality. A cast ingot at the steel mill is rolled down by successive passes to ultimately produce bars and billets with a fibrous grain which runs the length of the bar. These high-quality bars are then cropped to the proper length to make a particular forging. In *Figure 8-86*, a comparison is made between three methods of producing a crank. In addition to the proper placement of grain flow pattern, the die designer must also take into account all functional surfaces of the part to help eliminate

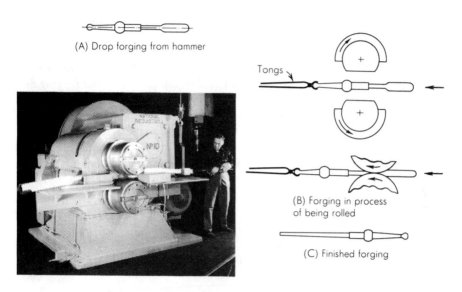

(A) Drop forging from hammer

Tongs

(B) Forging in process of being rolled

(C) Finished forging

Figure 8-85. Forging roll in operation. (*Courtesy, National Machinery*)

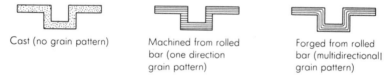

Cast (no grain pattern)

Machined from rolled bar (one direction grain pattern)

Forged from rolled bar (multidirectional grain pattern)

Figure 8-86. Three methods of producing a crank.

costly machining operations wherever possible. In some cases streamlining is necessary for appearance.

Above all else, overall economy must be considered in die designing for the forged part. In a costly aircraft or missile forging, quality is often of the utmost importance. Dies must be designed to impart the maximum grain flow benefit to the finish-machined part. On a hook to be used in overhead lifting, where human lives and expensive fabricated components are at stake, the ultimate in grain pattern should be developed in the dies. A small lever with a high factor of safety in its design can occasionally be produced with no regard for grain flow but a high regard for its final cost. *Figure 8-87* illustrates the grain flow patterns which are developed on various forgings made by different forge die designs.

A tensile specimen pulled to destruction will exhibit a higher ductility when the grain pattern runs parallel to the pulling force than a specimen of equal dimensions and hardness having the grain pattern transverse to the pulling force. The tensile strength is affected to a lesser extent on a straight no-shock load than when impact loads occur. In this high-shock area of application, grain flow plays an important part in lengthening the service life of the part. Such established data make it highly desirable for the forging die designer to work closely with the engineers who have designed a component. The forgings shown in *Figure 8-87* will all look the same when completed, but the grain pattern may be detrimental to the strength of the part. The hook and the bell crank can be produced more economically without bending, and the gear is produced more economically by not flattening the stock on its end prior to forging. However, if the part is to be highly stressed in the areas of transverse grain structures, it would be better to perform the bending or flattening operations. The illustrations of *Figure 8-88* depict the poorer grain patterns.

Another prime consideration in the design of tools is economy in manufacture. In this area the tool designer has two things to consider: The effect of the tool design on the part price, and the cost of the tools. Some very fine tools can be constructed to accomplish an operation in a minimum of time. One use's requirements may be high enough to justify the expense of

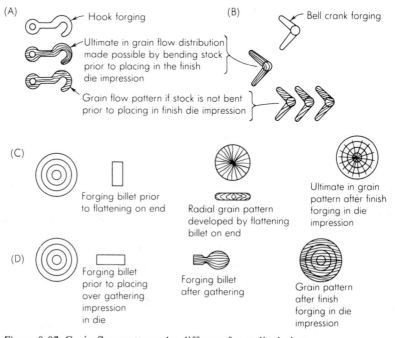

Figure 8-87. Grain flow patterns by different forge die designs.

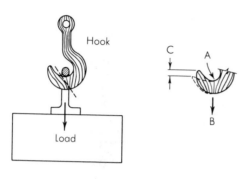

Load at B tends to deflect
unsupported point as shown
by distance C. This cannot
take place unless material
at A elongates. This
elongation is a tensile or
pulling type stress and
as can be seen, the grain
fibers are transverse to
this tensile stress, thus
reducing the ultimate
load that the point
can withstand.

Grain pattern on 1st and 2nd teeth
will withstand less stress
than the 3rd and 4th teeth.

Figure 8-88. Examples of poor grain patterns.

these tools, whereas another may be interested in obtaining a low tooling cost with not much regard to the part price. If customer *A* needs 10,000 pieces and the tools will cost $3000 and the pieces produced from these tools 30 cents each, the total cost will be $6000 or 60 cents per part. Customer *B* needs 1000 pieces of an almost identical part; if produced from the same tool design they will cost $3.30 a piece. Therefore, it would be to customer *B*'s advantage to get tooling at $1000 and parts from these tools at 75 cents each. Thus the cost per part would be only $1.75. Conversely, if customer *A*'s forgings were produced from the $1000 tools the cost per part would be 85 cents. This can be depicted graphically as shown in *Figure 8-89*.

The flattening out of the curve along the axes shows the tool designer two basic facts: with little or no tooling the piece price rises, and with too extensive a tool design no benefit on piece price is gained. The designer should acquaint himself with all available equipment and processes and in all cases evaluate his tool designs with overall economy in mind. Occasionally, the nature of the part is such that a saving can be made by upsetting a large section prior to finish impression forging rather than endeavoring to roll or draw a smaller section down from a large bar or billet. Possibly a hole could be drilled more economically than a set of piercing dies could be constructed for the quantity

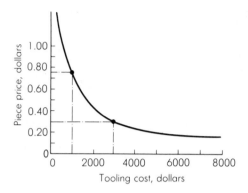

Figure 8-89. Relation of piece price to tooling cost.

involved. Coining or sizing a part by exerting extreme pressures on certain contours may eliminate a costly machining operation.

Forging Design

The tools necessary to produce a given forging cannot be made until the shape of the final forging has been determined. Therefore, it is essential that the tool designer has an understanding of the underlying principles of forging design. These will be considered in the following order: forging draft, parting planes, fillets and corner radii, shrinkage and die wear, mismatch of dies, tolerances, and finish allowances.

Forging Draft

Draft is the angle or taper that must be imparted to the surfaces of the forging that are contained in the die and are not self-releasing. For maximum production it is necessary to free the forging from the die cavity quickly. Forgings that are made in hammers and presses generally require a draft angle of 3° to 7° for external surfaces and 5° to 10° on internal surfaces. The application of draft can best be seen from the example shown in *Figure 8-90*. As the dies part, owing to the action of the hammer or press, the forging will stick in whichever die offers the maximum frictional resistance to release of the part contained within its cavity. Draft as provided to all *B* surfaces permits the forging to release itself from either die. The *A* surfaces are self-releasing and require no draft addition. In the case of upsetter work (*Figure 8-76*) the application of draft can be materially reduced since the stock is firmly held by the gripping dies as the plunger retracts. Draft on the head of an upset forging

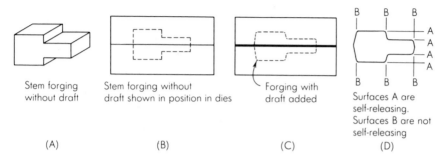

Stem forging without draft

Stem forging without draft shown in position in dies

Forging with draft added

Surfaces A are self-releasing. Surfaces B are not self-releasing

(A)   (B)   (C)   (D)

Figure 8-90. Application of forging draft.

can be reduced in many cases to as little as $1/2°$ (*Figure 8-91*). In addition to flat surfaces, a cylindrical, self-releasing surface is often encountered (*Figure 8-92*). As can be seen, the arc of the circle forms its own draft and will not present a sticking problem during forging. Generally, the draft on external surfaces can be less for internal surfaces. The reason is that as the metal is being forged, it is undergoing a shrinkage due to cooling, and all surfaces, internal as well as external, are shrinking towards the center of the part. Thus, when a hole or other depression is being forged into a part, the forging tends to grip the plug in the die which forms the depression. On external surfaces, the reverse is true (*Figure 8-93*).

Parting Lines

A parting line is the dividing plane between the two halves of a pair of forging dies. This line may be perfectly flat or multidirectional. A flat parting plane is more economical to produce, but very often the part which results from using a flat parting plane is not an economical forging. A straight parting line is shown in *Figure 8-90C* on the stem forging. However, another type of forging may not lend itself to this type of parting. *Figure 8-94* indicates an extreme amount of stock addition (shaded at *a*) which would result if a straight parting line were used. This would result in additional weight and an expensive machining operation to remove this excess; the only benefit that

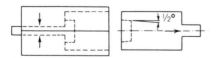

Figure 8-91. Draft on head of an upset forging.

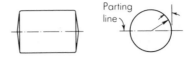

Figure 8-92. Draft formed by arc of
a circle.

Figure 8-93. Surfaces *a*
are pulling away from the
die, whereas surfaces *b*
are pulling toward the
die. Greater draft angle is
required on *b* surfaces
than on *a* surfaces.

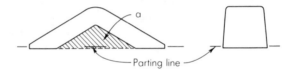

Figure 8-94. Stock addition resulting from a straight
parting line.

would be gained would be a slight reduction in the time necessary to machine
the die. A lock-type parting line would be preferable and would produce a
part which would have the desired shape. A lock-type parting line for the
same piece is seen in *Figure 8-95*. Other types of lock-type parting lines are
shown in *Figure 8-96*. In *Figure 8-96 A* the crank arm shown could have been
parted through the center line *x-x*, but then draft would have to be added to
faces *a* of the two large bosses. This would complicate the drilling operation
which must subsequently be performed because the drill would enter a slanted
surface and would tend to bend in the direction of deflection (*Figure 8-97*).
The boss is to enter a 0.813″ (20.64 mm) wide clevis slot in final assembly and
the remaining draft after drilling would make this impossible. A further
consideration in this parting line design was economy in forging. Had the
forging been parted through center line *x-x*, a separate bending operation
would have to be performed on the stock before placing it into the die cavity
as shown in *Figure 8-98*. In *Figure 8-96 B* a gear blank is shown with a straight
parting line at *x-x* and also an elevated parting line at *y-y*. It was necessary to

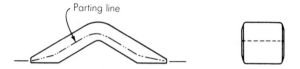

Figure 8-95. Use of lock-type parting line for the piece shown in Figure 4-87.

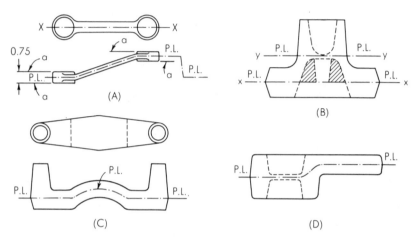

Figure 8-96. Various types of lock-type parting lines.

raise the parting plane to y-y in the hole to reduce the amount of stock to be removed in machining. The stock saving is indicated by the shaded areas. The plug in the top die which must form the hole is considerably stronger than if it were extended down to parting plane x-x. From the considerations for the designs of Figure 8-94, 8-95, 8-96A and 8-96B, the apprentice tool designer should readily understand the location of the parting planes shown in Figure 8-96C and 8-96D.

Fillets and Corner Radii

A fillet can mean any rounding of the apex of an internal angle and a corner radius the rounding of the apex of an external angle. Fillets are employed in practically all forging work because they not only increase the lift of the die but also result in sounder forgings. As can be seen in Figure 8-99 the material will progressively flow into the die cavity. The fillet radius has allowed the material to start its downward flow without nicking. If too small a fillet is used or if none is provided, the material would flow as seen in Figure

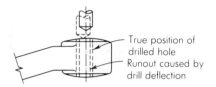

Figure 8-97. Drilling a hole in a casting.

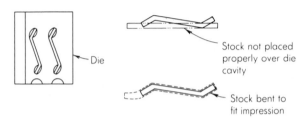

Figure 8-98. Bending operation required because of faulty parting line.

8-99B. The material has filled the cavity but has folded back against itself forming a defect. This condition is called a lap and is generally caused by using inadequate filleting in the die.

Corners must be rounded on forgings because the material flowing into the die cavity will trap air and oil at these points, making them very difficult to fill. Scale will adhere in the sharp corner of a die to a greater extent than when it is rounded and will halt the progressive flow of material, resulting in an unfilled cavity. *Figure 8-100* can be used as an approximate guide in establishing the necessary corner or fillet radii.

Shrinkage and Die Wear

Most steel forgings are formed at temperatures from 2100° F to 2400° F (1149° C to 1316° C). In this state the material is greatly expanded, and during the forging operation, is in the process of cooling and consequently shrinking. The die impressions are put in using a shrink scale which amounts to 3/16″ per foot when the material to be forged is steel. It is difficult to control both control both heating and cooling of a material within close limits during forging. Consequently it is difficult to have the forging leave the die impression at exactly the temperature which is equal to 3/16″ per foot shrinkage. Shrinkage variances usually amount to ±0.001 to 0.003″ per inch (±0.001 to 0.003 mm/mm) of width or length on a given forging. Die wear is that quantity of dimensional difference which comes about through normal abrasion of the impression.

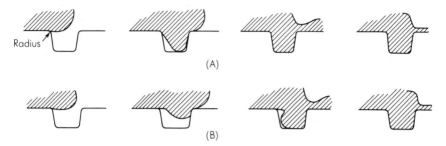

(A)

(B)

Figure 8-99. Progressive flow of metal into a forging die.

This usually amounts to between 1/64″ (0.4 mm) and 1/32″ (0.8 mm) on external surfaces of forgings weighing up to 10 lb. (4.5 kg) and a like amount on internal surfaces such as plugs. On forgings between 20 and 50 lb. (9.1 kg and 22.7 kg) this may increase from 1/16″ to 3/32″ (1.6 mm to 2.4 mm).

Mismatch

When a forging is parted in such a manner as to include a portion of the part in the top die and a portion in the bottom die, it is difficult to maintain alignment of the half impressions because either the upper or the lower die may shift during forging. This shift may occur sideways or endways as illustrated in *Figure 8-101*. Forgings produced from a shifted die will be mismatched. This shift is generally kept to a minimum as the forging die designer will add a finish allowance if the mismatch is detrimental to the workability of the part. On forgings up to 10 lb. (4.5 kg) the dies may shift 0.015″ (0.38 mm) during a production run, and forgings from 10 to 20 lbs. (4.5—9.1 kg) may shift 0.020″ (0.51 mm). A 0.005″ (0.13 mm) shift allowance should be added for each additional 10 lb. (4.5 kg).

Tolerances

Thickness tolerances apply to the forging dimensions which cross the parting plane of the dies. They are applied both as a plus and as a minus value. If an excessive flash is thrown between the die faces and becomes too cold, the flash will tend to keep the dies from fully closing. To prevent this, the flash thickness in the die is kept as thick as possible.

Commercial plus tolerances on thickness dimensions are approximately + 1/32″ (0.8 mm) on forgings up to 1 lb. (0.45 kg), + 1/16″ (1.6 mm) on forgings from 5 to 20 lb. (2.3 to 9.1 kg), and + 3/32″ (2.4 mm) from 20 to 50 lb. (9.1 to 22.7 kg).

Net wt. of forging, lb (kg)	Corners, in. (mm)	Fillets, in. (mm)
1 to 5 (0.45-2.27)	1/16 (1.6)	1/8 (3.2)
5 to 10 (2.27-4.54)	3/32 (2.4)	3/16 (4.8)
10 to 30 (4.54-13.61)	1/8 (3.2)	1/4 (6.4)
30 to 50 (13.61-22.68)	5/32 (4.0)	5/16 (7.9)

Figure 8-100. Corner radius design.

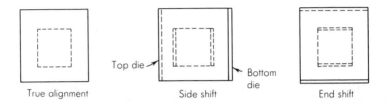

Figure 8-101. Shifting of dies.

Commercial minus tolerances on thickness dimensions are -0.010" (-0.25 mm) up to 1 lb. (0.45 kg), to -0.023" (-0.6 mm) from 5 to 20 lb. (2.3 to 9.1 kg), and -0.031" (-0.78 mm) from 20 to 50 lb. (9.1 to 22.7 kg).

The forging die designer can control the plus variations by providing ample flash thickness in the die, establishing a trimming operation to remove the majority of flash before finish forging, or by designing a set of coin dies to force the excess material down to size. The minus variations are ordinarily caused by the faces of the die pounding down and thus causing an undersize condition when the dies are closed. This can be corrected by providing ample striking surface on the faces of the dies. Assuring a smooth surface finish to the die face is another control that can be applied to reduce face pound-down. Scale falling into the die cavity can also cause an undersize condition, but the die designer has little control over this factor. It is customarily controlled by the inspection department during forging.

Finish Allowances

A finish allowance is an additional amount of material added to surfaces which cannot be controlled close enough by forging and must be subsequently machined. The finish allowance for a given forging is established by considering the effects of shrinkage, straightness, mismatch, the minus tolerance on thickness, and surface decarburization. Decarburization is the loss of carbon from the outer surfaces of a forging caused by heating. On some grades of steel, the loss of carbon may be very high; this should be checked with the steel supplier. If a heat treatment is to be given a forging, the outer layers will not respond to the treatment because the carbon content has been lowered. Therefore, to get full hardness even at the outer layers, it is essential to add sufficient stock to assure that virgin material is being subjected to the heat treatment. An example of the application of finish would be as seen in *Figure 8-102* where surfaces *A* and *B* require a hardened and ground finish.

Figure 8-102 shows that the forging designer has applied 1/16″ (1.6 mm) finish allowance all over specifying the forged stem (A) diameter 1.000″ (25.4 mm) nominal or 0.125″ (3.18 mm) total finish.

less 0.020″ (0.51 mm) scale pitting

less 0.020″ (0.51 mm) decarburization

less 0.014″ (0.36 mm) minus thickness tolerance

less $\dfrac{0.003″}{0.057″}$ $\dfrac{(0.08 \text{ mm})}{(1.45 \text{ mm})}$ shrinkage

The nominal diameter of 1.000″ (25.4 mm) less 0.057″ (1.45 mm) equals 0.943″ (23.95 mm) diameter or 0.034″ (0.86 mm) per side. This remaining 0.034″ (0.86 mm) has been added to guarantee a full clean-up in the event the stem forging is not completely straight as located for turning, and also to allow approximately 0.010″ (0.25 mm) for grinding after turning. The same approach was used in determining the 1/16″ (1.6 mm) allowance for surface *B*, except that thickness tolerance is not considered but end shift (mismatch) is a possibility.

Drop Forging Dies and Auxiliary Tools

Forging dies are made from high-carbon, nickel-chromolybdenum alloy steel blocks which have been forged to achieve the utmost grain refinement and resistance to shock. In addition to the qualities developed by

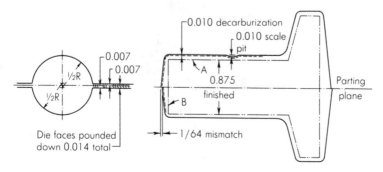

Figure 8-102. Application of finish allowance.

the hot working, a maximum resistance to wear is insured by a double heat treatment.

Before the forging impressions are sunk, shanks are shaped for locking the dies in the hammer, and the blocks are squared for perfect alignment, (*Figure 8-103*). The striking faces are planed, or if the forging to be produced necessitates a lock, that is, if the striking surfaces must be in two or more planes, the faces are shaped accordingly.

A blueprint or model of the part to be forged is translated into metal templates which, in turn, are used for laying out the blocks. Six types of impressions may be incorporated in a die for shaping the material progressively from bar form to the finished forging (*Figure 8-104* and *8-105*). Impression *a*, usually referred to as the swager, is used to reduce and draw out stock when differences in cross sections of the forging make this necessary. This operation is similar to that performed on the helve hammer (*Figure 8-77*). The swager is ordinarily used for forgings having only a short section requiring reduction of area, whereas the helve hammer is used for forgings for which most of the bar length must be reduced. Impression *b*, commonly called the edger or roller, distributes the stock so it will fill the next impression without excessive waste.

A bender, impression *c*, is included when curves or angles in the forging make it necessary to bend the stock before it will fit properly in the finishing impressions.

Impression *d*, known as the blocker, gives the forging its general shape and allows the proper gradual flow of metal necessary to prevent laps and cold shuts. Although this impression has the same contour as the finished part,

Figure 8-103. Preparation of die blocks.

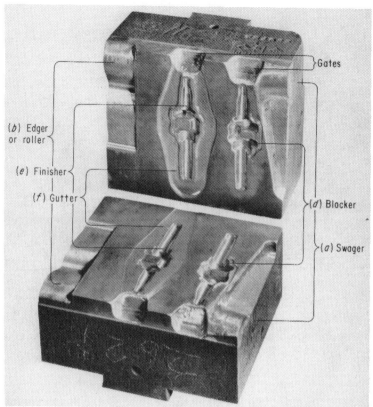

Figure 8-104. Finished die with bender, blocker, and finisher impressions, and cutoff.

large fillets and radii are added to permit the easiest flow of metal. The finisher impression, *e*, brings the forging to its final size. A gutter, *f*, is cut into the block around this impression to provide space for the excess metal, or flash, which is forced out of the finishing impression when the forging is hammered to size.

The cutoff, *g*, cuts the forging from the bar by shearing off the tong hold. It is principally used for small forgings on which no further processing is to be performed at the forging heat. It permits rapid and economical production.

The blocking and finishing impressions are cut in the block by highly skilled operators who use a milling machine especially designed for sinking dies. Cutters of various types are used in accordance with the shape of each section of the impression, but much of the accuracy of the die depends upon hand work performed after it is sunk.

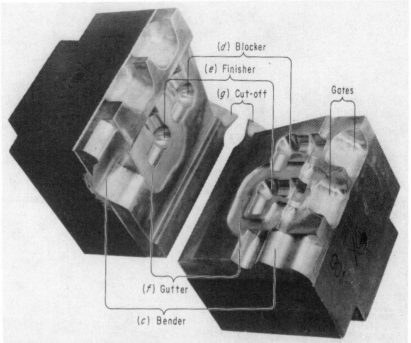

Figure 8-105. Finished die with roller, swage, blocker, and finisher impressions.

When the forging die impressions are completed, the blocks are clamped together in the position in which they will meet in the forging operation, and a lead-antimony alloy is poured into the finishing impression. The resulting lead cast is used to check the accuracy of the forging dimensions, and is sent to the customer for approval.

Since steel shrinks in cooling from its forging temperature, and the lead alloy does not, it is necessary to allow for this shrinkage in checking the lead cast. The correction amounts to 3/16″ per foot (15.6 mm/m), or 1/64″ per inch (0.016 mm/mm).

After the lead cast has been approved, the dies are finished by machining the gutter, flash relief, and roller, and the cast is used as a model from which auxiliary tools can be made accurately. These tools include trimming, punching, and restrike dies.

Figure 8-106 shows the forging dies and working steps involved in producing a large connecting rod. The essential steps encountered in the production of this forging will explain the method for producing a wide variety of forgings.

Before the impressions were sunk into the die faces, it was determined that the forging would best resist the stresses imposed upon it if the grain flow ran parallel with the length. It was, therefore, decided to place the impressions in the die with the length running from front to back, and the bar was to be placed in like manner.

The next consideration was that of establishing a basic parting plane which in this case lends itself to the preferred straight-line type of parting. Subsequent machining can be materially reduced by including plugs in the upper and lower dies to form that crankpin, piston pin holes, and the depressions which form the channel section. It is, therefore, proper to place the parting plane along the thickness of the connecting rod rather than the width.

After these factors have been established, critical surfaces that require a high degree of accuracy to be usable must have finish allowances added. In this case a 3/32" (2.4 mm) allowance was added to the top and bottom faces of the crank and piston ends. The hole in the crankpin end was elongated 3/16" (0.8 mm) to provide material for a slitting saw to part the rod from the cap portion and still end up with a round hole for the crankshaft bearing. An allowance of 3/32" (2.4 mm) was added to the sides of the two holes as well as to the ends of the bolt bosses where a good seat must be established for the heads of the bolt and nut which assemble the cap to the rod portion.

Draft must now be added to allow surfaces in the die so that the forging will be self-releasing. All external surfaces around the periphery have a 7° angle and the plugs which form the holes have a 10° angle. Before the preforming operations of rolling and swaging can be laid out on the die face, it is necessary to know the size of the original bar. From the forging design with finish, draft, and approximate flashing, the largest cross section of the rod perpendicular to the direction of stock placement is first determined. In this case it is the plane *A-A* in *Figure 8-106* at 5. This section is shown in *Figure 8-107A*. By approximation of areas, length times width, the area of stock to fill this section would be computed as follows (for U.S. customary units):

left flashing: 1 1/2" x 3/16" = 0.282 in²

right flashing: 1 1/2" x 3/16" = 0.282 in²

plug flashing: 2" x 3/16" = 0.375 in²

crankpin section *L*: 1" x 2" = 2.000 in²

crankpin section *R*: 1" x 2" = 2.000 in²

4.939 in² total

This area is slightly less than that of a 2 1/2" diameter bar. It is apparent that the 2 1/2" bar could not be laid across the length of the forging and

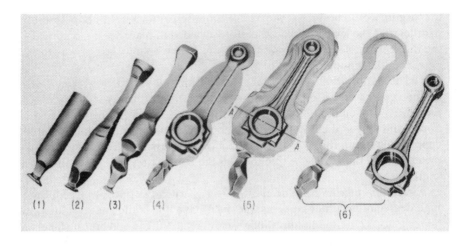

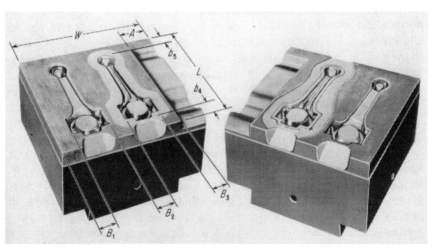

Figure 8-106. (*A*) Working steps in producing a connecting rod forging: (1) original bar; (2) drawn in a helve hammer; (3) result of rolling; (4) rough form after blocking; (5) finished forging as it leaves the hammer; (6) the flash and the trimmed forging. (*B*) Forging dies.

hammered until the cavity was filled, as the volume of material required on the balance of the rod is considerably less. The I-beam section may take as little as 1 1/8″ round at the small end when calculated as above. The rod forging will, therefore, require extensive reduction from a large bar diameter to a smaller bar diameter at various sections along its length. A swager such as is seen in *Figure 8-106* or a helve-hammer operation as seen in *Figure 8-77* can

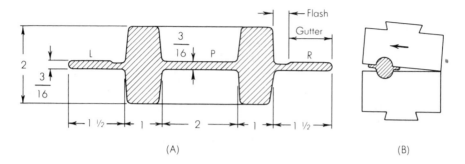

Figure 8-107. (*A*) Cross section of finished forged connecting rod; (*B*) shift in halves of a mating forging.

make this reduction. On this particular forging it was decided to use the helve hammer to reduce the steel, since a very small shipment of forgings was needed, and to add a swager to the die would increase the die width approximately 3″. This would necessitate ordering larger, costlier blocks and would add to the block preparation cost by increasing the time to face the blocks and also the time to sink the swaging impression in the face. A set of standard helve-hammer dies was used which did not allow the stock to be reduced beyond the 1 1/8″ round minimum requirement. After the swaging operation has been eliminated from the sequence of operations that must be performed on the forging die blocks, it is now possible to decide on the size of blocks required.

In addition to the widths and lengths of a forging, empirical data relating to minimum striking surface are necessary to establish the die block size. The first consideration is the width required. As shown in the inset of *Figure 8-106*, the roller is placed on the right side of the die. The width, A, required is approximately equivalent to the required steel size plus 3/4″ on sizes up to 2″ round, steel size plus 1″ on sizes from 2″ to 3″ round, steel size plus 1 1/4″ on sizes from 3″ to 4″ round, and steel size plus 1 1/2″ for diameters from 4″ to 5″. The next considerations are the distances b, b_1, b_2, etc. required between or around impressions. The area covered by the b allowances serves a dual purpose, the foremost being striking surface and the other being a provision for the excess metal or flashing. If too little striking surface is provided between impressions, rapid upsetting of the die face will result, causing an undersize condition on the forging thickness. The striking surface between cavities is reduced considerably by the flashing and guttering of the dies.

Flashing is that portion of excess metal adjoining the forging at the parting line. In die designer's terminology, it can also be considered the portion of the die which forms the excess. The amount of flashing that must be provided varies from one type of forging to another. However, on steel sizes to 1 1/2″ diameter, the dies can be flashed approximately 3/16″ wide around the periphery of the forging and about 1/32″ thick. On sizes between 1 1/2″ diameter and 2″ diameter, a 7/32″ wide by 3/64″ thick flash will usually suffice. Sizes from 2″ diam to 2 1/2″ diam: 1/4″ by 1/16″; 2 1/2″ to 3″ diam: 5/16″ by 3/32″; 3″ to 4″ diam: 3/8″ by 1/8″. In addition to the flash extension, a further provision must be made in the die for any excess material. This is called the gutter and may be incorporated in the die by use of *Figure 108.*

From the data just given it is now possible to determine the width of block required in six steps as follows:

1. Roller width = 2 1/2″ + 1″ = 3 1/2″
2. Blocker width = widest part of forging = 4″
3. Finisher width = widest part of forging = 4″
4. (b_3) flash + gutter = 1/2″ = 5/16″ + 1 1/4″ + 1/2″ = 2 1/16″
5. (b_2) flash + gutter = 1/2″ = 5/16″ + 1 1/4″ + 1/2″ = 2 1/16″
6. (b_1) taken same as b_2 or b_3 = 2 1/16″

Therefore, an 18″ wide block should be used and the impressions laid out on the die face in accordance with the six steps. It should be noted that in steps 4 and 5, an additional 1/2″ must be added to provide ample striking surface of the die faces. A blocking impression was added to this die to distribute the stock more evenly in the varying contours of the finish impression. If a part has very few sectional changes and is nearly symmetrical about its axis, the blocking impression can be eliminated from the die. However, generous filleting between sections is then necessary.

The overall length of this particular connecting rod is 18″. The distances b_4 and b_5 are also taken the same as b_2 and b_3. Thus the length of block will be approximately 22″ (18 + 2 1/16 + 2 1/16). Die block manufacturers generally stock large blocks in 2″ increments; thus it is desirable to keep the finish impression as close to the center of the block as possible so that the least

Steel size, in. (mm)	Depth of gutter, in. (mm)	Width of gutter, in. (mm)
1/2-1 1/2 (12.7-38.1)	1/8 (3.2)	1 (25.4)
1 1/2-2 (38.1-50.8)	3/16 (4.8)	1-1 1/4 (25.4-31.8)
2-2 1/2 (50.8-63.5)	3/16 (4.8)	1 1/4-1 1/2 (31.8-38.1)
2 1/2-3 (63.5-76.2)	3/16 (4.8)	1 1/4-1 1/2 (31.8-38.1)
3-4 (76.2-101.6)	1/4 (6.4)	1 1/2-1 3/4 (38.1-44.4)

Figure 8-108. Gutter design sizes.

amount of shifting can take place as the dies come together. If it is placed on the outer edge of the die, there is a tendency to pull the dies in that direction and thus cause a shift in the mating halves of the forging (*Figure 107B*). After the finish impression is sunk into the die face, the blocking impression is sunk. This impression is very similar to the finish impression, but differs in many respects. In the case of the connecting rod and similar forgings, the I-beam section is left solid but made with sufficient volume to fill the I-section when the stock is transferred to the finisher. This is done to distribute the stock to the proper areas and also to provide less resistance to the flow of material to deep sections.

The blocking impressions are seldom flashed or guttered, whereas the finish impression is usually flashed in both dies and gutters only in the top die. By flashing both dies, the final forging has a neater, more symmetrical appearance (*Figure 8-109*). By guttering the top die only, the forging will sit flush in the trimming die which must be used to trim the excess flash.

The roller can next be sunk into the die. The purpose of the roller is to gather steel prior to insertion in the blocking impression and also to break off scale which is formed on the bar during heating. Scale is detrimental to die life and should be kept out of the die impressions by all means possible.

As the steel comes from the helve hammer, it will be reduced in the center section only and will not be worked at the large ends of the bar. The roller is designed to make possible the use of a smaller bar than the 2 1/2″ previously figured. This is accomplished by filleting in from both sides of the large section, thus increasing the bar size slightly at the point where the most steel is needed. On the small end of the rod, less material is needed and here the roller should be designed to elongate and reduce the steel. The side view of the roller for the connecting rod is seen in *Figure 8-110*. After all impressions are sunk in the die, it is given a final dressing to fillet all sharp corners at the flash line to permit ease of metal flow (*Figure 8-111*). The radius is approximately 3/64″ on 1/16″ thick flashing, 1/16″ on 3/32″ flashing, and 3/32″ on 1/8″ thick flashing.

In addition to forging dies, auxiliary tooling is required to complete the forging. This tooling consists of a hot punching die and hot trimming die

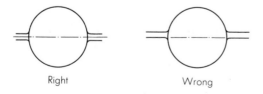

Right Wrong

Figure 8-109. Methods of flashing.

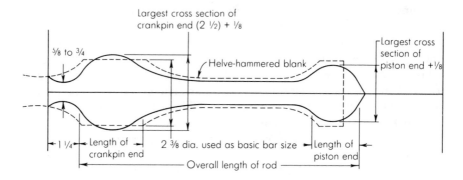

Figure 8-110. Side view of roller for gathering the material in a connecting rod.

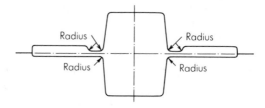

Figure 8-111. Filleting a die.

mounted side by side in the same press. The forging is punched first so that handling time is reduced. If it were trimmed first, the forging would fall through the trim die and would have to be picked up from the base of the shoe and relocated for punching. The small punch holder is merely bolted to the top die. Any rotation about its axis due to capscrew hole clearances would not affect the punched hole since it is a full diameter. However, the large hole punch is bolted and doweled in the top die. Since it is oval, any rotation would affect the concentricity of the hole with the outside wall. The trimming punch is also keyed in the top die to prevent shift in any direction.

The trim punch is made flat where it strikes the bosses, whereas the I-beam section is contoured directly into the face of the punch. This gives full bearing all over as the forging is pushed through the trimmer. Distortion is thus eliminated and the I-beam section will hold its true shape. When pushing a forging through a trimmer, particularly where the operation is performed hot and the metal is still plastic, it is desirable to contour the punch so that no nicks or flat spots occur in the finished forging. Both the trim and the punching die are relieved for the flash. Even though the gutter side is up, handling often results in bent flashing. This must be provided for in order that the forging rest properly in the impression. The built-up edge which remains also provides grinding stock if the trimmer needs re-conditioning. If the forging is trimmed prior to punching, the piercing die need not be relieved. For trim punch design, see *Figures 8-112* and *8-113*.

A hot trim die is made from unhardened medium carbon alloy steel for heavy trimming jobs and mild steel for light duty service. The cutting edge is built up with stellite, a material which maintains very high hardness characteristics at elevated temperatures. Stellite is available in the form of welding rod and is applied to the trimmer by this process. After the bead of stellite is placed around the impression, it is ground to fit the contour of the forging at the parting line. Stellited trimmers can be used for cold trimming if the base material is a medium carbon alloy. Pressures encountered in cold work are considerably higher and thus a substantial base material is needed for the support of the stellite. Cold trimming dies are also made from tough alloy tool-steels which can be hardened to 52-54 Rc and still maintain their cutting edges. Cold trimming punches are made from tough alloy tool-steels and hardened 45-47 Rc. The punch is made softer so that if it comes in contact with the edge of the trimmer, its edge will shear off rather than the trimmer. Hot trim punches are made from medium carbon alloys, hardened 38-40 Rc.

Initial clearance of about 0.015 to 0.020" (0.38 to 0.51 mm) is provided between the punch and the die. The trimming die is relieved on a 5° to 7° angle to give adequate clearance for the forgings to drop through freely. As the draft angles wear, the trimmer is opened up so that a wide trim flat is not left on the forging. However, if it is desired to hold a close trim dimension at the parting line, the trimmer must be watched very closely and, if worn beyond the acceptable tolerance, must be re-stellited and ground to size. The trimming

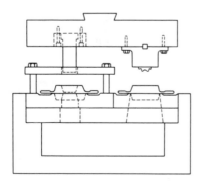

Figure 8-112. Punch-and-trim die.

die for the connecting rod is made in two pieces to facilitate shaping of the stock and to provide a means of adjustment (*Figure 8-114.*) A three-piece trimmer is also illustrated to show this principle.

Heavy mild steel plates, approximately 1 1/2 to 2″ (38 to 51 mm) thick, are used to back up a trim die, and a hole is usually burned out to a size slightly larger than the forging which the plate must accommodate.

Hot piercing dies are made of medium carbon alloy steels (36-49 Rc) and the punches are medium carbon alloy with a stellite facing around the periphery of the punch and on its face. Punching dies must be provided with a stripper plate (*Figure 8-115*) to release the forging from the punch. The stripper posts must be set far enough apart to admit the forging, and if punched prior to trimming, must be wide enough to admit the forging and the flash. Where production warrants the expense of combination dies, the trimming and punching can be combined in one die set (*Figure 8-115*).

The various die components called out are as follows: (1) shank, (2) trimmer punch, (3) trimmer, (4) bottom plate, (5) forging locator, (6) positive knockout, (7) piercing punch, (8) stripper plate, (9) legs (2), (10) knockout arms, (11) brackets (2), (12) knockout guides (2), (13) top shoe.

Figure 8-115 shows the die partially closed with the forging in position and about to be trimmed. As the press ram descends further, the punch will remove the web from the inside of the forging and the slug will drop through. On the return stroke, the stripper plate will strip the flash from the punch and the knockout on the locator will push the forging out of the trimmer. The stripper plate and positive knockout are actuated by the knockout arms. This die principle can be utilized on many types of forgings but, again, the production must warrant the more complex tooling.

When small, thin sections adjoin heavy sections, there are cooling stresses set up which tend to distort the forging. There are also problems of distortion of light sections in normal handling operations. Consequently, a cold restrike

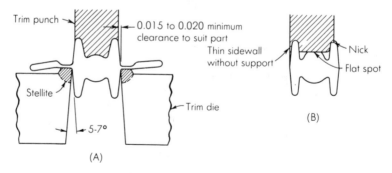

Figure 8-113. Trimmer punch design: (*A*) correct; (*B*) incorrect.

die is required to remove the major deformations encountered after forging and subsequent handling.

Hot restriking dies are made from medium carbon alloy steels and hardened to 36-40 Rc. They have half impressions sunk in the die faces about the same as the finish impression of the forging die, but the impression is not flashed. A set of restrike dies for a connecting rod was shown earlier in *Figure 8-80*. Cold restrike dies are made from a good grade of alloy tool-steel hardened to 50-52 Rc.

Forgings vary in size, shape, and operations. The connecting rod example required forging dies, trimming and punching dies, and restrike dies. Some connecting rods are furnished very close to size on the thickness so that a simple grinding operation is all that is required to finish the sides of the two bosses. Here a coining die is constructed such as the one shown in *Figure 8-83*.

Two or more small forgings may be made progressively from a long bar of steel so long as the forging temperature is adequate to keep the material plastic. In this case, a cutoff is added to the die (*Figure 8-104*). The cutoff consists of two sharpened edges which pinch the steel in half as the dies come together. These cutting edges can be milled into one corner of the die or separately inserted into the side of the die. In the photograph, the cutoff is milled into the die proper. *Figure 8-116* illustrates an inserted type of cutoff.

The bending of stock is illustrated in *Figure 8-98* and a bending impression on the die face in *Figure 8-104*. To incorporate a bender in a die, it

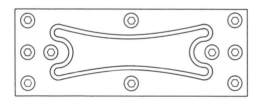

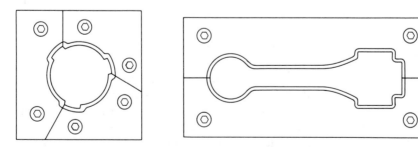

Figure 8-114. Trimming dies made in one, two, or three sections.

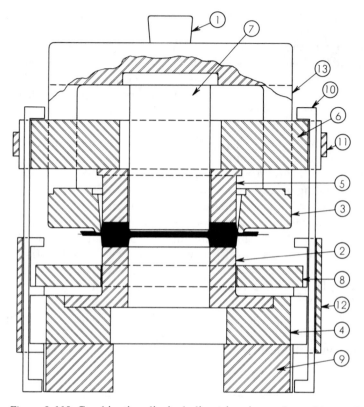

Figure 8-115. Combination die, including trimming and punching tools.

is necessary to leave a horn on one die which will pocket in the other die (*Figure 8-117*). The horn is reduced in size and the pocket increased to allow for the stock which must fit between it.

The forging tool designer is faced with a new problem on practically every die design, and many of the problems cannot be foreseen until after the work is set up in the forging equipment and the dies tried out.

No-Draft Forging Dies

No-draft forgings are used most often on nonferrous alloys. They are usually mounted in a die set and run in a forging press. Depending on the shape of the part, two or more sides of the die cavity are movable to allow the finished forging to be removed. This process uses higher-cost tooling than most conventional forgings and the production rates are much slower. In many cases, the parts are run once with the die not completely closed, excess

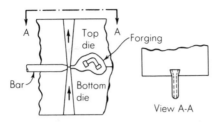

Figure 8-116. After the forging is cut off, the bar end is then started through another forging cycle. This can continue until the bar is no longer at forging temperature.

flash material is removed and then the part is reheated and run a second time with the die closed completely.

Although this is an expensive process, the advantage of secondary machining operations being eliminated due to the absence of the draft angles makes the overall cost of the part considerably less.

Flashless Warm Forming

Flashless warm forming is done on a special machine using lower heat and higher pressure than conventional forging processes. The billets of stock are cut precisely and the weight checked to determine the exact volume of material to fill the die cavity. This process provides a part to closer tolerances and with better surface finishes than a conventional forging. Warm forming also provides considerable material cost savings as there is no excess. In a conventional forging operation, excess material may be as high as 50%

Upset or Forging Machine Dies

A wide variety of work that can be accomplished in no other way is made possible by the use of the upset principle. Forging machines can form forged head configurations on short or extremely long bars. The heads so formed will exhibit the same desirable qualities of grain flow that are accomplished by press or hammer forging. The heads vary in size and shape and it is not always possible to form them in one blow. Occasionally the stock must be gathered in two or more blows or passes as they are generally called. The volume of a particular head may require the use of a great length of the original bar, and definite rules must be followed to eliminate the possibility of defects. Three basic rules which pertain to the gathering of stock are as follows:

1. The limit of length of unsupported stock that can be gathered or upset in

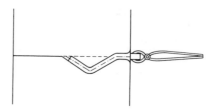

Figure 8-117. Incorporating a bender in a die.

one pass without injurious buckling is not more than three times the diameter.

2. Lengths of stock more than three times the diameter of the bar can be successfully upset in one blow by displacing the material in a die cavity no greater than 1 1/2 times the diameter of the bar, provided the stock extending beyond the die face is no greater than 1/2 of the diameter.

3. An upset requiring more than three diameters of stock in length and extending up to 2 1/2 bar diameters beyond the die face can be made if the material is confined by a conical recess in the punch which does not exceed 1 1/2 bar diameters at the mouth and 1 1/8 bar diameters at the bottom, provided the heading tool recess is not less than 2/3 the length of working stock or not less than the length of working stock minus 2 1/2 times its diameter.

These rules cover the absolute limits, whereas in actual practice, the dies are designed with broader limits. *Figure 8-118* illustrates the application and violation of these rules.

Figure 8-119 illustrates a simple bevel gear forging designed for manufacture on an upsetting machine. Upset dies are subject to practically the same die considerations as hammer or press forgings; i.e., mismatch, die wear, die closure (thickness tolerance), scale pitting, etc. When computing the thickness tolerance of the upset forging, the weight is taken as only that portion of the forging which is actually formed by the dies. In this case it is the beveled head only and the 1 1/4″ shank is excluded from the weight. If the volume of the head, in cubic inches, is calculated and multiplied by 0.284, the weight will be approximately 1.4 lb. This would require a thickness tolerance of +0.040″ -0.010″ since 0.284 lb is the weight of one cubic inch of steel.

In designing the set of upset dies, the first step is to find the theoretical weight of the head and add the plus variations of the head thickness tolerance to this weight. This amounts to an additional 0.1 lb. Thus 1.4 lb plus 0.1 lb = 1.5 lb.

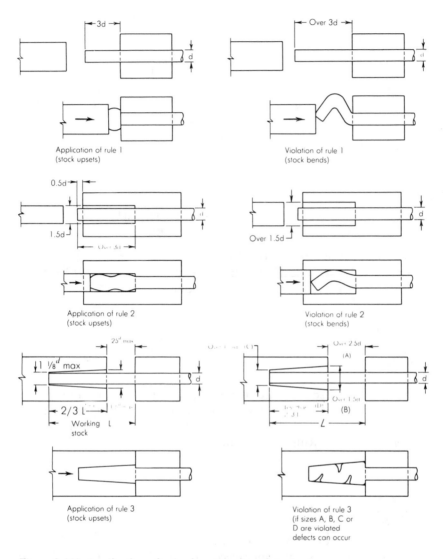

Figure 8-118. Application of rules for gathering of stock.

The next step is to determine the length of stock (1 1/4″) which will be required to fill the head. Therefore the stock length is 1.5 lb divided by 0.35 lb per inch = 4.28″ and 0.35 lb is the weight of 1 1/4″ diam steel stock for a 1″ length.

A layout of the completed forging (*Figure 8-120*) is now made with a set of hypothetical dies which would be required to make the forging in one blow.

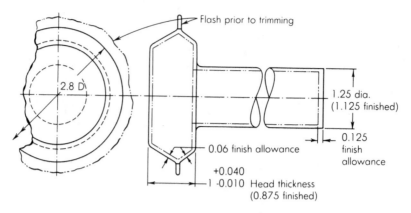

Figure 8-119. Bevel gear forging designed for production by upsetting.

The die designer can now consider the three rules for upsetting. If none of the rules is violated, the die design can be completed, since the head of the forging can be made in one blow. Rechecking the rules:

Rule 1.

$$\frac{4.28}{1.25} = 3.42$$

This head exceeds the three diameters of Rule 1. If this ratio had been between 2 and 2 1/2, the forging dies could be completely designed on the basis of one pass to forge. However, at times the 3.42 ratio will not result in a defective forging if Rule 2 is not violated.

Rule 2.

$$\frac{3.562}{1.250} = 2.85$$

The 2.85 ratio exceeds the limit of Rule 2 which requires that the 3 9/16" dimension of the dies could not exceed 1 1/2" x 1 1/4" or 1 7/8". If the cavity of the plunger were, say, 1 5/8" diam, the die could be completely designed on the basis of one pass to forge and the only consideration would be to limit the stock extension beyond the die face by no more than 5/8" or 1/2" diam.

It is not necessary to proceed with the checking of Rule 3 since Rule 2 has already been violated.

A subsequent pass or passes must be added to the original one-pass die and the resulting die design checked against the three rules for upsetting. To best arrive at this determination, begin with Rule 3 and consider a die design which

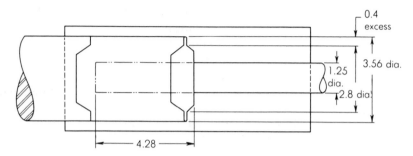

Figure 8-120. Layout for one-blow forging of a bevel gear blank.

will not violate this rule. The time-tested procedure called the progressive taper method is utilized. Gathering steel in a conical-type punch by this method can be seen in *Figure 8-121*.

The volume of stock contained in the punch must equal the original weight of 1.5 lb. The punch is designed as shown with an average taper of 1 1/2" diam at the midpoint. Two conditions have now been satisfied; the upset does not violate any part of Rule 3, and the weight of steel which the cone can contain is not less than the weight needed for the head.

The stock has now been increased in size so the die designer can return to the hypothetical one-pass die design and determine whether the conical pass just completed will meet all conditions. Again rechecking the rules:

Rule 1.

$$\frac{3.00 \text{ (overall length of cone)}}{1.50 \text{ (average bar diam)}} = 2$$

If Rule 1 is violated, additional cone passes should be designed until the stock is increased in size to the point where three diameters is not exceeded.

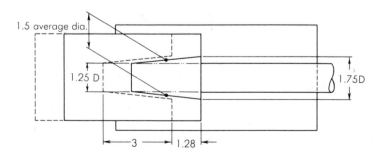

Figure 8-121. Gathering steel by use of a conical punch.

Rules 2 and 3 are not applicable unless Rule 1 has been violated. The dies and completed forgings from each pass are shown in *Figure 8-122*.

A trimming pass has been added to finish the forging in the same machine without additional equipment or handling.

Clearance between punch and die should be as shown. This space will prevent seizing of the dies due to unequal expansion, scale formations, and the like. An important consideration in upset work is the grip which must be provided on the stock to withstand the force of the heading punches. The

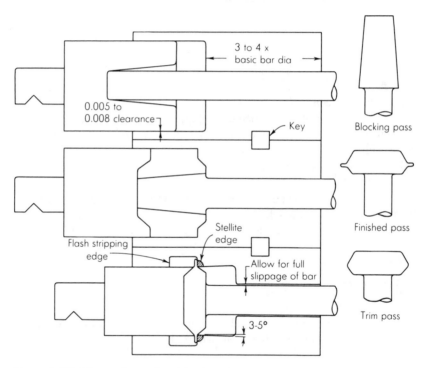

Figure 8-122. Dies and completed forgings from each pass.

length of grip in the dies should be equal to a minimum of three bar diameters and, when heating the bar, this part of stock should be kept as cold as possible. If the stock must be heated throughout, or the grip length is less than three diameters, corrugations can be bored in the dies to prevent slippage. The finish allowance on the machine part must be increased by the depth of the grooving if the die is corrugated (*Figure 8-123A*).

Figure 8-124 and *Figure 123B* show the design which is necessary throughout the grip section when corrugations are not required.

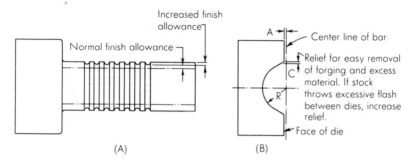

Figure 8-123. Finish allowance on machine part (A); design of grip section (B).

Stock size, in. (mm)	R	A	Total grip, 2A	C
1/2 (12.7)	0.250 (6.35)	0.002 (0.05)	0.004 (0.10)	1/64 (0.40)
5/8 (15.9)	0.3125 (7.938)	0.0025 (0.06)	0.005 (0.13)	1/64 (0.40)
3/4 (19.0)	0.375 (9.52)	0.003 (0.08)	0.006 (0.15)	1/64 (0.40)
7/8 (22.2)	0.437 (11.10)	0.0035 (0.09)	0.007 (0.18)	1/64 (0.40)
1 (25.4)	0.500 (12.7)	0.004 (0.10)	0.008 (0.20)	1/64 (0.40)
1 1/4 (31.8)	0.625 (15.88)	0.005 (0.13)	0.010 (0.25)	1/32 (0.80)
1 1/2 (38.1)	0.750 (19.05)	0.006 (0.15)	0.012 (0.30)	1/32 (0.80)
1 3/4 (44.4)	0.875 (22.22)	0.0065 (0.16)	0.013 (0.33)	1/32 (0.80)
2 (50.8)	1.000 (25.4)	0.007 (0.18)	0.014 (0.36)	1/32 (0.80)
2 1/2 (63.5)	1.250 (31.8)	0.0075 (0.19)	0.015 (0.38)	3/64 (1.19)
3 (76.2)	1.500 (38.1)	0.008 (0.20)	0.016 (0.41)	3/64 (1.19)

Figure 8-124. Die design dimensions.

The detailed analysis of the bevel gear forging shows the method used in designing a typical set of upset dies. By not exceeding any of the three rules, interesting variations of the upset principle can be made on a variety of other forgings which lend themselves to this method of manufacture. Two variations are in *Figure 8-125*: the sliding die upset and the bend and upset.

The die steels necessary for upset work vary with the severity of service and whether low or high production limits are required. Heavy dies with little change in section can be made from 0.50% carbon, 2.00% tungsten oil-hardening steels. If the die has light sections, warpage will become a problem during quenching and an air-hardening 5.00% chromium hot work die steel should be used. Long production runs warrant the use of the 5.00% chromium steel in almost all cases. This steel exhibits high resistance to thermal fatigue cracking and can be heat treated to a high strength and toughness. High thermal fatigue resistance constitutes a major requirement in the selection of

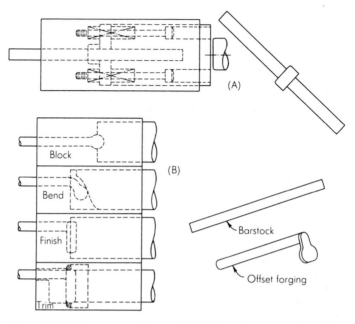

Figure 8-125. Two types of upset: (*A*) sliding-die; (*B*) bend-and-upset.

upset die steels since in service they must be cooled constantly with water or water-oil mixtures.

If extreme wear on long runs is present, hardened inserts can be incorporated into the die design to reduce die replacement costs. All dies, punches, and inserts should be hardened within a two-point range on the Rockwell C scale. The following ranges for each type of tool can be used as a guide in determining the heat treatment. However, dependability in service is still the criterion, and changes in specifications may be necessary after the tool is used.

Material	Part	Hardness RC
0.50C, 2.00W or 5.00CR	Gripper dies	45-50
0.50C, 2.00W or 5.00CR	Punches	42-48
2.00W or 5.00 CR	Inserts	45-50

BENDING, FORMING, DRAWING, AND FORGING DIES

Review Questions

Bending Dies

1. What is the basic principle involved in a bending die?
2. What causes springback?
3. Calculate bending forces for the following bends (assume S is 30 tons and width of $W = 2.50$)
   a. 90° bend in 5/16 metal, 1/2 IR, 12″ long.
   b. 60° bend in 1/4 metal, 3/4 IR, 16″ long.
   c. 30° bend in 3/16 metal, 3/16 IR, 9″ long.
4. When are edge-bending dies used? Why?
5. Design a simple V-bending die.
6. Design an edge-bending die.

Forming Dies

1. How does forming differ from bending?
2. Name five types of forming dies.
3. How does a solid form die differ from a pad die?
4. What are the two types of power sources for pressure pads?
5. What happens to metal during an embossing operation?
6. Name three types of embosses formed by an embossing die?
7. What is the smallest permissible radius that can be curled?
8. How are curls strengthened?
9. What is the bulging medium in a bulging die?
10. What is the difference in metal behavior in coining and swaging dies?
11. What is the range of swaging pressures?
12. What is the critical height of a hole flange?

Draw Dies

1. Will a smaller draw radius allow less or more material to flow into the draw form?
2. What is the primary use of an air cushion on a die in a drawing operation?
3. What is the primary function of a knockout die?
4. What is the minimum corner radius of a drawn cup?

5. State the formula for determining blank diameters for drawing operations, when d/r is between 15 and 20.
6. Why must a pressure pad or blankholder always be used when drawing thin stock?
7. What determines the shut height of a die?
8. Why must both punch and draw dies be vented?
9. Determine the drawing pressure given a 1/16" (1.6 mm) thick shell of 3" (76 mm) diameter, 2" deep, having a yield strength of 30,000 psi.
10. What is the range of blankholding pressure compared to drawing pressure?

Progressive Dies

1. What is a progressive die?
2. What operations can be performed in a progressive die?
3. When should a progressive die be used?
4. List the items that should be considered when designing a progressive die.
5. Why is piloting required?
6. Where are pilots located in the strip?
7. What is segmental-type die block construction? Give an example.
8. Sketch four types of shanks used on die sets.
9. Sketch the various pin and bushings available.
10. What is the feed level of the die?
11. Calculate the blank diameter for the part shown in *Figure A*.
12. What size hole should be pierced before the 0.218 diameter hole is extruded?
13. What is the percent of reduction from the blank diameter to the shell diameter.

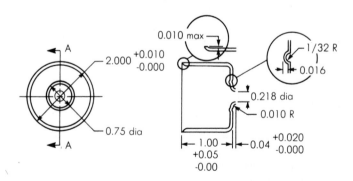

Figure A.

14. Design a progressive die stock strip to run the part (*Figure A*). The width of the strip should be the blank diameter plus 1.375 rounded to the next highest 1/4". The pitch should be the blank diameter plus 0.375". If the percentage of reduction is less than 48%, one draw and a restrike are required; if more than 48% two draws and a restrike are required. Add 0.25 to the blank diameter to calculate the radius of the relief cut-out. Notice that the inside radius at the top of the cut indicates the part is blanked by a pinch trim punch.

Extruding Dies

1. Discuss the basic principles of impact extrusion.
2. What affects the pressures required to extrude common metals?
3. Discuss why lubricants play an important role in successful impact extrusion.
4. Discuss the three basic methods of impact extrusion.
5. Discuss why coining is desirable at times before cold extrusion.
6. Why is tool design considered the most important factor of successful impact extrusion?
7. Discuss the important factors of punch design.
8. Discuss common characteristics of punch and pressure anvils.
9. What is the function of punch guides?
10. Discuss design considerations of diepots.

Forging Dies

1. What are the three classes of forging equipment (machines)?
2. What determines whether hot or cold forging is to be done?
3. Why are forgings inherently strong and tough?
4. What is forging draft?
5. Why are fillets and rounded corners specified for forged parts?
6. What kind of steel should be used for forging dies?
7. What is the limit of length of unsupported stock for an upset forging?
8. What is the size limit of a die cavity?
9. What is the first step in designing a set of upset dies?
10. What is the second step in designing a set of upset dies?

BENDING, FORMING, DRAWING, AND FORGING DIES

Answers to Review Questions

Bending Dies

1. Bending is a uniform straining of material, around a straight axis. Metal flow takes place within the plastic range of the metal, so that the bend retains a permanent set after removal of the applied stress.
2. The elastic stresses causes the metal to spring back after the bending pressure is released.
3. $P = \dfrac{KLSt^2}{W}$

   a. $P = \dfrac{1.33 \times 12 \times 30 \times 0.3125^2}{2.5} = \dfrac{46.75}{2.5} = 18.7$ tons

   b. $P = \dfrac{1.33 \times 16 \times 30 \times 0.25^2}{2.5} = \dfrac{39.9}{2.5} = 15.96$ tons

   c. $P = \dfrac{1.33 \times 9 \times 30 \times 0.1875^2}{2.5} = \dfrac{12.62}{2.5} = 5.05$ tons

4. In multiple station dies and when two or more bends are done at the same time. Because in edge bending the area not being bent is stationary, only the flange being bent moves, and edge bending requires only 1/2 the pressure of "V" bending.
5. Basic sample shown in *Figure 8-2A*.
6. Basic sample shown in *Figure 8-2B*.

Forming Dies

1. Forming is usually done along a curved axis instead of a straight axis as in bending.
2. (1) embossing, (2) curling, (3) beading, (4) twisting, and (5) hole flanging.
3. A solid form die has a solid punch and die and does not have to be mounted in a die set, and is used for simple forms with large tolerances.

   A pad for die has a punch, die, and pressure pad. It is usually mounted in a die set and can produce close tolerance parts.

4. (1) springs and (2) air cushion.
5. The metal is stretched into the die rather than being compressed.
6. (1) stiffening, (2) lettering, and (3) weld projection nibs.
7. Two times the material thickness.
8. Wires are rolled inside the curls.
9. Rubber, heavy grease, water, or oil.
10. In swaging, the part contours are not precise. The material is not fully confined as in a coining operation and overflow flash is usually produced.
11. From 25 to 2500 tons are used.
12. 2 1/2 times metal thickness.

Draw Dies

1. Less.
2. To provide pressure on the blankholders.
3. To eject the part from the die.
4. Four times stock thickness.
5. $2.00 \div 0.125 = 16$

$$D = \sqrt{d^2 + 4\ dh - 0.5r}$$

6. To prevent wrinkles and puckers.
7. The depth of draw and the type of draw die.
8. To let air escape from the die and to prevent vacuum from holding the part on the punch.
9. $P = \pi\, dts\ \dfrac{D}{d} - C$

$$D = \pi d^2 + 4dh$$

$$D = \sqrt{3^2 + (4 \times 3 \times 2)} = \sqrt{33} = 5.745$$

$$P = 3.1416 \times 3 \times 0.0625 \times 30{,}000 \left(\frac{5.745}{3} - 0.65 \right)$$

$$P = 17671.459 \times 1.265 = 22354.4 \text{ lbs or } 11.18 \text{ tons}$$

10. It varies from very little to a maximum of about one-third of the drawing pressure.

Progressive Dies

1. A progressive die performs a series of fundamental sheet-metal operations at two or more stations during each press stroke.
2. All sheet-metal operations that can be performed while the part is attached to a carrier strip.
3. When parts can be produced more economically than a series of single operation dies.

4. a. The die should be heavily constructed.
   b. The type of die set.
   c. The type of stripper (solid, stripping loaded or punch guide type).
   d. The type of stock guides (lifter pins or carrier rails).
   e. Pilots (locate from internal holes in the part or holes in scrap strip).
   f. Die block size and material.
   g. Shut height and press tonnage available.
   h. Operator safety.
5. To position the stock strip so each station will perform the operations in the proper location.
6. Piloting may be done directly using holes in the part or indirectly using holes pierced in the scrap strip.
7. It is using two or more pieces to make up the die block rather than a single piece. This is usually done for ease of construction, by breaking down a complex shape into two or more simple components of that shape. It also lowers maintenance costs; if broken, a single segment can be replaced rather than the entire die block.
   (See *Answer Figure A.*)
8. (See *Answer Figure B.*)
9. (See *Answer Figure C.*)
10. The height to the bottom of the stock as it is feeding through the die.
11. Stock is mild steel 0.016" thick

$$d/r = \frac{2.00}{0.125} = 16$$

$$D = \sqrt{d^2 + 4dh - 0.5r} = \sqrt{2^2 + 4 \times 2 \times 1 - 0.5 \times 0.125}$$

$$= 3.464 - 0.0625 = 3.401 \text{ dia.}$$

12. $J = \sqrt{\dfrac{TB^2 + 4TA^2 + 4HA^2 - 4HB^2}{9T}}$

$$J = \sqrt{\frac{0.016 \times 0.25^2 + 4 \times 0.016 \times 218^2 + 4 \times 0.04 \times 0.218^2 - 4 \times 0.04 \times 0.25^2}{9 \times 0.016}}$$

$$J = \sqrt{\frac{0.001 + 0.00304 + 0.0076 - 0.01}{0.44}}$$

$$J = \sqrt{0.01138}$$

$$J = 0.107 \text{ diameter}$$

13. $\% = \dfrac{3.401 - 2.00}{3.401} = \dfrac{1.401}{3.401} = 0.41 = 41\%$

14. Stock width = 3.401 + 1.375 = 4.776 rounded to 5.00

    Pitch = 3.401 + 0.375 = 3.776

    Radius of relief cut out = $\dfrac{3.401 + 0.25}{2} = \dfrac{3.651}{2} = 1.825\ r$

2 pc.   3 pc.   4 pc.

Answer Figure A.

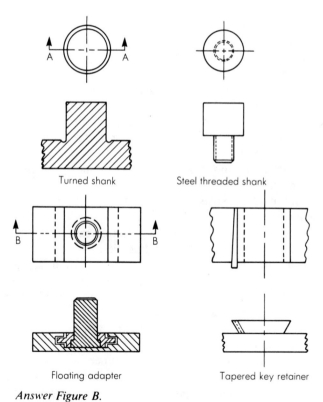

Turned shank   Steel threaded shank

Floating adapter   Tapered key retainer

Answer Figure B.

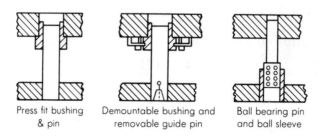

| Press fit bushing | Demountable bushing and | Ball bearing pin |
| & pin | removable guide pin | and ball sleeve |

Answer Figure C.

Extruding Dies

1. Sufficient pressure is applied to the material to cause plastic flow in all directions to fill the cavity between the punch and die.
2. a. The alloy used.
   b. Its microstructure.
   c. The restriction to flow.
   d. The severity of work hardening.
   e. The lubricant used.
3. Correct lubrication reduces considerably the pressures required and makes possible the cold extrusion of steel by preventing metal-to-metal welding of the part material to the punch and die.
4. a. In backward extrusion, metal flows in a direction opposite to the direction of punch movement.
   b. In forward extrusion, metal flows in the same direction as the punch movement.
   c. In combination extrusion, metal flows in both directions simultaneously.
5. Coining slugs to a close fit in the die and to locate the punch minimizes punch breakage in backward extruding dies.
6. Without a proper tool, various problems can arise: part not to specifications, excessive tool breakage, and excessive run time.
7. The most important part of design is end profile. This determines to a great extent punch life and part quality.
8. They must be free of grinding marks requiring a 4 μ in. finish and lapped in the direction of metal flow.
9. To minimize the column loading on the punch.
10. Diepots should have the same surface condition as punches. Carbide should be used for high run parts. The diepot is shrunk fit into a ring or die shoe to prevent bursting. The shrink fit of 0.004" per inch of diameter is usually used.

Forging Dies

1. (a) Hammer, (b) Forging machine or upsetter, and (c) Forging press.
2. The elasticity of the metal or its size.
3. Because the grain flow follows the shape of the forging.
4. Draft is the angle or taper in the side walls of the die to allow the part to be removed.
5. Fillets allow metal to flow into the cavity without folding. Round corners let air and oil escape as metals fills the cavity. Sharp corners may trap air or oil causing a void in the forging.
6. High carbon, nickel-chrome-molybdenum alloy steel that has been forged (for grain refinement and resistance to shock) and double heat treated for resistance to wear.
7. Not more than three times the diameter of the bar.
8. No greater than 1 1/2 times the bar diameter.
9. To find the theoretical weight of the head including the plus variations of the head thickness.
10. To determine the length of stock required to fill the head.

References

1. J. Verson, "Tooling for Cold Extrusion", American Machinist, October 7, 1957.
2. R. Quadt, "When to Use Cold Extrusion of Aluminum", Design Engineering, November 1956.
3. "Computations for Metal Working in Presses", E. W. Bliss Company, Canton, Ohio.
4. Tool & Manufacturing Engineers Handbook, SME, Third Edition, 1976.

9

DESIGN OF TOOLS
FOR INSPECTION AND GAGING

A drawing or blueprint of a workpiece may be used as a guide for its manufacture only if it completely defines and limits the size, shape, and composition of the workpiece by dimensions and specifications. Tolerances may be given, directly or indirectly, to every dimension or specification.

Tolerance is the total amount by which a dimension may vary from a specified size. In providing tolerances, the designer recognizes that no two workpieces, distances, or quantities can be exactly alike. The designer specifies an ideal, unattainable condition (nominal dimension), and then states what degree of error can be tolerated.

Every dimension of every workpiece must be specified as being between two limits. No proper dimension can ever be given as a single fixed value, because a single value is unattainable. The designer guards against this pitfall in several ways. Critical dimensions on a drawing or blueprint have the tolerance given as part of the dimension (for example, $0.125 \pm 0.002''$ [3.18 ± 0.05 mm]). Less critical dimensions are provided tolerance by a general note in the title block (for example, unless otherwise specified, all dimensions are $\pm 0.010''$ [0.25 mm]). Tolerance may also be specified indirectly in the bill of material. If the designer calls for the use of purchased material or parts, and does not further specify tolerance, it must be assumed that the vendor's manufacturing tolerances are acceptable.

In establishing tolerances, the designer may consider a number of factors including use, appearance, and cost. The workpiece will have an intended use and the specified tolerance must be compatible with it. Appearance is sometimes a factor, because a workpiece dimension with no actual function may have aesthetic requirements. For no other reason than the cost of materials, the size of a workpiece must be limited.

Cost is perhaps the governing factor in deciding tolerances, and as such, is often more important than use. As tolerances become smaller, the cost of meeting them increases rapidly. The designer often must weigh the need for a close tolerance against its cost, and must also consider the capability of the machines that will produce the workpiece. If the use of the workpiece requires extremely close tolerances to assure proper mating allowances, it may be less expensive to combine easily held tolerances with selective assembly. It may be possible in this way to obtain the necessary mating allowances for function without increasing the cost with close tolerances.

In a conventional blueprint or drawing, the workpiece is shown in a fixed relationship to reference planes at right angles to each other. The location of each element of the workpiece is defined by stating its distance (dimension) from each reference plane. Each dimension has a tolerance.

The reference planes in a single view may be compared with the X and Y axes of a Cartesian coordinate system. The location of the element is anywhere within a rectangular area formed by the minimum and maximum values of the X and Y dimensions as established by the tolerances. Engineering intent can often be expressed more exactly by stating the nominal location of the element and by stating how far from the true position (TP) the element may vary. In this case, a positional tolerance applied to the nominal location takes the place of the X and Y tolerances as shown in *Figure 9-1*. As will be seen later, the workpiece designer's intent can be further shown by modern geometric dimensioning and tolerancing methods.

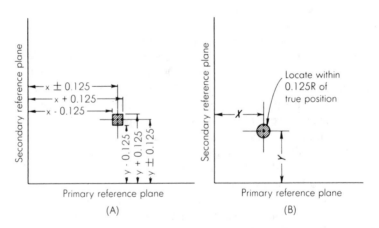

Figure 9-1. Coordinate dimensioning and positioning tolerance: (*A*) the element so dimensioned may be located anywhere within the shaded rectangle; (*B*) the element so dimensioned may be located anywhere within the shaded circle.

Conversion Charts

Since most machine movements are made in two directions at 90° to one another, X and Y, the chart shown in *Figure 9-2* may be used to determine the two-sided tolerance approximations of positional tolerance diameters. Mathematically, they can also be approximated by multiplying the position tolerance ⊕ by 0.7. For example, 0.7 x 0.010 = 0.007 (±0.0035).

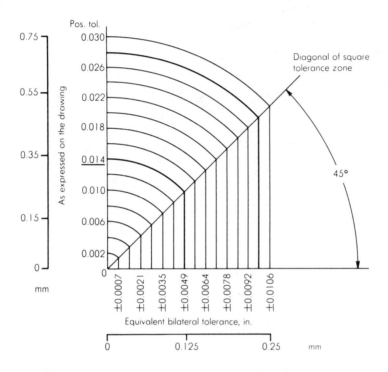

Figure 9-2. Conversion chart: positional tolerance to bilateral. (*Courtesy, The Sandia Corp.*)

The chart in *Figure 9-3* may be used to determine whether an actual hole center (as measured on an open setup) is within the positional tolerance specified on the drawing, and also to determine the actual amount of difference (radius) from TP. For example, on a drawing a hole is located by dimensions labeled TP with a positional tolerance. The location of the actual center of the hole on the part is determined by coordinate measurements. The difference in the actual measurements and the corresponding TP dimensions are located on the chart using the values in the upper and right-hand borders.

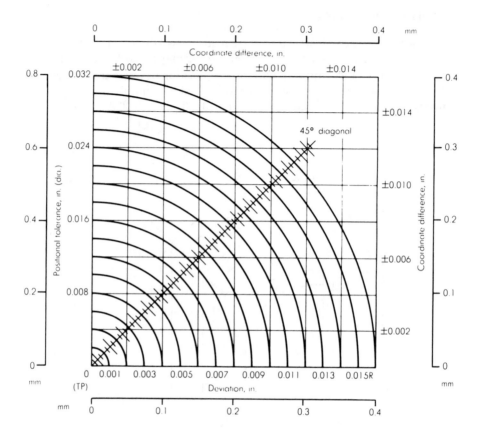

Figure 9-3. Conversion chart: bilateral to positional tolerance. (*Courtesy, The Sandia Corp.*)

It can now be shown whether the location is within the required positional tolerance limits using values in the left-hand border.

To determine the actual difference from TP, a radius may be drawn through the point, intersecting the 45° diagonal line. The amount of difference may now be read from the scale on the 45° line using the values in the lower border. Mathematically, it can also be approximated by multiplying the total ± tolerance by 1.4. For example, 1.4 x 0.007 (±0.0035) = 0.010.

Basic Principles of Gaging

Gage Tolerances. Since it is not possible to produce many parts with exactly the same dimensions, working tolerances are necessary. For the same reason, gage tolerances are necessary. Gage tolerance is generally determined from the amount of workpiece tolerance. A 10% rule is generally used for determining the amount of gage tolerance for fixed, limit-type working gages. When no gage tolerance is specified, the gagemaker will use 10% of the working tolerance as the gage tolerance for a working gage. Working gages are those used by production workers during manufacture.

The amount of tolerance on inspection gages—those used by the inspection department—is generally 5% of the work tolerance. Tolerance on master gages—those used for checking the accuracy of other gages—is generally 10% of the gage tolerance. Where tolerances are large, gages used by the inspection department are not different from the working gages.

Four classes of gagemakers' tolerances have been established by the American Gage Design Committee and are in general use. These four classes establish maximum variations for any desired gage size. The degree of accuracy needed determines the class of gage to be used. *Figure 9-4* shows these four classes of gagemakers' tolerances.

Above	To and Including	CLASS			
		XX	X	Y	Z
0.010″	0.825″	0.00002″	0.00004″	0.00007″	0.00010″
.254mm	20.95mm	0.00051mm	0.00102mm	0.00178mm	0.00254mm
0.825″	1.510″	0.00003″	0.00006″	0.00009″	0.00012″
20.95mm	38.35mm	0.00076mm	0.00152mm	0.00229mm	0.00305mm
1.510″	2.510″	0.00004″	0.00008″	0.00012″	0.00016″
38.35mm	63.75mm	0.00102mm	0.00203mm	0.00305mm	0.00406mm
2.510″	4.510″	0.00005″	0.00010″	0.00015″	0.00020″
63.75mm	114.55mm	0.00127mm	0.00254mm	0.00381mm	0.00508mm
4.510″	6.510″	0.000065″	0.00013″	0.00019″	0.00025″
114.55mm	165.35mm	0.00165mm	0.00330mm	0.00483mm	0.00635mm
6.510″	9.010″	0.00008″	0.00016″	0.00024″	0.00032″
165.35mm	228.85mm	0.00203mm	0.00406mm	0.00610mm	0.00813mm
9.010″	12.010″	0.00010″	0.00020″	0.00030″	0.00040″
228.85mm	305.05mm	0.00254mm	0.00508mm	0.00762mm	0.1016mm

Figure 9-4. Standard gagemakers' tolerances.

Class *XX* gages are precision smoothed (lapped) to the very closest tolerances practicable. They are used primarily as master gages and for final close tolerance inspection.

Class *X* gages are precision lapped to close tolerances. They are used for some types of master gage work, and as close tolerance inspection and working gages.

Class *Y* gages are precision lapped to slightly larger tolerances than Class *X* gages. They are used as inspection and working gages.

Class *Z* gages are precision lapped. They are used as working gages where part tolerances are large and the number of pieces to be gaged is small.

Going from Class XX to Class Z, tolerances become increasingly greater, and the gages are used for inspecting parts having increasingly larger work tolerances.

To show the use of the 10% rule in connection with *Figure 9-4*, assume a gagemaker is to choose the correct tolerance class for a working plug gage that is to be used on a 1.0000″ (25.400 mm) dia. hole having a working tolerance of 0.0012″ (0.030 mm). One-tenth of the work tolerance would indicate a gage tolerance of 0.00012″ (0.003 mm), as noted in the table, a class Z gage. If the work tolerance were only 0.0006″ (0.015 mm) on the 1.0000″ (25.4 mm) diam hole, then a class X gage would be indicated, with a tolerance of 0.00006″ (0.0015 mm). If the work tolerance, however, were 0.015″ (0.38 mm), then the gage tolerance indicated by the 10% rule would be 0.0015″ (0.38 mm). As this is larger than the maximum tolerance class, a class Z gage would be needed, and the gage tolerance would be 0.00012″ (0.003 mm).

The smaller degree to which a gage tolerance must be held, the more expensive the gage becomes. Just as the production cost rises sharply as working tolerances are reduced, the cost of buying or manufacturing a gage is much higher if close tolerances are specified. Gage tolerances should be realistically applied from the work to be gaged.

Allocation of Gage Tolerances

After deciding the tolerance for a specific gage, the direction, plus or minus, of that allowance must be decided. Two basic systems, and many changes of them, are used in making this decision.

The Bilateral System

In the bilateral system, the go and no-go gage tolerance zones are divided into two parts by the high and low limits of the workpiece tolerance zone. The division is illustrated by *Figure 9-5A*, which shows the black rectangles representing the gage tolerance zones are half plus and half minus in relation to the high or low limit of the work tolerance zone.

Referring to *Figure 9-4*, assume that the diameter of the hole to be gaged is 1.2500 ± 0.0006″ (31.75 ± 0.015 mm). The total work tolerance in this case is 0.0012″ (0.030 mm), since the hole size may vary from 1.2506 to 1.2494″ (31.765 to 31.735 mm). Using 10% of the total work tolerance as our gage tolerance, the gage tolerance is then 0.00012″ (0.003 mm). From *Figure 9-4*, this diameter would require a Class Z gage tolerance. The diameter on the go-plug gage for this example would be 1.2494 ± 0.00006″ (31.735 ± 0.0015 mm), and the diameter of the no-go gage would be 1.2506 ± 0.00006″ (31.765 ± 0.0015 mm).

One disadvantage of this system is that parts that are not within the

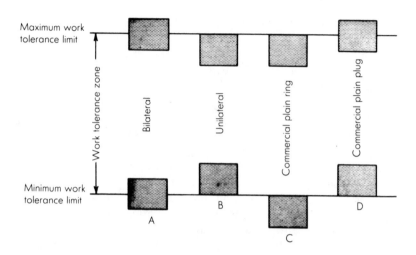

Figure 9-5. Different systems of gage tolerance allocation.

working limits can pass inspection. Using the above example, if the hole to be gaged is reamed to the low limit (1.2494″ [31.735 mm]) and if the go-plug gage is at the low limit (1.24934″ [31.7332 mm]), then the go-plug gage will enter the hole and the part will pass inspection, even though the diameter of the hole is outside the working tolerance zone. A part passed under these conditions would be very close to the working limit, and the tolerance on the mating part should not be such as to prevent assembly. Plug gages using the bilateral system could also pass parts in which the holes were too large. A common misconception is that gages accept good parts and reject bad. With the bilateral system, however, parts can also be rejected as being outside the working limits when they are not.

The Unilateral System

In the unilateral system (*Figure 9-5B*), the work tolerance zone entirely includes the gage tolerance zone. This makes the work tolerance smaller by the sum of the gage tolerance, but guarantees that every part passed by such a gage, regardless of the amount of the gage size variation, will be within the work tolerance zone.

If the diameter of the hole is $1.2500 \pm 0.0006''$ (31.750 ± 0.0015 mm), again using 10% of the working tolerance as the gage tolerance, the go gage diameter would be $1.24940 + 0.00012''$ ($31.7348 + 0.003$ mm), and the no-go gage diameter $1.25060 - 0.00012''$ ($31.7652 - 0.003$ mm).

This system of applying gage tolerance, like the bilateral system, may reject parts as being outside the working limits when they are not, but all parts passed using the unilateral system will be within the working limits. The unilateral system has found wider use in industry than the bilateral system for plain plug and ring gages.

One partial solution to the problem of gages rejecting parts that are within working limits is to use working gages with the largest unilateral gage tolerance practical, and inspection gages with the smallest unilateral gage tolerance practical. Thus, no piece can pass inspection that is outside tolerance, and the possibility of the inspection gage turning down acceptance work is reduced because of its small tolerance.

Figure 9-5C shows the commercial practice of allocating the plain ring gage tolerances negatively with reference to both the maximum and minimum limits of the workpiece tolerance. In another practice (*Figure 9-5D*), the no-go gage tolerance is divided by the maximum limit of the workpiece tolerance, and the go gage tolerance is held within the minimum limit of the workpiece tolerance.

The final results of these allocation systems will differ greatly. The choice of the system to be used, modified, or unmodified, must be determined by the product and the facilities for producing it. The objectives in choosing an allowance system should be the economic production of as near 100% usable parts as possible, and the acceptance of the good pieces and rejection of the bad.

Gage Wear Allowance

Perfect gages cannot be made. If one did exist, it no longer would be perfect after just one checking operation. Although the amount of gage wear during just one operation is difficult to determine, it is easy to measure the total wear of several checking operations. A gage can wear beyond usefulness unless some allowance for wear is built into the gage.

The wear allowance is an amount added to the nominal diameter of a go-plug and subtracted from that of a go-ring gage. It is used up during the gage life by wearing away of the gage metal. Wear allowance is applied to the nominal gage diameter before gage tolerance is applied.

The amount of wear allowance does not have to be decided in relation to the amount of work tolerance, although a small work tolerance can restrict wear allowances. When specifying a wear allowance, the material from which the gage and work are made, the quantity of the work, and the type of gaging operation to be performed must be taken into consideration. It is important to establish a specific amount of wear allowance. When the gage has worn the established amount, it should be removed from service without question. This avoids any controversy as to whether a gage is still accurate.

One method, which uses a percentage of the working tolerance as the wear tolerance, can be explained with the following example: For a $1.500 \pm 0.0006''$ (38.1 ± 0.015 mm) diameter hole, the working tolerance is $0.0012''$ (0.03 mm). The basic diameter of the go-plug gage would be $1.49940''$ (38.0848 mm). Using 5% of the working tolerance ($0.00006''$ [0.0015 mm]) as the wear allowance, and adding this to the basic diameter, the new basic diameter would then be $1.49946''$ (38.0863 mm). The gagemaker's tolerance of 10% of the working tolerance ($0.00012''$ (0.003 mm)) would then be applied in a plus direction as allowed by the unilateral system, with a resultant go plug diameter of $1.49946 ^{+0.00012}_{-0.00} $ ($38.0863 ^{+0.003}_{-0.0}$) . *Figure 9-6* shows the wear allowances and gage tolerances used by the manufacturers of U.S. weapons.

Other manufacturing companies do not build a wear allowance into their gages, but set up a standard to determine when a gage has worn beyond its usefulness. The gage is allowed to wear a certain percentage above or below its basic size before being taken out of service. Gages should be inspected regularly for wear. No set policy for wear allowance or gage inspection is practical for all industries. In operations where an extremely high degree of accuracy must be maintained, the amount of allowable wear is smaller and inspection must be more frequent than in operations where tolerances are greater.

A gagemaker normally makes the gage to provide maximum wear even if no wear allowance is designed into the gage. That is, he makes the go end of a plug gage to its high limit, and he makes the go component of a ring gage to its low limit.

The no-go gage will slip in or over very few pieces and so wear very little, and as any wear on a no-go gage puts that gage farther within the product limits, no wear allowance is applied. When the no-go gage begins to reject work actually well within acceptable limits, it must be retired.

Gage Materials

For medium production runs, hardened alloy steel is used for wear surfaces of gages. For higher-volume production runs, gage wear surfaces are usually chromium-plated. When a high degree of accuracy is needed, the production run is long, and wear is excessive, tungsten carbide contacts are often used on gages. Worn gaging surfaces can be ground down, chrome-plated, reground, lapped to size, and put back into service.

Gaging Policy

Gaging policy is the standardization of the methods for determining gage tolerances and their allocation, and of fixing wear allowance. It is a guide to

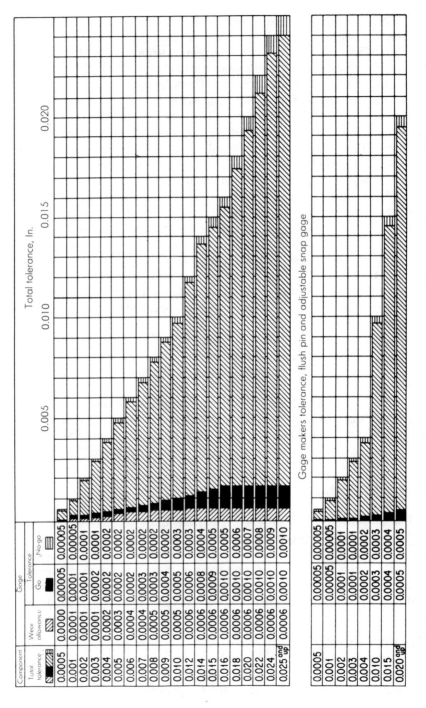

Figure 9-6. Wear allowances and tolerances used as standard practice by plants in U.S. Ordnance production.

determine when gages are required, when and how they are to be inspected, and what types of gages should be used.

There is no one policy in universal use. Gage users should have their own policy—the one best suited to their work. A gage policy is helpful in eliminating controversy over the method of gaging, gage tolerances and allocation, and gage wear.

The policy should establish a general rule for determining the amount of gage tolerance. The 10% rule, used in conjunction with the four standard classes of gage tolerances, is widely used in industry.

The size and application of wear allowance is also part of the gage policy. How often gages are inspected will depend upon how often they are used, what materials they are used to check, and how close the tolerances are. When a high degree of accuracy is needed, the gage policy must be strict.

Gage Types and Applications

Measurement compares amount or length with a known standard. Dimensions are proven by end measurement; the end of a specified length is measured against the end of the standard; the last drop of a specified amount of liquid brings the fluid level up to a mark on the side of the vessel. Almost all dimensional measurement is end measurement. End measurement is determined by origin; that is, both the standard and the length or amount being measured must start at a common point.

Every workpiece must be measured by three planes located at right angles to each other. Every dimension has as its origin one of the three planes. Given a horizontal reference plane, two vertical reference planes can be constructed at right angles to each other, and any structure can be defined by dimensions starting at the reference planes.

Surface Plate

The surface plate is used as the main horizontal reference plane. It is the primary tool of layout and inspection. Surface plates are usually made of cast iron or granite, with one surface finished extremely flat. When accuracy of 0.00001" (0.00025 mm) or finer is required, toolmakers' flats or optical flats are used instead. A number of accessories such as squares, straight edges, parallels, angle irons, sine plates, leveling tables, height gages, length standards, and indicating equipment are used with the surface plate.

The workpiece being inspected is carefully placed on the surface plate with its primary reference plane on the plate surface. If the workpiece has elements extending below its established reference plane, it may be necessary to mount the workpiece on adjustable supports (jacks) or parallels to make the two

Figure 9-7. Typical surface plate setup.

reference planes parallel. Vertical reference planes are placed on the surface plate as required.

To avoid increased error, all dimensions on a drawing or blueprint should start from a reference point or plane rather than from the end of another dimension. If, however, a workpiece feature is measured from a point rather than a reference plane, its location should be proven in relation to that point rather than the reference plane. A typical surface plate setup is shown in *Figure 9-7*.

Templates

A template represents a specified profile, or it may be a guide to the location of workpiece features with reference to a single plane. A straightedge may be used as a template to check flatness. To control or gage special shapes or contours in manufacturing, special templates are used for comparison by

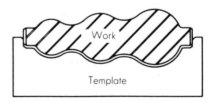

Figure 9-8. Gaging the profile of a workpiece with a template.

eye to insure uniformity of individual parts. These templates are made from thin, easily machinable materials, some of which may be hardened later if production requirements demand longer use. *Figure 9-8* shows an application of the contour template to inspect a turned surface. Templates of this type are also used widely in the sheet metal industry and where production is limited. Templates are satisfactory when the part tolerance will permit this type of inspection.

To inspect or gage radii or fillets visually, a standard commercial type of template can be obtained from various gage manufacturers. These templates or gages are used by inspectors, layout men, toolmakers, diemakers, machinists, and patternmakers. The five different basic uses are shown in *Figure 9-9*. These gages are usually made to nominal sizes in increments of 1/64″ for inch gages (0.5 mm for millimeter gages). Special gages of this type might readily be designed and made for specific jobs.

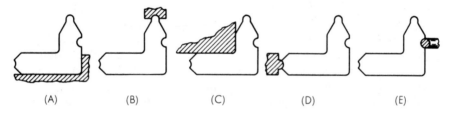

| (A) | (B) | (C) | (D) | (E) |

Figure 9-9. Commercial radius gage and applications: (*A*) inspection of an inside radius tangent to two perpendicular planes; (*B*) inspection of a groove; (*C*) inspection of an outside radius tangent to two perpendicular planes; (*D*) inspection of a ridge segment; (*E*) inspection of roundness and diameter of a shaft.

A screw pitch gage is used to determine the pitch of a screw. Individual gages can be obtained for most of the commonly used thread forms and sizes. To determine the pitch, the gage is placed on the threaded portion as shown in *Figure 9-10*. A screw pitch gage is seldom designed for special use, since it does not check thread size and will not give an adequate check on thread form for precision parts.

Plug Gages

A plug gage is a fixed gage, usually made up of two members. One member is called the go end, and the other the no-go or not-go end.

The actual design of most plug gages is standard, being covered by American Gage Design (A.G.D.) standards. However, there are many cases where a special plug gage must be designed.

A plug gage usually has two parts: the gaging member, and a handle with

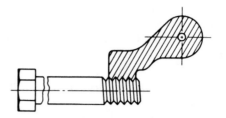

Figure 9-10. Screw pitch gage and method
of application.

the size, go or no-go, and the gagemaker's tolerance marked on it. There are
generally three types of A.G.D. standard plug gages. First is the single-end
plug gage. This type has two separate gage members, a go and a no-go, each
having its own handle (*Figure 9-11A*). The second is called a double-end plug
gage. This type consists of two gage members, a go and a no-go, mounted on a
single handle, with one gage member on each end (*Figure 9-11B*). The third
type, called the progressive gage, consists of a single gage member mounted
on a single handle (*Figure 9-11C*). The front two-thirds of the gage member is
ground to the go size and the remaining portion is ground to the no-go size.
The go and no-go sizes are therefore put together in the same gage member.

Standard A.G.D. plug gages generally have three methods of mounting the
gage members on the handle. The smaller sizes are usually wiretype gages.
The gage member is simply a straight blank or nib with no shoulder, taper, or
threads. The gage member on this type is held in the handle by a setscrew or
with a collet chuck built into the handle, as shown in *Figure 9-12A*. The

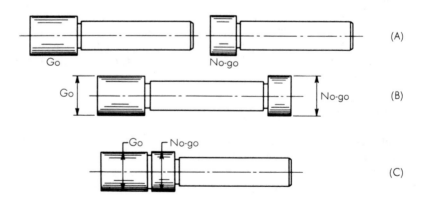

Figure 9-11. AGD cylindrical plug gages used to inspect the diameter of holes.

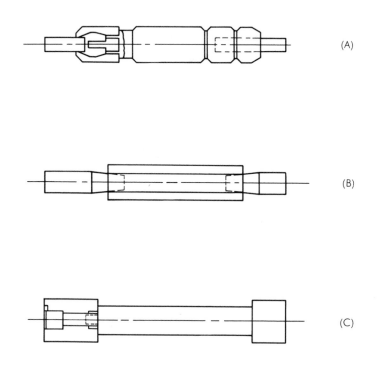

Figure 9-12. AGD cylindrical plug gages (*A*) reversible wire type, (*B*) taper lock design (*C*) trilock design.

advantage of this type of mounting is that the gage members are reversible when one end becomes worn, thus increasing the life of the gage.

Figure 9-12B shows another type of mounting called the taper-lock design. In this type, the gage member is manufactured with a taper ground on one end. This taper fits into a tapered hole in the handle much the same as a taper-shank drill fits into a drill press spindle. This type is not reversible.

The third type of mounting is called the trilock design. This design is usually for larger gages. The gage member has a hole drilled through the center and is counterbored on both ends to receive a standard socket head screw. Three slots are milled radially in each end of the gage member. The gage is held to the handle by means of a socket head screw with the three slots engaging three lugs on the end of the handle as shown in *Figure 9-12C*. This type of gage is reversible.

It is sometimes necessary to design special plug gages. There are many types, depending on the job requirement. A square, hexagonal or octagonal

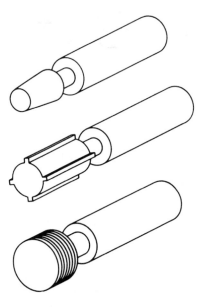

Figure 9-13. Special plug gages used to
inspect the profile or taper of holes.

hole requires a special plug gage. Internal splines require spline plugs designed in accordance with ANSI B92.1. Internal threads require thread plugs designed in accordance with ANSI B1.2, etc. *Figure 9-13* shows several of these special plug gages.

Ring Gages

Ring gages are usually used in pairs, consisting of a go member and a no-go member as shown in *Figure 9-14*. They are fixed gages, and their design is also covered by A.G.D. standards.

In sizes up to 1.51″ (38.4 mm) the design is a plain ring, knurled on the O.D., and lapped in the I.D. to a close tolerance. The no-go member is identified by a groove around the O.D. of the gage.

In sizes up to 1.51″ (38.4 mm), the gage has a flange to reduce weight and increase rigidity. The no-go in this type is also identified by a groove around the O.D. of the gage.

Special ring gages are required occasionally. An example would be the inspection of a shaft having a key way, where a key-shaped segment added to a ring gage would allow gaging of shaft size and key-slot width and depth at the same time. External splines require spline rings designed in accordance

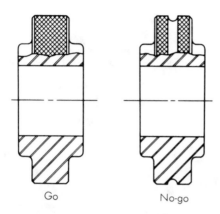

Go   No-go

Figure 9-14. Ring gage set used to inspect
the diameter of shafts.

with ANSI B92.1. External threads require thread rings designed in
accordance with ANSI B1.2, etc. *Figure 9-15* shows several of these special
ring gages.

Snap Gages

A snap gage is a fixed gage arranged with inside measuring surfaces for
calipering diameters, lengths, thicknesses or widths. Snap gages come in a
variety of types.

A plain adjustable snap gage is a complete external-caliper gage used for
size control of plain external dimensions. It has an open frame, with gaging
members provided in both jaws. One or more pairs of the gaging members can
be set and locked to any predetermined size within the range of adjustment.
The design of most of these types is covered by A.G.D. standards. They are
shown in *Figure 9-16*.

A plain, solid, snap gage, shown in *Figure 9-17*, is a complete external-
caliper gage used for size control of plain external dimensions. It has an open
frame and jaws, the latter carrying gaging members in the form of fixed,
parallel, nonadjustable anvils.

The thread-roll snap gages shown in *Figure 9-18* are complete external-
caliper gages employed for size control of thread pitch diameter, lead of a
thread, and thread form. They have open frames and jaws in which gaging
members are provided. One or more pairs of the gaging members can be set
and locked to a predetermined size within range of the thread to be checked.
Gaging members vary with the different diameters and pitch or thread.

Figure 9-15. Special ring gages to check profile or
taper on parts. (*Courtesy, Hemco*)

A form and groove or blade-type snap gage as shown in *Figure 9-19* is a
complete external-caliper gage used for size control of grooves or for
checking close shoulder diameters. It has an open frame and adjustable-blade
anvils.

Many companies have master drawings of snap gages. The designer need
only fill in the dimension needed to gage a part. These master drawings are
then sent to gage manufacturers for filling the order. There are special cases
when a single-purpose fixed-snap gage is desired. For example, the outside
diameter of a narrow groove (as shown in *Figure 9-20*) is checked with a
special double-end snap gage designed to inspect the diameter.

A snap gage may at times be better than a ring gage as an inspection tool.
Figure 9-21 illustrates how a ring gage may accept out-of-round workpieces
that would be rejected by a snap gage.

Flush-Pin Gages

The flush-pin gage is a simple mechanical device used to measure linear
dimensions. The important parts of the gage are the body and a sliding pin or
plunger. The indicating device is a step, ground either on the plunger or on the
flush-pin body, equal to the total tolerance of the dimension. When the gage is
mounted on the workpiece, the position of the plunger can be checked
visually or by fingernail touch.

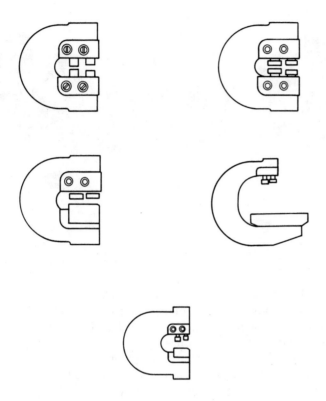

Figure 9-16. AGD adjustable snap gages.

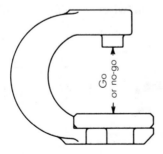

Figure 9-17. Plain snap gage.

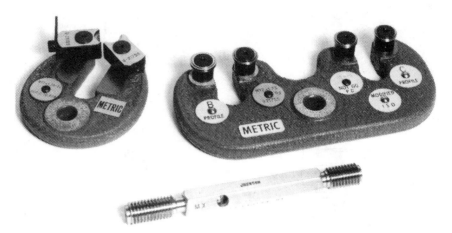

Figure 9-18. Thread snap gages. (*Courtesy, Johnson Gage Co.*)

Figure 9-22 shows a slotted workpiece being checked with a flush-pin gage. Also shown are the relative positions of the plunger at the high and low limits of the depth. The flush pins principle applied in this way is simple in operation, is rugged and foolproof, does not require a master for presetting, and is economical when compared to micrometer or dial gaging methods. The dimension could also be checked with a depth micrometer. However, the depth micrometer requires a greater degree of operator skill and there is the possibility of misreading the instrument.

Figure 9-23 shows an inspection fixture containing two flush-pin gages. Also shown is the workpiece being checked. The first dimension being checked (X) is the distance from the center of the spherical radius to a flat surface. The dimension (Y) is between the center of the same spherical radius and over the outside of a roll (wire). A standard tooling ball in the base of the

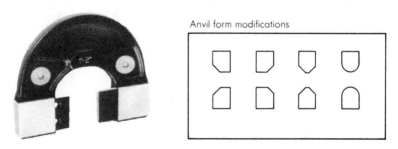

Figure 9-19. Adjustable form and groove snap gage shown with typical anvil form modifications. (*Courtesy, Standard Gage Co.*)

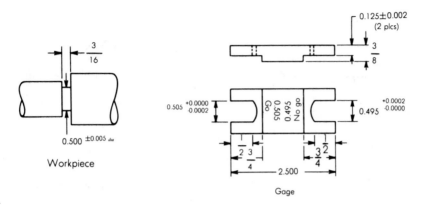

Workpiece

Gage

Figure 9-20. Special snap gage.

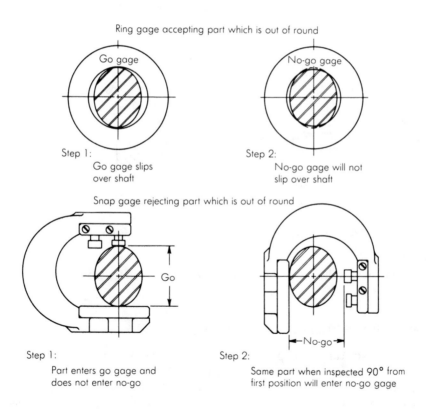

Ring gage accepting part which is out of round

Go gage

No-go gage

Step 1:
Go gage slips
over shaft

Step 2:
No-go gage will not
slip over shaft

Snap gage rejecting part which is out of round

Go

No-go

Step 1:
Part enters go gage and
does not enter no-go

Step 2:
Same part when inspected 90° from
first position will enter no-go gage

Figure 9-21. Gage comparison.

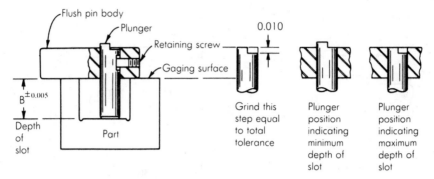

Figure 9-22. Basic application of flush pin gage indicating various positions of plunger.

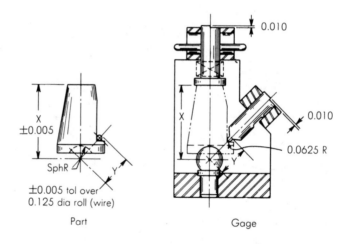

Figure 9-23. Workpiece with flush pin-type inspection fixture.

fixture locates the radius of the workpiece and provides the origin for both dimensions.

Amplification and Magnification of Error

A surface cannot be perfectly flat. Two parts can not be exactly alike. Measured distances or quantities can not be exactly equal. The inability of an inspection device to detect error proves only that the capability of the inspection device is limited. Error or deviation is ever present, and, unless specified or necessary limits are exceeded, are not in themselves a disqualifying factor. The designer specifies what degree of error can be allowed. The inspector confirms that the error in the workpiece is within tolerance. Accuracy of measurement is limited by the accuracy of the standard used for comparison and by the skill of the person making the comparison.

As tolerances become smaller, the primary gaging methods are not precise enough to detect the degree of error. An inspector may, for instance, find that qualification of a workpiece no longer depends on whether or not a gage enters a hole, but rather on how much pressure is needed to insert the gage or

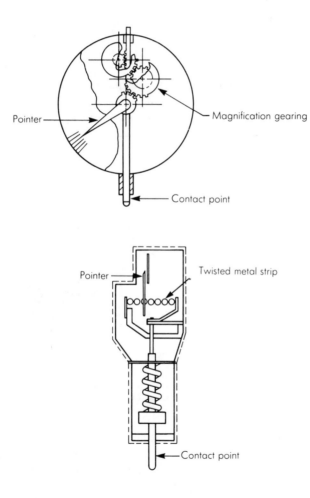

Figure 9-24. (Above) Basic operating principles of dial indicator. (Below) This Comparator employs 100% mechanical amplification without gears or pivots.

whether the gage enters with too little pressure. Inspection becomes more dependent on human judgment and is therefore less reliable.

As changes in dimension become too small to be easily measured, it becomes necessary to amplify or magnify them prior to measurement. This can be done mechanically, pneumatically, optically, or electronically. Some examples of mechanical amplification are shown in *Figure 9-24*.

Dial Indicators

The dial indicator (*Figure 9-25*) is perhaps the most widely used instrument for precise measurement. Basically, it consists of a probe, rack, pinion, pointer, dial, and case. The probe, which is attached to the end of the rack, is placed on the workpiece. A change in workpiece size changes the position of the probe, which in turn moves the rack. The rack movement turns the pinion, which through a gear train causes the dial pointer to move. The graduated dial is calibrated for direct reading of variation from the nominal dimension. Several amplification factors are involved.

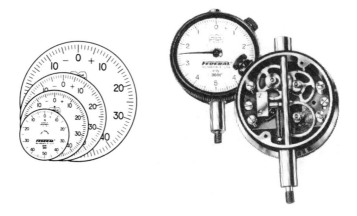

Figure 9-25. Dial indicators. (*Courtesy, Federal Products Corp.*)

Dial indicators are commercially available from many sources. Standard models vary greatly in size, amplification ratio, mounting facilities, and precision. Inexpensive models are available for production use as well as very precise models for use in the gage crib.

An indicator gage has one primary advantage over a fixed gage: it shows how much a workpiece is oversize or undersize. When using an indicator as part of a gaging device, a master block to the nominal dimension to be checked must be used to preset the indicator to zero. Then, in applying the gaging device, the variation from zero, the nominal dimension, is read from the dial scale.

Extremely precise dial indicators are part of the standard inspection equipment found in a gage crib, together with surface plates, parallels, V blocks, and vernier height gages. *Figure 9-26* shows standard inspection tools being used to check runout of a workpiece mounted between centers. As the

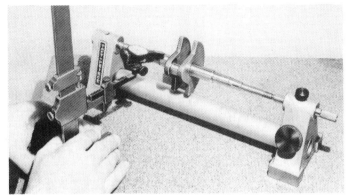

Figure 9-26. Checking runout with standard inspection tools. (*Courtesy, Taft-Peirce*)

part is manually turned, runout of each of the diameters can be read directly as Full Indicator Movement (FIM). *Figure 9-27* shows standard inspection tools being used to check squareness of a workpiece. In checking squareness or runout, the indicator is zeroed with the probe placed on the workpiece. If

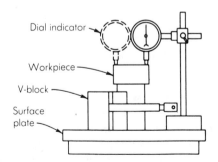

Dial indicator

Workpiece

V-block

Surface plate

Figure 9-27. Squareness gaging fixture composed of standard inspection tools.

the same standard components are used directly as a height gage, the gage must first be zeroed to a known standard such as a gage block or master.

If inspection is frequently needed, it may be expensive to tie up quality control equipment. *Figure 9-28* shows a commercially available gaging fixture complete with surface, indicator, and stand. Also available are attached centers and rolls for checking concentricity of centered or uncentered parts.

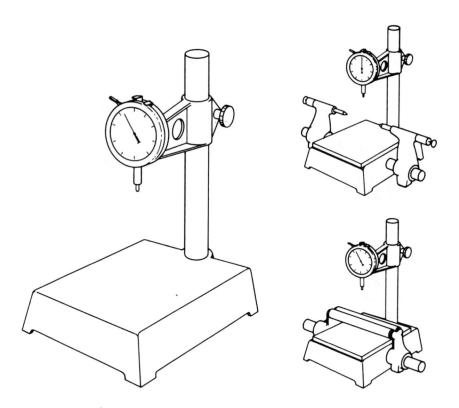

Figure 9-28. Gaging fixture with accessories.

Dial indicators are often used with commercially available plug and snap gages as shown in *Figure 9-29* and *Figure 9-30*. *Figure 9-31* shows three different types of depth gages commercially available. *Figure 9-32* shows several indicator-type bore gages, together with a gage for measuring large internal diameters. *Figure 9-33* shows an indicator-type sweep gage used as part of a special gaging fixture.

Figure 9-29. Comtorgage for checking IDS (A); indicating thread plug gage—portable style (B); indicating thread plug gages—bench style (C); indicating spline gage for checking internal splines (D). (*Courtesy, Ex-Cell-O Corp.*)

The pantograph mechanism in *Figure 9-34* is widely used as a motion transfer mechanism with indicator type gages. It is used as an extension of the indicator spindle to work around projections or recesses on the workpiece without losing accuracy in the gage. It also serves as a shock absorber.

(A)

(B)

(C)

Figure 9-30. Dial snap shown with narow anvils (A); indicating thread gage shown with attachments for checking geometrics (B); indicating spline gage for external splines (C). (*Courtesy, Standard Gage Co.*) (*Courtesy, Johnson Gage Co.*) (*Courtesy, Bendix Automation & Measurement Division*)

Pneumatic Amplification Gages

There are two general types of air gages. One type is operated by varying air pressure (back pressure), and the other is operated by varying air velocity (flow) at constant pressure.

Figure 9-35 shows a differentially controlled, constant amplification pressure-type gage. With this equipment, filtered and regulated air is directed through two master jets to cause the basic restriction in the flow of air. From the master orifice, the air flows directly to the gaging plug, where it is further restricted by the workpiece or setting master, which builds up back

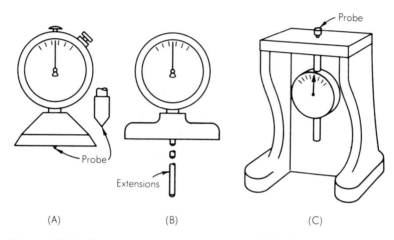

Figure 9-31. Indicator-type depth gages: (*A*) special indicator for checking shallow recesses; (*B*) depth gage with probe extensions; (*C*) bench mounted depth gage.

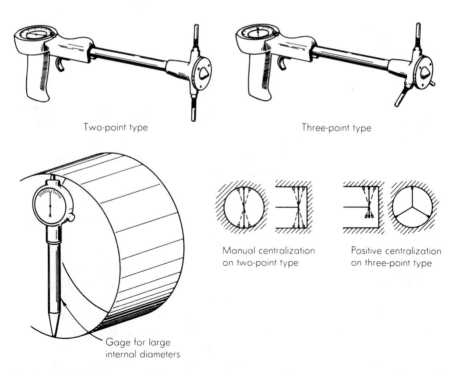

Figure 9-32. Indicator-type bore gages.

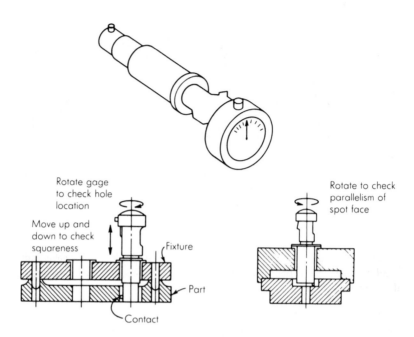

Figure 9-33. Special gaging fixture using indicator-type plug gage as component.

pressure inside the bellows of the differential pressure meter. At the same time, air passing through the equalizing jet flows directly to the zero setting valve. By closing or opening this valve, the pressure in the sealed chamber surrounding the bellows is regulated to match the back pressure from the gaging plug to set the indicator to zero with a master ring.

Any change in pressure caused by differences in the sizes of the gaged workpieces is detected by the differential meter. The system shown in *Figure 9-35* uses one master gage for setting to zero. After the zero setting has been obtained, the actual size of the part can be read on the indicating dial.

This type of gage can also be used for checking external dimensions by replacing the plug gage with either a ring gage or a probe (air cartridge) as shown in *Figure 9-36.* The ring gage operates exactly like the plug gage.

The probe operates in a different manner, but on the same basic principle. As the plunger within the probe rises or lowers, because of workpiece variations, it blocks the flow of air escaping through a jet in the side of the probe. Zero setting registers when the escape jet is blocked halfway, so that the mean amount of air escaping is read as the mean pressure reading. Air probes are widely used on fixtures as shown in *Figure 9-37.*

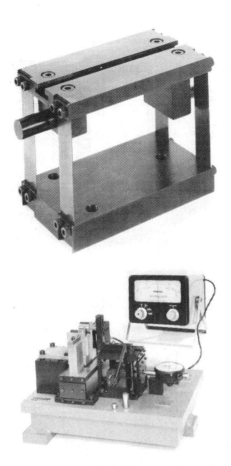

Figure 9-34. (*Top*) Pantograph mechanism. (*Bottom*) Pantograph as used with gaging fixture. (*Courtesy, Federal Products Corp.*)

A schematic of another pressure-type air gage is shown in *Figure 9-38*. It is shown with an electric circuit which allows it to work as a controller for automatic inspection and machine tool size control. In operation, clean regulated compressed air is fed to the gaging circuit and to the pressure-electric relays from the nozzles against the workpiece. The pneumatic system is calibrated by placing two masters on the gaging member one at a time and adjusting the two restrictions until the pointer on the bourdon indicator shows the respective minimum and maximum part tolerance limits. The restrictions exhibit linear pressure characteristics which give an exact indication of part size, regardless of the position of the pointer.

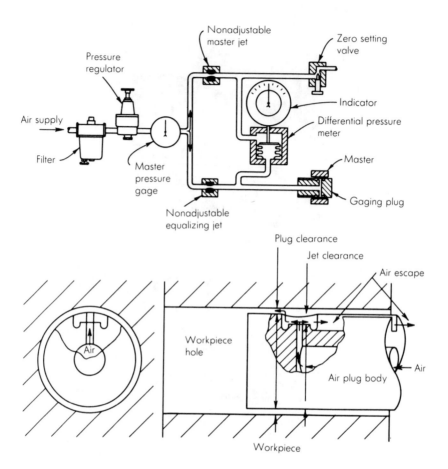

Figure 9-35. Pressure-type air gaging system.

At the same time, the pneumatic size signal is converted to electrical outputs by the pressure—electric relay. The schematic shows an SPDT switching element (up to four can be combined in a single relay). The reference pressure is set by the firing point adjustment screw.

Before firing, the reference pressure is present in both the *A* and *B* chambers. Due to the larger area of diaphragm *A*, the switch leaf is held firmly against one set of contacts. When the gaging pressure rises above the reference pressure, the control diaphragm closes the nozzle, which stops any further flow of the reference pressure to chamber *A*. The pressure in chamber *A* falls off rapidly to zero because of the large vent. The reference pressure then snaps

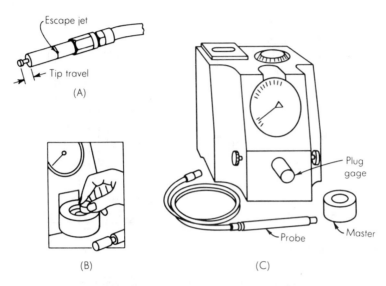

Figure 9-36. Pressure-type air gage: (*A*) probe for inspecting internal diameter; (*B*) ring gage for inspecting outside diameter; (*C*) complete gage with accessories.

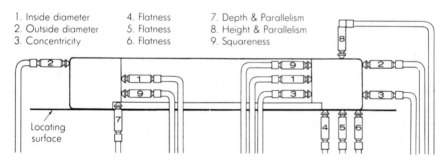

Figure 9-37. Typical air gage cartridge application. (*Courtesy, Dearborn Gage Co.*)

the switch leaf to the left, breaking one of the contacts and making the other contact.

A flow air gage is shown in *Figure 9-39*. Filtered air is reduced in pressure by a pressure regulator and flows upward through a transparent tapered tube. A float is suspended in the air column in the tube and moves up and down according to the flow of air. From the top of the tube, the air flows through a hose and exhausts through the clearance between the gaging member and the

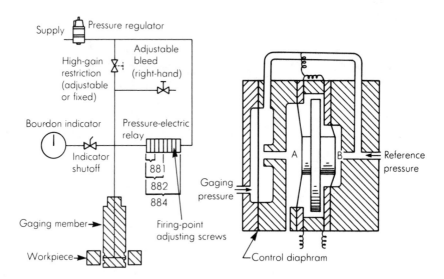

Figure 9-38. Pressure type air gage with integral electric circuit.

workpiece. Since the rate of air flow is proportional to the clearance, it is indicated directly by the position of the float in the column.

Proper setting masters are used to set the upper and lower limits of the float travel representing the maximum and minimum part tolerance limits. The proper float movement between these limits is then regulated by the external float positioning and calibration adjustments.

For parts with a surface finish greater than 50 μin. (1.27 μm), and for other special applications, air gage contacts such as shown in *Figure 9-40* are specified.

Figure 9-41 shows a special air-electric gage designed for 100% inspection.

Figure 9-42 shows air gage application data for a wide variety of measurements.

Electronic Gages

Electronic gages measure by changing physical displacement into an output voltage which is proportional to the displacement. A typical electronic gage consists of a transducer to convert the measurement into a voltage. The voltage is then amplified, rectified, and displayed on a suitable voltmeter, which is calibrated to display the actual workpiece dimension or difference from a predetermined basic specification. The transducers are generally a version of one of the three gage heads shown in *Figure 9-43*. The most common transducers are the linear-variable-differential-transformer (LVDT)

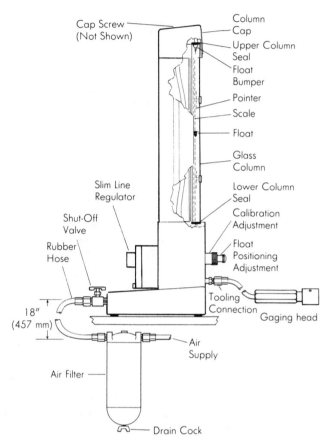

Figure 9-39. Flow type air gage.

and the E-transformer shown in *Figure 9-44*. A schematic diagram of an electronic gaging system is shown in *Figure 9-45*.

The high sensitivity and extremely fast response time of electronic gaging makes it suitable for high-speed sorting and classifying operations. A typical application is shown in *Figure 9-46*. *Figure 9-47* shows electronic gage application data for a wide variety of measurements.

Optical Projection Gaging

Optical projection is a method of measurement and gaging using a precision instrument known as an optical projector or comparator. Optical gaging is gaging by sight rather than feel or pressure. The measurement or

Figure 9-40. (*Top, Balljet Spindles*) Recommended for gaging inside diameters in soft or porous parts; when surface finish is 65 μin (1.65 μm) rms or rougher; for narrow lands; for gaging bores to hole edge; for diameters larger than 1/2" (12.7mm). Standard diameters from 0.375 to 2.00"(9.52-50.8mm) for thru or blind holes. Calibrate with gage blocks and standard calibrator. (*Middle, Leafjet Spindles*) Semistandard contact gaging elements for checking hole diameters 7/16" (11.1mm) and larger. Used also for checking to extreme bottom of blind holes and to extreme edge of holes and grooves. Two or more leaves hinged on thin flexing steel section; no sticking. (*Bottom, Bladejet Spindles*) For use on hole diameters 5/8-4" (16 - 102 mm); for checking blind or thru holes in laminated and sintered parts; for inspecting gun bores and holes in which oil grooves, keyways, or slots prevent using standard Balljet or Leafjet spindles. (*Courtesy, Bendix Automation & Measurement Division*)

gaging is performed by placing a workpiece in the path of a beam of light and in front of a magnifying-lens system, thereby projecting an enlarged silhouette shadow of the object upon a translucent screen as illustrated in *Figure 9-48*.

Measurement of a workpiece normally requires a chart gage with reference lines in two planes. In gaging applications, a precisely scaled layout of the contour of the part to be gaged, usually with tolerance limits, is drawn on the screen. *Figure 9-49* shows a workpiece with the chart gage used for its inspection.

Optical gages are almost unaffected by wear. There is no wear to a light beam, and any fixture wear can be compensated for by repositioning to the setting point. Little operator skill is required; no touching skill or sensitivity is required. Dimensions can be changed on the screen easily and quickly. The chart provides exact duplication. Several dimensions can be checked at the same time, eliminating too much handling of the part.

Most optical projectors have standard magnification of 10 power up to 100 power. The effective area that can be projected through a given lens system can be determined by dividing the screen diameter by the magnification of the lens being used. For example, a comparator with a 14" (356 mm) diameter

Figure 9-41. Thirty seven dimensions on camshafts are checked simultaneously by this gage which utilizes the Federal air-electric measuring system. A green master light at the top of the panel lights to show when all dimensions are good. Lights on individual modular units signal out of tolerance dimensions and show whether they are "scrap" or "salvage". (*Courtesy, Federal Products Corp.*)

screen, using 10-power magnification, will project a complete area of a 1.400″ (35.56 mm) diameter specimen at one setting.

A chart gage is an accurately scribed, magnified, outline drawing of the workpiece to be gaged, containing all the contours, dimensions, and tolerance limits necessary. The chart is made of glass or plastic. For very short and quick checks, drafting paper can be used. However, paper should not be used consistently because it shrinks with changes of weather. Chart layout lines should be dark black, sharply defined, and from 0.006″ to 0.020″ (0.15 mm to 0.51 mm) wide for best legibility. Dimensions extend normally to the center of

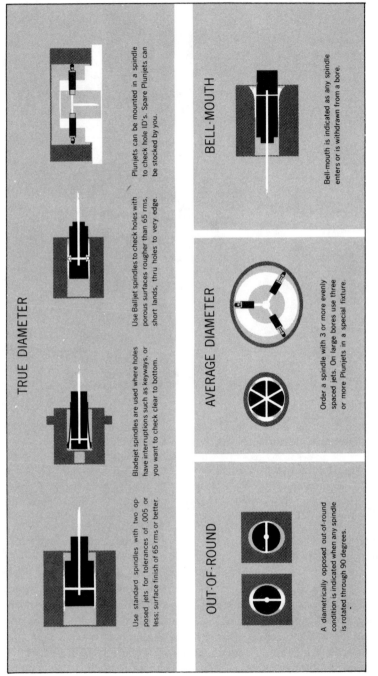

Figure 9-42. Application data - air gages. (*Courtesy, Bendix Automation and Measurement Division*)

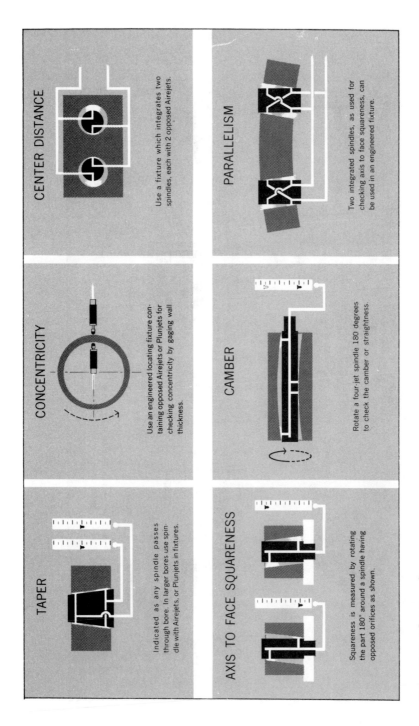

Figure 9-42. (Continued)

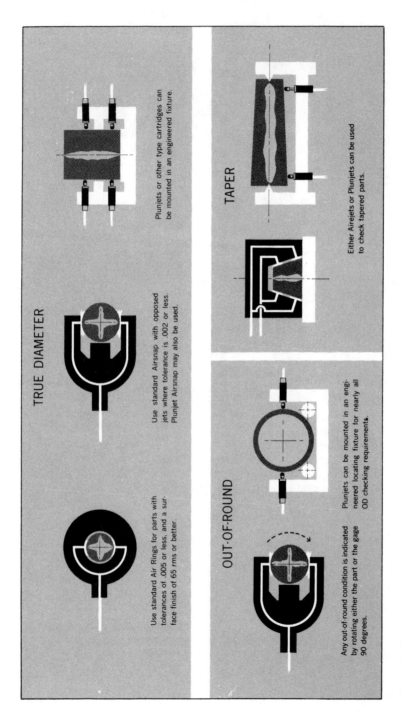

Figure 9-42. (Continued)

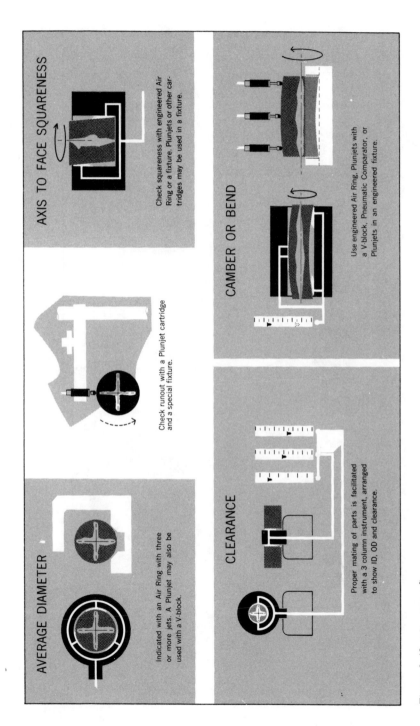

Figure 9-42. (Continued)

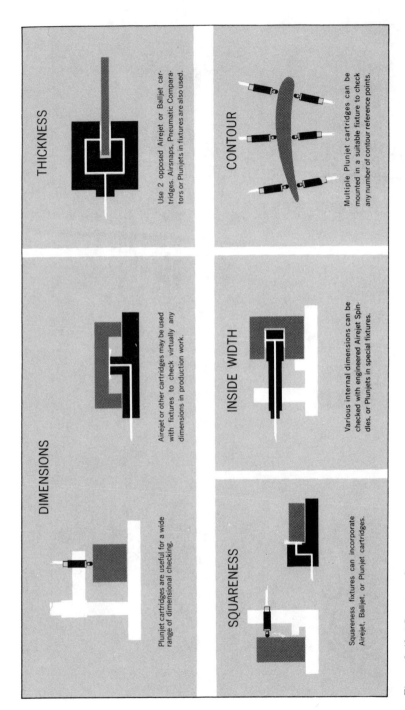

Figure 9-42. (Continued)

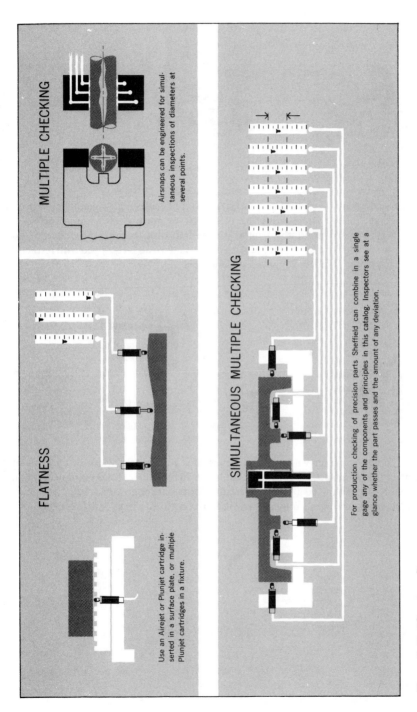

MULTIPLE CHECKING

Airsnaps can be engineered for simultaneous inspections of diameters at several points.

FLATNESS

Use an Airejet or Plunjet cartridge inserted in a surface plate, or multiple Plunjet cartridges in a fixture.

SIMULTANEOUS MULTIPLE CHECKING

For production checking of precision parts Sheffield can combine in a single gage any of the components and principles in this catalog. Inspectors see at a glance whether the part passes and the amount of any deviation.

Figure 9-42. (Continued).

Lever type

Single scale Amplifier with Digital Display Unit provides high magnification over a long range. Silicon wafers differing in size by as much as .020″ are checked for thickness to tens of millionths. Digital Unit permits fifth decimal place determination at any point within ±.010″ range of Amplifier. Gage Head is equipped with retraction lever, has approximately 2.5 grams gaging pressure.

Reed spring type

Dual input Amplifier with two fixtures is used to compare width of an armature contact to slot in spool. Proper size contact provides .0002″ total clearance. Gaging accuracy is to within .000010″. Gage head polarities are opposite so Amplifier ignores combined size, responds only to size difference between pieces being matched.

Cartridge type

Figure 9-43. Transducers and their application. (*Courtesy, Federal Products Corp.*)

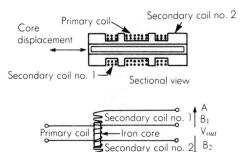

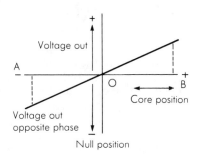

LVDT type

E-transformer type

Figure 9-44. Transducers. (*Courtesy, Federal Products Corp.*)

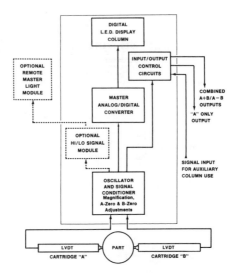

Figure 9-45. Schematic diagram of an electronic gage and principle of operation. As shown in the block diagram at left, each cartridge contains a linear variable differential transformer (LVDT) which converts any movement of the gaging cartridge contact into an electrical signal output to the instrument's 51-diode display column.

In operation, a high-frequency oscillator supplies excitation from the instrument to the LVDT's primary coil. This causes the LVDT's secondary coils to supply a stepless linear signal input to the instrument's signal conditioner which compares the inputs to the original oscillator frequency and rejects all extraneous signals.

As shown, the signal conditioner also provides means for adjusting magnification or span as well as zero position for inputs from both A and B gaging cartridges.

Pure gaging signals are then demodulated, amplified, and fed through the instrument's analog-to-digital converter to the light emitting diode (LED) display column which provides instantaneous visual indication of the part's size or condition in relation to its specified tolerance spread.

The *Combined Output* circuit converts linear analog dimensional data into voltage over the full range of the instrument's calibrated scale. Output signal from the instrument is proportional to the amount of the combined dimension being displayed. Signals from cartridges +A, +B, A+B, or A-B are obtained by setting the mode selector switch at front of the instrument's amplifier module.

The *A Only Output* circuit provides an analog voltage that is proportional to the value of the A cartridge channel regardless of what operating mode the selector switch is in. The output voltage is approximately 0.5 volts/0.001" (0.025 mm) for all magnifications.

The *Signal Input* circuit provides means for displaying any analog signal of ±10 volts from other instruments in other locations. When connecting line from the remote signal source is plugged in, the signal from the instrument's gaging circuit is interrupted and the incoming information replaces it on the front LED display column. (*Courtesy, Dearborn Gage Co.*)

the lines. When maximum and minimum tolerance lines are used, the magnification should be high enough to maintain a minimum of 0.020" (0.51 mm) spacing between the lines. For closer tolerance checking, special lines or bridge arrangements, based on gaging to the edge of the lines, are often used. *Figure 9-50* shows two portions of a chart gage. The first portion (A) uses maximum and minimum tolerance lines. The second portion (B) has a close tolerance bridge arrangement.

Workholding equipment includes vises for flat workpieces and staging centers for round workpieces such as shafts and cylindrical parts with machined centers. V-blocks, in a range of sizes, are mounted on bases to clamp to projector work table slots. Diameter capacities run from almost zero to 5" (127 mm) diameter.

Coordinate Measuring Machines

Coordinate measuring machines (CMMs) are available in many configurations, from single-axis, manually-operated digital readout machines to the multi-axis direct computer controlled (DCC) machines shown in *Figure 9-51* and *Figure 9-52*.

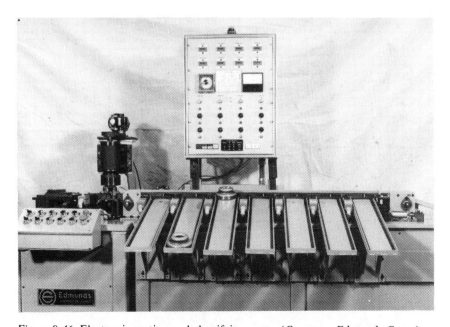

Figure 9-46. Electronic sorting and classifying gage. (*Courtesy, Edmunds Gages*)

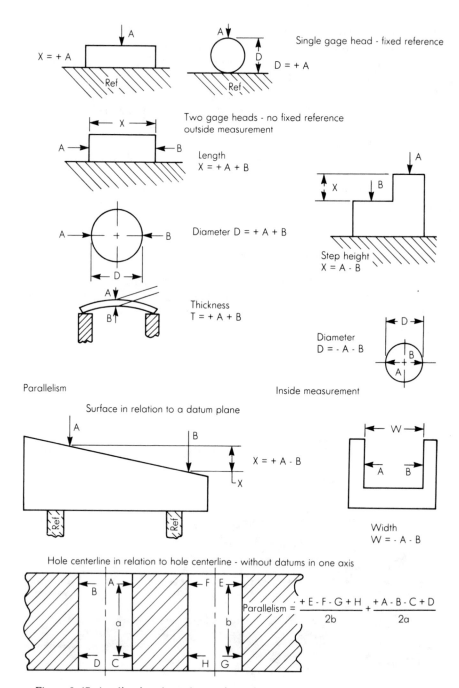

Figure 9-47. Application data-electronic gaging.

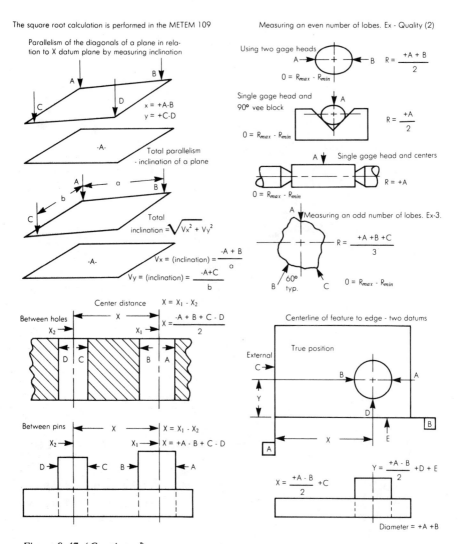

Figure 9-47. (Continued)

A system diagram of a measurement processor for automatic direct computer control of a CMM is shown in *Figure 9-53*.

CMMs are presently being used to check a wide variety of characteristics on an equally wide variety of parts such as engine blocks, cams, gears, etc. By substituting an LVDT probe for the hard ball probe, complete profiles can be measured automatically.

Figure 9-54 shows application data and geometric capabilities of a measurement processor.

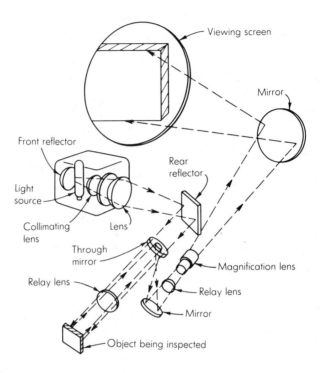

Figure 9-48. Optical projection gaging principle.

Gaging Geometric Dimensioned and Toleranced Parts

Modern geometric dimensioning and tolerancing (geometrics) is rapidly being adopted as the universal engineering drawing language, since it allows the workpiece designer to more clearly express his intent and improve the understanding of the manufacturer.

In order to properly gage geometric dimensioned and toleranced parts, it is first necessary to understand the terms and symbols endorsed by the American National Standards Institute (ANSI Y14.5) and the International Standards Organization (ISO 1101) as shown in *Figure 9-55*.

Additional Symbols and Definitions

Maximum Material Condition (MMC)—a modifier showing that the feature size contains the maximum amount of material. Examples: The

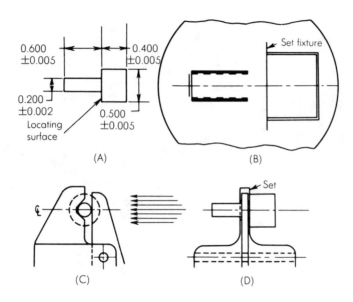

Figure 9-49. Optical gage setup: (*A*) workpiece; (*B*) chart gage; (*C*) side view of holding fixture; (*D*) front view of holding fixture.

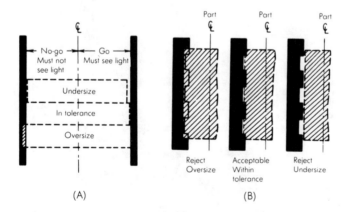

Figure 9-50. Chart gage segments: (*A*) maximum and minimum tolerance lines; (*B*) close tolerance bridge arrangement.

minimum diameter of a hole and maximum diameter of a shaft. MMC is designated by the symbol Ⓜ.

Regardless of Feature Size (RFS)—another modifier which requires that

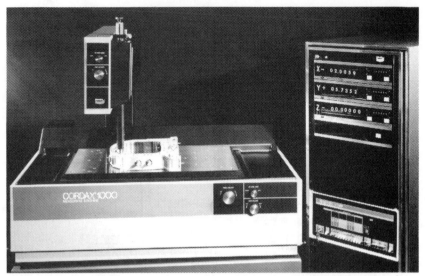

Figure 9-51. Coordinate measuring machine. (*Courtesy, Bendix Automation & Measurement Division*)

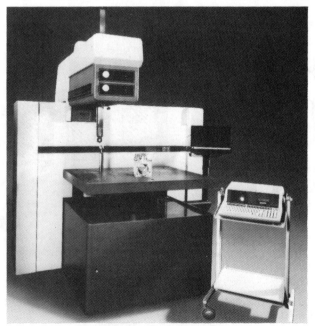

Figure 9-52. Coordinate measuring machine. (*Courtesy, Bendix Automation & Measurement Division*)

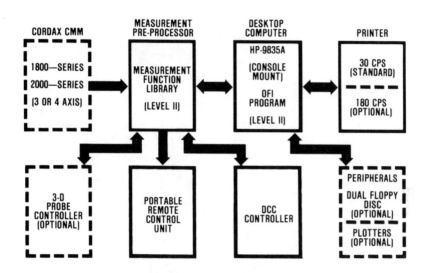

Figure 9-53. Measurement processor for a direct, computer-controlled coordinate measuring machine.

the tolerance of form, runout, or position must be met regardless of where the feature lies within its size tolerance. In the ISO system, there is no symbol for RFS, and it is assumed unless MMC is specifically designated. In the USA system, the RFS principle is used automatically when a geometric tolerance other than position is applied. For a tolerance of position, RFS is specified on the drawing by the symbol (S).

Least Material Condition (LMC)—a modifier opposite to MMC—where the feature size contains the minimum amount of material. Examples: The maximum diameter of a hole and minimum diameter of a shaft. It is used to permit additional tolerance when the accuracy of location can be relaxed as the feature size approaches MMC. LMC is designated by the symbol (L).

Projected Tolerance Zone—used to control the 90° angle of a hole into which a pin, stud, screw, etc., will be assembled. It does this by extending a tolerance zone beyond the surface of the part to the functional length of the assembled pin, screw, stud, etc. A projected tolerance zone is indicated by the symbol (P).

Basic Dimension—a theoretical value used to describe the exact size, shape or location of a feature. A tolerance is always required with a basic dimension to show the permissible variation from the stated value. A basic dimension is symbolized by boxing (1.000).

Datum Identification Symbol—identifies the feature of a part from which functional relationships are established. It is symbolized by boxing the letter as shown in *Figure 9-56.*

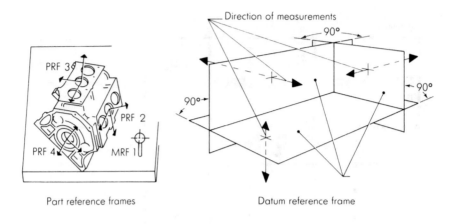

Part reference frames Datum reference frame

BASIC GEOMETRIC CAPABILITIES

INPUT FEATURES	RELATIONSHIP		FEATURE CONSTRUCTIONS			
	SHORTEST DISTANCE	ACUTE ANGLE $\angle$	POINT	NORMAL LINE	PARALLEL LINE //	NORMAL PLANE $\perp$
POINT & POINT		N.A.	MID-POINT	$\eta = 2$	N.A.	MID-PLANE
POINT & LINE		N.A.	N.A.	N.A.	N.A.	N.A.
POINT & PLANE		N.A.	N.A.	N.A.	N.A.	N.A.
LINE & LINE			INTERSECTION		N.A.	MID-PLANE
LINE & PLANE			INTERSECTION	N.A.	PROJECTION	
PLANE & PLANE			N.A.	N.A.	INTERSECTION	N.A.

Figure 9-54. Coordinate measuring machine application data and geometric capabilities.

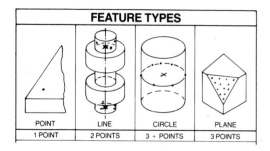

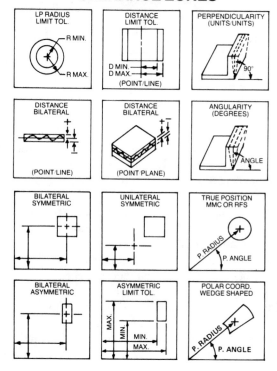

Figure 9-54. (Continued)

Characteristic	ANSI-Y14.5	ISO 1101
Flatness	▱	▱
Straightness	—	—
Roundness (Circularity)	○	○
Cylindricity	⌀	⌀
Profile of any line	⌒	⌒
Profile of any surface	⌓	⌓
Perpendicularity (Squareness)	⊥	⊥
Angularity	∠	∠
Parallelism	//	//
Runout (Circular)	↗	↗
Runout (Total)	Total ↗	↗↗
Position	⊕	⊕
Concentricity	◎	◎
Symmetry	≐	≐

Figure 9-55. Geometric characteristics and symbols.

Figure 9-56. Datum identification symbol.

Each feature requiring identification as the datum uses a different letter(s). To eliminate confusion, the letters I, O, and Q are not used as reference letters. Where the alphabet is used up, double letters, such as AA, BB, ZZ, etc., are used.

Feature Control Symbol (Frame)—also a boxed expression containing the geometric characteristics symbol and the form, runout, or location tolerances, plus any datum references and modifiers for the feature or datum. Proper symbols and application are shown in *Figure 9-57.*

Note: In the past, USA datum references preceded the tolerances, see *Figure 9-58.* However, this practice is being phased out in favor of the international sequence where the datum reference follows the tolerance, see *Figure 9-59.*

For convenience or conservation of drawing space, the feature control symbol and datum identification symbol may be combined as shown in *Figure 9-60.*

A common datum such as the axis or centerplane of a part can be established by the use of two datum letters separated by a dash, see *Figure 9-61.*

Feature control symbols are placed on drawings in accordance with standard drawing practices as noted in *Figure 9-62.*

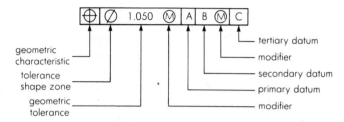

geometric
characteristic

tolerance
shape zone

geometric
tolerance

tertiary datum

modifier

secondary datum

primary datum

modifier

Figure 9-57. Feature control symbol (frame).

Figure 9-58. Datum references preceeding tolerance.

Figure 9-59. Datum reference following tolerance.

Figure 9-60. Combined feature control symbol and datum identification symbol.

Figure 9-61. Common datum for axis or centerplane.

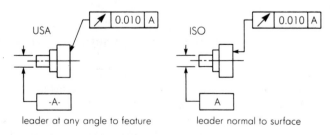

USA

ISO

leader at any angle to feature

leader normal to surface

Figure 9-62. Feature control symbols.

The Tolerance Zone which is shown in the feature control symbol represents the total allowable change from the desired form, attitude, runout or location of the feature. Where the specified tolerance is to indicate the diameter of a cylindrical zone, the diameter symbol ϕ is placed ahead of the tolerance in the feature control symbol as shown in *Figure 9-63*. Otherwise, the tolerance zone represents the total distance between two parallel lines, planes or geometric boundaries.

Datum Targets—may be used wherever datum orientation is required on castings, forgings, weldments, etc., or to ensure repeatability between manufacturing and inspection.

Older datum targets are designated by a circle divided into four quadrants, with the datum letter of the datum plane in the upper left and the number of the target in the lower right of the circle. The newer datum target is shown with a circle divided in half. The lower half denotes the specific datum and the upper half, when used, shows the datum size. These symbols and their proper applications are shown in *Figure 9-64*.

Three Plane Concept

A datum is a point, line, plane, cylinder, axis, etc., which is assumed to be exact for purposes of computation or reference and from which location of other features of a part may be determined. Data are established by, or relative to, actual part features and therefore include all the irregularities of the part feature. When more than one datum feature is required on a part, it is advisable to establish a three-plane datum system consisting of three planes at right angles to each other. Where a primary datum plane is established by three points (minimum) somewhere on the most influential datum surface, a secondary datum plane is established by two points (minimum) on the second most influential datum surface, and a third datum plane is established by one point (minimum) on the least influential datum surface.

Where the size of a part such as a cylinder or width is used as a datum on a RFS basis, a datum axis or datum centerplane must be established by making contact with the feature surface extremities by using precision expanding chucks, mandrels, etc.

Virtual Condition—the boundary described by all the effects of the maximum material condition of a feature plus any applicable geometrical tolerances.

Full Indicator Movement (FIM)—the total indicator movement reading observed when properly applied to a part feature; the same as full indicator reading (FIR) and total indicator reading (TIR).

General Rules—to provide users of geometrics with a better understanding of the system and provide for proper applications, five general rules have been established by ANSI Y 14.5

Figure 9-63. Tolerance zone.

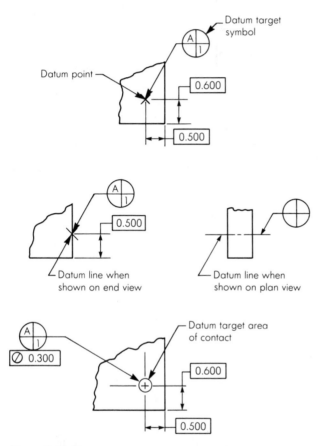

Figure 9-64. Datum targets.

Rule 1. Unless otherwise specified, the limits of an individual feature of size controls the form as well as the size.

- a) No element of the actual feature shall extend beyond a boundary of perfect form at the maximum material condition (MMC).
- b) The actual size of the feature measured at any cross-section shall be within the least material condition (LMC) limit of size.
- c) Form control provisions do not apply to commercial stock. *Figure 9-65* shows the allowable changes of form permitted by this rule.

Rule 2. For a tolerance of position, (formerly called True Position

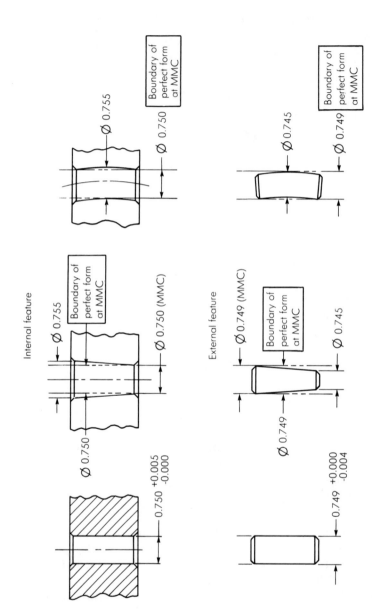

Figure 9-65. Rule 1.

Tolerance), MMC or RFS is specified on the drawing with respect to the individual tolerance, datum reference, or both, as applicable. See *Figure 9-66*.

Rule 3. For other than a tolerance of position, where no modifier is specified, RFS applies with respect to the individual tolerance datum reference, or both. Where MMC is required, it must be specified on the drawing. See *Figure 9-67*.

Rule 4. Each tolerance of attitude (orientation), runout, or location and datum reference specified for a screw thread applies to the pitch diameter. Where an exception to this practice is necessary, a notation must be added beneath the feature control symbol or datum identification symbol, *Figure 9-68*.

Rule 5. Datum features of size which are connected by a separate tolerance of form, attitude, or position, and are noted within the same feature control symbol, apply at their virtual condition. When this is not intended, a zero tolerance at MMC should be specified for the appropriate datum features. For example, see *Figure 9-69*.

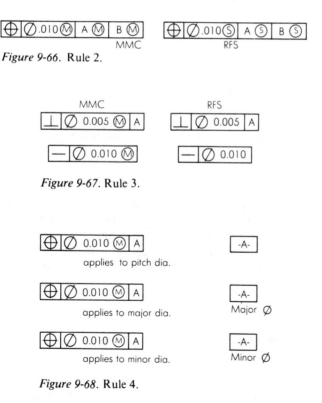

Figure 9-66. Rule 2.

Figure 9-67. Rule 3.

Figure 9-68. Rule 4.

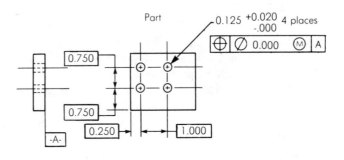

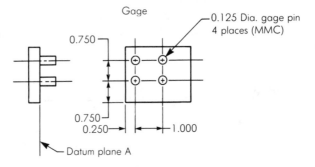

Note: Go plug gaging is not rec'd. with this gage.

Figure 9-69. Positional tolerance of zero at MMC.

Flatness

Flatness is a condition in which all elements of a surface are in the same plane. The flatness tolerance specifies a tolerance zone bordered by two parallel planes within which the entire surface must lie. No datum is needed or proper with a flatness tolerance. When checking flatness, all elements of the concerned surface must also be within the specified size limits of the part to be acceptable. An example of flatness tolerance and its meaning is shown in *Figure 9-70*.

Gaging Methods

Various methods of checking flatness are available. Method selection depends on the accuracy required, the size of the part, and the time available to make the check. Flatness cannot be easily checked by functional methods.

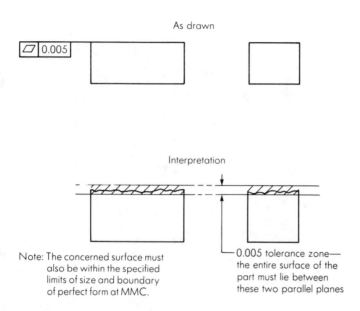

As drawn

☐ 0.005

Interpretation

Note: The concerned surface must
     also be within the specified
     limits of size and boundary
     of perfect form at MMC.

—0.005 tolerance zone—
the entire surface of the
part must lie between
these two parallel planes

Figure 9-70. Flatness tolerance and its meaning.

One of the most widely used methods for checking flatness is the direct contact method. With this method, the part being checked is brought into direct contact with a reference plane of known flatness which has been covered with a thin coating of prussian blue. The high spots on the part are indicated by the transfer of the blue dye. This method, however, does not lend itself to geometrics since the results are not measurable.

Another common method for checking flatness is with an indicator, stand, leveling device, and a surface plate of known flatness, as shown in *Figure 9-71*. With this method, the surface to be checked is first adjusted with the leveling device as shown by the indicator to establish the most accurate plane possible parallel to the surface plate. The entire surface is then explored with the indicator to determine the full indicator movement (FIM) which is the measure of flatness.

Figure 9-72 shows a relatively fast and accurate method for checking flatness productively. The gage consists of a surface plate with an air probe (shown), electronic probe, or mechanical indicator set into a recessed mount with the measuring contact projecting past the plate surface. To check flatness, the complete surface area to be measured is then passed over the extended contact point which measures and displays the differences of the individual elements on the surface to determine the full indicator movement (FIM).

Figure 9-71. Leveling a part. (*Courtesy, Federal Products Corp.*)

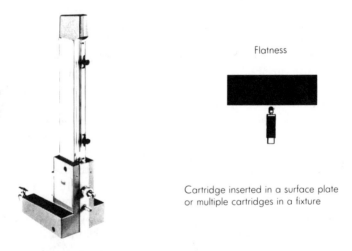

Flatness

Cartridge inserted in a surface plate
or multiple cartridges in a fixture

Figure 9-72. Surface plate and recessed probe.

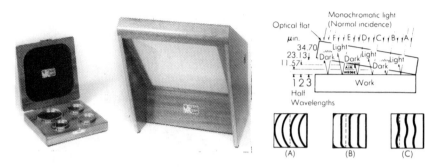

Figure 9-73. (*Left*) Optical flats and monochromatic light source. (*Right*) Rays fall where wedge thicknesses are just one, two, three, etc., half-wavelengths, and are reflected partly from the work. At reflection, each of these particular rays interferes with itself, in accordance with optical laws, thus cancelling its own light and appearing from above as a narrow dark band. Since each dark band is like a contour line, it defines a path across the wedge wherever its thickness is exactly uniform. (*Courtesy, Van Keuren Co.*)

The combination of an optical flat and a monochromatic light source as shown in *Figure 9-73* is also used for checking flatness of relatively smooth surfaces on parts being produced on a production line. Differences from perfect flatness on the surface being checked cause differences in the parallelism and straightness of the interference fringes, which are proportional to the flatness error when the optical flat is applied to the surface of the part under the monochromatic light.

These differences are easily converted to a measure of flatness.

The plano-interferometer shown in *Figure 9-74* checks flatness in a similar manner.

For large, accurate surfaces, the inspector must use a precision level or an autocollimator such as those shown in Figure 9-75. While each of these instruments has a different principle of operation, the use of the instruments for the measurement of flatness is basically the same. In each case, the instrument is used to measure a change in angular rotation of a carriage resting on two feet as the carriage is moved along several tracks on the surface in equal distances. This rotation is changed into linear rise and fall, the tracks related to one another as the surface area is mapped to indicate the changes from a reference plane. *Figure 9-76* shows the checking of flatness with an autocollimator.

Along with any of the above checks, the surface must also be checked separately to assure that it is within the specified limits of size and the boundary of perfect form at MMC.

Figure 9-74. The 12-inch aperture Plano Interferometer is an optical measuring instrument used as a simple and efficient means for measuring the flatness of 12-inch diameter reflective surfaces and providing a permanent record of the observed fringe pattern. There is no physical contact between the master reference and the surface to be measured. (*Courtesy, Davidson Optronics, Inc.*)

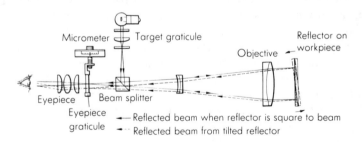

Figure 9-75. A precision level (*above*); an autocollimator (below). (*Courtesy, L. S. Starrett Co.*)

Straightness

Straightness is a condition where a portion of a surface or an axis is a straight line. The straightness tolerance specifies a tolerance zone bordered by two parallel straight lines within which the entire axis or element must lie. No datum is needed or proper with a straightness tolerance. When checking straightness of a surface element, all elements of the surface must also be within the specified size tolerance and the boundary of perfect form at MMC. However, when checking the straightness of an axis or a center plane either RFS or MMC, virtual condition results and the boundary of perfect form may be exceeded up to the stated tolerance, but each cross sectional element must still be within the specified part size. Examples of straightness tolerances and their meanings are shown in *Figure 9-77*.

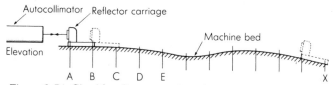

Figure 9-76. Checking flatness and straightness with an autocollimator.

Gaging Methods

Straightness can be checked by various methods depending on the accuracy required, and the part size.

One common method of checking straightness is with a straight edge or surface plate of known straightness and gage blocks or thickness stock. With this method, the part to be checked is supported at two places equidistant from the surface, and the differences from parallel measured with gage blocks or an indicator set up to determine the FIM or equivalent. Cylindrical parts and narrow surfaces are sometimes placed directly on the surface plate and the differences checked with thickness stock. For very small tolerances, straightness can be checked by placing a toolmaker's knife edge directly on the part as shown in *Figure 9-78*. Differences in straightness are indicated by the presence and width of light gaps. Deciding straightness by this method, however, does not generally provide measureable results.

A more practical check can be made with a measuring machine as shown in *Figure 9-79*. The tracking accuracy of these machines permits checking, with a high degree of resolution, the locations of many separate points along the surface and provides numerical information on any out-of-straight condition.

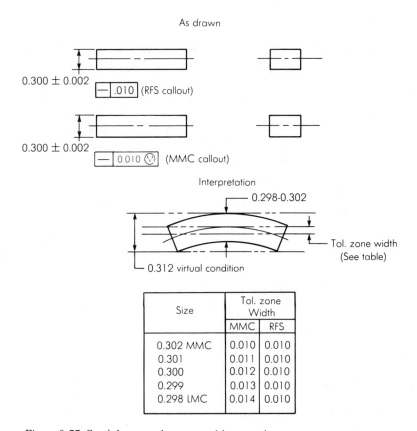

As drawn

0.300 ± 0.002

$\boxed{-\ \boxed{.010}}$ (RFS callout)

0.300 ± 0.002

$\boxed{-\ \boxed{0.010\ \ⓥ}}$ (MMC callout)

Interpretation

0.298-0.302

Tol. zone width
(See table)

0.312 virtual condition

Size	Tol. zone Width	
	MMC	RFS
0.302 MMC	0.010	0.010
0.301	0.011	0.010
0.300	0.012	0.010
0.299	0.013	0.010
0.298 LMC	0.014	0.010

Figure 9-77. Straightness tolerance and its meaning.

When these machines have computer-assist, precision setups are not needed and the condition can be displayed directly.

Straightness can also be measured by comparison to the changed path of a precision slide. *Figure 9-80* shows an arrangement for checking straightness where the part is mounted on the slide and passed under an indicating device which displays variations from the straight. *Figure 9-81* shows an arrangement holding the part is held motionless while an indicating device mounted on a precision slide moves over the part. In both cases, the part must be aligned with the slide track to determine the FIM, unless the equipment is equipped with an auto-level device, which eliminates the need.

Large, accurate straightness measurements are made with precision levels, electronic levels, autocollimators, or alignment telescopes using the same methods described for checking flatness.

When straightness is specified MMC, it can be checked with a functional

As drawn

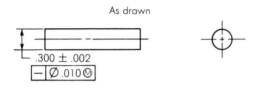

.300 ± .002

| – | Ø .010 Ⓜ |

Interpretation

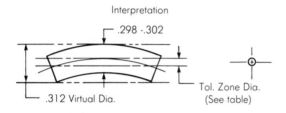

.298 -.302

.312 Virtual Dia.

Tol. Zone Dia.
(See table)

Size	Tolerance Zone Dia.
.302 MMC	.010
.301	.011
.300	.012
.299	.013
.298 LMC	.014

Note: A functional gage to check this part is shown in Figure 9-82.

Figure 9-77. (Continued)

gage. A simple gage to check the straightness of an axis (MMC) on the sample part is shown in *Figure 9-82*. A gage to check the straightness of a centerplane (MMC) would basically consist of two parallel surfaces 0.312″ (7.92 mm) apart through which the part must pass to be acceptable.

Circularity (Roundness)

With respect to a cylinder or cone, circularity is having all points of the surface intersected by any plane at right angles to a common axis, equal distances from the axis.

With respect to the sphere, circularity is having all points of the surface intersected by any plane passing through a common center, equal distances from that center.

The circularity tolerance specifies a tolerance zone bounded by two concentric circles within which each circular element of the surface must lie.

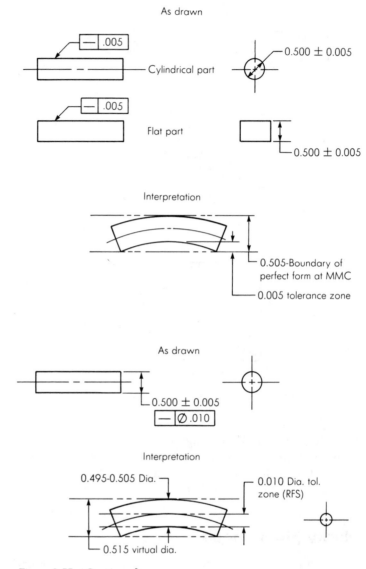

Figure 9-77. (Continued).

No datum is needed nor proper with a circularity tolerance. When checking circularity, all elements of the surface must also be within the specified size tolerance to be acceptable. Examples of circularity tolerances and their meaning are shown in *Figure 9-83*.

Figure 9-78. Checking straightness with a toolmaker's straightedge.

Figure 9-79. Checking straightness with a measuring machine. (*Courtesy, Federal Products Corp.*)

Gaging Methods

Although the common method for checking circularity use V-block or checking from centers, they are not recommended in this text. The many possible sources of error such as lobing, angle of V-block, center misalignment, etc., contribute to an inaccurate reading of the conditions. Circularity

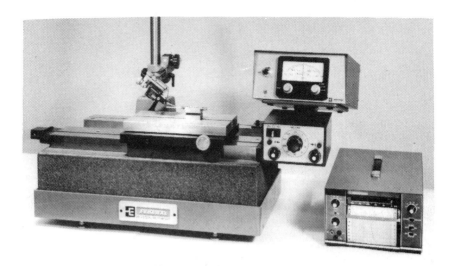

Figure 9-80. Checking straightness with part mounted on slide - indicator fixed. (*Courtesy, Federal Products Corp.*)

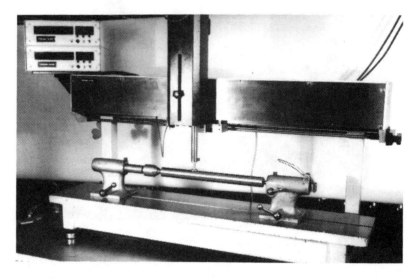

Figure 9-81. Checking Straightness while the part is held motionless.

should be checked with a precision rotating spindle, rotating table, or circular tracing instrument as shown in *Figure 9-84*. Measurement is made by centering the part to be measured on the table, establishing an axis, and placing a stylus in contact with the surface of the circular cross-section. The

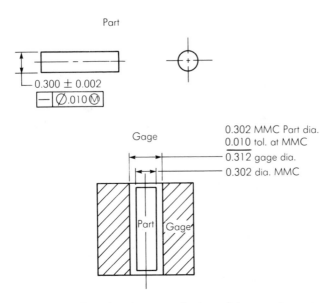

Part

0.300 ± 0.002

$\boxed{-\ \varnothing.010\ \text{Ⓜ}}$

Gage

0.302 MMC Part dia.
0.010 tol. at MMC
0.312 gage dia.
0.302 dia. MMC

Part Gage

Figure 9-82. Functional gage to check straightness of an axis MMC.

stylus contacts the surface normal to the axis that is being examined to pick up, magnify, and display departures from roundness for determination of any out-of-round. These instruments can be equipped with auto-centering capability, eliminating the need to precisely align the part before checking. Properly equipped, these same instruments can also be used for the complete analysis of circular sections. *Figure 9-85* shows a special roundness gage for checking crankshaft journals.

Along with any of the above checks, elements of the surface must be measured separately to assure that they are within the specified size tolerance and the boundary of perfect form at MMC.

Cylindricity

Cylindricity is having all points on the surface of a cylinder equal distances from a common axis. Cylindricity tolerance specifies a tolerance zone bounded by two concentric cylinders within which the surface must lie. No datum is needed nor proper with a cylindricity tolerance. When checking cylindricity, all elements of the concerned surface must also be within the specified size tolerance and the boundary of perfect form at MMC. An example of cylindricity tolerance and its meaning is shown in *Figure 9-86*.

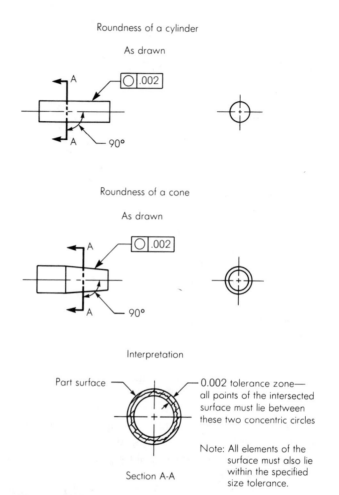

Figure 9-83. Roundness tolerance and its meaning.

Gaging Methods

Cylindricity is checked with the same equipment that is used to check roundness, except that roundness readings must be taken at a number of sections along the entire length of the part and placed in order to establish a common axis from which a tolerance zone can be established and measurements made to determine whether or not they fall within the tolerance zone specified. As with the circularity checks, elements of the cylindrical surface must be measured separately to assure that they are within the specified size tolerance and boundary of perfect form at MMC.

Figure 9-84. Checking roundness and cylindricity. (*Courtesy, Federal Products Corp.*)

Figure 9-85. Roundness gage for checking crank shafts. (*Courtesy, Federal Products Corp.*)

As drawn

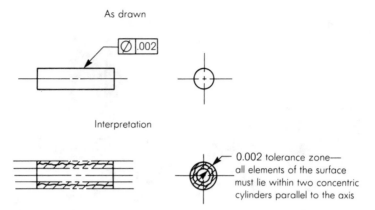

Interpretation

0.002 tolerance zone—
all elements of the surface
must lie within two concentric
cylinders parallel to the axis

Note: All elements of the specified surface must be within the specified size tolerance
& boundary of perfect form at MMC.

Figure 9-86. Cylindricity tolerance and its meaning.

Profile of a Line or Surface

Profile tolerancing is used to specify an allowable deviation from a desired profile where other geometric controls cannot be used. The profile tolerance specifies a uniform boundary along the desired profile within which the elements of the surface or line must lie. Datums may or may not be necessary to establish proper relationship of the profile to mounting surfaces for assembly purposes, etc. Most profiles are defined using basic dimensions, and as such are not covered by Rule 1, which requires that form control must be contained within the part-size tolerance. However, if, for whatever reason, conventional size dimensions and tolerances are associated with the profile tolerancing, Rule 1 *would* apply and the profile would be required to be within the specified size tolerance. Examples of profile tolerances and their meaning are shown in *Figure 9-87*.

Gaging Methods

Many profiles are checked with contour gages having the opposite form of the nominal contour of the part, or limit gages representing the go and not go sizes of the part as determined by the maximum and minimum material condition resulting from the compound effects of the form and size tolerances. Neither of these methods are suitable for geometrics.

One suitable method would be to compare the part with a master part conforming exactly with the basic dimensions of the part. A method to do this is shown in *Figure 9-88*.

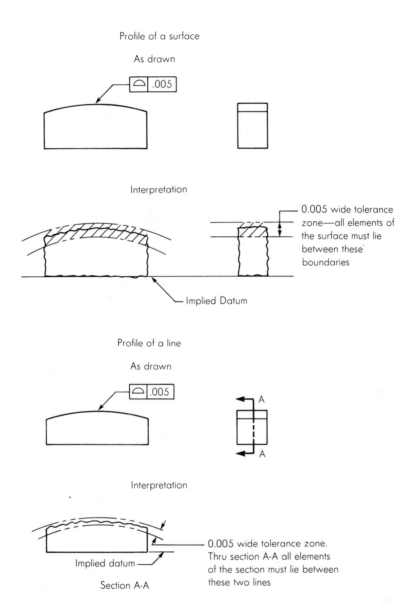

Figure 9-87. Profile tolerance and its meaning.

Another method has been used to check cams and other profiles which can be rotated. Modern technology, however, now makes it possible to read these profiles directly with long-range digital reading sensors, compare the readings

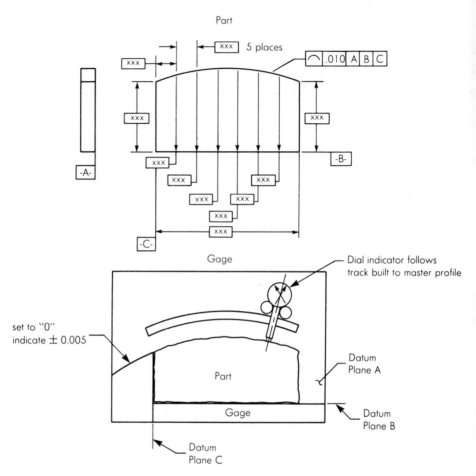

Figure 9-88. Checking profile - indicator following a master profile.

with a perfect master in the memory of a microcomputer and print out any out-of-tolerance conditions along with their location.

Figure 9-89 shows another generally accepted method which uses a fixture with several indicating devices mounted side by side in a plane which contains the profile to be inspected. The indicators must be set with a master representing the basic profile. Conformance of the part is easily determined by reading part differences directly from the indicators or display unit.

Perhaps the best way to check small complex profiles is with an optical projector using one of the methods previously described in the section on optical projectors.

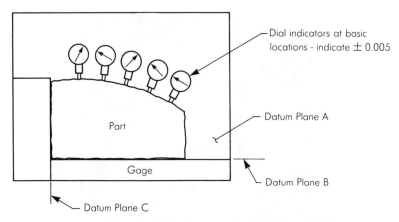

Note: Dial indicators to be set to "0" with master part built to basic DIMS.

Figure 9-89. Checking profile-fixed indicators set to master part.

Profiles can also be checked on an optical comparator with a contour projector tracer attachment as shown in *Figure 9-90*.

The contour tracer shown in *Figure 9-91* is also useful for checking profiles involving small length displacements such as a thread, etc. It works on the same principle as the instrument shown in *Figure 9-81* for checking straightness, except on a large scale. As the stylus moves across the profile, the movement is magnified and traced on a timed chart which can be easily interpreted. For very fine displacements, the actual instrument (shown in *Figure 9-81*) could be used.

Perpendicularity (Squareness)

Perpendicularity is the condition of a surface, median plane, or axis which is exactly 90° from a datum plane or datum axis. A perpendicularity tolerance always requires a datum and is specified by one of the following:

1. A tolerance zone bordered by two parallel planes perpendicular to a datum plane or datum axis within which the surface or median plane of the considered feature must lie.
2. A tolerance zone bordered by two parallel planes perpendicular to a datum axis within which the axis of the considered feature must lie.
3. A cylindrical tolerance zone perpendicular to a datum plane within which the axis of the considered feature must lie.
4. A tolerance zone defined by two parallel lines perpendicular to a datum plane or datum axis within which all elements of the surface must lie.

Figure 9-90. Optical comparator with a contour projector tracer attachment. (*Courtesy, Optical Gaging Products, Inc.*)

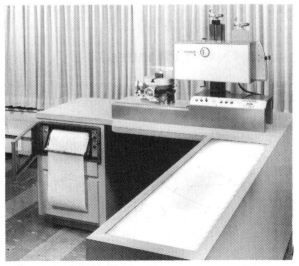

Figure 9-91. Contour reader and console assembly. (*Courtesy, PMC Industries*)

A perpendicularity tolerance applied to a surface also controls the flatness of the surface to the extent of the stated tolerance and requires the surface to be within the stated limits of size. Examples of perpendicularity tolerances and their meanings are shown in *Figure 9-92*.

Gaging Methods

The most common method of checking perpendicularity of surfaces is by direct comparison with gages of known squareness such as the precision square. To make this check, the square and the part are placed in contact with each other while resting on a surface plate. Out-of-squareness is determined by measuring the gap between the square and sections along the part with a thickness gage as shown in *Figure 9-93*.

Another direct contact method utilizes a direct reading cylindrical square, which is a cylindrical square with one face-off angle and with dotted curves and graduations etched on the surface indicating the amount of out-of-squareness for the bounded area. In use, the part and the cylindrical square are placed on a surface plate and brought together while rotating the

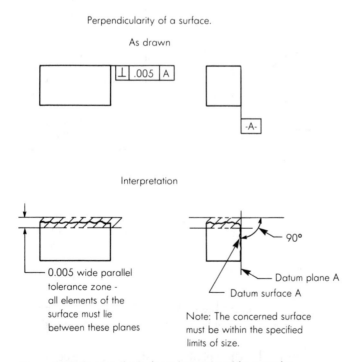

Figure 9-92. Perpendicularity tolerance and its meaning.

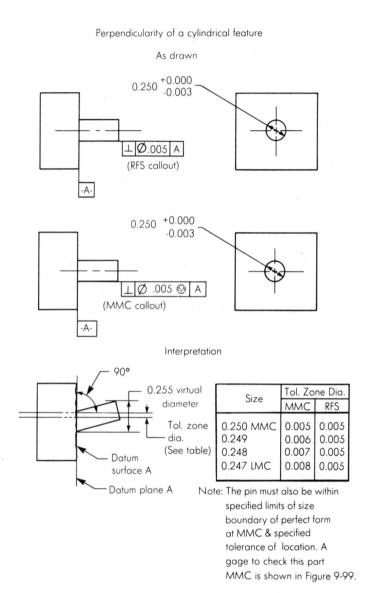

Perpendicularity of a cylindrical feature

As drawn

$0.250 \, ^{+0.000}_{-0.003}$

⊥ | Ø.005 | A
(RFS callout)

-A-

$0.250 \, ^{+0.000}_{-0.003}$

⊥ | Ø .005 Ⓜ | A
(MMC callout)

-A-

Interpretation

90°

0.255 virtual diameter

Tol. zone dia.
(See table)

Datum surface A

Datum plane A

Size	Tol. Zone Dia.	
	MMC	RFS
0.250 MMC	0.005	0.005
0.249	0.006	0.005
0.248	0.007	0.005
0.247 LMC	0.008	0.005

Note: The pin must also be within specified limits of size boundary of perfect form at MMC & specified tolerance of location. A gage to check this part MMC is shown in Figure 9-99.

Figure 9-92. (Continued)

cylindrical square to produce the smallest light gap. The number of the top most dotted curve in contact with the part surface indicates the squareness error.

Another widely used surface plate method for checking perpendicularity is as follows: check whether two preselected points on the vertical surface of the

Perpendicularity - non-cylindrical

As drawn

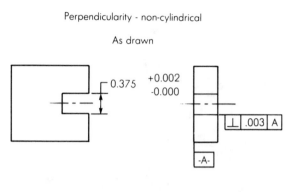

Interpretation

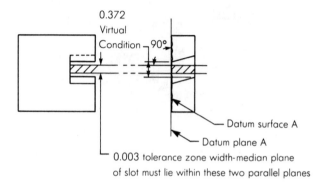

0.372
Virtual
Condition — 90°

Datum surface A

Datum plane A

0.003 tolerance zone width-median plane
of slot must lie within these two parallel planes

Note: Slot must also be within specified limits of size boundary of
perfect form at MMC & specified tolerance of location.

Figure 9-92. (Continued)

part are in a common plane at right angles to the surface plate that the part
and gage are resting on. Measurement is made by contacting the part, with the
spherical base of the comparator square and the probe of the indicator
mounted in the stand as shown in *Figure 9-94*. Before the measurement is
made, the indicator is set to zero with the aid of a cylindrical square as shown
in *Figure 9-94*.

With this method, multiple checks must be made at varying heights to get
an indication of the complete surface condition.

The squareness gage, shown with its master in *Figure 9-95*, measures in a
similar manner except that the part surface can be completely indicated in a
vertical manner due to the precision guideways built into the gage. Accuracy
is maintained by comparing and adjusting the squareness to the master gage.

Large, accurate perpendicular measurements are generally made with an

Figure 9-93. Checking squareness with a square.
(*Courtesy, L.S. Starrett Co.*)

Figure 9-94. Transfer inspection of squareness using cylindrical and comparator
squares. (*Courtesy, Taft-Peirce*)

Figure 9-95. Squareness gage and master.
(*Courtesy, PMC Industries*)

autocollimator, optical square, and reflector stand using a setup as shown in *Figure 9-96.* Sometimes, special fixtures are used such as those shown in *Figure 9-97*, which permit the concerned feature to be searched with indicating depth gages.

None of the above methods, however, are applicable to the other examples shown.

The examples specifying perpendicularity, cylindrical size feature RFS, and noncylindrical feature RFS cannot be checked by functional gaging. They would also be difficult to check economically by other means, since a number of direct or differential measurements would have to be made, depending on the feature basic form errors, just to determine the attitude of the axis or median plane. For small quantities, the part could be checked by standard layout methods with the help of a simple staging fixture to quickly arrange the datum surface so that the concerned feature was properly lined up with the surface plate to easily check with a height stand and indicator as

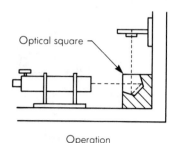

Operation

Light beams from the autocollimator which is aligned
with the datum plane are turned at an exact right angle
with an optical square (pentaprism) which permits the
measurement of perpendicularity by observing the
reflection from a mirror moved along the second plane.

Figure 9-96. Checking perpendicularity with an
autocollimator.

shown in *Figure 9-98.* For production quantities, fixtures are generally
designed with air or electronic probes properly positioned and connected to
determine the squareness of the axis or median plane to the datum surface.

The example specifying perpendicularity, cylindrical size feature, and
MMC could be checked with a functional gage as shown in *Figure 9-99.*

Angularity

Angularity is a condition of a surface or axis at a specified angle (other
than 90°) from a datum plane or axis. Angularity tolerance specifies a
tolerance zone defined by two parallel planes at a specified basic angle from a
datum plane or axis, within which the surface or axis of the feature must lie. A
datum is always required, and the desired angle is always shown as a basic
angle. When checking angularity, all elements of the concerned feature must
also be within the specified size limits of the part to be accepted. On surfaces,
the angularity includes a control of flatness to the extent of the angularity
tolerance. Examples of angularity tolerances and their meaning are shown in
Figure 9-100.

Gaging Methods

Angularity tolerance is another RFS tolerance which cannot be checked
with functional gaging.

Most short-run parts are checked for angularity with standard surface

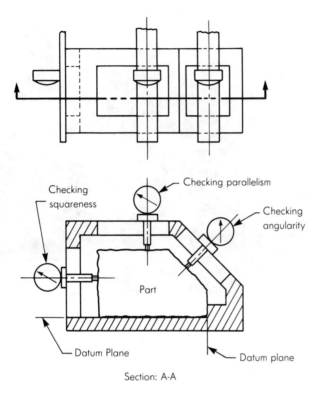

Section: A-A

Entire concerned part surface is searched by sliding indicating depth
gage around qualified surface. Dimensions can also be checked at
the same time by setting the depth gage to a proper master.

Figure 9-97. Checking squareness, parallelism, and angularity
with indicating depth gages.

plate methods. Sine plates or simple staging fixtures are used to place the
datum surface of the concerned feature in proper alignment with the surface
plate. A height stand and indicator as shown in *Figure 9-101* are used to check
the surfaces.

For production quantities, fixtures are generally designed with specifically
located and interrelated air or electronic probes which determine and then
display any out-of-tolerance condition of the feature angle being checked.

Large, accurate angularity measurements are generally made with an
autocollimator or special fixtures and indicating depth gages in a manner
similar to those shown for checking perpendicularity in *Figure 9-96* and
Figure 9-97.

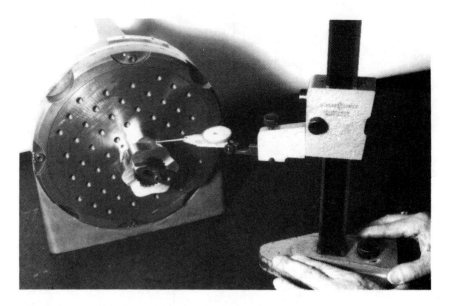

Figure 9-98. Checking perpendicularity with a height stand and an indicator. (*Courtesy, Scherr-Tumico*)

Parallelism

Parallelism is a condition of a surface, line, or axis which is equidistant from a datum plane or axis at all points. A parallelism tolerance always requires a datum and is specified by one of the following:

1. A tolerance zone bounded by two parallel planes or lines parallel to a datum plane or datum axis within which the elements of the surface or the axis of the considered feature must lie.
2. A cylindrical tolerance zone whose axis is parallel to a datum axis within which the axis of the considered feature must lie.

The parallelism tolerance applied to a surface also controls the flatness of a surface to the extent of the stated tolerance and requires the surface to be within the stated limits of size. Examples of parallelism tolerances and their meaning are shown in *Figure 9-102* and *Figure 9-103*.

Gaging Methods

Parallelism of a surface is generally checked by placing the datum of the part on a surface plate and searching for any out-of-parallel conditions with a height stand and indicator.

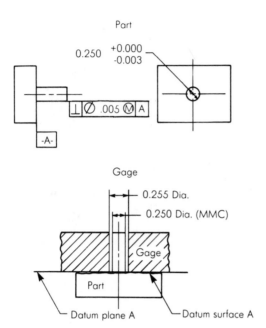

Note: The pin must also be within specified limits of size, boundary of perfect form at MMC, and specified tolerance of location.

Figure 9-99. Functional gage to check perpendicularity. Cylindrical size feature, MMC.

Many times, a part must be placed in a special fixture to check perpendicularity, angularity, location, etc. In these cases, a check for parallelism is generally incorporated in the same fixture as shown in *Figure 9-97*.

When parallelism concerns cylindrical size features or datums, the inspection becomes more difficult especially when the feature, datum or both are RFS. Here again, standard layout methods using a surface plate, height stand and indicator can be used for small quantities as long as the inspector understands the meaning expressed by the feature control symbol.

Figure 9-104 shows a design approach that can be used to check the parallelism of these cylindrical size features.

Runout

Runout is the deviation from the desired form of a part surface of revolution when the part is rotated 360° about a datum axis. The runout tolerance specifies the maximum full indicator movement (FIM) allowed

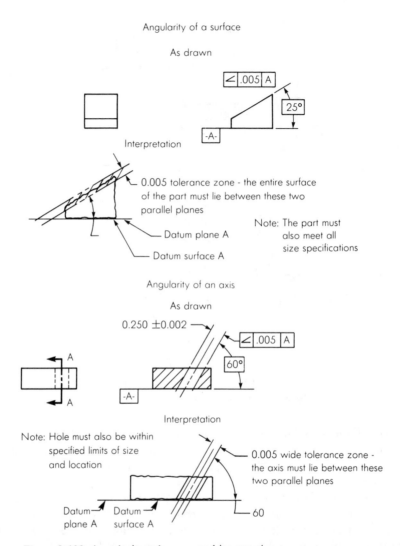

Figure 9-100. Angularity tolerance and its meaning.

during the 360° rotation. Runout tolerance is always applied on an RFS basis and always requires a datum. It is used to maintain surface-to-axis control on a part. See *Figure 9-105*.

Circular runout controls only circular elements of a surface individually and independently from one another, but total runout provides composite control of all surface elements at the same time. When checking runout, all elements of the concerned surface must be within the specified size limits of

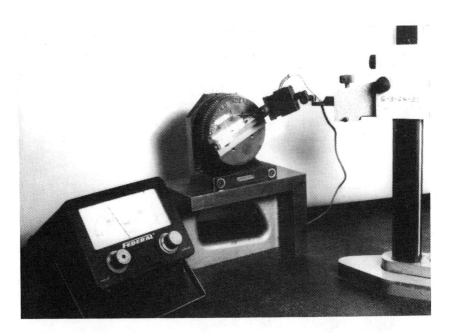

Figure 9-101. Checking angularity with a height stand and indicator. (*Courtesy, Federal Products Corp.*)

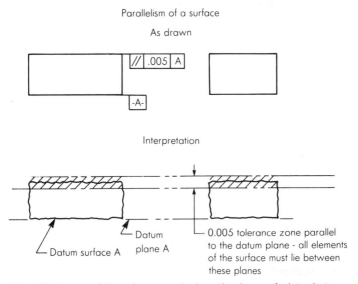

Figure 9-102. Parallelism tolerance and its meaning.

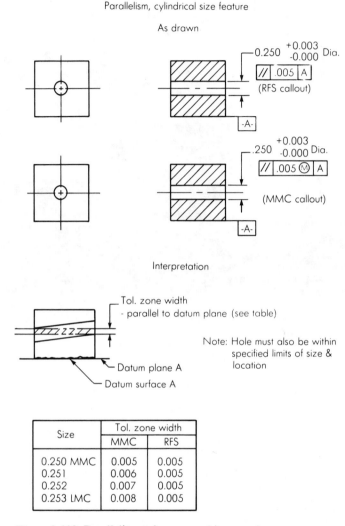

Figure 9-103. Parallelism tolerance and its meaning.

the part to be acceptable. Examples of runout tolerances and their meaning are shown in *Figure 9-106*.

Gaging Methods

Checking runout on a part on which the datum axis is established by two part centers is relatively simple and easily understood. Low production parts

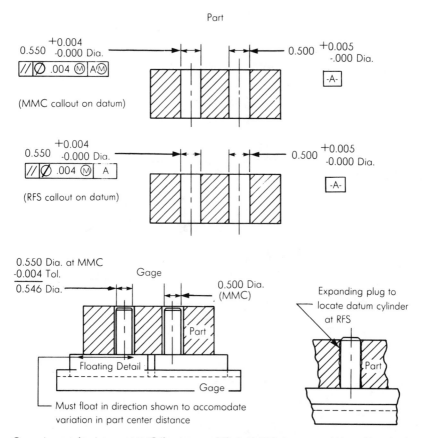

Figure 9-104. Gage for checking parallelism - cylindrical size feature.

are generally mounted between the centers of a bench center and rotated 360° with a dial indicator in contact with the surface to be checked as shown in *Figure 9-25*. For high production, special fixtures with several air, electronic, or mechanical indicators are generally more practical (see *Figure 9-107* and *Figure 9-108*).

Parts where the datum axis is established by two functional diameters are generally checked in the same manner, except that the datum is established by closing in on the datum features with expanding chuck-like devices as shown in *Figure 9-109* and taking the readings in the same manner as from centers. Another method widely used with electronic gaging allows the part to be mounted in any suitable fashion while indicating the datums as well as the

Figure 9-105. Runout symbols.

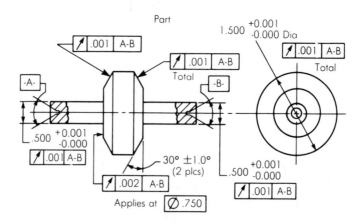

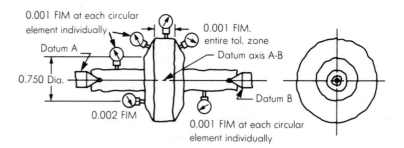

Notes:

1 Indicators to be normal to surface being checked.

2 All features concerned must be within their specified limits of size & boundary of perfect form at MMC.

Figure 9-106. Runout tolerance and its meaning.

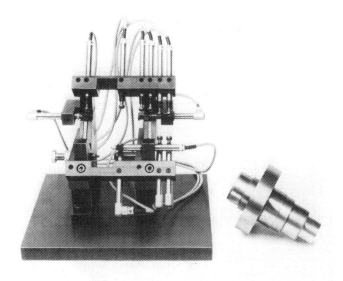

Figure 9-107. Checking runout.

Figure 9-108. Checking runout with non-contact gaging. (*Courtesy, Dearborn Gage Co.*)

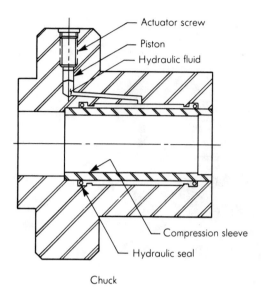

Chuck

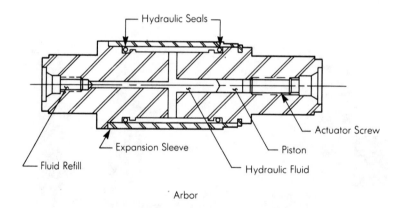

Arbor

Figure 9-109. Hydraulically actuated arbors and chucks.

features to be checked. The readings are then adjusted electronically within the equipment and the runout can be displayed directly.

Methods for checking runout on parts where the datum is established from a diameter, or a combination of datum diameter and functional face, are shown in *Figure 9-110*, *Figure 9-111*, and *Figure 9-112*. The cylindrical datums must be established by chuck-like devices or established electronically as noted above since runout is always RFS.

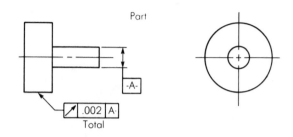

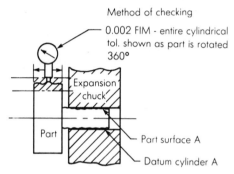

Method of checking

0.002 FIM - entire cylindrical
tol. shown as part is rotated
360°

Notes:
1 Datum cylinders axis must be established RFS.
2 All concerned features including datum must be within the specified
limits of size and boundary of perfect form at MMC.

Figure 9-110. Checking runout from an O.D. datum.

When checking runout, all features of the part must also be measured separately to ensure that they are within their specified size limits and within the boundary of perfect form at MMC.

Position—formerly called True Position Tolerance

Position is a tolerance of location which is used to specify how far a feature may vary from the theoretically exact location on the part drawing.

A position tolerance defines a zone within which the axis or centerplane of a feature can vary from this theoretically exact position. Position tolerances are applied only to size features, and datums are always required. A position tolerance is mainly used to maintain surface-to-surface control or axis-to-axis control on an MMC basis. True position tolerance theory and bonus tolerance are shown in *Figure 9-113.*

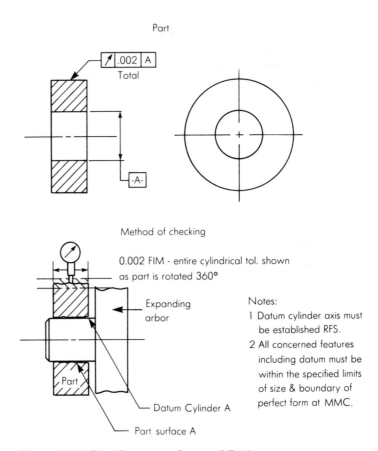

Part

.002 | A
Total

-A-

Method of checking

0.002 FIM - entire cylindrical tol. shown
as part is rotated 360°

Expanding
arbor

Part

Datum Cylinder A

Part surface A

Notes:
1 Datum cylinder axis must
  be established RFS.
2 All concerned features
  including datum must be
  within the specified limits
  of size & boundary of
  perfect form at MMC.

Figure 9-111. Checking runout from an I.D. datum.

Gaging Methods

When position tolerance is applied in an MMC basis, it allows functional gaging to be used.

Figure 9-114 shows two identical parts containing clearance holes, assembled with two 0.5 diameter bolts. Each part can be dimensioned and toleranced as shown, with MMC specified after the hole location tolerance. Also shown is a hole relation gage for each of the parts. Hole relation gages check hole-to-hole relationship, not hole location to some other part feature. The gage could contain fixed pin gage features in place of the separate gage pins shown which fit tight into nominally located bushings in the gage.

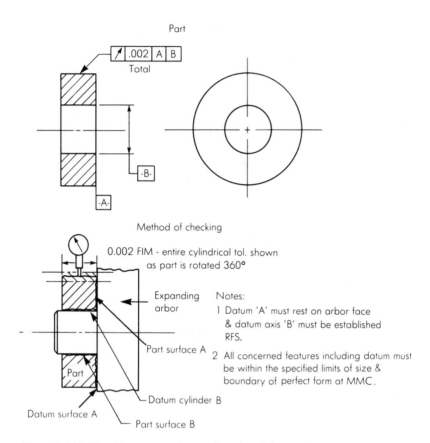

Figure 9-112. Checking runout from a functional face and ID.

The locational and squareness tolerance is the actual difference in size between the gage pin and the clearance hole feature. Since feature size will vary from hole to hole, from part to part, and from process to process, the true tolerance is a variable since the 0.5 diameter bolt or gage pin and the interchangeable design requirement is still a constant. The positional tolerance is 0.1.

Substituting RFS for the MMC callout would make gaging difficult, since the gage pin must always be 0.01 smaller than the actual hole size. Eleven or more gage pins would be required for each hole (0.5, 0.501, etc.) so that, for instance, if a particular hole measured 0.512, the 0.502 gage would be used.

Figure 9-115 shows how the true tolerance varies with hole size. All

Perfect hole location at MMC

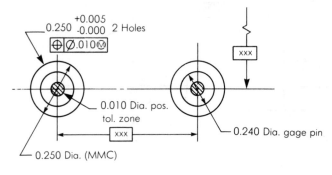

0.250 $^{+0.005}_{-0.000}$ 2 Holes

0.010 Dia. pos. tol. zone

0.240 Dia. gage pin

0.250 Dia. (MMC)

Holes offset at MMC

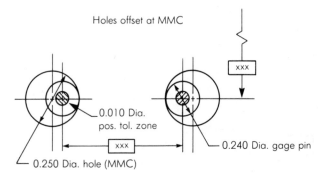

0.010 Dia. pos. tol. zone

0.240 Dia. gage pin

0.250 Dia. hole (MMC)

Holes offset at LMC

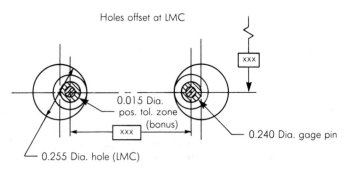

0.015 Dia. pos. tol. zone (bonus)

0.240 Dia. gage pin

0.255 Dia. hole (LMC)

Figure 9-113. Position tolerance theory - bonus tolerance.

tolerances greater than 0.01 diameter are bonus tolerances and are not specifically allowed in any system other than the positional tolerancing system.

Figure 9-116 shows the tolerances and gages for the manufacture and assembly of two different parts. One part contains clearance holes and the

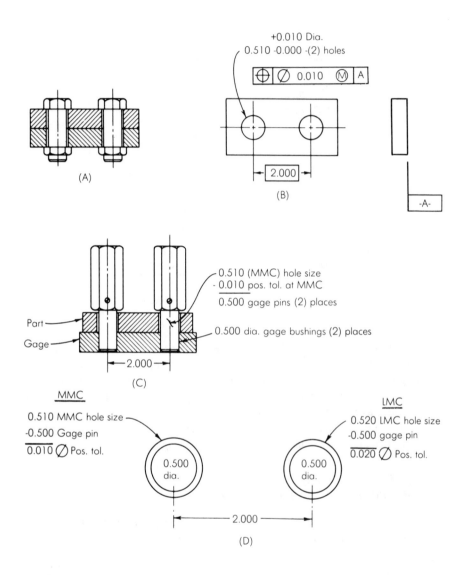

Figure 9-114. Two parts with clearance holes, assembled with bolts.

other part contains holes tapped for 0.5 diameter bolts. Shown also is the feature relation gage required for the tapped part. Gage thickness or bushing height must be at least the maximum thickness of the untapped

Feature diameter	Positional and squareness tolerances diameter
0.510 (MMC or most critical size)	0.010 at MMC (tightest tolerance)
0.511	0.011
0.512	0.012
0.513	0.013
0.514	0.014
0.515	0.015
0.516	0.016
0.517	0.017
0.518	0.018
0.519	0.019
0.520 (LMC or least critical size)	0.020 at LMC (loosest tolerance)

Figure 9-115. Variable tolerances allowed by MMC.

part, to guarantee that the bolts will be properly located and square for assembly. The two, go-thread gages simulate the bolts at assembly. Gage bushing size is determined by adding the 0.01 diameter positional tolerance specified for the tapped features to the bolt or tap size. Stepped gage pins with go threads may be used to take advantage of standard bushing size as long as a 0.01 difference in size is maintained between the gage bushing and that portion of the gage pin that lies within the bushing.

Basic Design Rules for Positionally-Toleranced Parts

Two basic rules govern the design of gages for positionally toleranced parts. These principles apply regardless of the number of features that make up an interchangeable pattern.

1. For parts with internal features, the nominal gage feature size is directly determined by subtracting the total positional tolerance specified at MMC from the specified MMC size of the feature to be gaged for location.
2. For parts with external features, the nominal gage feature size is directly determined by adding the total positional tolerance specified at MMC to the specified MMC size of the feature to be gaged for location.

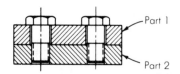

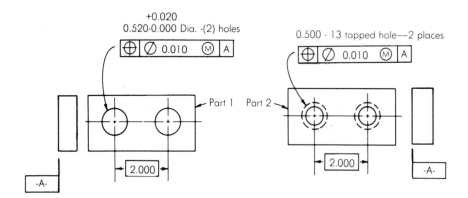

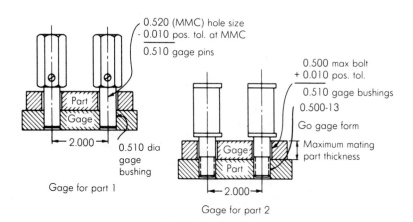

Figure 9-116. Tolerances and gages for the manufacture and assembly of two different parts.

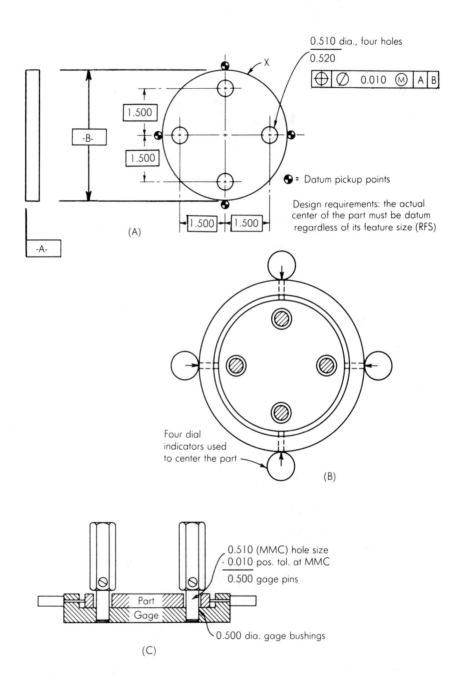

Figure 9-117. Workpiece with four holes that must be located from the specific center of the workpiece.

Part

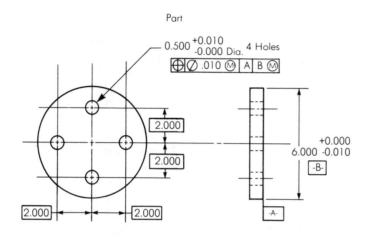

Gage

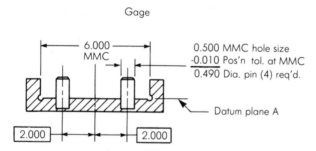

Figure 9-118. "Shake" gage to check a less critical (MMC) O.D. datum feature.

Figure 9-117 shows a workpiece with four holes that must be located from the specific center (datum) of the workpiece regardless of the actual workpiece size (RFS). The design specification drawing includes the exact pickup points so that the same center can be repeatedly found. Also shown is a hole relation gage which uses four dial indicators to determine and correctly position the datum for the hole gaging operation.

Figure 9-118 shows a similar workpiece with the less critical MMC requirement on the datum diameter requiring that the holes be located from the center of the datum diameter when the datum is at MMC. Also shown is the design of a gage to check the part functionally.

Figure 9-119 shows another workpiece in which the holes never need to be

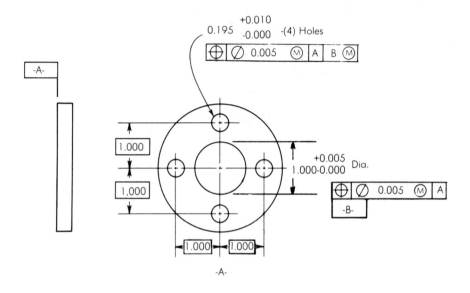

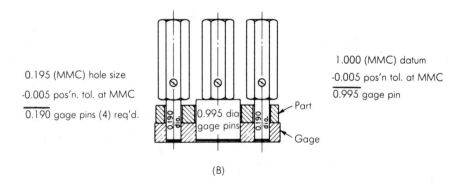

(B)

Figure 9-119. Gaging workpiece features not specifically located relative to the datum: (*A*) workpiece; (*B*) hole location (shake) gage.

exactly located from the center of a specified datum feature. A gage fit allowance has been specified that is directly reflected in the size of the datum gage feature since the gage is 0.995 in diameter and differs from the MMC size of the part diameter (1.000) by 0.005. Quite a large allowance could be specified if the datum was merely a convenient starting place for manufacturing.

A pattern of interchangeable features (holes) is the most critical feature on the part shown in *Figure 9-120*, but is not locationally critical in relation to any single datum feature. The 0.3 minimum breakout specification is the result of a stress analysis, and is an end-product requirement. No single datum is specified, and the 0.3 minimum specification can be readily gaged with a tubing micrometer or a fork gage as shown.

Figure 9-121 shows a workpiece with seven holes. The specified positional tolerance includes the location and angularity tolerances for each radial hole. Also shown is the gage required for checking the part. In use, all seven gage pins must go through the part at one time. In designating the datum, if RFS callouts had been used instead of MMC for the diameter and width of the slot, the gage would be required to center on the two datum features. As a result, one gage pin could be used to individually qualify each hole in reference to the datum.

Position tolerance is often used to control the location of Coaxial Mating part features, as shown in *Figure 9-122*, along with the gages to check them.

Figure 9-123 shows the method used to check bore alignment functionally when specified MMC. Another common application of position tolerance MMC is on noncylindrical mating parts such as shown in *Figure 9-124*.

Occasionally, a projected tolerance zone will be specified to prevent interference at assembly as shown in *Figure 9-125*, along with its interpretation.

A case where it was desired to place all the usable size and location tolerances into the size limit by stating the position tolerance as zero (0.000) at MMC is shown in *Figure 9-126*, along with the functional gage to check the part. It should be noted that the gage eliminates the need for a go-plug gage check.

Concentricity

Concentricity is another tolerance of location which is used to maintain axis-to-axis control on an RFS basis. It is a condition in which two or more solid features, in any combination, have a common axis. A concentricity tolerance specifies the diameter of a cylindrical tolerance zone within which the axis of the feature or features must lie. The axis of the tolerance zone must coincide with the axis of the datum feature or features which are always

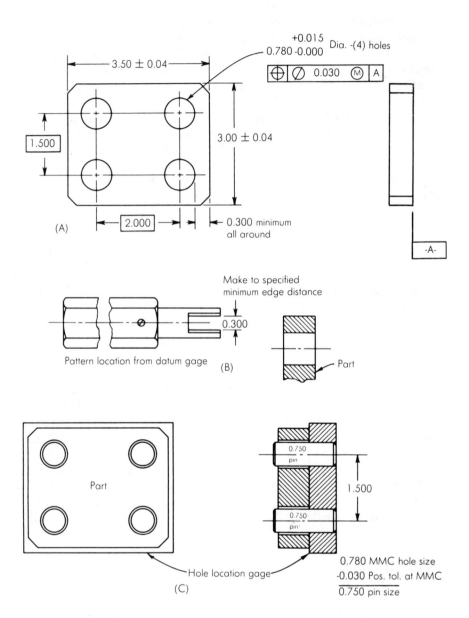

Figure 9-120. Gaging an end product: (*A*) workpiece; (*B*) gage for checking required edge distance; (*C*) gage for checking hole location.

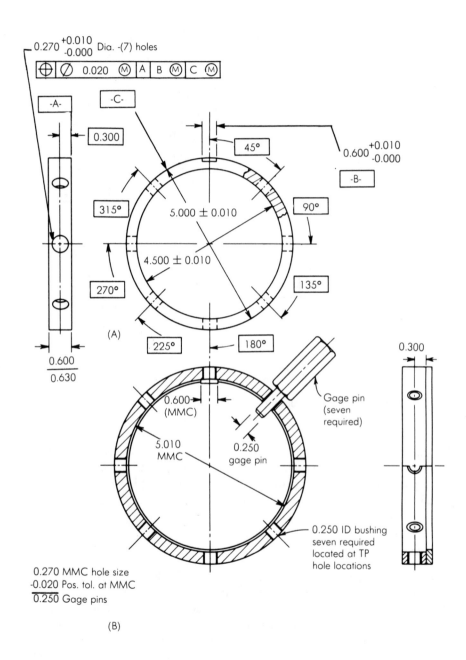

0.270 $^{+0.010}_{-0.000}$ Dia. -(7) holes

⊕	⌀	0.020	Ⓜ	A	B	Ⓜ	C	Ⓜ

-A-

-C-

0.300

45°

0.600 $^{+0.010}_{-0.000}$

-B-

315°

5.000 ± 0.010

90°

4.500 ± 0.010

270°

135°

(A)

225°

180°

0.300

0.600
0.630

0.600
(MMC)

Gage pin
(seven
required)

5.010
MMC

0.250
gage pin

0.270 MMC hole size
-0.020 Pos. tol. at MMC
0.250 Gage pins

0.250 ID bushing
seven required
located at TP
hole locations

(B)

Figure 9-121. Gaging radial hole patterns: (*A*) workpiece; (*B*) hole location and sqaureness gage.

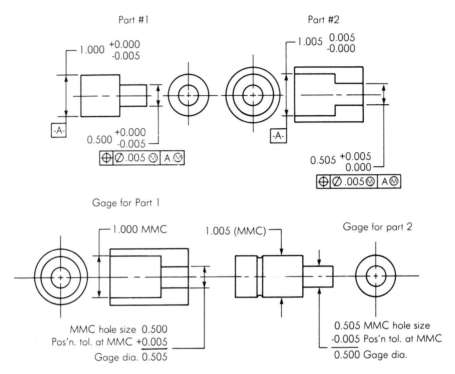

Figure 9-122. Gaging coaxial mating parts at MMC.

required. When checking concentricity, all elements of the concerned surfaces must also be within the specified size limits of the part to be acceptable. An example of a concentricity tolerance and its meaning is shown in *Figure 9-127.*

Gaging Methods

Since concentricity tolerance is always on an RFS basis, it requires that the datum axis be established by chuck-like devices which close in on the part, or by electronic means previously described under runout, which allows the part to be mounted in any suitable fashion while indicating the datums as well as the features to be checked. The results are electronically interpreted and any eccentricity is displayed directly on the readout. All features must also be measured separately to assure that the part conforms to the specified size tolerances.

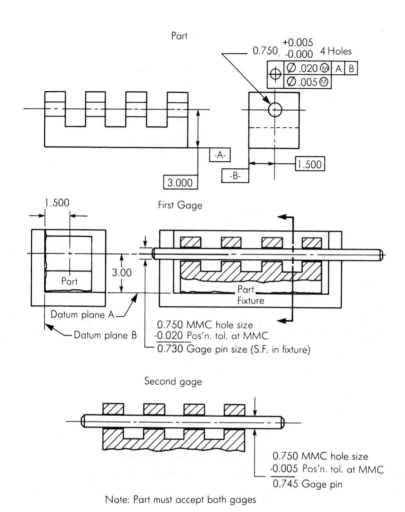

Figure 9-123. Checking bore alignment at MMC.

Symmetry

Symmetry is still another tolerance of location. It is a condition in which a feature or features are symmetrically disposed about the centerplane of a datum feature. A symmetry tolerance specifies a tolerance zone bounded by two parallel planes equally placed about the centerplane of the datum feature, which is always required. Positional tolerance, RFS, may be applied instead of a symmetry tolerance with identical meaning. When checking symmetry,

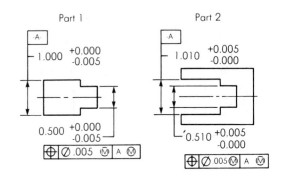

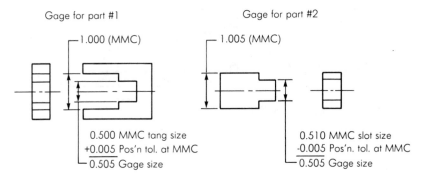

Figure 9-124. Checking non-cylindrical mating parts at MMC.

all elements of the concerned feature must also be within the specified limits of the part to be acceptable. An example of a symmetry tolerance and its meaning are shown in *Figure 9-128*.

Gaging Methods

Symmetry tolerances generally will be specified on noncylindrical parts on an RFS basis since MMC principles are usually specified as position tolerances. Symmetry can be checked with standard surface plate methods. However, for high production the multi-sensor type electronic gaging methods mentioned previously for checking other RFS features are recommended.

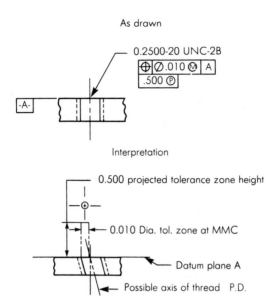

Figure 9-125. Projected tolerance zone.

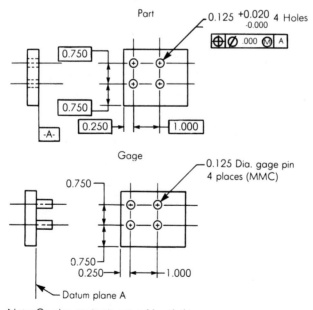

Note: Go plug gaging is not rec'd. with this gage.

Figure 9-126. Positional tolerance of zero at MMC.

As drawn

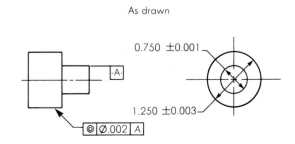

0.750 ±0.001

1.250 ±0.003

⊚ ⌀.002 A

Interpretation

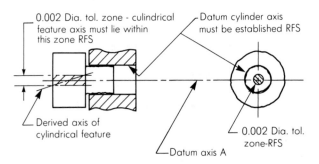

0.002 Dia. tol. zone - culindrical
feature axis must lie within
this zone RFS

Datum cylinder axis
must be established RFS

Derived axis of
cylindrical feature

0.002 Dia. tol.
zone-RFS

Datum axis A

Note: Part must also be within specified limits of size.

Figure 9-127. Concentricity - a tolerance of
location and its interpretation.

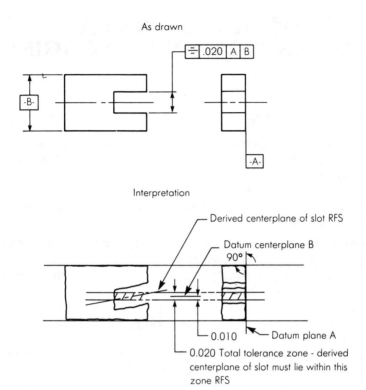

Note: Part must also be within specified limits for size.

Figure 9-128. Symmetry - A tolerance of location and its interpretation.

DESIGN OF TOOLS FOR INSPECTION AND GAGING

Review Questions

1. What is usually the governing factor in the specification of tolerance?
2. Explain the difference between coordinate dimensioning and position tolerancing.
3. What are the four classes of gagemaker's tolerance?
4. What is the usual gagemaker's tolerance for working gages? For inspection gages?
5. Using the 10% rule, what is the correct gagemaker's tolerance for a plug gage to check a 5.000 $^{+0.002''}_{-0.000''}$ diameter hole?
6. What is the purpose of a master gage? What tolerance is applied to master gages?
7. What are the two general types of air gages?
8. What precautions are required when using an air plug to check a bore with a 90 μin. finish?
9. Name the two ANSI Y14.5 accepted modifiers and show their symbols.
10. What is the symbol for a datum target?
11. What is the MMC of a 0.250″ $^{+0.004''}_{-0.000''}$ diameter hole? The LMC?
12. Discuss the difference in checking a cylinder for roundness (circularity) and checking it for cylindricity.
13. Which of the following geometrics do not require a datum: flatness, straightness, profile, perpendicularity, roundness, cylindricity, angularity?
14. What is the virtual condition of the following part? (See *Figure A*)

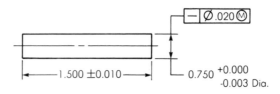

Figure A.

15. Design a simple, functional gage to check the part shown in *Figure A*.
16. What is the purpose of the projected tolerance zone?
17. Design a simple function gage for the following part. (See *Figure B*)

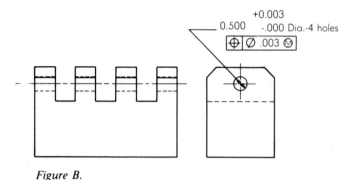

Figure B.

DESIGN OF TOOLS
FOR INSPECTION AND GAGING

Answers to Review Questions

1. Cost.
2. With coordinate dimensioning, the location of each element of a workpiece is defined by stating its distance (dimension) from each reference plane and assigning an amount by which the distance may vary (tolerance). With position tolerancing, the nominal (exact) location of each element is specified along with a statement of how far from this position (true position tolerance) the element may vary.
3. XX, X, Y, Z.
4. 10% of the part tolerance, 5% of the part tolerance.
5. Y. Determined as follows:
   Total part tolerance = 0.002″
   10% of total part tolerance = 0.0002″
   According to *Figure 9-4*, since 5.000″ lies between 4.510″ and 6.510″, this row is examined to find the proper tolerance. Since the "Z" tolerance is greater than 0.0002″ the next tighter tolerance which is equal or less than 0.0002″ must be used. The "Y" tolerance satisfies this requirement since it is 0.00019″.
6. Master gages are used to check the accuracy and set other gages. They are generally built to 10% of the gage tolerance.

7. Pressure (back pressure) and Velocity (flow).
8. Contact gaging spindles must be used to eliminate any error introduced by reading roughness as well as size. this error is usually considered negligible for finishes of 50 μin. or less.
9. Maximum material condition regardless of feature size. Ⓜ
   Regardless of feature size. Ⓢ
10. (See *Answer Figure A.*)
11. 0.025″, 0.254″
12. Whereas the roundness tolerance specifies a tolerance zone bounded by two concentric circles within which each circular element of the surface must lie, cylindricity specifies a tolerance zone bounded by two concentric cylinders within which the entire surface must lie.
13. Flatness, straightness, roundness, and cylindricity.
14. 0.770″ diameter.
    Determined as follows:
     0.750″ part diameter at MMC
    +0.020″ tolerance at MMC
     0.770″ virtual condition.
15. (See *Answer Figure B.*)
16. To assure assembly between mating parts.
17. (See *Answer Figure C.*)

Answer Figure A.

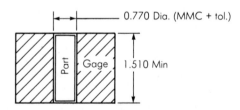

Answer Figure B.

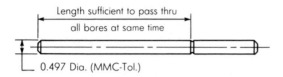

Answer Figure C.

10

TOOL DESIGN FOR
THE JOINING PROCESSES

The joining processes are generally divided into two classes: mechanical and physical. Mechanical joining does not ordinarily involve changes in composition of the workpiece material. The edges of the pieces being joined remain distinct.

In the physical joining process, parts are made to join along their contacting surfaces through the application of heat or pressure or both. Often a filler material is added, with the edges losing their identity in a homogeneous mass.

Two pieces of wood nailed together are joined mechanically. The same two pieces of wood could be joined physically by an adhesive. At the exact center of the joint, only the adhesive would be found. The adhesive would penetrate into the pores of the wood for some distance, and the workpiece edges would no longer exist as true entities but would have become a blend of wood and glue.

Joining processes may require tooling to hold the parts in correct relationship during joining. Another function of the tooling is to assist and control the joining process. Often, several parts may be joined both mechanically and physically. Thus, two workpieces may be bolted together to assure alignment during subsequent welding. Mechanical joining at times may be considered as the tooling method for the final physical joining.

Tooling for Physical Joining Processes

Physical joining processes generally cannot be performed without tooling, because the high temperatures required usually make manual positioning impractical. The tooling must hold the workpieces in correct relationship during joining, and it must assist and control the joining process by affording adequate support. Tooling used for hot processes must not only withstand the temperatures involved, but in many cases must either accelerate or retard the flow of heat. Hot fixtures must be designed so that their heat-expanded dimensions remain functional.

Design of Welding Fixtures

The purpose of a welding fixture is to hold the parts to be welded in the proper relationship both before and after welding. Many times a fixture will maintain the proper part relationship during welding, but the part will distort after removal from the fixture. Good fixture design will, of itself, largely determine product reliability. Major fixture design objectives, some basic and some special, are as follows: (1) to hold the part in the most convenient position for welding, (2) to provide proper heat control of the weld zone, (3) to provide suitable clamping to reduce distortion, (4) to provide channels and outlets for welding atmosphere, (5) to provide clearance for filler metal, (6) to provide for ease of operation and maximum accessibility to the point of weld.

Other factors that also will influence fixture design are: (1) cost of tool, (2) size of the production run and rates, (3) adaptability of available welding equipment, (4) complexity of the weld, (5) quality required in the weldment, (6) process to be employed, (7) conditions under which the welding will be performed, (8) dimensional tolerances, (9) material to be welded, (10) smoothness required, (11) coefficient of expansion and thermal conductivity of both workpiece and tool materials.

The tool designer must be familiar with the gas, arc, and resistance welding processes. Each of these processes will require individual variations of the general design factors involved. For instance, heat dissipation is not a critical factor in some of the welding processes. Expansion is not a problem if outer ends of the workpiece are not restricted.

Gas Welding Fixtures

The general design of a gas welding fixture must take into consideration the heating and cooling conditions. A minimum of heat loss from the welding area is required. If the heat loss is too rapid, the weld may develop cracks. Heat loss by materials, particularly aluminum and copper, must be carefully controlled. To accomplish this, large fixture masses should not be placed close to the weld line, however, the part may distort. The contact area and clamps should therefore be of the minimum size consistent with the load transmitted through the contact point. In welding copper and aluminum, the minimum contact surface often permits excessive heat loss, and prevents good fixture welds. This necessitates tack welding the fixtured parts at points most distant from the fixture contact points, with the rest of the welding done out of the fixture. With this method, excessive distortion may result, and subsequent stress relieving of the part may be required.

One of the simplest fixtures for gas welding is a gravity-type fixture shown in *Figure 10-1*. This design eliminates excess fixture material from the weld area to minimize heat loss, while providing sufficient support and locating points. The design also permits making welds in a horizontal position, which is generally advantageous.

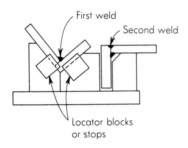

Figure 10-1. Simple welding fixture using gravity to help locate parts.

Figure 10-2 shows another simple form of gas welding fixture which holds two flat sheets for joining. C-clamps hold the workpieces to steel support bars. Alignment is done visually or with a straight edge. A heat barrier of alumina-ceramic fiber is placed between the workpieces and the steel bars. Holddown plates are used to keep the workpieces flat and to prevent distortion. If the parts to be welded have curved surfaces, the supporting bars and holddown plates may be machined to match the part.

Simple parts may be properly located or positioned in a fixture visually. As

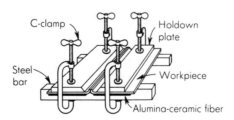

Figure 10-2. Workpieces with simple fixturing for gas welding operations.

workpiece shape becomes complex, or the production rate increases, positive location is desirable. The same locating methods used in workholder design can readily be adapted for the design of welding fixtures.

The selection of material for gas welding fixtures is governed by these factors: (1) part print tolerances; (2) material heat resistance; (3) heat transfer qualities, and (4) the fixture rigidity required to assure workpiece alignment accuracy. The fixture material should not be affected in the weld zone and should prevent rapid heat dissipation from the weld area. Some of the fixture materials commonly used are cast iron, carbon steel, and stainless steel.

Arc Welding Fixtures

Arc welding concentrates more heat at the weld line than gas welding. The fixtures for this process must provide support, alignment, and restraint on the parts, and also must permit heat dissipation.

Some of the more important design considerations for arc welding fixtures are as follows: (1) the fixture must exert enough force to prevent the parts from moving out of alignment during the welding process, and this force must be applied at the proper point by a clamp supported by a backing bar; (2) backing bars should be parallel to the weld lines; (3) backing bars should promote heat dissipation from the weld line; and (4) backing bars should support the molten weld, govern the weld contour, and protect the root of the weld from the atmosphere.

Backing bars are usually made from solid metal or ceramics. A simple backup could be a rectangular bar with a small groove directly under the weld. This would allow complete penetration without pickup material by the molten metal. In use, the backup would be clamped against the part to make the weld root as airtight as possible. Some common shapes are shown in *Figure 10-3*. *Figure 10-4* shows a backing bar in position against a fixed workpiece.

The size of the backup bar is dependent upon the metal thickness and the material to be welded. A thin weldment requires larger backup to promote heat transfer from the weld. A material with greater heat-conducting ability requires less backup than that required for a comparable thickness of a poor conductor.

Figure 10-5 shows backing bars designed for use with gas, which may be used to blast the weld area (*A*), flood the weld area (*B*), or may be concentrated in the weld area (*C*). Backup bars may be made of copper, stainless steel (used for tungsten inert gas), titanium ceramic or a combination of several metals (sandwich construction).

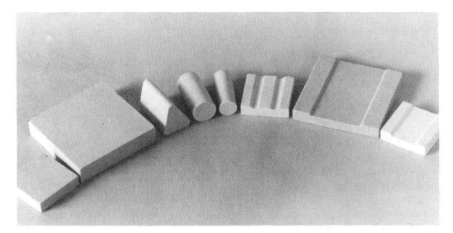

Figure 10-3. Typical backing bars. (*Courtesy, Alloy Rods Division, Chemetron Corp.*)

Figure 10-4. Workpiece with simple fixturing for arc welding operations. (*Courtesy, Alloy Rods Division, Chemetron Corp.*)

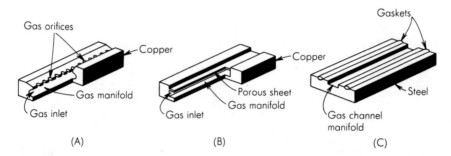

Figure 10-5. Backing bars with provisions for (*A*) directed gas flow, (*B*) diffused gas flow, and (*C*) pressurized gas.

Resistance Welding

One of the simplest and most economical processes for joining two or more metal parts is resistance welding. In resistance welding, fusion is produced by heat generated at the junction of the workpieces by local resistance to passage of large amounts of electric current and by the application of pressure. *Figure 10-6* illustrates the elements of a resistance welding machine. Resistance welding processes used for low-cost high-production are shown in *Figure 10-7* and include the following: spot welding, projection welding, seam welding,

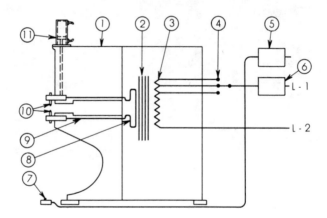

Figure 10-6. Elements of typical resistance welder: (1) housing; (2) low-voltage, high-current transformer; (3) primary coils; (4) tap switch; (5) welding timer; (6) power interrupter; (7) foot switch; (8) secondary loop; (9) bands from electrodes to secondary; (10) electrodes; (11) cylinder which exerts pressure on work.

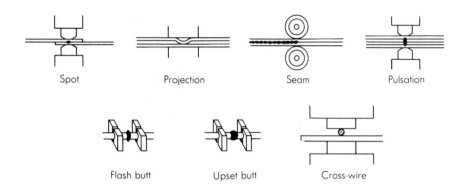

Figure 10-7. Resistance welding methods. (*Assembly and Fastener Engineering.*)

pulsation welding, flash butt welding, upset butt welding, and cross-wire welding.

Spot Welding. Spot welding, compared to riveting, may be considerably faster and less expensive since there is no need to drill holes and insert rivets. *Figure 10-8* illustrates typical spot-welded joints and electrode shapes which can be produced on standard welders. *Figure 10-9* illustrates the principle of series welding where two welds are made with each stroke of the welder without any markings, indentations, or discoloration on one side of the assembly. *Figure 10-10* illustrates the principle of indirect welding ordinarily used where the welding current must pass through the side or ends of parts due to design. The welding electrode tip size directly affects the size and shear strength of the weld. Tip area is a function of the workpiece material gage. For thin sheets (up to 0.250"—6.35 mm) the diameter of the electrode tip can be calculated by using the formula $d = 0.1 + 2t$ where t is the material thickness. For thick material (1/4-2"—6.0-50.0 mm) $d = t$.

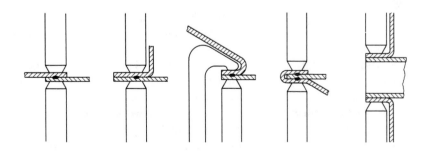

Figure 10-8. Typical spot-welded joints. (*Machine Design.*)

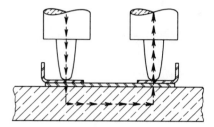

Figure 10-9. Assembly showing series-welded joint. (*Machine Design.*)

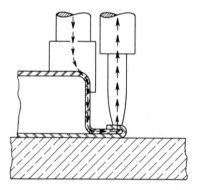

Figure 10-10. Spot-welded assembly showing a typical joint design for an indirect weld. (*Machine Design.*)

Projection Welding. In projection welding, embossments or projections are formed on one or both workpieces for localization of heat. Dimpled workpieces are placed between plain, large-area electrodes. Projection welding provides increased strength with reduced electrode maintenance.

Seam Welding. In seam welding, the material to be welded passes between two rotating disk electrodes. As the current is turned on and off a continuous tight seam is produced.

Pulsation Welding. In pulsation welding, the current is applied repeatedly to make a single weld while the pressure is applied. This process will produce a better weld for heavier material.

Flash Butt Welding. In flash butt welding, the work is clamped in dies, the current is turned on, and two joints are brought together by means of cam control to establish flashing and upsetting followed by discontinuation of the welding current. *Figure 10-11* illustrates a flash welding fixture.

Upset Butt Welding. Upset butt welding differs from flash butt welding

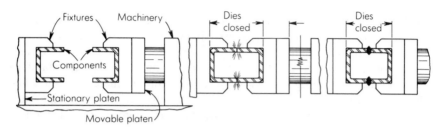

Figure 10-11. Flash butt welding.

in that pressure is applied continuously through the clamping dies after the welding current is applied so that heat is developed entirely from the resistance effect of the current.

New Welding Processes. Newer joining techniques growing in popularity are ultrasonic welding, high-frequency resistance welding, foil-seam welding, magnetic-force welding, percussion welding, friction welding, thermo-pressure welding, diffusion-bond welding, electro-slag welding, electron-beam welding, plasma-arc welding and laser welding.

Function of Electrode Holders and Electrodes. In resistance welding, the parts are positioned between electrodes which exert heavy pressure, conduct the current into the materials to be welded, and dissipate

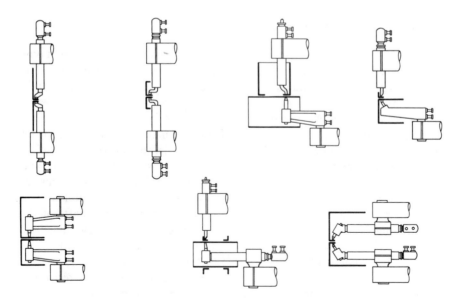

Figure 10-12. Typical standard electrode tips and operations.

the heat from the outer surface of the materials being welded. Holders and adapters are mounted in the machine so that the position of the electrode can be adjusted to suit a particular workpiece. Wherever possible, the electrode tips should be water cooled. *Figure 10-12* illustrates typical standard electrode tips. The design of welding electrodes and the material from which they are made are of great importance. For increased life, the design must provide sufficient strength with adequate heat conduction and cooling. Electrode nose shapes are shown in *Figure 10-13*.

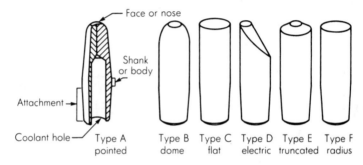

Figure 10-13. Standard types of electrode face or nose shapes. (Type D was formerly called offset.)

Resistance Welding Fixtures

There are two general types of fixtures for resistance welding. The first type is a fixture for welding in a standard machine having a single electrode. The second type is a fixture and machine designed as a single unit, usually to attain a high-production rate.

Certain design considerations apply to fixtures for resistance welding: (1) keep all magnetic materials, particularly ferrous materials, out of the throat of the welding machine; (2) insulate all gage pins, clamps, locators, index pins, etc.; (3) protect all moving slides, bearings, index pins, adjustable screws, and any accurate locating devices from flash; (4) give consideration to the ease of operation and protection of the operator; (5) provide sufficient water cooling to prevent overheating; and (6) bear in mind that stationary parts of the fixture and work are affected by the magnetic field of the machine. Workholder parts and clamp handles of nonmagnetic material will not be heated, distorted, or otherwise affected by the magnetic field.

There are other considerations that will affect the design of resistance welding fixtures and the machine if high-production is required:

1. The fixture loop or throat is the gap surrounded by the upper and lower

arms or knees containing the electrodes and the base of the machine that houses the transformer. This gap or loop is an intense magnetic field within which any magnetic material will be affected. In some cases, materials have actually been known to melt or puddle. Power lost by unintentional heating of fixture material will decrease the welding current and lower the welding efficiency. This power loss may sometimes be used to advantage; e.g., if the current is burning the parts to be welded, the addition of a magnetic material in the throat will increase the impedance, lower the maximum current, and halt the burning of parts.

2. The throat of the machine should be as small as possible for the particular job.
3. Welding electrodes should be easily and quickly replaceable. Water for cooling should be circulated as close to the tips as possible. Provide adjustment for electrode wear. If the electrodes tend to stick, knockout pins or strippers may be specified. Current-carrying members should run as close to the electrodes as possible, have a minimum number of connections or joints, and be of adequate cross-sectional area.
4. Provide adjustment for electrode wear.
5. Check welding pressure application.
6. Have knockout pins or strippers if there is a tendency of the electrode to stick to the electrode face. These may be leveraged or air operated.

General Fixture Design Considerations. Simple fixtures may have the part located visually with scribed lines as a guide. This is quite similar to locating parts for gas welding. For higher production, a quicker locating method is needed. A locating land may be incorporated in the fixture to accurately establish the edge position of the part to be welded (*Figure 10-14*). In some cases, setup blocks may be used in place of a locating land (*Figure 10-15*).

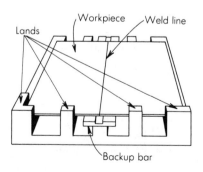

Figure 10-14. Locating lands.

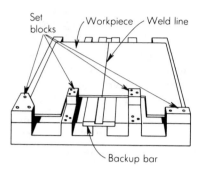

Figure 10-15. Set block locators.

When welding a variety of similar parts with different dimensions, setup blocks have a distinct advantage over the land method of locating. With proper design, setup blocks can be interchangeable to accommodate varying workpieces. Dowel pins may be used as locators (*Figure 10-16*).

Other means of locating are V-blocks, adjustable clamps, rest buttons and pads, spring plungers, and magnets where applicable.

Clamping Design Considerations. Clamps used in welding fixtures must hold the parts in the proper position and prevent their movement due to alternate heating and cooling. Clamping pressure should not deform the parts to be joined. Clamps must be supported underneath the workpiece (*Figure 10-17*). Owing to the heat involved, deflection by clamping force could remain in the part.

Quick-acting and power-operated clamps are recommended to achieve fast loading and unloading. C-clamps may be used for low production volume. Power clamping systems may be direct acting or work through lever systems (*Figure 10-18*).

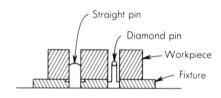

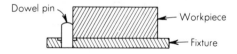

Figure 10-16. Dowel pin locators.

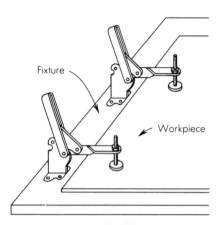

Figure 10-17. Typical clamp installation with the fixture supporting the workpiece directly beneath the clamps.

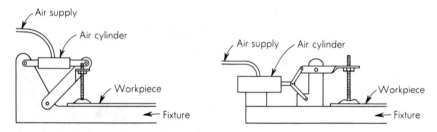

Figure 10-18. Air-actuated clamping methods.

In heavier plate applications, urethane tip or spring-loaded clamp spindles are recommended to compensate for plate thickness variations.

Where weldments change in size and quantity, and where weldment tolerances are not critical, welding platens and their stock tooling can be used (*Figure 10-19*).

Where weldment tolerances are critical, specially designed fixtures are used. These fixtures use all standard jig and fixture components (*Figure 10-20*).

Laser Welding Fixtures

Because of the laser's high heat intensity, it can be used for welding. Since the laser delivers its energy in the form of light, it can be operated in any transparent medium without contact with the workpiece. In welding, the

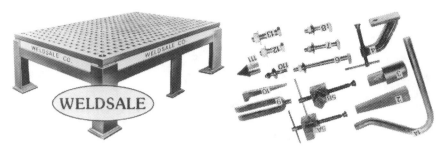

Figure 10-19. Welding platen and tooling. (*Courtesy, Weldsale Company*)

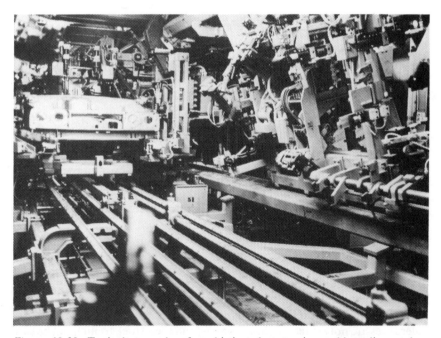

Figure 10-20. Typical example of sophisticated, extensive weld tooling and fixturing. Shown is Gilman framing, assembly robot line. Welding done on auto frames and car roofs, fully automated and computerized. (*Courtesy, The Milwaukee Journal.*)

power is delivered in pulses rather than as a continuous beam. The beam is focused through a lens, on to the workpiece, where the weld is to be made, and the intense heat produces a fusion weld (*Figure 10-21*). Laser welding is limited to depths of approximately 0.175-0.200″ (4.45-5.08 mm). Additional energy only tends to create gas voids and undercuts in the workpiece.

The laser should be mounted in a firm structure to prevent vibration. The

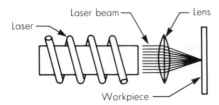

Figure 10-21. Laser welding.

design should allow sufficient room between the work table and the focusing lens to accommodate various types of positioning devices.

The tooling used to position parts for laser welding is similar to that used in other areas of welding. However, laser tooling is scaled down to accommodate smaller piece parts. In numerical control operations where more than one fixture is used, working heights of fixtures must be controlled in relation to each other to prevent accidental defocusing. This accidental defocusing causes poor quality. Because the laser energy is delivered to the workpiece without mechanical force, no fixturing is required on small piece parts. Use of mechanical force for these applications could induce strains and deformations.

Tooling for Soldering and Brazing

Soldering and brazing differ from welding in several respects. The metal introduced to the workpiece for the joining operation is nonferrous, usually lead, tin, copper, silver, or their alloys. The workpiece or base metal is not heated to the melting point during the operation. The added metal is melted and usually enters the joint by capillary action. Lead-tin soldering is called soft soldering and is conducted below 800° F (427° C). Silver soldering is called hard soldering and requires temperatures from 1100° —1600° F (593° —871° C). Copper brazing requires temperatures from 1900° —2100° F (1038° —1149° C).

The success of these processes depends on chemical cleanliness, temperature control, and the clearance between the surfaces to be joined. Cleanliness is usually obtained by introducing a flux which cleans, dissolves, and floats off any dirt or oxides. The flux also covers and protects the area by shielding it from oxidation during the process. It may to some extent reduce the surface tension of the molten metal to promote free flow. The worst contamination is usually due to oxidation during the process. Many joining operations are conducted in a controlled atmosphere with a blanket of gas to shield the operation.

Temperature control, although influenced somewhat by fixture design, is dependent primarily on the heat application method. In a simple low-production process, the workman may hold a torch closer to the workpiece for a longer period of time. In a very precise high-production process, the instrumentation of a controlled atmosphere-type furnace may be adjusted.

Clearance between the surfaces being joined determines the amount of capillary attraction, the thickness of the alloy film, and consequently the strength of the finished joint. The best fitting condition would have about 0.003-0.015″ (0.08-0.38 mm) clearance. Larger clearances would lack sufficient capillary attraction, while smaller clearances would require expensive machining or fitting.

Many soldering and brazing operations are conducted without special tooling. As with mechanical joining methods, many workholding devices can be used to conveniently present the faces or areas to be joined. An electrical connecting plug can be conveniently held in a vise while a number of wires are soldered to its terminals. In many high-production assembly operations, parts are manually mated with a preformed brazing ring between them. They are then placed directly on the endless belt of a tunnel-type furnace, or on a gas-heater ring fixture.

If the shape of the workpiece is such that it will not support itself in an upright or convenient position, a simple nesting fixture may be required. *Figure 10-22* shows a simple nesting fixture in which two workpieces and a

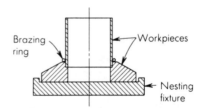

Figure 10-22. Simple nesting fixture
with work in place.

brazing ring have been placed. The fixture can be mounted on a table while an operator applies heat with a hand torch. The same fixture could be mounted on a powered rotating base in the flame path of a fixed torch, while a feed mechanism would introduce wire solder at a predetermined rate (*Figure 10-23*). The same fixture could be attached in quantity to the belt of a tunnel furnace. A number of the fixtures could be attached to a rack for processing in a batch furnace.

Tooling for Induction Brazing. *Figure 10-24* shows a nest-type fixture to hold mating workpieces within the field of an induction work coil. *Figure 10-25* shows the same fixture as altered to permit use of an internal induction heating coil. If the external coil is used, the fixture designer must provide some method of moving the fixture or coil while workpieces are loaded and unloaded.

Induction coils (inductors) provide a convenient and precise way of quickly and efficiently heating any selected area of an electrically conductive part or assembly of such parts to any required depth to provide a specified

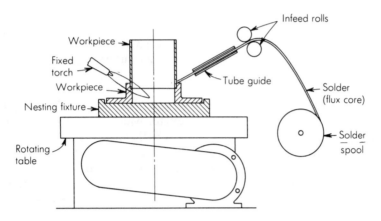

Figure 10-23. Soldering machine using simple nesting fixture.

brazed joint. Correct selection must be made of frequency, power density, heating time, and inductor design.

Induction Heating Theory. The flow through an electrical conductor results in heating as the current meets resistance to flow. Thus, I^2R losses (I, current; R resistance) may be as low as in the case of current flowing through copper wire (resistance is low) or high when this same current reaches and flows through a heating element (resistance is high). This high loss is the resistance heating obtained with conventional electric heaters.

Induction heating is also resistance heating with current flow meeting

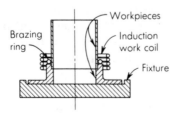

Figure 10-24. Nesting fixture for brazing with an external inductor.

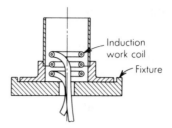

Figure 10-25. Nesting fixture for
brazing with an internal inductor.

resistance. A workpiece so heated has been made the secondary of a simple transformer, the primary being the inductor which generally surrounds the part and through which alternating current is flowing. The flow of alternating current in a primary induces by electromagnetic forces (magnetic flux) a flow of current in the secondary. Alternating current tends to flow on the surface, and there is a relationship between frequency of the alternating current, the depth to which it flows, and the diameter of stock which can be heated efficiently.

Considering a piece of plain carbon steel being heated for brazing or surface hardening, the depth at room temperature in which there is instantaneous flow of current is related only to the frequency:

$$D = 4/F$$

Where:

D = depth in inches

F = frequency in cycles/sec.

The depth D increases with temperature. For heat to be generated, the $I^2 R$ must have a time factor t to become $I^2 Rt$.

Additional depth results from the current following the path of least resistance (cooler underlying metal), and from heat flow by conduction. This depth may be approximated by the equation $D = 0.0015t$ where D is depth in inches and t is time in seconds. The higher the frequency for a given heating time, the shallower will be the depth of heat. The converse is true. For a given frequency, the depth of heat is also directly proportional to the time (square root).

An examination of the equation $D = 4/F$ will give an answer to the relationship between frequency and diameter of stock which can be heated efficiently. Large diameters may be heated efficiently with low frequency. Sixty cycles is used efficiently on 10" (254 mm) diameter steel workpieces.

The theoretical depth of current penetration mentioned above, increases with temperature and must never exceed the radius of the stock being heated. For practical purposes, this depth D ($D = 4/F$ at room temperature) should be two to three times the radius for through heating and ten times the radius for surface hardening.

Power density and heating time are closely related to the design of the inductor. It is upon the inductor design that the success of any induction heating application is largely dependent.

The purpose of the inductor is to set up a magnetic flux pattern in which the work to be heated is correctly positioned so that the required heating is accomplished (*Figure 10-26*).

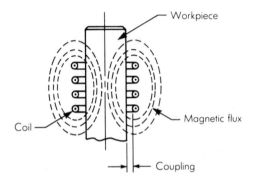

Figure 10-26. Magnetic flux in induction heating.

Coil design is influenced by the application, as is the selection of frequency, power density, and heating time.

Regardless of the material used in the induction brazing process, the inductor must be designed to heat the area of the joint sufficiently to cause the bonding material to flow. Temperatures can range from a few hundred degrees for soft solder and epoxies to 2100°F (1149°C) for copper. No attempt will be made here to discuss joint design, placement of material, type of material, or fluxing. These factors first must have been resolved and proper consideration given to their relationship to the induction heating method, which enables the precise heating of the joint area only. Neither will any discussion be given to frequency selection other than to comment that extremely small diameters usually require the high frequencies of a vacuum tube oscillator, and either oscillators or motor generator sets are suitable for parts of medium cross section. When attempting, however, to heat heavy sections and to avoid overheating of sharp corners on small or medium-size workpieces, the use of motor-generators at 3000 or 10,000 cycles is preferred.

Copper is normally used for the coils with appropriate water cooling. This can be accomplished by using copper tubing, solid copper to which tubing has been attached, or solid copper which has been drilled out or machined to provide water passages.

An inductor should have the following characteristics: (1) proper physical shape to surround the section to be heated; (2) large enough diameter to permit loading and unloading or, in the case of a continuous operation, to permit clear passage; (3) proper support or rigidity to maintain its designed shape; (4) insulation to avoid electrical breakdown (spacing or an air gap between the turns and the workpiece is usually sufficient); and (5) ability to withstand operating conditions (exposure to water, dirt, and flux must be taken into consideration).

The air gap or space between the work and the inside of the inductor should be kept low for reasons of efficiency. A gap of one-eighth inch (3.2 mm) is reasonable. For irregularly shaped sections where efficiency is not too important, gaps of one-quarter inch (6.4 mm) and greater can be tolerated.

The area to be heated determines the length or width of the inductor. For most joint designs on small parts, a single turn is sufficient. On larger diameters, it is necessary to heat a wider band requiring either several turns or a wider inductor. An inductor that is too narrow will simply require a longer heating time to allow the heat flow to cover the area. An inductor that is too wide will heat more metal than necessary and, therefore, will be less efficient. A single-turn inductor should be no wider than the bore diameter; for greater coverage, multiturn inductors are used.

The electrical characteristics of the high-frequency power source determine the number of turns in an inductor. Generally, the high-frequencies of a vacuum tube oscillator (200-3,000 kHz) usually require multiturn coils, whereas the motor-generator sets (1-10 kHz) can operate into either multiturn coils or single-turn coils through the use of variable-ratio transformers. Turns of the inductor should be kept as close as possible without touching. Typical inductors are shown in *Figure 10-27*.

Number of Inductors. The number of inductors to be used at a time is, of course, influenced by production requirements. The number can be extended to as many as it may appear convenient to load. Instead of a large number, a continuous coil can be used, through which parts are moved progressively. This method is generally limited to a workpiece having a joint on or near its end. The determination of how many parts are to be processed at a time is related to heating time and power density. The time generally runs from 10 or 15 seconds on small parts to as long as two minutes on large parts where the heat must penetrate. The power density is approximately 0.5-1.5 kW/in² (775-2325kW/m²) of surface area. An average value of one kW/in²

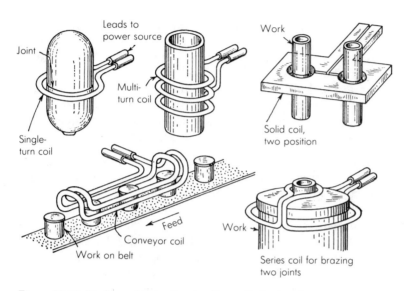

Figure 10-27. Designs of induction-heating coils for brazing.

(1550kW/m²) is recommended. The actual value depends on the ratio of surface area to volume.

The amount of energy (kilowatt seconds) is determined by the volume of metal to be heated and the temperature necessary to flow the bonding material. Since the exact amount of metal heated in a brazing operation depends on many factors such as conductivity of the work and heating time, only a rough approximation can be attempted. The following procedure is recommended (for U.S. customery units):

1. Estimate the volume of metal which would be expected to be brought to brazing temperature.
2. Convert the volume to weight in pounds based on steel weighing 0.3 lb/in³.
3. For 400° F solder, convert to kilowatt seconds by multiplying by 100.
4. For 1200° F brazing material, convert to kilowatt seconds by multiplying by 300.
5. For 1600° F brazing material, convert to kilowatt seconds by multiplying by 500.
6. For 2000° F brazing material, convert to kilowatt seconds by multiplying by 700.
7. Estimate the surface area of the joint. Based on one kW/in.², divide the area into the above value of kilowatt seconds to determine the approximate brazing time in seconds.

8. Determine from production requirements (allow for handling each part) the number of parts to be processed at a time.
9. Total power required will depend on the above value and kilowatts per piece based on one $kW/in.^2$ of surface area.
10. Power available is determined by the equipment to be used. The maximum number of parts which can be processed at once can be approximated by dividing the kilowatt rating of the equipment by the surface area of each part in inches (one $kW/in.^2$).

Material. A solid-type inductor is preferred because of its greater rigidity. Accidental contact while loading parts will not bend it out of shape. Less expensive coils can be made of copper tubing which, except for extremely low power (one kW per coil), should not be less than 0.19″ (4.8 mm) in diameter. Both round and rectangular sections are used; the rectangular section is especially adaptable when a wide, single turn is needed. Its wide flat area can be adjacent to the work for more uniform heating. A one-inch (25.4 mm) wide inductor can be made of 0.25″ by one-inch (6.35 by 25.4 mm) tubing or from a wide flat strip brazed to a round tube for cooling.

Leads. The leads to the inductor should be kept as short as possible and close together. In some situations coaxial leads may be necessary (450 kHz and up). If the leads are not a continuation or part of the inductor and a brazed joint is used, they must not reduce the flow of current to the inductor. Silver solder is recommended, and carefully prepared joints (even mitered) are essential. Insulating materials should be used between leads if they are not self-supporting. The connection to the power source will be determined by the design of the output transformer.

Cooling. Inductors must have sufficient cooling to avoid overheating. Inductors for brazing operations can be made by forming a copper tube on a suitable mandrel. The tube must not collapse to restrict internal coolant flow.

Cooling water may be brought in from the output transformer through the connectors carrying the current or by separate insulated connections to a water supply.

Coil Support. Solid inductors are usually self-supporting, as are single-turn inductors. Support of multiturn inductors or an array of inductors can be accomplished with any nonmetallic material. Many plastic materials can be used. Coils can be attached by studs (preferably copper) brazed to the inductor. Use of copper or bronze-brass studs is highly recommended, as cooling water will cause rust problems with steel or any ferrous metal and hinder their removal.

Heating Internally. When tubular sections are to be joined it is often more convenient to heat from the inside. The factors of coil design are essentially the same as for ID inductors except that the leads must be brought to the coil axially instead of radially (*Figure 10-25*). Since heating from the ID is considerably less efficient (the magnetic flux is weak outside the confines of the coil), two to three times as much energy (kilowatt seconds) may be required. This can be obtained by using more power or time or both. When high production requirements preclude the use of longer heating cycle, the ID coil efficiency can be improved by the use of an inside core which increases the magnetic flux outside of the coil. Basic design elements are as follows:

1. For vacuum tube oscillators, compacted ferromagnetic stock may be used. This material, which can be purchased in round bars, is then made a core inside the inductor. The turns can be held tightly on the surface or wound in grooves machined in the core. No insulation is required, and one lead may be brought through a hole in the center of the core.
2. For motor-generators, iron laminations are used. They are made from 0.007" (0.18 mm) thick strips of transformer steel, 0.25" (6.35 mm) wide, which has been cut to the correct coil length. The strips are wound together and inserted in a multiturn coil to form a core. If a single-turn inductor is used, laminations are cut in a *C*-shape and stacked tightly around the turn as shown in *Figure 10-28*.

Figure 10-28. Special-purpose inductor.

The design of inductor coils for brazing operations is relatively simple. Electrical matching to the proper power source is no problem. The physical configuration is dictated by the shape and size of the area to be heated. The inductor is a primary winding for a transformer with the workpiece acting as a secondary. The coil need only be formed to surround or—for ID heating—be adjacent to the surface to be heated. A wide latitude of shape is permitted. A square part can be heated with an inductor formed on a square mandrel (*Figure 10-29*) or a round mandrel. Irregularly-shaped parts can be heated with round inductors. If a corner of a workpiece over-heats, the inductor can be modified to merely provide a larger air gap at the corner.

Occasionally, a joint must be heated by proximity. An ID coil cannot be

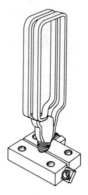

Figure 10-29. Inductor formed on
rectangular mandrel.

used and an ID coil would necessitate heating entirely too much metal. An inductor for heating such an area is shown in *Figure 10-30*. Such an inductor is insufficient and can be improved by use of laminations as shown. The *C*-shape is open toward the work.

Generally, in inductor should have the same contour as the area to be heated. With a reasonably uniform coupling, the part can be loaded with no difficulty. When there are large changes in diameter in the section to be heated, the larger diameter will tend to overheat unless some compensation is made. This can be done by having fewer turns per linear inch in that area, by increasing the coupling, or by using two turns in series (electrically) on the small diameter (*Figure 10-31*). The intensity of the magnetic field in which the surface finds itself should be as uniform as possible.

Tooling for Thermal Cutting

Three basic types of thermal cutting systems are used in metal fabricating operations; oxy-flame, plasma-arc, and laser.

Laser cutting is primarily used on different types of NC and CNC punching machines. Machines that are equipped with lasers generally have a large working table area (60 x 50 in. or 152.4 x 127.0 cm nominal sheet size). The workpiece is held in position by several hydraulic holders or clamps. All of the holes or slots are punched, and the outside perimeter is then laser cut. Many of the machines have nesting capabilities, which allow more than one part per material sheet.

Plasma-arc cutting is used on punching machines in the same manner as a laser. Plasma-arc cutting is also used on thermal machining centers. These

Figure 10-30. Inductor with C-shaped laminations added.

Figure 10-31. Inductor with loops in series.

machining centers carry one or more torches on a traveling bridge. The torches are guided by an optical scanner, which follows a paper template or can be operated by tape control or computer. The material to be cut is placed on watertables. The watertable is used to control smoke, noise, and ultraviolet radiation. A dye in the water eliminates ultraviolet emissions.

Oxy-flame or oxyacetylene cutting machines employ the same method as the plasma-arc machining centers, except that the cutting tables do not contain water. The cutting table is made up of a number of cross-supports on which the workpiece is set. These cross-supports are held in position by a square or rectangular container into which the drop-outs may fall. After several cutting operations, the cross-supports must be replaced because they are damaged during the cutting operation.

Some fixturing is used on plasma-arc and oxy-flame machining centers. These fixtures are constructed to support the material to be cut, allowing for proper heat dissipation to prevent distortion. The fixtures use standard jig and fixture components. The key to designing fixtures for cutting operations is to allow clearance for the cutting media below the piece part. If there is insufficient clearance, fixture damage may result (*Figure 10-32* and *10-33*).

Without further discussion regarding torch cutting, it is to be understood that this chapter is dealing with basic low-carbon steel plate. Added precautions must be taken in cutting high-alloy steels; due to the hardening ability of the workpiece at the line of cut, suitable machining allowance or post heat treatment must be considered.

Tooling for Mechanical Joining Processes

Many workpieces can be held for mechanical joining without tooling. A

Figure 10-32. A steel plate circle cut with specialized fixturing. (*Courtesy, Ryerson*)

Figure 10-33. 45 bevel and square cuts made on ½ inch aluminum plate a speed of 110 inches per minute with cutting fixture. (*Courtesy, Linde Division, Union Carbide Corporation*)

workman can often manually align two workpieces and insert a fastener. This method has several limitations. The workpiece must be small or light enough to be positioned manually, and the forces incurred in the joining process must be relatively small. The complexity of an assembly so joined is also limited by the number of components that a workman can conveniently handle. An elementary workholder can often be used to advantage in even the most simple joining operation.

A universal-type vise can be of great value in mechanical joining. A primary workpiece can be held in any position while other workpieces are fastened to it. Several workpieces may be held in alignment between the jaws, while the workman applies a fastener. Clamping pressure can be used to counteract the joining forces, such as the torque applied to a threaded fastener.

Threaded Fasteners

Threaded fasteners are used for a wide variety of applications, the bolt and nut being the most common. Tooling for threaded fasteners is as varied as their application. There are, however, design principles that apply to all cases.

A primary application of threaded fasteners is joining and holding parts together for load-carrying requirements, especially when disassembly and reassembly may be required. Typical assemblies for several types of threaded fasteners are illustrated in *Figure 10-34*. Threaded fasteners are also used extensively for assemblies subject to environmental conditions such as high temperatures and corrosion.

Advantages of threaded fasteners include their commercial availability in a wide range of standard and special types, sizes, materials, and strengths. Extensive standardization efforts have made most threaded fasteners interchangeable.

Bolts and Studs. Bolts are externally threaded fasteners generally assembled with nuts (*Figure 10-34A*). While most bolts are headed, some are not. The means of distinguishing between bolts and screws are discussed in ANSI Standard B18.2.1, "Square and Hexagonal Bolts and Screws, Inch Series." Studs are cylindrical rods threaded on one or both ends or throughout their lengths (*Figure 10-34C*).

Bolts with hexagonal heads, frequently called hex heads, are the most commonly used. These heads have a flat or indented top surface, six flat sides, and a flat bearing surface. The flat sides facilitate tightening the bolts with wrenches. Hex heads are often used on high-strength bolts and are easier to tighten than bolts with square heads. They are generally available in standard strength grades and to special strength requirements for specific applications.

Round-head bolts have thin circular heads with rounded or flat top surfaces and flat bearing surfaces. When provided with an underhead configuration that locks into the joint material, round-head bolts resist rotation and are tightened by turning their mating nuts. Included in this classification, even though the configurations differ, are countersunk and T-head bolts.

Variations of round-head bolts include those with square, ribbed, or finned necks on the shanks below the heads to prevent the fasteners from rotating in their holes.

Square-head bolts have square-shaped, external wrenching heads. They are available in two strength grades. Lag bolts, sometimes called lag screws, usually

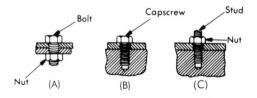

Figure 10-34. Typical assemblies using threaded fasteners: (A) bolt and nut; (B) capscrew; (C) stud.

have square or hex heads, gimlet or cone points, and thin, sharp coarse-pitch threads. They produce mating threads in wood or other resilient materials and are used in masonry with expanding anchors.

Battery bolts have square heads and are generally stainless steel or lead or tin coated for clamping onto battery terminals. Fitting-up bolts have square heads and coarse-pitch, 60° stub threads. They are used for the preliminary assembly of structural steel components. T-bolts are square-head bolts used in the T-slots of machine tools.

Bent bolts are cylindrical rods having one end threaded and the other end bent to various configurations. These include eyebolts, hook bolts, and J-bolts. Other bent bolts, such as U-bolts, have both ends threaded. The ends of bent bolts are usually square (as sheared).

Studs are unheaded, externally threaded fasteners. They are available with threads on one or both ends or continuously threaded. Studs with collars and threaded on one or both ends are also available. Heat-treated and/or plated studs are available to suit specific requirements. They are also made with chamfered or dog-point ends.

An advantage of studs for some applications, such as the assembly of large and heavy components, is their usefulness as pilots to facilitate mating of the components, which expedites automatic assembly. For many applications, studs provide fixed external threads, and nuts are the only components that must be assembled.

Nuts. Nuts are internally threaded fasteners that fit on bolts, studs, screws, or other externally threaded fasteners for mechanically joining parts. They also serve for adjusting, transmitting motion, or transmitting power in some applications, but they generally require special thread forms.

Hex and square nuts, sometimes referred to as full nuts, are the most common. Hex nuts are used for most general-purpose applications. Square machine screw nuts are usually limited to light-duty and special assemblies. Regular and heavy square nuts are often used for bolted flange connections.

Single-thread nuts, sometimes called spring nuts, are formed by stamping a thread-engaging impression (arched prongs) in a flat piece of metal (*Figure 10-35*). These nuts are generally made from high-carbon spring steel (SAE 1050-1064), but are also available in corrosion-resistant steel, beryllium copper, and other metals.

Stamped nuts are hex fasteners stamped from spring steel or other metals, with prongs formed to engage mating threads. Like single-thread nuts, they rely on spring action for clamping and resistance to loosening, but they have more prongs to engage the threads on the mating fastener. Applications include replacement for full nuts in low-stress uses and as retaining nuts against full nuts (*Figure 10-36*). Stamped nuts are made with integral washers, in closed top or bottom styles, and as wingnuts.

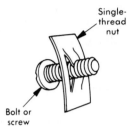

Figure 10-35. Single-thread nut.

Screws. Screws are externally threaded fasteners capable of insertion into holes in assembled parts, of mating with preformed internal threads, or of cutting or forming their own threads. Because of their basic design, it is possible to use some screws, which are sometimes called bolts, in combination with nuts.

Screws are available in a wide variety of types and sizes to suit specific requirements for different applications. Major types discussed in this section include machine screws, capscrews, setscrews, sems (screw and washer assemblies), and tapping screws.

Machine screws are usually inserted into tapped holes, but are sometimes used with nuts. They are generally supplied with plain (as sheared) points, but for some special applications they are made with various types of points. Machine screws have slotted, recessed, or wrenching heads in a variety of styles and are usually made from steel, stainless steel, brass, or aluminum. Many machine screws are made from unhardened materials, but hardened screws are available.

Capscrews are manufactured to close dimensional tolerances and are designed for applications requiring high tensile strengths.

The shanks of capscrews are generally not fully threaded to their heads. They are made with hex, socked, or fillister slotted heads (*Figure 10-37*). Low-head capscrews are available for applications having head clearance problems. Most capscrews are made from steel, stainless steel, brass, bronze, or aluminum alloy.

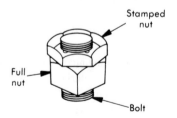

Figure 10-36. Stamped nut applied and tightened after full nut is in place.

Setscrews are hardened fasteners generally used to hold pulleys, gears, and other components on shafts. Hardness of the shaft is an important consideration in selecting a proper setscrew. They are available in various styles, with square-head, headless-slotted, hex-socket, and splined (fluted) socket styles being the most common. Holding power is provided by compressive forces, with some setscrews providing additional resistance to rotation by penetration of their points into the shaft material.

Sems (screw and washer assemblies) is a generic term for preassembled screw and washer fasteners. The washer is placed on the screw blank prior to roll threading and becomes a permanent part of the assembly after roll threading, but is free to rotate. Sems are available in various combinations of head styles and washer types. Washers commonly used include flat (plain), conical, spring, and toothed lock washers.

Sems are used extensively in the manufacture of automobiles and appliances and other mass production industries because they are suitable for automatic assembly operations. These fasteners permit convenient and rapid assembly by eliminating the need for a separate washer assembly operation. They also ensure the presence of the proper washer in each assembly, and prevent the loss of washers during maintenance.

Tapping screws will cut or form mating threads when driven into holes. Self-drilling, self-piercing, and special tapping screws are also available. They are made with slotted, recessed, or wrenching heads in various head styles and with spaced (course) inch or metric threads. Tapping screws are generally used in thin materials.

Advantages of tapping screws include rapid installation because nuts are not needed, and access is required from only one side. Mating threads fit the screw threads closely, with no clearances necessary. Underhead serrations or nibs on some screws increase locking action and minimize thread stripout.

Captive screws remain attached to panels or assembly components after they have been disengaged from their mating parts. Advantages include fast assembly and disassembly and prevention of damage to other assembly components caused by lost or loose screws.

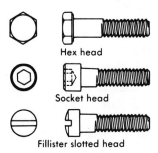

Hex head

Socket head

Fillister slotted head

Figure 10-37. Capscrews with various heads.

Threaded Assembly. The most common method is to place two workpieces in their correct relative location, drill a hole through them, insert a bolt in the hole, and torque a nut onto the bolt. The hole may have been drilled in both pieces prior to mating. In many cases no tooling will be required. If it is convenient for a workman to hold the workpiece together while inserting the bolt, adding the nut, then tightening the nut, a simple workholder may be advantageous.

The workholder must locate the workpiece so that the holes are conveniently positioned, and should support the workpiece against the torque and thrust loads imposed in tightening the fasteners.

One of the two workpieces being assembled may be tapped to receive the threaded fastener, or a self-tapping fastener may be used. Power tools may be used to drive the fasteners. The fixture design principles for nut runners are as follows: (a) location—the fixture must align the workpiece precisely with the nut runner; and (b) support—the fixture must withstand the weight of the workpiece plus the thrust and torque loads imposed.

Figure 10-38 shows a template nesting fixture resting on the table of a power screwdriver. It is designed to hold a metal cabinet while two side shields are assembled to it by twenty self-tapping screws.

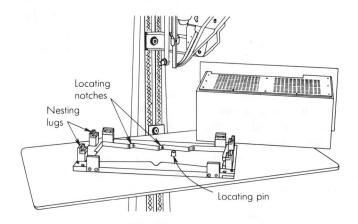

Locating
notches

Nesting
lugs

Locating pin

Figure 10-38. Partial nest-type workholding fixture for assembly of cabinet.

A round locating pin is attached to the machine table exactly under the driver. The fixture base is a template and has ten radiused notches that can be manually held against the locating pin. When the components of the cabinet are placed in the fixture, each template notch is directly beneath a predrilled hole in the components into which a screw is to be inserted. As each notch is held against the locating pin, the predrilled hole is placed directly beneath the driver, and the machine is cycled to insert and drive one screw. After ten screws have been driven to secure one side shield, the cabinet is inverted and another ten screws are driven to fasten the other shield.

The fixture design principles are again the same. The fixture must precisely locate the workpiece relative to the driver. In this case ten precise location points are involved. The fixture must also establish the exact height of the workpiece. The fixture must support the workpiece in resistance to the torque and thrust loads imposed.

For high-production volumes, several machines may be arranged to simultaneously insert and drive screws into a single-fixtured workpiece.

Rivets

Perhaps the most widely used pin-type fastener is the rivet—a pin headed on one end, the other end being plastically deformed after insertion to prevent retraction. The riveting process is extremely varied, and may be used to assemble the parts of a timepiece or the structural members of a bridge. Rivet diameters may vary from perhaps 0.015—5.0" (0.38—127 mm). The holes may be drilled or pierced before or during the operation.

It is important that the correct size and shape riveting tool be used, and that excessive driving pressures be avoided. Excessive pressure in driving may result in any of the following problems: (1) bulging of the edge of the piece being riveted; (2) buckling or other distortion, particularly if thin material is used; (3) weakening or fracturing of metal near the hole.

Figure 10-39 shows an L-shaped workpiece clamped in a fixture while a second workpiece, a channel, is being riveted to it. The portable riveting yoke literally squeezes the rivets to deform them. The holes are drilled and the rivets are inserted prior to the operation. In practice, it would be necessary to employ stops or other means to locate the channel with reference to the primary workpiece.

Figure 10-40 shows the sequence of a punching and riveting operation used in mass production. The two workpieces are placed between the tools, and the machine is cycled to automatically pierce a hole, insert a rivet, and then head the rivet.

Eyeleting. Eyelets, like rivets, are used extensively as low-cost fasteners for light assembly work. For high-production operations, costs can be reduced to a minimum with the use of eyelet-attaching machines. These machines are actually small power presses with hopper feed mechanisms. *Figure 10-41* illustrates typical tooling of an automatic eyelet-setting machine in loading and clinching positions.

Tubular Riveting. A tubular rivet is a cross between a solid rivet and an

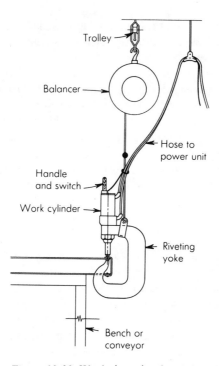

Trolley

Balancer

Hose to
power unit

Handle
and switch

Work cylinder

Riveting
yoke

Bench or
conveyor

Figure 10-39. Workpiece simply supported
for riveting.

eyelet. The straight end of the rivet has a center hole which permits this part of
the rivet to clinch easily when it is struck by the contoured riveting tool.
Figure 10-42 illustrates tooling for a typical tubular riveting operation.

Spin Peening. Some materials to be assembled are brittle, while other
assemblies require slender unsupported rivets or eyelets which will not
withstand the single impact required to rivet or eyelet without distortion or
cracking. For delicate assembly operations spin-peening machines may be
used. These machines deliver innumerable light blows while the hammer
spins. The peening and spinning action of these machines is either mechanical
or pneumatic.

Clinch Allowance. Clinch allowance is that part of a rivet or eyelet
which extends beyond the combined thickness of the assembly before the rivet
or eyelet has been set. *Figure 10-43* shows the relation of rivet or eyelet length
to material thickness and clinch allowance. The proper rivet or eyelet length
should be approximately equal to the material thickness plus the clinch
allowance. Both rivets and eyelets are available commercially in 0.03″ (0.8
mm) length increments. Two rules of thumb have been developed: (1)

maximum length of clinch for full-tubular and bifurcated rivets should be figured at 100% of shank diameter; (2) maximum length of clinch for semitubular rivets should be 50-70% of shank diameter to prevent buckling and ensure a tight set.

The hole diameter and the method of producing the hole in the material also affect clinch diameter. The minimum clinching radius for tubular rivets or eyelets can be determined by multiplying the wall thickness by three. The

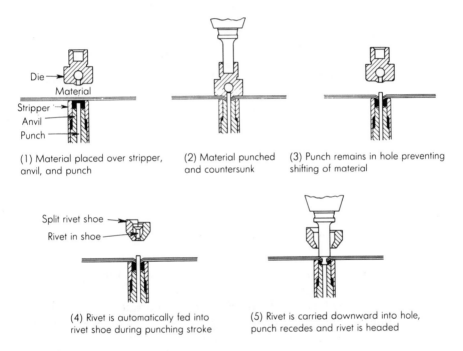

(1) Material placed over stripper, anvil, and punch

(2) Material punched and countersunk

(3) Punch remains in hole preventing shifting of material

(4) Rivet is automatically fed into rivet shoe during punching stroke

(5) Rivet is carried downward into hole, punch recedes and rivet is headed

Figure 10-40. Sequence of punching and riveting operations.

clinching contour of riveting tools must be free of nicks and circular grooves. After hardening, the contour must be highly polished.

Riveting Equipment

Rivets can be deformed or set in many ways. Pressure may be applied continuously or by a series of hammer blows. The rivets may be manually or automatically inserted. The holes may be drilled prior to or during the riveting sequence. The tool used to apply the deformation pressure may be a

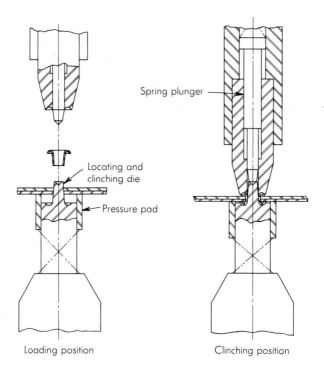

Spring plunger

Locating and clinching die

Pressure pad

Loading position   Clinching position

Figure 10-41. Eyelet curling.

common hammer, a pneumatic hammer (riveter or rivet gun), a portable squeezing yoke, or a stationary machine.

The final shape of the rivet becomes that of the tool (die) used to apply the deformation pressure. The shape may be flat (reflecting the contour of a hammer or a conventional bucking bar) or may be curved as shown in *Figures 10-41* and *10-42*. The rivet die may be a simple bucking bar, and interchangeable rivet set placed in the nozzle of an air hammer, or a complete forming die placed in a standard hydraulic or mechanical press.

Workholders or fixtures used to hold and locate workpieces being assembled by riveting are generally of three types: stationary, portable, and self-contained (riveting die fixture).

Pneumatic Hammer. Conventional riveting is performed by placing the rivet in a predrilled hole, holding a pneumatic hammer against the head of the rivet, holding a bucking bar against the end of the rivet, and then cycling or activating the hammer. The initiating pressure is exerted by the hammer while the formation pressure is exerted by the bucking bar. Pneumatic hammers (rivet guns) are commercially available in a wide range of types and sizes. The nozzle of the rivet gun receives and retains the rivet set or die. Rivet sets are commercially available in a wide variety of shapes to mate with the many types of rivets commonly used.

Portable Yoke Type. This type of riveting equipment is commonly

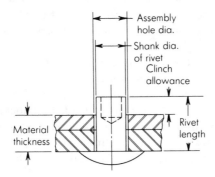

Figure 10-42. Tubular riveting.

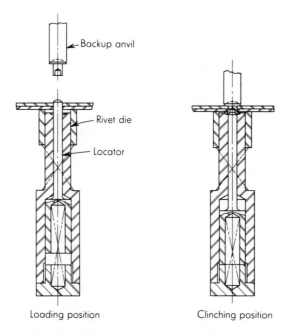

Loading position   Clinching position

Figure 10-43. Determination of rivet length for tubular rivets.

used for squeezing large rivets (0.19—1″ [4.8—25.4 mm]) diameter, and consists of a yoke with a cylinder (air or oil) to provide the squeezing action. Equipment size can be minimized by using hydraulic cylinders and high pressures (5000 lb/in.^{2} [34.5 MPa]). Equipment weight is minimized by using special high-strength heat-treated steel. The cylinder advances an anvil to the rivet with a primary pressure of approximately 1000 lb/in.^{2} (6.9 MPa). When resistance is encountered, the pressure increases to 5000 lb/in.^{2} (34.5 MPa) for the rivet upsetting portion of the action. *Figure 10-39* shows a hydraulic riveting yoke.

Stationary Machines. Stationary machines are often used for riveting. The workpieces being assembled are located with reference to each other (mated) and are placed between the upper and lower elements of the machine. Small stationary machines have a spring-loaded pin locator in the rivet die. Predrilled holes through the mated workpieces engage the pin locator as shown in *Figure 10-44*. The upper riveting die (backup anvil) pushes the rivet into the holes, depresses the spring pin, and upsets the driven head on the lower anvil. The rivets are fed from a hopper to a feed track. The lower end of the feed track locates the rivet directly above the pin locator. The end of the track is split to allow the upper die to pick off the rivet for insertion.

Large stationary machines often combine both riveting and hole piercing.

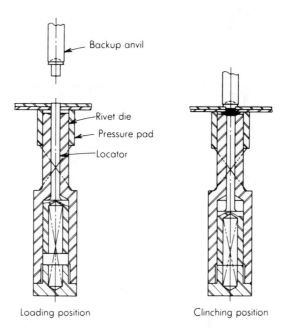

Figure 10-44. Stationary riveting machine operation.

Figure 10-40 illustrates the sequence of operations. The workpieces need not be predrilled, but must be located with reference to each other. The locating and holding fixture may be part of the machine. Extremely large stationary machines can hold and precisely locate large aircraft sections while thousands of the holes are pierced and rivets are driven. The entire production sequence is automatic, with the machine motions governed by tape control or CNC (computer numerical control). Closed-circuit TV is used to monitor the operation.

Stationary Holding Fixtures. Stationary holding fixtures are those in which two or more parts are located and pinned or clamped in position. They are usually free-standing fixtures where portable riveting equipment is used. Air-operated riveting hammers are used for smaller rivets such as those used in aircraft or appliance work. Larger rivets used in structural work are deformed with hydraulic or air-operated riveting yokes (rivet squeezing action). *Figure 10-45* illustrates a simple holding fixture where riveting equipment is brought to the work.

Portable Fixtures. Portable riveting fixtures are those used to locate two or more parts for transport to a stationary-type rivet unit or machine. The fixtures also hold the workpieces during the riveting sequence.

Riveting Die Fixture. A riveting die fixture is completely self-contained for use in a punch press. This fixture consists of locating elements for two or more parts, and rivet buttons for driving one or several rivets in one stroke or hit of a press. The locating elements may be movable so that parts can be positioned outside of the press and rivets can be placed in location. The assembled parts are placed in location in the riveting die, and all rivets are driven at once by the action of the press. *Figure 10-46* shows a riveting die fixture.

Riveting Fixture Design. Beyond the primary requirements that the fixture precisely locate and hold workpieces, ease of loading and unloading is extremely important. Many parts can be riveted without clamping if the weight of the parts alone will keep the riveting surfaces in contact with one another. Light-gage panels do require clamps because of the tendency to warp or twist. Clamps for low-volume production are usually hand-operated cam or toggle type. For high-production, air-operated clamps of the same type are used. Locating pins and sheet holders (*Figure 10-47*) are used in preference to clamps whenever possible. The area of application must be accessible to the riveting tool. Tilting fixtures are often used for this reason.

Stapling

Stapling is a joining operation using preformed U-shaped wire staples. Staples are made in a variety of shapes, wire sizes, leg lengths, and crown

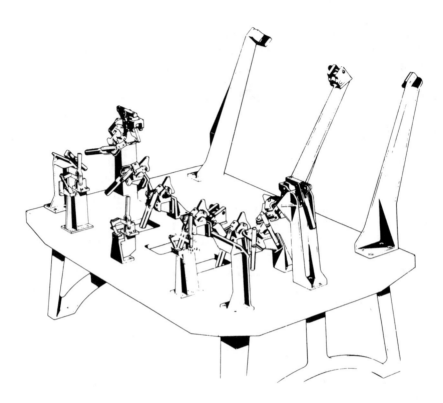

Figure 10-45. Stationary riveting fixture.

sizes. Staples are available in blunt-end, chisel-point, or divergent-point styles, and are cohered into strips and sticks. The sticks are loaded manually into staplers. *Figure 10-48* illustrates staple nomenclature. Gun tackers and hammer tackers are used to drive wire staples. Wire staples can also be driven and clinched by low-cost stapling machines.

Wire Stitching

Wire stitching is the process of joining two or more pieces of material with wire fed from a coil, cut to length, U-formed, driven through, and clinched by a specially designed machine called a stitcher. *Figure 10-49* illustrates the principle of wire-stitching machines. For low-cost, high-volume production, automatic wire-stitching machines are ideal for fastening components together. These machines are available to perform the following operations:

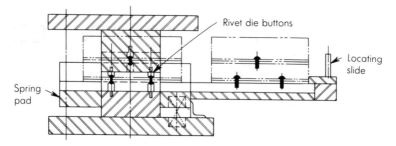

Figure 10-46. Punch press riveting.

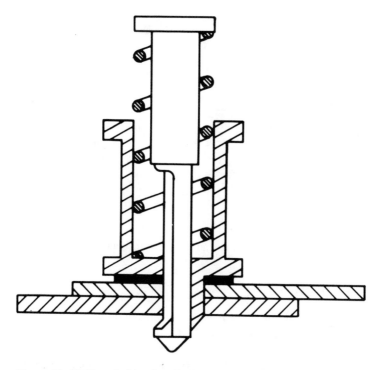

Figure 10-47. Sheet holder for aligning holes during riveting.

(1) carton or industrial-type stitching; (2) carding, bagging, and labeling; (3) book stitching; and (4) metal stitching. These machines are similar, varying in sizes from small bench models to large floor models. Each machine draws wire from a coil, cuts it to proper length, forms it into a stitch, drives it through the material, and clinches it with every stroke of the machine.

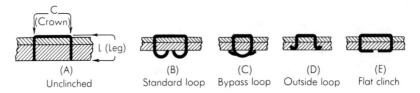

Figure 10-48. Staple nomenclature.

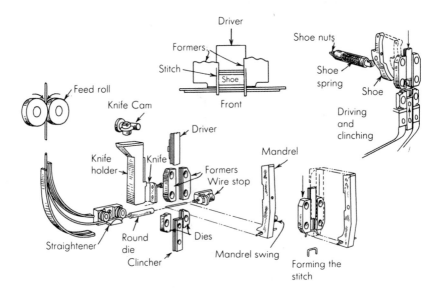

Figure 10-49. Metal-stitching nomenclature and principles.

Metal Stitching

Metal stitching is one of the newest methods of fastening thin-gage metals, and fastening metals to nonmetals. Metal stitching will fasten the work more economically since there is no need for other operations such as punching or drilling prior to fastening. Standard stitching machines can be tooled to form loop or flat clinches. Flat clinches are used when the stitched joint must carry heavy loads, such as aircraft constructions where joints are subject to severe stresses. The flat clinch is formed by an upward movement of the clinching die, folding the legs flat against the bottom of the material. Loop clinches provide only point contacts with the material and, therefore, provide less strength than flat clinches. Loop clinching is formed by curling the legs of the wire and stationary solid dies. *Figure 10-50* illustrates flat and loop clinches.

Special Applications. Metal stitches are used extensively by manufacturers of electronic components. Variable-resistor control and switch shafts are assembled into mounting bushings by wire retaining or snap rings. The rings are made of round or flat spring-tempered steel wire. The specially tooled machine draws the wire from a coil, cuts it to proper length, forms it to a U-shape, and drives it against a grooved die to form the round retaining ring into the circular groove of the shaft. *Figure 10-51* illustrates the parts and clinching dies of C-ring, retaining-ring assembly.

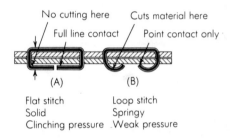

Figure 10-50. Types of metal stitches.

Staking

In staking, two or more parts are joined permanently by forcing the metal edge of one member to flow either inward or outward around the other parts. Staking is an economical method of fastening parts. The operation is completed with a single stroke of an arbor press, kick press, punch press, air, or hydraulic press. *Figure 10-52* illustrates locking a ring to a shaft with center punch staking. *Figure 10-53* illustrates joining together a bushing, a bracket, and a shell with a ring-staking punch. *Figure 10-54* illustrates the principles of inward staking by forcing metal of a ring against the knurled portion of a shaft. To provide added rigidity and torque in assembly, spot staking may be used. Spot staking is essentially the same as ring staking except that three or more equally spaced chisel edges of the staking punch force metal into splined portions of the parts to be assembled. *Figure 10-55* illustrates a typical splined design and spot-staking punch to force metal into this spline. In some cases, it is desirable to use a combined spot- and ring-staking punch, as is shown in *Figure 10-56.*

To facilitate efficient metal flow, the tips of punches must be free of nicks and circular grooves, and, after hardening, these surfaces should be highly polished. Polishing the punch tips in the direction of metal flow, instead of circular polishing, is desirable.

Figure 10-51. C-ring retaining-ring assembly.

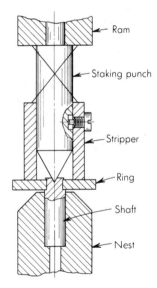

Figure 10-52. Staking by center punch.

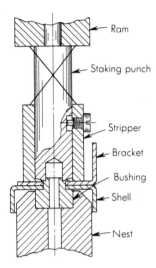

Figure 10-53. Staking by ring punch.

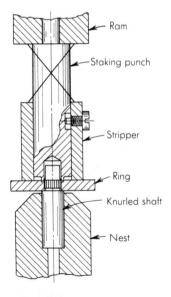

Figure 10-54. Inward staking
by ring punch.

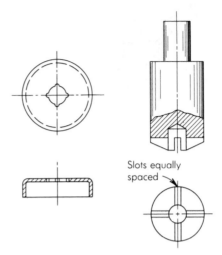

Figure 10-55. Spot-staking punch and
workpiece with splined hole.

Tooling for Adhesive Bonding

The design of tooling for adhesive joining does not differ greatly from that
of tooling used in the physical joining process. In most cases, all standard jig
and fixture components can be used.

Tooling can range in size from fixtures to hold small desk tops to fixtures
which hold wing sections of a Lockheed C-5A transport plane.

The type of adhesive used is the main element in the design of fixtures. Nine
different types of adhesives are used today, all having different characteristics.

Plural components (Class IA) can cure at room temperature. Heating will
increase cure, therefore, fixtures must withstand heat to start chemical
reaction.

Heat-activated (Class IB) and film adhesives (Class V) require heat to start
chemical reaction.

Moisture cure (Class IC) adhesives cure under atmospheric moisture. Due
to this moisture, tooling must be designed of nonrusting material.

Evaporative adhesives (Class II) also require heat for curing. Again,
fixtures must be able to withstand temperatures without distortion.

Hot-melt adhesives (Class III) are applied in a molten state, which permits
high-speed production. For higher production rates, these fixtures must be
easily loaded and unloaded with the use of hydraulic or pneumatic clamping.

Delayed-tack adhesives (Class IV) are heat-activated to produce a
tackiness that is retained upon cooling for periods up to several days. Fixtures

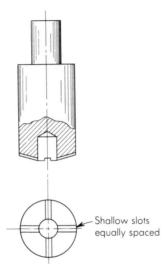

Figure 10-56. Combined spot-
and ring-staking punch.

must control the piece part configuration for periods of days without losing
its intended holding pattern.

Pressure-sensitive adhesives (Class VI), just as the name implies, apply
masking tape, surgical tape, and labels through the use of pressure pads or
pressure rollers to form a bond. Pressure-sensitive adhesives, used on cartons
for food packaging, are becoming more common as application equipment
becomes available (*Figure 10-57*).

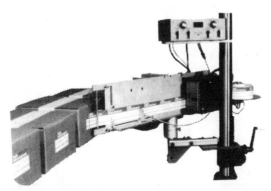

Figure 10-57. Application equipment for pressure-
sensitive adhesives. (*Courtesy, Weber Marking Systems,
Inc.*)

An understanding of adhesives is increasingly important because the use of plastics is on the rise. Since many composites are compromised when subjected to drilling or localized compression, adhesives provide a reliable solution.

Curing Methods. Conventional methods of adhesive curing include convection and infrared oven curing. These techniques can be troublesome due to temperature variation and abuse baking.

Cure response has been improved by the following methods:
1. Induction Curing
2. Dielectric Curing
3. Radiation Curing (ultraviolet, microwave, electron beam, infrared)

Induction Curing utilizes a heating coil to energize the workpiece, rapidly increasing its temperature. The heat is conducted into the adhesive, initiating or accelerating the cure.

Dielectric Curing involves use of a varying electric field. Electric energy is converted into heat for the curing process.

Radiation Curing includes use of microwaves (waves shorter than visible light) and infrared radiation (wavelengths longer than visible light) to effect an adhesive cure. Ultraviolet radiation and electron beams can also be used, although electron beam curing is primarily applied to laminating and coating operations.

High processing speeds and productivity gains are possible with these methods, but capital costs to implement the processes are high.

Adhesive Selection. The principal advantage of adhesive bonding is improved efficiency of production, and new developments have made it possible to select the adhesive that best meets individual job requirements for toughness and resistance to environmental factors.

The properties of the surfaces to be joined must be considered as well as the properties of the adhesive since the substrate must uniformly coat the entire surface to create the strongest possible bond. The safety concerns of toxicity and dermatological hazards should also be addressed when selecting the appropriate adhesive.

Joint Design. There are eight possible types of loading on a bondline.
1. Shear.
2. Peel.
3. Tension/Compression.
4. Cleavage.
5. Creep.
6. Vibrational Fatigue.
7. Mechanical Shock.
8. Thermal Shock.

Adhesives are generally weak in peel and cleavage and are stronger when subjected to shear and tension/compression. These characteristics should be considered when designing the workpiece.

Bibliography

Daryl J. Doyle, *Criteria for Proper Adhesive Selection: From Application to Viscosity*, SME Technical Paper AD90-450.

Kieran F. Drain and Kevin J. Schroeder, *New Developments in Structural Adhesives*, SME Technical Paper AD90-127.

Tool and Manufacturing Engineers Handbook, Fourth Edition, Society of Manufacturing Engineers, 1989.

Welding and Brazing, Eighth Edition, American Society for Metals, 1971.

TOOL DESIGN
FOR THE JOINING PROCESSES

Review Questions

1. Why is it important to provide for proper heat control in the weld zone?
2. What are some of the major design objectives for a welding fixture?
3. Before designing a welding fixture, what factors should be considered?
4. When are parts that have been welded together stress-relieved?
5. What is the advantage of a gravity-type fixture as shown in *Figure 10-1*?
6. What function does the holddown plate serve in a gas welding fixture?
7. What factors govern the selection of materials used in the design of a gas welding fixture?
8. What materials are commonly used in gas welding?
9. What are some of the design considerations for arc welding fixtures?
10. What is the design function of a backing bar?
11. What determines the size of the backup bar?
12. What materials are used for backup bars?
13. What important design considerations should be incorporated into the resistance welding fixture?
14. Would setup blocks be used for high-production multiple-part fixtures? Why?
15. What joining materials are used in brazing or soldering?
16. Why is it important to use inert gas for shielding?
17. What is meant by a physical joining process in a welding fixture?
18. Why has alumina-ceramic fiber replaced asbestos in heat shielding?
19. Why are quick-acting and power clamps preferred over C-clamps?
20. Where would rest buttons, spring plungers, and magnets be used in weld

fixturing?

21. How does soldering or brazing differ from welding?

22. Hard soldering or silver soldering requires a heat range of what temperature?

23. Besides cleanliness, all forms of soldering require wide gap joints. True or False?

24. Brazing flux tends to clean the work area, increase fluidity, and prevent what?

25. Induction coils should be fastened with cast-iron bolts. True or False?

26. Why do induction coils require water cooling?

27. What is meant by nesting in cutting operations?

28. A watertable in cutting is used for washing off the workpiece. True or False?

29. What is the name of a pin with a head on one end, fastened by deforming the pin end?

30. What is eyeleting?

31. In what areas are metal stitches used extensively?

32. Two or more parts are permanently joined by forcing the metal edge of one member to flow either inward or outward. What is this method called?

33. Adhesives are generally strongest under what two types of loading?

TOOL DESIGN
FOR THE JOINING PROCESSES

Answers to Review Questions

1. Excessive distortion may result and subsequent stress relief of the part may be required.

2. Provide support, alignment, and restraint of the parts. Also must permit heat dissipation.

3. Holding the part, proper heat control, suitable clamping, outlets for welding atmosphere, provision of clearance for filler material, and accessibility to the point of weld.

4. When excessive distortion occurs.

5. This design eliminates excess fixture material from the weld area. This also permits welds in a horizontal position.

6. Keep workpieces flat and prevent distortion.

7. Part tolerances, heat resistance and heat transfer, and fixture rigidity.

8. Cast iron, carbon steels, and stainless steel.
9. Prevent parts from moving out of alignment, use of backing bars, heat dissipation from weld lines, support molten weld, govern weld contour, and protect weld root.
10. Complete penetration without pickup of the backup material.
11. Metal thickness and the material to be welded.
12. Copper, stainless, beryllium, titanium, ceramic, or a combination of several metals.
13. Keep magnetic material out of machine, insulate gage pins, clamps, locators, and index pins. Protect slides, bearings, and index pins from flash; ease of operation, and water cooling.
14. Yes. Positive locating and interchangeable to accommodate varying workpieces.
15. Lead, tin, copper, silver, or their alloys.
16. To prevent oxidation during the weld process.
17. That parts are made to coalesce along their contacting surfaces through the application of heat and/or pressure and often the addition of a filler material, with the edges losing their identity in a homogenous mass.
18. Asbestos was banned by the federal government. OSHA regulations declare asbestos as hazardous to health.
19. Speed for production.
20. Heavy plate work and irregular shapes.
21. Low temperature and nonfusion process.
22. 1100-1600° F (590-870° C).
23. False.
24. Oxidation.
25. False.
26. Intense heat produced would distort and damage coils.
27. Arrangement of pieces cut to result in minimum waste.
28. False.
29. Rivet.
30. A joining device whereby a portion of the fastener is force outward to form a bond.
31. Electronics.
32. Staking.
33. Shear and tension/compression.

11

MODULAR TOOLING AND
AUTOMATED TOOL HANDLING

Special Design Considerations

Accessibility. Since CNC machining centers can perform many operations such as milling, drilling, tapping, turning, and boring, the fixtures must be designed for all these operations. This requires cutter accessibility to all parts of the workpiece that need machining operations, as shown in *Figure 11-1*. Additionally, the closer a clamp is placed to a machined feature, the more the machining operation is restricted. Clamping can affect the tool diameter, cycle time, finish, and accuracy.

Probably the most important objective is to keep the fixture and clamping to a low profile to prevent interference with the ideal programmed pattern for the cutting tool. The programmer must raise the tool above the clamp, move it over the clamp, and then drop it back down before resuming a cutting operation. Keeping the clamping as low as possible reduces the amount of travel to jump over the clamp, which results in cycle time savings. It also permits the tool to be chucked as short as possible.

Accuracy. Most CNC machining centers are capable of holding extremely close tolerances, especially where all machining can be done with the same setup. For example, a machine may position to within 0.0005″ (0.0127 mm) of the programmed coordinates and return to the same position (repeatability) over and over within half that amount, or ±0.00025″ (0.0064 mm). Fixtures for CNC have an equal role in transferring this accuracy to the workpiece. While closer tolerances increase the cost of a fixture, they can also deliver a more accurate, consistent product.

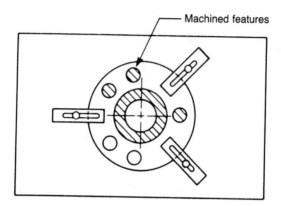

Figure 11-1. Cutter accessibility.

Particular attention should be paid to flatness and parallelism when designing fixtures for CNC. *Figure 11-2* illustrates the effect on an 18″ (45.72 cm) workpiece when held between vice jaws that are misaligned a total of 0.003″ (0.076 mm). If a slot were milled down the center, it would be out of parallel by 0.009″ (0.23 mm). This could be enough to scrap the workpiece.

Rigidity. Accuracy, surface finish, and productivity are affected by rigidity. The cutting tools produce severe shock, pressure, and vibration on the workpiece, which must be alleviated by good fixturing.

At least one or two solid surfaces should be designed into the fixture, as shown in *Figure 11-3*, to take the shock of the cutting tool. Clamps should be strong enough to hold the workpiece securely without distorting it.

Speed and Ease of Workpiece Changing. The tool designer must also consider loading and unloading the workpiece. Usually the lot size and total quantity of parts to be machined will dictate the amount of money that can be spent on the fixture. Using standard fixture components can lower the build cost substantially.

In many cases, a simple clamp will work fine. This type of clamp can be manually guided when close positioning is required.

Many of the more sophisticated mechanical, pneumatic, and hydraulic clamping systems can also be used. These clamps move out of the way when released and permit easy removal of the workpiece. The fluid-operated systems provide the additional advantage of being activated jointly by a single lever, button, or programmed sequence.

Realizing and acting on these needs, however, may not eliminate two relatively serious problems:

1. How to supply the large number of individual fixtures for the many different workpieces in a timely manner.

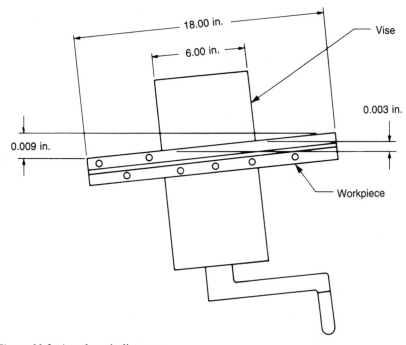

Figure 11-2. Angular misalignment.

2. Where to store the fixtures after completing the workpiece(s) that they were used to hold.

Dedicated fixtures for each workpiece are not feasible. The process and time involved to provide a dedicated fixture from inception to the time it is put on the job may be factors.

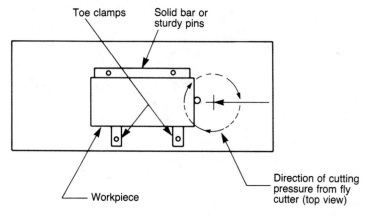

Figure 11-3. Rigidity in tool design.

As for storing large quantities of dedicated fixtures for possible reuse in the future, there is no practical solution. Either they are scrapped out and new ones are built as needed, or they are stored in space that could be better utilized producing parts.

Modular tooling systems were designed to solve both of these problems at the same time. These systems are called by other names, but regardless of the name, they are all kits of tooling components that can be used together in various combinations to locate and clamp workpieces for machining, assembly, and inspection operations.

The components of a typical starter kit are illustrated in *Figure 11-4*. A kit consists of mounting plates, angle plates, locators, clamps, and mounting accessories. Adapters are also available to permit the use of many standard and power workholding devices. The method of assembling these components varies between systems.

Fixtures made from modular tooling kits can also be used on standard machines and NC machines as well as CNC machining centers. They can also be positioned on U.S. or metric machine tables.

Modular tooling systems are invaluable when confronted with short lead time or small production quantities that do not warrant the design and construction of a special jig or fixture.

Constructing Modular Tooling Fixtures. The first step in assembling a jig or fixture is to select a base large enough to handle the workpiece. Next, the main structure is constructed with riser blocks and reinforced with

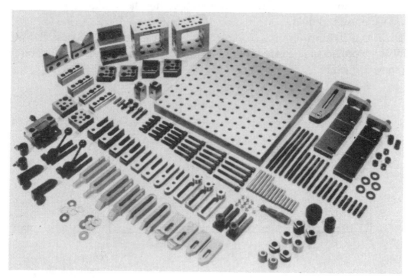

Figure 11-4. Starter kit of modular fixturing components. (*Courtesy, Fritz Werner Machine Tool Corp.*)

stop-thrust elements. Finally, the more specialized elements are added to properly locate and clamp the workpiece for machining.

An erector set fixture is shown in *Figure 11-5* without a workpiece in place and in *Figure 11-6* with a workpiece in place.

Construction with Sample Parts. An ideal method of constructing a jig or fixture with a modular tooling kit is to build it around a sample part. Simply position the sample part on the base and add locators, supports, and clamps as needed. This method reduces the construction time to a fraction of the time it would take to design and build a dedicated jig or fixture to do the same job. The use of a sample part can also expedite frequent assembly and disassembly of jigs and fixtures.

Construction with a Template. When no sample part is available, a jig or fixture can be assembled around a template of the part. Templates are also useful to check for interference that could occur when loading or unloading a part.

Machining on Modular Fixtures. Jigs and fixtures assembled from modular tooling seldom need machining. Occasionally, limited machining may be required to produce a special component. Excessive machining, however, should always be avoided, since it will eliminate the economic advantages of the system. If, for some reason, the jig or fixture is impossible to construct without considerable machining, modular tooling should not be used.

The Tool Assembler. Even though modular tooling is easily assembled, the need for tooling knowledge and experience is not diminished. Jigs and fixtures constructed with modular tooling kits must be strong enough to withstand machining forces imposed on them, they must be built to utilize adjustable components, and must often be built to accommodate an in-process part. The tool assembler must have the imagination and experience to foresee potential problems and plan accordingly. The selection of a tool assembler requires careful consideration.

Pallet/Fixture Changers. The prevalence of Just-In-Time (JIT) production methods has made smaller lot sizes necessary and, consequently, more frequent changeover is required. The result is machines that may sit idle 30% to 40% of the time.

This has lead to the use of manual pallet/fixture changers in vertical machining centers. One pallet slides away from the spindle to allow off-line loading, unloading, cleaning, and setup work while the machine continues making parts on another pallet (*Figure 11-7*). This process is effective on small or large lot sizes and allows changeover in less than one minute. The operator is then free to spend more time inspecting work while parts are still being produced.

New machining centers are now outfitted with pallet/fixture changers as standard equipment.

Figure 11-5 and *Figure 11-6*. Erector-set fixture shown in *Figure 11-5* with a workpiece in place. (*Courtesy, Flexible Fixturing Systems, Inc./Erwin Halder, Ltd.*)

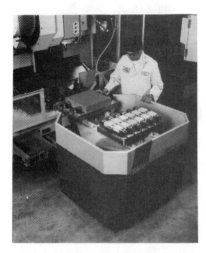

Figure 11-7. Pallet changer. (*Courtesy, SMW Systems, Inc.*)

Advantages of Modular Tooling

Reduced Lead Time. Reduced lead time is the major advantage of a modular tooling system. Jigs and fixtures can usually be assembled in a few hours time with readily available components, virtually eliminating lead time. The tooling can often be assembled in less time than it takes to prepare a tape for the part.

Adaptability. Tooling changes to accommodate new products or revisions to existing products are fast and easy with modular tooling. For companies experiencing frequent product changes, new products will not be held up waiting for tooling, since changes to existing tooling can be made immediately without interrupting production. Sometimes tool trials show the need for revisions that are difficult and time consuming to make on a dedicated fixture. These changes can be made easily and quickly with modular tooling.

Even when a dedicated fixture is planned, modular tooling can be used to establish the basic design and tool clearances.

Reusability. Although modular tooling kits may seem expensive, they usually pay for themselves in one to two years. At the completion of a production run, the modular jig or fixture can be completely dismantled and the components returned to the kit for reuse, whereas conventional tooling is usually stripped of any reusable parts and then scrapped at a fraction of its original cost. An example of this is shown in *Figure 11-8*. The fixture on the left will be useless at the end of the production run, whereas the modular fixture will be completely torn down and the components returned to the kit for reuse. Storage of the

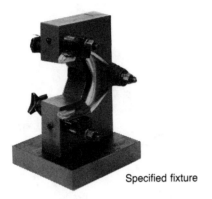

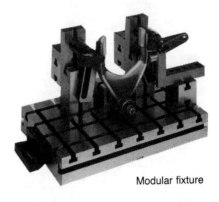

Specified fixture　　　　　　　　　　　　　　Modular fixture

Figure 11-8. Two methods of constructing a milling fixture. (*Courtesy, Flexible Fixturing Systems, Inc./Erwin Halder, Ltd.*)

dedicated fixture for possible revision and reuse at a later date may be considered, but storage costs, when added to revision costs, usually makes this an expensive proposition.

Backup Ability. Modular tooling can be swiftly assembled to temporarily replace a dedicated jig or fixture while it is being repaired, reworked, or revised. Just having backup tooling available for emergencies may make modular tooling well worth the investment.

Modular Tooling Systems

Typically, all of the major components of a modular tooling system feature a grid pattern of accurate locating holds or precision-spaced T-slots which are used to accurately position components to locate and clamp workpieces for machining. The major components include subplates, riser/tooling blocks, four-sided, six-sided, and two-sided tooling blocks, angle plates, and tooling cubes.

Subplates. Subplates or tooling plates (*Figure 11-9*) are being used at an increasing rate by progressive job shop and production manufacturers to increase productivity while lowering tooling costs. Subplates are adaptable to any machine table or pallet system and greatly increase the number of locating and clamping points available on the table surface.

Subplates are machined for an exact fit to the customer's machining center or machine, and subplate sets are available for machines using shuttle tables. The use of matched sets greatly reduces the need to indicate individual tables, thereby increasing the production capacity of multi-pallet machines.

Riser/Tooling Blocks. Most machining centers have unusable dead space between the centerline of the spindle and the top of the machine table or pallet. This dead space varies between machines, but in each case it places limitations on the machine.

Figure 11-9. Subplates. (*Courtesy, Mid-State Machine Products*)

To work around these limitations, operators will sometimes move the workpiece to the edge of the table to get more vertical quill movement. This arrangement, however, prevents machining the back of the workpiece without resetting the job, thereby eliminating the cost saving potential provided by the indexable pallet.

Another method used by operators is to elevate the workpiece on blocks that are high enough to eliminate the dead space. This method is costly in terms of setup time. Riser/tooling blocks (*Figure 11-10*) offer a workable alternative to the time-consuming double setups and eliminate the need for unstable parallels. The riser/tooling block is mounted squarely in the center of the machining center pallet. The dead space is eliminated by the additional height of the block, thereby allowing the operator to make full use of the machine's indexing capacity and to machine up to five sides of a workpiece in one setup. The riser/tooling block's heavy-duty construction, along with qualified dowel holes or T-slots, assures setups will be solid and repeatable.

Figure 11-10. Riser/tooling blocks. (*Courtesy, Mid-State Machine Products*)

Four-sided Tooling Blocks. Four-sided tooling blocks (*Figure 11-11*) are designed for use on horizontal machining centers and provide four identical surfaces for attaching workpieces or other components. When mounted on a rotary table or fourth axis, they can be indexed 90° to present four work setups to the cutting tool in rapid succession. In some cases, even the top surface is used to locate the workpieces for machining.

Six-sided Tooling Blocks. Six-sided tooling blocks (*Figure 11-12*) are used in the same manner as the four-sided blocks, but with the addition of two identical mounting surfaces.

Two-sided Tooling Blocks. Two-sided tooling blocks (*Figure 11-13*) are set up on the machine table in the same manner as the four-way blocks, but provide for more Z-axis travel. When mounted on a rotary table or fourth axis, they can be indexed 180° to present two work setups to the cutting tool in rapid succession. Two-sided tooling blocks are the logical solution to mounting workpieces that are too large to clear the spindle or the coolant-chip shield if mounted on a column subplate. The open frame design provides an access

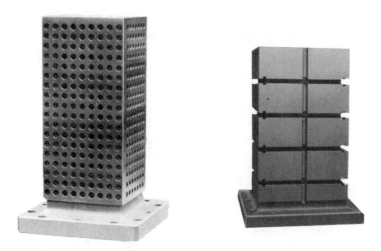

Figure 11-11. Four-sided tooling blocks. (*Courtesy, Mid-State Machine Products*)

opening so the spindle can reach the back of the workpiece after indexing. Operations can then be performed on the front and back of a workpiece using the same setup.

Angle Plates. Angle plates (*Figure 11-14*) are ideal for machining operations where a four-sided tooling block is neither required nor economically feasible. They are ideal for applications that require the fixture be mounted near the front edge of the pallet, for extra thick workpieces that take up most of the pallet, and for workpieces that require their centerline to be on the centerline of the pallet.

Tooling Cubes. Tooling cubes (*Figure 11-15*) were designed for use on many of the newer machining centers that are equipped with pallet changers having limited weight capacities. The inside of the tooling cube is hollow and necked down at both ends to reduce weight while still providing sufficient rigidity to resist intense machining forces. Tooling cubes can be located quickly against edge stops with dowel pins or parallels. They are usually bolted to the pallet through the center with threaded alloy bolts using either standard straps or a made-to-fit cover.

Modular Tooling System with Self-Adjusting Fixturing Elements. The major components of this system include grid plates, pallets, consoles, tombstones, double angle plates, and angle plates in various sizes.

Mounting holes are multipurpose with 0.5″ (12.7 mm) diameter bushings at the top, and 1/2-13 threaded inserts at the bottom, or with 0.63″ (16 mm) diameter bushings at the top and 5/8-11 threaded inserts at the bottom. They are located on a grid pattern with either 2.0000 ±0.0016″ (0.0152 mm) or 2.0000

Figure 11-12. Six-sided tooling blocks. (*Courtesy, Mid-State Machine Products*)

Figure 11-13. Two-sided tooling blocks. (*Courtesy, Mid-State Machine Products*)

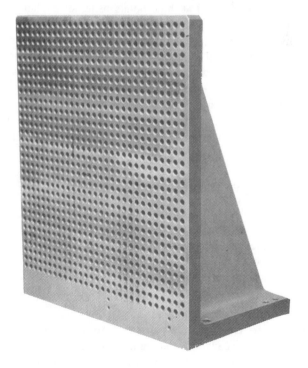

Figure 11-14. Angle plates. (*Courtesy, Mid-State Machine Products*)

Figure 11-15. Tooling cubes. (*Courtesy, Mid-State Machine Products*)

$\pm 0.0004''$ (0.01 mm) spacing between centers. Components can be attached using dowel pins to locate them and bolts to clamp them with, or by using two shoulder bolts which both locate and clamp.

With conventional clamping, castings and forgings—which are often badly warped—are usually clamped at three points using strap clamps. The warped undersurface locates unevenly at these points, as shown in *Figure 11-16a*. When clamping pressure is applied, severe strains are produced, as shown in *b*. After machining, the surface will spring out flat when the clamps are released, as shown in *c*, resulting in scrap or rework.

With this system, the hardened steel elements "float" in their sockets to adjust to workpiece irregularities and surface profile, as is shown in *d*. The upper arms of the clamps adjust in the same manner during clamping, as shown in *e*, without causing distortion of the workpiece. When the workpiece is removed, the machine surface does not change.

Quick Change Tooling

Machine tool productivity is greatly enhanced by sharp reductions in setup time, workholding, and tooling. Vast changes have been made to meet the demands of flexible manufacturing cells, automated parts handling systems, and computer integrated manufacturing. Quick change tooling is one means of increasing productivity and decreasing setup time by presetting jobs and utilizing off-line setup methods, such as pallet/fixture changers discussed earlier in this chapter.

Components of Quick Change Tooling. Quick change tooling consists of two parts: the clamping unit and the cutting head. The clamping unit mounts to the machine tool and acts as a receptacle for the interchangeable cutting unit (*Figure 11-17*).

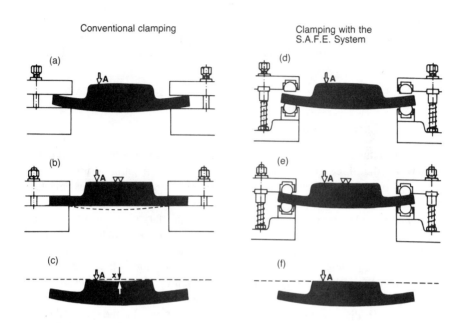

Figure 11-16. Modular tooling system with self-adjusting fixturing elements. (*S.A.F.E. System, courtesy Enerpac Group, Applied Power, Inc.*)

Figure 11-17. Quick change clamping unit and cutting head.

To change tooling, the machine operator simply releases the locking system, changes the cutting unit, and locks the new tool in position. The operator then makes the offset adjustment according to the previously recorded data and continues machining the part. The total machine downtime is only about 30 seconds.

Strategies for Machining

Understanding the impact of quick change tooling on productivity requires a review of the three basic functional areas of the manufacturing process: inventory planning and control; preproduction planning; and in-process machining.

Inventory Planning and Control. The primary objective of inventory planning and control is to maintain enough finished goods inventory to meet customer demand while keeping levels low enough to minimize costs. By producing smaller lots more frequently, inventory carrying costs can be reduced, as can shelf-life and problems such as rust, contamination, and deterioration.

Preproduction Planning and Setup Reduction. The type of planning just discussed necessitates setup reduction. In the preproduction planning phase of the manufacturing process, all elements are identified, organized, and scheduled in advance of the production run. The objective is to eliminate as many setups as possible, and to improve machine and operator efficiency during setup and in-process machining.

A setup reduction program can reap as much as a 75% reduction in setup time, as well as drastically reduce tool change time during in-process machining.

A setup reduction program generally consists of the following elements:
1. Quick-change tooling. This procedure involves changing an entire pre-gaged cutting unit, as opposed to changing an individual insert or tool.
2. Tool kitting. The identification, organization, and assembly of all tooling necessary to complete a production run or shift of operation.
3. Pregaged tooling. Once all tools are fitted with new cutting edges and assembled in the tool taxi, the "F" and "C" dimensions are measured and recorded in advance of production.
4. Preproduction tool maintenance. All tool maintenance is performed in advance of the production run to avoid catastrophic tool failure and in-process tool maintenance.
5. Advanced cutting tool materials. Materials such as ceramics and poly-crystalline diamonds permit longer and faster production runs.
6. Tool management software. This provides computerized tracking of tools and assists in the kitting process.

Preproduction planning is the least difficult and least expensive phase of the manufacturing process. Yet, it can provide a significant payback through reduced machine downtime.

In-process Machining. Most of the factors that inhibit productivity can be eliminated through effective preproduction planning. While setup reduction programs most often address downtime resulting from tool change, tool maintenance, and part setup, the downtime incurred to adjust tool position for part accuracy is much less understood. This process is referred to as tool offsetting and, by traditional methods, requires an eight-minute, 17-step sequence of operations to adjust tool position and produce dimensionally accurate parts. Quick change tooling reduces this process to a four-step procedure that takes about 30 seconds to perform.

Quick Change Workholding, Toolholding, and Part Registration

To achieve maximum machine productivity, presetting should encompass three functions. The first area is workholding in which chucks, chuck jaws, fixtures, pallets, etc., must be changed each time a new lot of parts is to be run.

The second area is toolholding, again where part changeover necessitates the exchange of an entire toolholder but is also comprised of changes for tool wear and breakage during normal operation. This includes simple procedures, such as the indexing of an insert, in which tolerances of milled pockets and tools may not lend themselves to precision tool setting (*Figure 11-18*).

Typically, after indexing an insert, the tool is backed off a predetermined amount, a trial cut is performed, the dimension is checked, an offset is entered into the machine (either manually or automatically), and a final cut is taken. Since cutting pressures can affect metal removal, the subsequent part cycled, taking a cut at full depth will also require inspection to verify that the compensation factor was correct. In the event that the offset was incorrect, at best, the process has to be repeated. At worst, the workpiece must be scrapped. Taking into account that many machines are equipped with multiple tools and multi-spindles, and that tools frequently require dual offsets (radial and depth, for example), it is no wonder that production rates scarcely reach 40% of machine uptime.

The third area to be addressed is that of part registry, which can be considered a natural extension of the capabilities to preset workholding devices. The concept of part registry can be defined as the accurate positioning of workpieces within a workholding device to a specified location, precisely simulating machine mounting conditions, yet prior to setting within the machine tool (*Figure 11-19*).

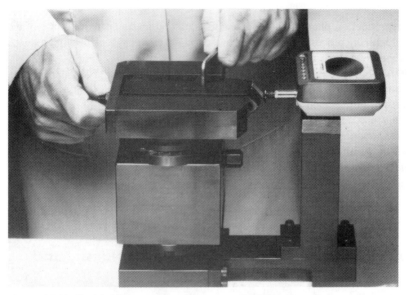

Figure 11-18. Tool holder presetting. (*Courtesy, ITW Woodworth*)

Figure 11-19. Part registry. (*Courtesy, ITW Woodworth*)

Part registry has grown in importance as the complexity and quality requirements (spatial dimensions, true positioning) of machined parts has increased. This has had an impact on the setup time of high-value, intricate, and fragile workpieces, where the loading, fixturing, and handling of a part are recognized as the critical elements of producing a component to specification. In addition, consideration should be given to applications involving heavy, un-wieldy parts in which machine loading is time-consuming and adds operator fatigue to the equation of machine productivity. Part registry also lends impetus to the trend toward chucking a part once for multiple machine processing and transfer between machines for improved quality and productivity.

Automatic Identification

The term Automatic Identification (Auto ID) is applied to bar coding and other forms of keyless data entry. The goal of Auto ID is reliable identification of physical objects.

Accuracy, speed, and reliability are the reasons for implementing an Auto ID system. Automatic identification systems are fast. Not only is the input of data quick and dependable, but the information is provided in a virtual instant to almost any database or software package. This allows the user to make immediate and informed decisions. Without real-time information provided by Auto ID systems like bar coding, MRP, MRP II, JIT, CIM, and other quality control programs can only be partially realized.

Bar coding is easy to learn and use, and relatively inexpensive to implement. By reading bar codes, data can be acquired 100% faster and 10,000 to 15,000 times more accurately than by any other means of manual input.

Bar Codes. The concept behind bar code technology is identification. Any bar code symbolizes a distinctive mark of identification for products, compa-rable to that of a finger print. Bar coding can be applied to almost any aspect of the manufacturing enterprise. Quality control and tracking product life from production to customer can be streamlined and made more efficient with the use of bar coding.

The symbol itself, the information contained in the symbol, or the language used may be referred to as the bar code. The symbol is the actual physical bar code that appears on an object being scanned (*Figure 11-20*). The code refers to the information encoded in the symbol. The symbology is the language spoken by the symbol. Each bar code symbology has its own unique characteristics. For example, some symbologies can only encode numbers (numeric); others include letters and numbers (alpha-numeric). In addition to what a bar code symbology can encode, the symbology can also vary in density and character length of the printed code.

Figure 11-20. Bar code symbol.

Bar Code Symbology For Manufacturing. The term symbology denotes each particular bar code scheme. To encode data, each symbology uses a unique set of rules that determines the type of information (numeric, alphanumeric, or ASCII characters) that can be placed in the symbology. Each one-dimensional bar code is a design of wide and narrow bars and spaces called elements that represent the information. Over 60 different types of symbologies have been developed.

Machine Tool Setups and Regrind Crib

Effective Tool Management, Reporting, and Tracking. Effective tool management is a critical factor for any industrial manufacturer. The accurate reporting of who regrinds tools, what departments or operators are using the various tools, and tracking of tools for inventory purposes can be made faster and more effective with bar coding.

Each tool type, identification number, description, and location is entered into a database. Identification numbers (bar codes) are assigned to each user/department of every tool or setup of items released from tool inventory. A bar code system has the capability to track items released to departments, work centers, job orders, or even operators. A numbering system, or the modification of one, would place the tools into logical groupings. For example, a two digit prefix may indicate the tool category, such as 02 for straight-shank drill, 04 for reamers, and so on. The use of suffixes could be used for reporting purposes.

A bar coded tool reporting system can effectively manage the regrinding, usage, and reporting of hundreds of tools and setups. Bar codes can be attached directly to large items. Small tools (e.g., drill bits) can have a bar code attached to the container/bin where they are stored. A preprinted bar code menu sheet identifies what has been taken or disbursed. Operators, departments, or work centers may have a bar code associated with them to record normal issuance and returns. This allows better management for supervisors/foremen to balance tool inventory and regrinding accordingly.

Additionally, the tracking of information found on traditional grinding crib time cards can be scanned off of a preprinted bar code menu sheet. This speeds up the collection of information by the employees who regrind tools, allowing them to spend more time on the actual regrinding operation.

Information obtained can be reported in a very timely manner in the form of a tool usage report. The report helps to determine what to buy and when, and what to scrap. The report can also establish a reorder point for each tool, and generate a report of all tools at or beyond their reorder levels.

Timely information is an essential component to improving daily tool management. In the few cases when a tool is not in the crib, all current locations and quantities of the tool type can be quickly ascertained. The system can provide invaluable assistance in reducing hidden inventory that exists in every shop and ensure the proper inventory for daily and weekly workloads.

Automated Guided Vehicles in Tool Control

Automated guided vehicle systems (AGVS) allow the transfer of materials to remote locations or through complex paths under computer control. The benefits of AGVS include inventory control, increased equipment and space utilization, manufacturing flexibility, higher productivity, and easier management control of operations.

Basic Functions. There are five basic AGVS functions:
(1) Guidance.
(2) Routing.
(3) Traffic Management.
(4) Load Transfer.
(5) System Management.
Guidance allows the vehicle to follow a predetermined route, which is optimized for the material flow pattern of a given application.

Routing is the vehicle's ability to make decisions along the guidance path in order to select optimum routes to specific destinations.

Traffic management is a system or vehicle's ability to avoid collisions with other vehicles while maximizing vehicle flow and load movement through the system.

Load transfer is the pickup and delivery method for an AGVS, which may be simple or integrated with subsystems.

System management is the method of system control that dictates system operation.

Manufacturing Applications. AGVS can be used to supply loads to and from machine tools. Parts can be routed to each machining operation to offer improved materials flow in simple to advanced manufacturing and assembly operations.

For metalworking and machining applications, AGVS can transfer parts between machining areas. The parts may be brought to an inspection station during routing between any of the machining areas. Different parts have different machining requirements and associated routings, but AGVS can readily accommodate these variable routing requirements.

Costly tooling demands high utilization, making timely, safe delivery of the correct workpieces extremely important. Workpiece delivery systems may be in a stand-alone configuration, where vehicles automatically pick up and deliver materials to machining centers and metalworking tools. When the operation is complete, automated guided vehicles (AGVs) pick up the materials and move them automatically to the next workstation or back to storage. AGVS are often used in group technology and flexible manufacturing cells to present fixtured parts.

Where total integration is required, the system may consist of AGVs directly interfacing with computer numerical control (CNC) machines via special input output stations and controlled by the same computer that operates the CNC equipment.

References

Boyes, William E., *Handbook of Jig and Fixture Design*, 2nd ed., Society of Manufacturing Engineers, Dearborn, MI, 1989.

Brown, Charles R., "Strategies for Innovative Machining," *SME Tech Paper TE91-145*, 1991.

DeLonghi, Charles S., "Quick Change Tooling for Today...and Tomorrow," *SME Tech Paper TE91-144*, 1991.

Diehl, Werner K., "The Productive Advantages of Preset Workholding, Toolholding and Part Registration to Reduce Setup Time," *SME Tech Paper TE88-125*, 1988.

Miller, Richard K., *Automated Guided Vehicles and Automated Manufacturing*, 1st ed., Society of Manufacturing Engineers, Dearborn, MI, 1987.

SMW Systems, Inc., *Here's How to Double the Output of Your Machining Center*, Santa Fe Springs, CA. Video.

MODULAR TOOLING AND AUTOMATED TOOL HANDLING

Review Questions

1. What three factors are affected by rigidity?
2. Name four components of a typical modular tooling starter kit.

3. Modular tooling systems are best suited to what types of production needs?
4. Should modular tooling be used in situations where considerable machining will be required?
5. Why are pallet/fixture changers particularly useful for Just-In-Time production?
6. Name the four major advantages of modular tooling.
7. Why are riser blocks used?
8. How does quick change tooling increase productivity and decrease setup time?
9. Name one form of Auto ID.
10. List the five basic functions of an automated guided vehicle system.

MODULAR TOOLING AND AUTOMATED TOOL HANDLING

Answers to Review Questions

1. Accuracy, surface finish, and productivity.
2. Mounting plates, angle plates, locators, and clamps.
3. Small production quantities and short lead times.
4. No.
5. Because JIT methods require smaller lot sizes and more frequent change-over.
6. Reduced lead time, adaptability, reusability, and backup ability.
7. To elevate the workpiece high enough to eliminate the dead space.
8. By presetting jobs and utilizing off-line setup methods.
9. Bar coding.
10. a. Guidance.
    b. Routing.
    c. Traffic management.
    d. Load transfer.
    e. System management.

12

THE COMPUTER IN TOOL DESIGN

One way to improve the competitive posture of American industry is through emerging technologies such as computer and control technology. The use of computers in production ranges from conceptual stages to final design, including manufacturing, quality control, and shipment of the product to the customer.

This has led to two basic uses of computers: first is the use of computers to assist in design, referred to as computer aided design (CAD); second is the use of computers in manufacturing stages, referred to as computer aided manufacturing (CAM). Recently, integration of CAD/CAM for industrial applications has led to a new technology called computer integrated manufacturing (CIM).

This chapter discusses the use of computers in design stages, with particular emphasis on tool design. It includes an overview of CAD in tool design, CAD hardware and software, impact of CAD on tool design, benefits, and future trends in CAD technology.

Overview of CAD in Tool Design

Computer aided design (CAD) is a term with several meanings. CAD may refer to computer aided drafting or graphics. This is, in most cases, called computer aided drawing and drafting (CADD). CAD may also mean computer aided analysis. This is referred to as computer aided design (CAD). Yet with the recent advent of artificial intelligence (AI) and other expert systems, the term CAD may suggest fully automated design, where the engineer specifies only the functions of a part, a description of the working environment, and operating conditions, and the computer arrives at a satisfactory (or even optimal) design. The term CAD encompasses all of these concepts.

Computer aided drawing and drafting (CADD) is the use of computers to assist in creating blueprint drawings. This is usually a two-dimensional representation of the product, showing the three views (top, elevation or front, and a side view) as normally produced with manual drawing. The views are usually associated with dimensional data as well as other manufacturing information such as machining symbols, welding symbols, tolerances, etc.

Computer aided design includes analysis of design as well as drafting. A part may be designed and tested for dynamic behavior and/or capability to carry certain loads. CAD systems may incorporate complex engineering routines to carry out the analysis. Some CAD systems are interfaced with finite elements programs and other *computer aided engineering* (CAE) systems to perform analyses of a product. CAD systems are generally interactive and have animation capability.

Both CADD and CAD systems are used in tool design. For example, in die design, a computer graphic model can be constructed using CADD. Analysis and animation of workability of the die is performed using a CAD system. CAD also performs analysis of the influences of changes in dimensional parameters, loading conditions, and working conditions, such as lubrication. This is called *parametric analysis*.

A CAD system, whether computer aided design or computer aided drawing and drafting, consists of three basic components: (1) hardware, (2) software, and (3) users.

Hardware components of a typical CAD system include a host computer, a graphics display, input devices such as a keyboard, a digitizer, a mouse system, etc., and output devices such as a plotter, a printer, etc. *Figure 12-1* depicts hardware components of a CAD system.

Software components of a CAD system consist of a set of programs that allow the system to perform design, analysis, and drafting functions. CAD software (*Figure 12-2*) consists of interactive computer graphic (ICG) software to create drawing images, applications software to carry out analysis, an operating system, a database, and a user's interface.

The user is the tool designer who creates a design. The user should be trained to operate as well as interact with the CAD system through input devices which create and modify images on the system's display, run the application models to carry out the required analysis, and store the model in the computer's memory or in secondary storage devices such as hard disks. It is often unnecessary for a single user to conduct both drafting and analysis functions. In most cases, the analysis is carried out by an analyst.

Broad-based emergence of CAD on an industry-wide basis did not begin to materialize until the 1980s. However, CAD as a concept is not new. Although it has changed drastically over the years, CAD originated more than 30 years ago during the mid 1950s. Some of the first computers included graphics displays. Currently, a graphics display is an integral part of every CAD system. Graphics displays (*Figure 12-3*) represent the first real step toward bringing together the worlds of tool design and the computer, quickly followed by the development of

Figure 12-1. The hardware components of a typical CAD system.

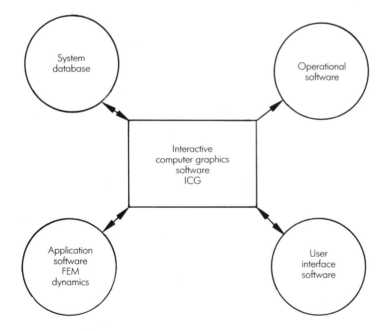

Figure 12-2. CAD software.

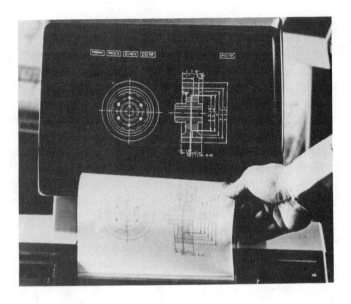

Figure 12-3. Graphics display.

plotters, depicted in *Figure 12-4*. With the advent of the digitizing tablet in the early 1960s (*Figure 12-5*), CAD hardware as we know it today began to take shape. Development of computer graphics software followed soon after these hardware advances.

Early CAD systems were large, cumbersome, and expensive. So expensive, in fact, that only the largest companies could afford them. During the late 1950s and early 1960s, CAD was looked upon as an interesting but impractical novelty that had only limited potential in tool design applications. However, with the introduction of the silicon chip during the 1970s, computers began to take their place in the world of tool design.

Integrated circuits on silicon chips allowed the production of microprocessors, which are the core of minicomputers. A minicomputer is a full-scale computer packaged in a small console no larger than a television set. These minicomputers had all of the characteristics of full-scale computers, but they were smaller and considerably less expensive. In the 1980s, the introduction of microcomputers served as the prime contributor to CAD development. Continuing advances in large-scale integration (LSI) and very large scale integration (VLSI) technology for manufacturing integrated chips allowed microcomputers to continuously decrease in cost and size and increase in power and speed.

The 1970s brought continued advances in CAD hardware and software technology. So much so that by the beginning of the 1980s, making and

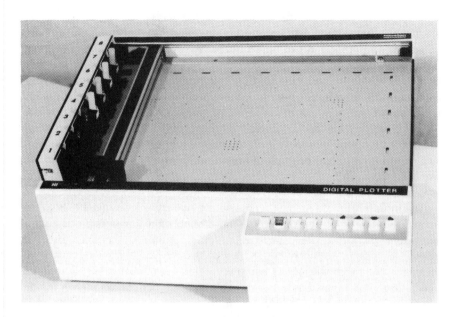

Figure 12-4. Digital plotter. (*Photo by Deborah Groetsch*)

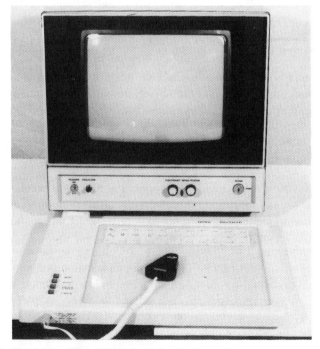

Figure 12-5. Digitizing tablet. (*Photo by Deborah Groetsch*)

marketing CAD systems had become a growth industry. Also, CAD had been transformed from its status of impractical novelty to one of the most important inventions to date. In 1980, development of CAD systems accelerated, and they became available in sizes ranging from microcomputer systems to large mini-computer and mainframe systems. Now, virtually no design organization, regardless of the size, can afford to work without a CAD system. There are several CAD systems that can operate by microcomputer. These CAD systems can be used for general purpose design as well as tool design. New CAD systems are designed to interface with downstream production functions such as manufacturing. The use of computers in the manufacturing stage is referred to as computer aided manufacturing (CAM).

The introduction of computer communications networks in the 1980s led to networking several computers such as mainframe, minicomputers, and micro-computers. This facilitated exchange of information. At the same time, it demanded standardization of CAD files so that several CAD programs from different vendors could interact with each other. Networking also allowed for integration of various hardware components and peripherals. This led to successful installation of the CAD system. It provided communication with other departments such as manufacturing. Networking also allowed for sharing the resources and peripherals, making it easy for small organizations to own a CAD system.

CAD Hardware

The most obvious differences CAD has made in the world of tool design are the instruments and techniques used by the designer. Traditionally, the tools of the designer have included such things as pencils, scales, paper, templates, and various other manual devices (*Figure 12-6*). Modern tool designers use computer aided design systems based on primarily interactive computer graphics. They usually run on a system consisting of computers, keyboards, graphic displays, digitizers, pucks, and plotters (*Figure. 12-7*).

Components of a CAD hardware system include a host computer and input and output devices. *Figure 12-8* shows the configuration of a computer-based CAD system. Traditionally, a CAD system is based on the host computer, whether it is a mainframe, a minicomputer, a microcomputer, or a workstation. The host computer is the "brain" of the system, while the other devices are peripherals to the computer to facilitate communication between the designer and the CAD system. The input/output devices include text input devices such as keyboards, cursor control devices such as mice and pucks, graphic display units, digitizers, plotters, printers, and other peripherals. Each component has its own function and influence in the performance of the CAD system.

Figure 12-6. Traditional designer's and drafter's tools. (*Photo by Deborah Groetsch*).

Figure 12-7. Standard hardware items. (*Courtesy, Modern Machine Shop, October, 1987*)

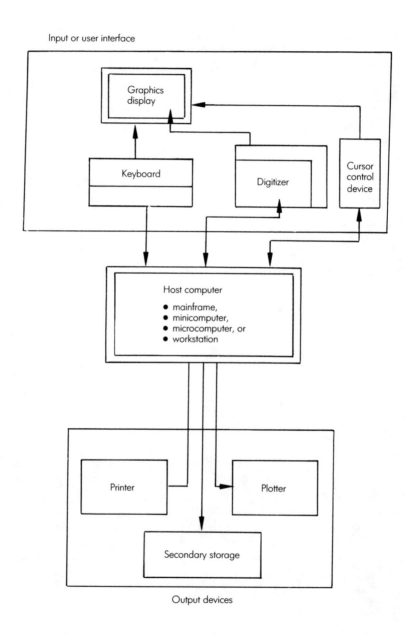

Figure 12-8. Components of a CAD hardware system.

Host Computer

The host computer is often called the central processing unit (CPU). Its capability depends on the type of host computer. There are four types commonly used as host computers for CAD systems. These include mainframes, minicomputers, microcomputers, and workstations. A CAD system could be based on one of the above or a combination through a local area network to increase the capability of the system. Selection among these systems depends on the intended application and the financial capability of the organization.

A mainframe computer is usually a multiple user computer connected to several terminals. It generally possesses a large memory and many hard disks for storage. A mainframe-based CAD system would typically consist of one or more design/drafting stations (*Figure 12-9*). A design/drafting station includes at least a display unit, a keyboard, a digitizer, and/or a cursor control device such as a mouse or joystick.

There are two types of graphics terminals: the *dummy terminal*, which relies on the computer for computation and works as the user interface; and the *intelligent terminal*, which has a microcomputer to carry out some computations. The computer is also connected to plotters, printers, digitizing boards, secondary storage devices, and other peripherals. The design/drafting station may be equipped with other input devices including lightpens and programmed function keyboards (PFK). The advantage of using mainframe computers in CAD systems is the high processing speed, high performance, and capability of accommodating large CAD systems. Another advantage is accessibility to accounting, planning, management databases, and other computer-based functions. Thus, simultaneous engineering concepts can be applied.

Minicomputers are smaller and less powerful than mainframes. They can be used as stand-alone or multi-user systems. A minicomputer-based CAD system is similar to a mainframe-based system, but accommodates fewer terminals. To reduce interactive time, minicomputer-based CAD systems may be linked in a network with mainframe computers. Typical minicomputer systems used in CAD systems are the PDP series and superminicomputers such as VAX 11/780 with 32 bits word and virtual memory operating systems. These minicomputers enabled the rapid growth of CAD systems throughout the 1970s and 1980s.

Microcomputers are sometimes called personal computers (PCs). They possess a dedicated central processing unit (CPU), a cathode ray tube (CRT) display or monitor, and input devices (keyboard, mouse, etc.) to allow graphic input. The IBM PC (*Figure 12-10*) and Apple Macintosh (*Figure 12-11*) are examples of microcomputers. The power of a microcomputer depends on its processors and the operating system. CAD software for PCs ranges from two-dimensional drafting to three-dimensional modeling and application. Microcomputer-based CAD systems are becoming popular because they are fast,

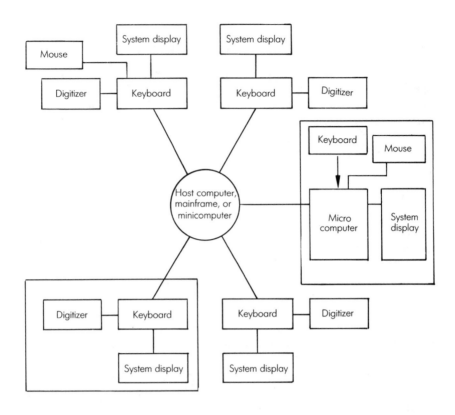

Figure 12-9. Workstations interfaced with a host computer.

Figure 12-10. IBM PC microcomputer. (*Courtesy, Machine Design, June 18, 1987*)

small in size, and accurate. Microcomputers of a 32-bit word length are now available with speeds up to 40 MHz, a large memory (up to 16 MB), disk drives as large as 1.0 GB, and super high resolution VGA monitors. This makes microcomputers suitable for CAD applications. Moreover, several microcomputers can be connected in a network. A file server can control the network and communicate with other peripherals such as plotters and printers.

Workstations are more powerful than microcomputers and typically have more memory, a large hard disk storage, and higher resolution color monitor graphics. They are usually single-user, but can be connected to a network for multi-user capability. Most workstations are 32 bits architecture UNIX operating systems. Workstations seem to be the next generation for CAD systems. *Figure 12-12* shows a PC SUN Microsystem workstation connected with a digitizer and mouse.

Figure 12-11. Apple Macintosh microcomputer.(*Courtesy, Machine Design, June 18, 1987*)

Figure 12-12. SUN Microsystem workstation. (*Courtesy, Machine Design, June 18, 1987*)

The System Display

The system display, usually called a monitor, is an output device for displaying graphic images and text. Images appearing on the system display are referred to as softcopies since they will disappear when the host computer is turned off. The system display replaces paper and pencil sketches and preliminary drawings traditionally used for documenting design concepts and ideas.

As a designer develops a particular tool, an image of the tool appears on the screen. Any image that appears can be stored under a file name or number so that it is not lost when the computer is turned off. Each successive stage in the design of a tool can be saved under a different file so that the designer has documentation of all phases of the design. These files are the computer equivalent of the check print file in manual documentation.

There are various display technologies used to develop graphics displays. Available technologies are all based on conversion of the computer's output signal to the display device into visible images. These technologies include the CRT and the laser flat panel display, or plasma display.

The laser panel display uses laser beams to trace an image on a film. The plasma screen is constructed from two plates of channelled, gas-filled glass. One plate has horizontal channels etched on it and the other has vertical channels. The two plates are assembled together and a small cavity is formed at the points of intersection. When the plasma beam is applied to the screen, the intersection points glow and a point appears on the screen. A line or curve can be drawn by sending beams that glow desired points and, thus, form the image. The laser and plasma screen are not as widely used as the CRT, which is the most dominant type of system display and has been used most often to produce a wide range of graphics.

The structure of a CRT is similar to that of a television set. The display device is a picture tube consisting of a phosphor-coated screen, a cathode that emits a high-speed electron beam, focusing systems, and a deflection system. The structure of the CRT is shown in *Figure 12-13*. The CRT operates by generating an electron beam from the heated cathode. The electron beam passes through the focussing and deflection units and strikes the phosphor-coated screen at high speed. Electrons energize the phosphor coating, causing it to glow at the points where the beam makes contact. By controlling the beam direction and intensity in proportion to the graphics information generated by the computer, the beam generates a picture on the screen. Graphics are generated on the screen based on one of the two common scanning technologies, random and raster. The type of graphics display is classified based on its scan technique.

The graphics display that operates using a *random scan* (also referred to as stroke writing, vector writing, and directed beam) generates the image by drawing vectors or line segments on the screen in random order in accordance with the user's input and the software. The beam, in this case, operates like a pencil to create the image. The *raster scan* approach divides the screen into a matrix of discrete points, called pixels. The number of pixels and spacing between pixels determines the quality of the display. Some screens have an array of 1024 x 1024 pixels. The pixels on the screen can be made to glow at different intensities to generate various levels of shade. A picture is generated by scanning the screen from right to left and top to bottom at frequencies between 30 and 60 times/second. Color screen scans allow the pixels to have different colors as well as brightness.

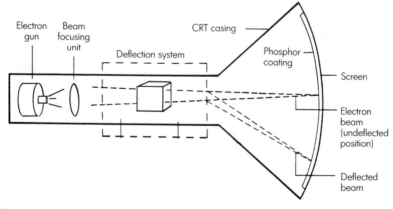

Figure 12-13. Schematic diagram of CRT structure.

These scanning techniques are used in the majority of CAD displays. Existing CRT displays are refresh displays, direct-view storage tube (DVST), and the raster scan. The refresh display is the oldest of the three systems. It utilizes the random scan approach to generate the image on the CRT screen. The image displayed on the screen must be regenerated at a rate of 30 to 60 times per second in order to avoid flickering of the image. The refresh display usually has high resolution (4096 x 4096) and can generate high-quality pictures. It is also suitable for animation. However, the need to refresh the picture places a limitation on the image size that can be displayed without flickering. A display processor unit is usually used to overcome this problem.

Direct-view storage tubes (DVST) use the random scan approach to generate the image on the CRT screen. The screen is coated with phosphor, which has a long-lasting glowing effect. This enables the screen to retain the image that has been projected to it, and makes the system work without the need for image refreshing. However, the image display is slow and any correction in the image requires clearing the screen before displaying the new image. DVST has historically been the lowest-cost terminal and is capable of displaying large amounts of data. Thus, it has been extensively used with CAD systems. However, it cannot produce colors or animation, and does not allow for use of a lightpen as an input device.

The Keyboard

Keyboards in CAD systems are input devices. Keyboards are used to input strings, numericals, pick a position on the screen, and locate an object. Conventional keyboards consist of 101 keys, a string input key, a numerical input key, and special function keys (*Figure 12-14*). This type of keyboard is employed to create/edit programs or perform word processing. Keyboards for CAD are similar to the conventional keyboard with some modifications. Programmable function keys (PFK) are added to program special functions, such as macros, thus eliminating extensive typing of commands or entering data for the CAD system. Programming of the function key is software-dependent and may be assigned different functions in different phases of software. Other modifications include attaching cursor control devices such as a mouse or track ball. These devices are used for locating position, moving an object on the screen, picking a menu item, activating certain objects on the screen, etc. The addition of a mouse and track ball makes the keyboard the most flexible input device for a CAD system.

Some CAD systems have their own function keyboards. Each key is programmed with a particular function. The function is controlled by the computer and a key could have different functions during each phase of the

Figure 12-14. Conventional keyboard. (*Courtesy, Machine Design, June 18, 1987*)

software. Thus, many alternative functions can be programmed which far exceed the number of keys. To help the user, the keyboard is designed to illuminate the function keys available for each design activity.

The Digitizer

The digitizer is commonly used for CAD geometric input. The digitizer consists of two parts (*Figure 12-15*): the digitizing tablet and an electronic handheld device such as a stylus or a puck. A stylus is a pen-like device while the puck is shaped like a box with a rectile on which cross-hairs are engraved. The digitizing tablet is a flat surface over which the puck can slide or the stylus can be used to locate a position or a function menu icon. Most digitizing tablets have a designated area for coordinate inputs, and the rest of the area for menu input command. The menu input area is used for activating systems commands. Digitizing tablets are made of different sizes ranging from 11 x 11″ to 36 x 36″ (28 cm x 28 cm to 91 cm x 91 cm). Resolution of the digitizer tablet is important when a puck is used. Typical resolution is 0.005″ (0.127mm) or 200 dots per inch (200 dots per 25.4 mm).

Figure 12-15. Digitizer and puck (input devices). (*Photo by Deborah Groetsch*)

There are several types of digitizing tablets. They can be classified by sensing method and technology into the following categories:

a) electromagnetic;

b) magnetorestrictive or strain wave;

c) acoustics.

In the electromagnetic tablets, the pointing element (puck or stylus) generates an out-of-phase magnetic field which is sensed by a wire grid in the tablet surface. This determines the coordinates, or location, of the element. In electrorestrictive tablets, a current pulse is sent from the wire grid in the pad which is picked up by the pointing element to determine the coordinates of the position. The acoustic digitizing tablet consists of two linear microphones positioned perpendicular to each other and intersecting at the origin. The pointing element for the acoustic tablet has a sound generator. The position of the pointing element is determined by measuring the sound travel between the pointing element and the microphone. The acoustic digitizer tablet has the ability for three-dimensional applications, while the other two tablets are limited to two-dimensional applications.

The puck is actually a handheld cursor which can be moved to locate certain positions on the digitizer. The puck has at least one push button which is pressed upon locating the position of its cross-hairs to send graphics information to the computer. Additional buttons may be available on the puck. These buttons may be programmed by the CAD software for other functions such as menu selections.

The digitizer is used for creating graphic images which are relayed to and displayed on the system display. Tool designers use the puck and the digitizing tablets in much the same way as they have traditionally used the pencil and sketch pad. The designer, or user, moves the pointing element (puck or stylus) to the desired location, then interrupts the computer to accept the coordinate value for this position, or activates the system command by interacting with the function menus.

Function menus are paper or polyester overlays which contain word or symbolic identifiers that correspond with stored data or design functions. A stored function might be the ROTATE function. The word "rotate" would appear in a specific location on the function menu. Each time the tool designer activates the ROTATE menu position, the image on the system display will rotate a given number of degrees.

The Mouse System

The mouse is an input device used to move the cursor on the screen of the display unit. It is very popular in CAD systems because of ease of use with icons, pop-up and pull-down menus, and the ability to locate a position on the screen. The mouse (*Figure 12-16*) is a small handheld box which can be slid on a flat surface (mouse pad) to move the cursor to a certain location or locate an object on the screen. The computer is interrupted to accept the data by pushing one of the mouse's pushbuttons. A mouse may have one to three pushbuttons.

There are two basic types of mice: mechanical and optical. The mechanical mouse has a rubber ball or two tracking wheels on the bottom surface. When the mouse is moved over a flat surface, the ball or wheels rotate and turn two orthogonally located potentiometers. The potentiometers signals are encoded into digital values which can be stored by the computation system when a pushbutton is depressed. Using this value, the program can determine the location of the cursor and retain the coordinates of the location, or execute a function if the location was on a menu function or an icon.

The optical mouse is similar in function to the mechanical mouse except it contains a light source and light sensors on its lower surface. It works by moving the mouse on a special pad which has a grid pattern. The movement of the mouse over the pad is determined by a light beam modulation and optical encoding techniques. The mouse has pushbuttons to input the signal when it is depressed.

Figure 12-16. Mouse (input device). (*Courtesy, Machine Design, June 18, 1987*)

Trackball

A trackball is similar to the mechanical mouse except it is fixed and has a ball on the top surface. It works by placing a hand on the ball and rotating it to move the cursor to a desired position. The trackball has one to three buttons used to input data. Trackballs exist as stand-alone devices or as attachments to the keyboard.

Other Input Devices

There are several other input devices used with CAD systems. These devices include lightpens, joysticks, and thumbwheels, which are all position locators. The lightpen is a handheld optical detector which is used to locate a position on the display screen. Joysticks and thumbwheels are cursor control devices which operate similarly to the mice and trackballs. Generally, they are not as popular as the mouse in the operation of CAD systems.

Documentation Output Devices

Several types of output devices are used with CAD systems for producing hardcopies of design documentation, such as working drawings and parts lists. Printers and plotters are used for this purpose. They can produce black and white or colored hardcopies. Dot matrix printers and laser printers are used to produce graphics output. Dot matrix printers have low quality prints, hence, they are used for producing quick, low-quality hardcopy for off-line editing. Printers in general are limited to small size documents. Therefore, plotters are frequently used for producing final hardcopy of the design document.

There are several types of plotters used in tool design in conjunction with CAD systems. They include pen and electrostatic plotters. Pen plotters may use one of several types of pens such as wet ink, ball point, or felt tip. They can produce black and white or color plots by scribing the drawing on paper, vellum, or mylar.

Pen plotters are of two types: flat-bed and drum. A flat-bed plotter uses a flat drawing surface to which the paper is attached. It uses an orthogonal carriage arrangement to control the movement of the pen that control the plot. One disadvantage of the flat-bed plotter is that paper is limited to the size of the drawing surface. With the drum plotter, paper is attached to a drum that rotates back and forth, thereby providing movement in the longitudinal axis. Hence, the length of paper is unlimited in this direction. The pen mechanism moves in the traverse direction to provide movement in the other axis. The hardcopy plots produced by pen plotters are usually of higher accuracy and quality than the corresponding image on the CRT. Speeds for pen plotters begin around 16″ per second (406 mm/sec.) and go much higher. A plotting speed of 22″ per second (559 mm/sec.) is common. Because of their accuracy, flat or drum plotters are the most commonly used in CAD operations.

Electrostatic plotters use a series of wire styli which apply electrostatic charges to charge-sensitive paper as it moves underneath the printing head. The paper then passes through toner and hot rolls to fix the toner to the paper. Resolution of electrostatic plotters is typically 200 dots per inch. Compared to pen plotters, electrostatics plotters are faster (up to 100 times) but their output and resolution are poorer.

When a tool designer has completed the design of a tool and has a finished image of the tool displayed on the screen, actual drawings can be produced by commanding the system to PLOT. Upon receiving this command, the plotter will produce a replica of the tool image that appears on the system display. *Figure 12-17* and *Figure 12-18* are drawings produced on a CAD plotter.

CAD Software

Tool designers use CAD systems to perform two primary tasks: (1) to produce detailed drawings for their tool design and (2) to carry out design analysis.

A CAD system, therefore, should contain at least two sets of programs used to perform these tasks. The first set is designed to produce detailed drawings.

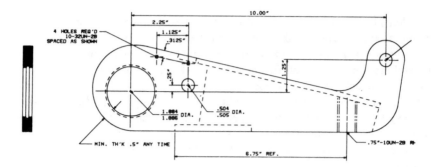

Figure 12-17. Drawing produced on a CAD plotter.

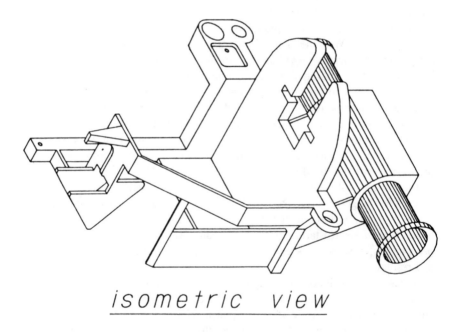

Figure 12-18. Hard copy from a CAD plotter.

This set is referred to as graphics software. In the early days of using these programs, they were difficult and cumbersome to write and use.

Recently, with the advent of graphics displays, the graphics software has been designed to work interactively. Graphics software that facilitates interaction with the user is referred to as an interactive computer graphics (ICG) system. ICG software is designed to generate images on the CRT screen, to manipulate the image, and to accomplish interaction between the user and the system. The second set of programs is used to carry out stress analysis, kinematic simulation, and other functions such as dimensional tolerancing. This set is referred to as applications programs.

Application programs may include finite-element analysis software to perform stress analysis and kinematic simulation software to test the performance of the designed tool. Operating system programs contain utilities that deal with file manipulation. The operating system is usually hardware-dependent. In addition to the aforementioned software, the CAD system needs a database that contains mathematical, numerical, and logical definitions of the application programs. The database for tool design may contain information related to dimensioning and tolerancing, dimensioning symbols, and welding symbols. Components of modular fixtures may also be included in the database.

A CAD system that contains graphics software, applications software, and a database is typically large, complex, and may be developed over several years. The user always needs professional training to become productive and experienced. The user should be trained to both operate and interact with the system in order to create and modify images, construct models, and store work in the computer memory and/or secondary mass storage device such as hard disk or magnetic tape. Interaction between the user and ICG system is accomplished through a user-interface dialogue. This interface is a collection of commands that helps the user to interact with the system. The language of the interface should be simple, efficient, complete, and easy to understand. Interface commands are usually written in a natural English-like language with simple grammar and a minimum number of easy-to-grasp rules. The user interface should allow the user to undo mistakes if needed and to edit the program to modify or add to the design.

Software Modules.[1] A typical CAD software system consists of several modules, which can provide several functions. Existing graphics software systems as investigated by Ibrahim Zied and reported in his book *CAD/CAM Theory and Practice*, have a common structure which can easily be divided into five major modules. These modules are: operating systems, graphics, applications, programming, and communications.

The *operating system* (OS) *module* provides the user with utility and system commands dealing with their accounts and files. The user can manipulate files (copy, delete, save, etc.), manage directories and subdirectories, create accounts, and program the system.

The *graphics module* provides the user with various functions to create detailed drawings, to edit and modify existing designs, and to construct geometric models to be used by the application module.

The *applications module* is used to carry out design analysis. Typical analyses includes mass and property calculation, assembly analysis, tolerancing analysis, finite-element analysis, mechanism analysis, and animation techniques.

The *programming module* is designed to provide system-dependent language for graphics and standard programming language for analysis. Programming language can be used to create graphics, customized menus, user-defined geometry such as surfaces, and to perform analysis and simulation.

The *communications module* is important for integrating the CAD system with other applications systems such as CAM. It is also crucial when the system is installed in a network and there is a need to transfer the CAD database of a model for analysis. This module serves the important purpose of translating databases between CAD and other applications systems.

Based on the above system structure, CAD systems have been carefully developed over the years. Early CAD systems automatically displayed two-dimensional representations of the part being designed. It is similar to the manual board, except the drawings are produced faster and more efficiently. These systems were not capable of interpreting the three-dimensionality of objects. The user had to read to 2-D representation and interpret it to 3-D. The systems also needed human assistance to verify that the drawing was correct.

Recently, CAD systems became capable of defining objects in three dimensions. The user defines 3-D models and the computer generates orthogonal views, perspective drawings, and close-up details of the object.

Graphics models in general are represented either in a form of modeling called wireframe or solid models. Wireframe models describe part contours using connecting lines for display purposes. *Figure 12-19* shows a wireframe

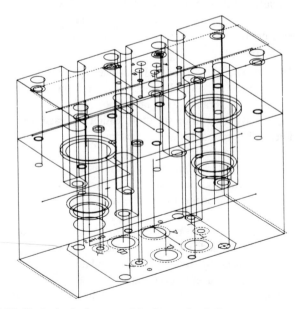

Figure 12-19. Typical wireframe model of a mechanical component.

representation for a 3-D object. Wire frame models provide accurate data about dimensions of the part. But they cannot present continuous surfaces. They have no automatic capability to remove hidden-lines feature. This could lead to an ambiguous representation of 3-D models when the part is complex.

In solid modeling representation, images are displayed as solid objects to the viewer (*Figure 12-20*). The model can include wireframe and surface definitions. It can determine if an entity or point falls inside, outside, or on the surface of the part. Solid modeling systems require more computer time and disk space than wire frame systems. But when mass property or cross section capability are required, solid modeling systems can compute this element. Solid models are also useful in kinematic studies and finite element analysis. They can be used to study fitting of parts in complex assembly.

Most CAD systems can provide useful functions other than image representation. These include automatic dimensioning, text writing, automatic cross hatching, and automatic generation of bill of material.

Figure 12-20. Solid model representation. (*Courtesy, Machine Design, June 18, 1987*)

CAD Systems for Tool Design

There are several CAD systems on the market that can be employed for tool design. The system selected for presentation in this section is a microcomputer-based system. The hardware consists of an IBM PC/AT compatible host computer, a 13-inch high resolution color monitor, a keyboard, a mouse, a digitizer, and a printer. The CAD software is VersaCAD Advanced, developed by T&W Systems Inc. The system offers such versatility in application that it can be considered CAD/CAF/CAM (computer aided design/computer aided fixturing/computer aided manufacturing). It is also easy to understand and use.

VersaCAD uses a system of 256 levels to prepare drawings with each level acting as a separate drawing sheet. The outline of the workpiece can be drawn on one level, as shown in *Figure 12-21*, while the holes are produced on another level. The milled opening in the center of the part can then be drawn on a third level. This system of drawing levels is very useful for complicated parts, or when only certain areas of the part are to be machined. Thus, the development of CAM programs can be simplified by calling up the appropriate level where machined details are shown.

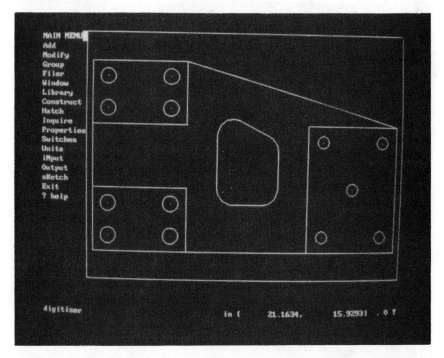

Figure 12-21. VersaCAD system of 256 levels. (*Courtesy, Modern Machine Shop, October 1987*)

The software provided is a two-dimensional system, which can construct axonometric (isometric, dimetric, and trimetric) drawings as well as the standard two-dimensional views. However, a three-dimensional system, Versa-CAD 3D, is available for use with this system. This wire-frame system has full shading capability.

There are other optional software programs that can greatly reduce the time required to prepare drawings:[2]

1. VersaLIST bill of materials software program automatically sorts, totals, and reports on parts, costs, and labor information.

2. VersaDATA is available as an option with this system. This software package allows the user to extract data from VersaCAD Advanced and incorporate it into software programs like Lotus or Symphony.

3. IGES (Initial Graphics Exchange Specification) software is an ANSI (American National Standard Institute) program which allows translation of drawings between VersaCAD Advanced and other CAD.

4. A new addition to VersaCAD Advanced is a mechanical drafting program that includes geometric dimensioning and tolerancing, dimensioning, symbols, and welding symbols. This program will automatically size the lines and symbols used for the drawing to their proper thickness and proportion based on the size of the drawing sheet. The geometric dimensioning and tolerancing may be controlled by the operator or the program can help the user through this form of dimensioning by prompting each entry in the feature control frame. These prompts are in the form of questions and answers that the user chooses to construct the feature control frame.

   By incorporating the basic rules of geometric dimensioning and tolerancing into this program some, if not most, of the more common mistakes in constructing the feature control frame can be eliminated. Including the new dimensioning symbols will also help the user properly construct engineering drawings based on ANSI Y14.5M-1982. The surface texture and welding symbols are also included in this program.

5. FixturePro package is used to design the appropriate workholder for an initially designed workpiece. This software system allows the designer to select from over 4,000 drawings of over 1,800 different standard jig and fixture components. Each component is shown in several different positions and can be rotated or positioned to suit the workpiece.

   The software follows the sizes and specifications of components established by the Tooling Components Manufacturers Association. Thus, any standard component can be used; the user is not locked into using any particular brand of components.

   Typical fixturing for a workpiece originally designed with VersaCAD Advanced software is shown in *Figure 12-22*. The fixture consists of a baseplate, strap clamps, locating pins, rest buttons, and keys to locate the fixture on the machine table. The side view shows how the part is supported and located.

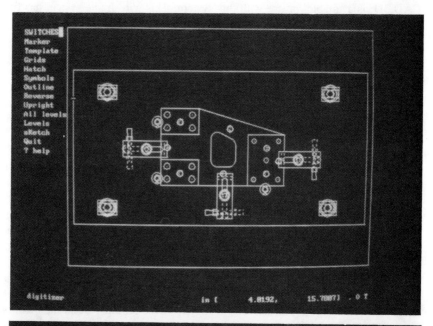

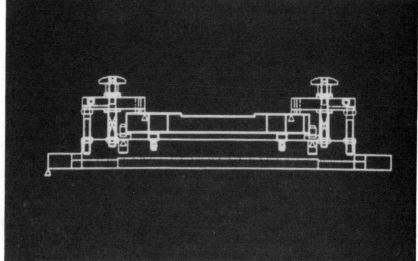

Figure 12-22. Typical fixture design (top); one strap removed from side view to clarify the workpiece (bottom). (*Courtesy, Modern Machine Shop, October, 1987*)

Impact of CAD on Tool Design

As previously mentioned, the major impact of CAD on tool design is in the areas of tools and techniques used to accomplish the design process. You have already seen that such design tools as pencils, scales, sketch pads, and pocket calculators have become virtually obsolete as they have been replaced by computer tools such as keyboards, system displays, digitizers, and plotters. In addition, the techniques used in designing a tool and documenting the design have undergone significant change.

In designing a tool, the designer goes through several stages. In reality, these stages overlap, but for the purpose of instruction, it is helpful to separate them into five distinct stages (*Figure 12-23*):

1. Identify the problem.
2. Develop tentative solutions or designs.
3. Accept the best design.
4. Make models or prototypes for testing.
5. Produce working drawings.

CAD will affect all five stages of the design process. In the initial phase of problem identification, the computer gives designers easy access to data concerning similar problems that have already been confronted, designed for, and solved. It is very common in tool design to discover that new problems are

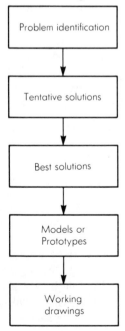

Figure 12-23. Five stages of tool design.

really just old problems that have reintroduced themselves. The computer gives designers immediate information about past problems encountered and the tools that were designed to solve them. This allows designers to avoid the problem of redundant effort or "reinventing the wheel" each time a design problem occurs. This may be in the form of a group technology (GT) program. GT is a manufacturing philosophy that groups parts into families to speed production and reduce costs. The system allows better retrieval of the design data already in existence.

In stage two, tentative solutions, designers use the keyboard, system display, digitizer, and function menu to begin roughing out design ideas and performing calculations. Eventually these new techniques will completely replace the rough sketches, calculations, and preliminary drawings traditionally accomplished using a pencil, sketch pad, and calculator.

In stage three, selection of the best design, one of the preliminary designs will be accepted, or a compromise that includes characteristics of several preliminary designs will be adopted. Tentative drawings will be called up from storage and displayed. Corrections and revisions will be made as needed and copies of the finished preliminary drawings will be plotted and circulated for final input.

In stage four, models or prototypes, the use of CAD technology makes its greatest contribution. Rather than making and testing actual models or prototypes of a tool, designers create three-dimensional models of tools in the computer. The model is displayed on the CRT and tested through computer simulation (*Figure 12-24*).

In stage five, working drawings, the files containing all the preliminary drawings and the three-dimensional model of the tool are given to drafting personnel for preparation of completely dimensioned and annotated working drawings of the tool. If the tool involves several parts, a parts list will be produced along with the working drawings.

Benefits of CAD

Users of CAD systems have discovered great increases in designer productivity and productivity of downstream production functions. This is because the designer can produce clear, accurate, and easy to handle designs. Many organizations realize they cannot survive the competition without switching from traditional pencil and paper design techniques to CAD/CAM technology.

The benefits of CAD are realized on both departmental and organizational levels. At the departmental level, benefits include production of designs with fewer errors, faster and easier analysis, greater accuracy in design calculations, reduction in lead times, and, consequently, higher productivity.

Figure 12-24. Testing a design through computer simulation. (*Courtesy, Machine Design, June 18, 1987*)

These benefits are often mirrored at the organizational level. They include production of high-quality products, increased customer satisfaction, more predictable turn around, gains in world competition, reduction in the design and manufacturing cycle, better production performance on reliability, reduction in inventory, and, consequently, improved cash flow.

Future Trends in CAD for Tooling

CAD is a constantly changing technology. The future of CAD in tool design will be one of continuous change and improvement. Hardware development has produced equipment that is vastly superior to its predecessors, and further improvements can be expected. Continuous improvement of microprocessor technology will lead to small, very powerful computers. Microcomputers with a large memory, vast hard disk capacity, and very fast processors will continue to support advances in CAD technology.

Improvements in hardware design facilitate the production of small, powerful computers called "workstations." Currently available workstations are capable of processing speeds of 27 MIPS (27 million instructions per second), double the speed of some existing mainframe computers. Parallel processing technology can lead to production of superspeed workstations and microcomputers.

In addition, it is expected that computer communication networks will be standardized and improved, making high-speed processing common practice. Networking of microcomputers and workstations will speed the continuing shift from mainframe computers. Thus, it is expected that even small companies will own a network of microcomputers and workstations. This will facilitate integration between CAD/CAM and other business functions. Hence, communication between large companies and suppliers will become easier and faster. A tool designer using CAD can design a tool and send it to a supplier for production in almost no time, but this demands standardization of CAD files. Standardization of CAD systems is expected to meet the demand of networking and exchange of information.

Graphic displays with high resolutions (up to 2048 x 2048) are also expected. The demand for faster, better plotters will fuel the growth of laser and electrostatic plotters, which can produce color plots at very high speed and with great accuracy. Input devices will shift toward mouse and trackball systems, due to the advances in pop-up menus and object-oriented software. Digitizers and specialized keyboards may become obsolete.

CAD software development will continually advance to accompany hardware. CAD standardization will become a requirement to facilitate interface not only with other CAD systems, but also with CAM and other business functions software. Communication modules software will be further developed to meet the needs of computer communication networks.

Advances in CAD hardware capabilities will lead to advances in CAD solid modeling software, which allows a tool to be designed and assembled with the workpiece on the machine tool. The manufacturing process will then be simulated for verifying the design. Databases for CAD systems will, consequently, grow to include graphic files for all equipment involved in the manufacturing process.

The process of Rapid Prototyping and Manufacturing (RPM) can be expected to gain greater use in tool design. Through RPM, designers can avoid the time-consuming, labor-intensive construction of conventional models. Physical, three-dimensional models or prototypes can be created through the process of stereolithography.

A CAD-generated computer model is mathematically "sliced" into thin layers, typically 0.005″ to 0.010″ (0.127 mm to 0.254 mm). This data is used to direct an ultraviolet laser beam over the surface of liquid polymer resin, which undergoes polymerization or partial solidification. Once a given layer has been polymerized in all locations corresponding to a single cross-section of the part, the system recoats another layer with fresh resin, and the process repeats, layer by layer, until the prototype or model is completed. After removal, cleaning, and finishing, the designer is able to hold an actual part within a few hours or several days, as compared to the weeks or months required to construct a model by conventional methods.

At the prototype model stage, design errors and omissions can be detected and ideas for improvements are readily generated. Using CAD and RPM technologies, the designer can modify the CAD design, make a new model, and proceed through several iterations within a week. Many deficiencies can be detected and corrected to produce a design of higher quality.

Artificial intelligence (AI) and expert systems software will become more advanced and will impact CAD and CAM software capabilities. The designer can input to the system a part description, and the CAD system will produce a complete design. Then, the information will be sent to the CAM system to produce a process plan, schedule the manufacturing system to produce the part, command the production line to produce, inspect the part, assemble the product, and package it for shipping or inventory. Advances in integration and AI technology will be the driving force in the factory of the future.

References

1. Zied, Ibrahim, *CAD/CAM Theory and Practice*, McGraw-Hill, Inc., New York, 1991. *(Reprinted with permission of McGraw-Hill, Inc.)*

2. Hoffman, Edward G., "Microcomputer CAD/CAM for All Shops," *Personal Computer Solutions for CAD/CAM*, Society of Manufacturing Engineers, 1989, Reprinted from *Modern Machine Shop*, 1987.

Bibliography

David Bedworth, Mark Henderson, and Philip Wolfe, *Computer-Integrated Design and Manufacturing*, McGraw-Hill, Inc., New York, 1991.

William Beeby and Phyllis Collier, *New Directions Through CAD/CAM*, Society of Manufacturing Engineers, 1986.

Thomas J. Drozda, *CAD/CAM for Production Tooling*, First Edition, Society of Manufacturing Engineers, 1989.

Mikell Groover, and Emory Zimmers, Jr., *CAD/CAM Computer Aided Design and Manufacturing*, Prentice Hall, Inc., Englewood Cliffs, NJ, 1984.

F. H. Mitchell, Jr., *CIM Systems, An Introduction to Computer Integrated Manufacturing*, Prentice Hall, Inc., Englewood Cliffs, NJ, 1991.

Elsayed A. Orady, *Design and Development of an Expert System for Simultaneous Engineering*, SME Technical Paper MS90-431.

Elsayed A. Orady, *Design and Development of Manufacturing Evaluation System for Rotational Parts*, Proceedings of The 15th Annual Conference of AMSE Association, Oct. 27-29, 1989, pp. 51-56.

Elsayed A. Orady and A. Shaout, *Computer Assisted Tools for Simultaneous Engineering*, Proceedings of IPC '91 Conference, April 8-11, 1991, pp.329-339.

THE COMPUTER IN TOOL DESIGN

Review Questions

1. Explain the acronyms CADD and CAD.
2. What are the three basic components of a typical CAD system?
3. What are the hardware components of a typical CAD system?
4. What technological development was the most important in allowing a role for the computer in tool design?
5. What are the two most obvious changes the computer has brought to the world of tool design?
6. What are the types of host computers used with a CAD system?
7. What does the abbreviation CRT stand for?
8. Classify the following pieces of CAD hardware as either input or output devices:
    a) System display (monitor),
    b) Puck,
    c) Keyboard,
    d) Mouse,
    e) Digitizer,
    f) Plotter.
9. What are the two basic scanning techniques used in graphic displays?
10. Define the term "softcopy."
11. How does a programmable function keyboard differ from a conventional keyboard?
12. What are the basic elements of a digitizer?

13. How does the tool designer use a digitizer?
14. What is a function menu?
15. What are the different types of mice used with a CAD system?
16. What is the difference between a trackball and a mouse?
17. What are the two types of pen plotters?
18. Compare pen plotters and electrostatic plotters from the point of view of speed and quality of output.
19. What are the basic modules of CAD software?
20. What are the two basic types of graphics representation models?

THE COMPUTER IN TOOL DESIGN

Answers to Review Questions

1. CADD stands for computer aided drafting and drawing, while CAD refers to computer aided design. CADD involves creating documentation (drawings, parts lists, specifications) for the design process. CAD involves using the computer as a tool for enhancing the design process (developmental stages, calculations, testing).
2. Hardware, software, and well-trained users.
3. Host computer, system display (monitor), keyboard, digitizer, cursor control device, and plotter.
4. Integrated silicon chips in the form of large-scale integrated chips (LSI) and very large scale integrated chips (VLSI).
5. Tools and techniques for designing and drafting.
6. Mainframe computer, minicomputer, microcomputer, and workstation.
7. Cathode ray tube.
8. a) Output device,
   b) Input device,
   c) Input device,
   d) Input device,
   e) Input device,
   f) Output device.
9. Random scan and raster scan.
10. Softcopy refers to images displayed on a system display or text display screen that disappear when the system is shut down.
11. A programmable function keyboard (PFK) is similar to a conventional keyboard but has additional keys that can be programmed for special functions, such as macros, to eliminate extensive typing of commands and data.

12. A digitizing tablet and a handheld electronic device such as a stylus or puck.
13. A digitizer is used to create images in much the same way as are pencil and sketchpad.
14. Paper or polyester overlays that contain word or symbolic identifiers corresponding with stored data or design functions.
15. Mechanical and optical.
16. The trackball is a mouse turned upside down where the user moves the cursor by rotating the ball on a fixed position.
17. Flatbed and drum.
18. Electrostatic plotters are about 100 times faster than pen plotters. The quality of output of pen plotters is much higher than that of electrostatic plotters.
19. a) Operating system module,
    b) Graphics module,
    c) Applications module,
    d) Programming module,
    e) Communications module.
20. Wireframe and solid models.

INDEX

V

Vacuum chucking, 224-225
Velocity, 64
Vise jaw fixtures, 297
Vises, 200, 213-223, 297
V locators, 153-157

W

Washers, 234
Weak links, 129
Wear resistance, 27

Wedge action clamp, 195
Wedge roller workholders, 213
Welded tool bodies, 230
Welding, 76-77, 636, 638, 642-643
Wire stitching, 674
Wood, 41-42
Wooden drill jigs, 261-264
Workholding principles, 137-139
Workpiece surfaces, 142

Y

Yield strength, 26